D0354290

RENEWALS 458-4574

DATE DUE

MAR 2			
DEC 11			
DEC 15			

GAYLORD PRINTED IN U.S.A.

WITHDRAWN
UTSA Libraries

Art and Affection

A LIFE OF VIRGINIA WOOLF

PANTHEA REID

NEW YORK OXFORD
OXFORD UNIVERSITY PRESS
1996

For

John Irwin Fischer

and in memory of Amanda Walker, 1953-1988

Oxford University Press

Oxford New York
Athens Auckland Bangkok Bogotá Bombay
Buenos Aires Calcutta Cape Town Dar es Salaam Delhi
Florence Hong Kong Istanbul Karachi
Kuala Lumpur Madras Madrid Melbourne
Mexico City Nairobi Paris Singapore
Taipei Tokyo Toronto

and associated companies in
Berlin Ibadan

Copyright © 1996 by Panthea Reid

Published by Oxford University Press, Inc.,
198 Madison Avenue, New York, New York 10016

Oxford is a registered trademark of Oxford University Press

All rights reserved. No part of this book may be reproduced,
stored in a retrieval system, or transmitted, in any form or by any means,
electronic, mechanical, photocopying, recording, or otherwise,
without the prior permission of Oxford University Press.

Library of Congress Cataloging-in-Publication Data
Reid, Panthea.
Art and affection: a life of Virginia Woolf / by Panthea Reid.
p. cm. Includes bibliographical references and index.
ISBN 0-19-510195-2
1. Woolf, Virginia, 1882–1941—Biography.
2. Women novelists, English—20th century—Biography.
PR6045.072Z8654 1996
823'.912—dc20
[B] 95-26568

1 3 5 7 9 8 6 4 2

Printed in the United States of America
on acid-free paper

Library
University of Texas
at San Antonio

Preface

In the fall of 1904, the Stephen children—Vanessa, Thoby, Virginia, and Adrian—made their now-celebrated move from their prestigious home in London's South Kensington district to a seedy address in the Bohemian, intellectual section of London known as Bloomsbury. Their exodus signaled for them the end of Victorian proprieties and repressions and the beginning of modernist freedoms. Virginia later claimed that the move heralded a renaissance of explorations in living and working. When the actual move took place, however, she did not participate, as her doctor and her sister thought her to be in too precarious a condition to join in the excitement.

Virginia was a manic-depressive, but at that time the illness had not yet been identified and so could not be treated. For her, a normal mood of excitement or depression would become inexplicably magnified so that she could no longer find her sane, balanced self. After her mother died, when Virginia was thirteen, she was disconsolate, unable to write for the only extended time in her life. Then after her half sister died, when Virginia was fifteen, she was destabilized for a brief period. When her father died, in February 1904, Virginia at twenty-two felt sad and regretful but also rather relieved to be rid of the increasingly ill-tempered and demanding old man. The repercussions of this loss were, however, only delayed. The Stephen children took a European holiday designed to suit the painterly interests of Virginia's older sister, Vanessa, and, in effect, to celebrate their father's death. But on their return in May, Virginia's irritation, guilt, and grief escalated into a full-blown manic episode. Some alien force seemed to have taken over her mind and being, changing the dynamics of her emotional makeup so that she transformed her beloved sister into her worst enemy.

That summer, while Virginia was recovering, Vanessa planned the Stephens' migration from South Kensington to Bloomsbury, and in the fall she shouldered the enormous task of orchestrating the move. Although Virginia thought herself well again, after only a few days at their new address, Vanessa sent her—following doctor's orders—to stay with their maiden aunt in Cambridge. On a black-bordered postcard she explained to Virginia, whom she addressed as "My own baby," that sheer boredom at their aunt's would "prevent another breakdown."

Vanessa also had more selfish motives: Her feelings had been hurt by Virginia's manic accusations, and she wanted to be free in London to conclude the move, decorate their new home according to her own tastes, and devote any spare time she might have to painting. Again and again, Virginia protested that she could stretch her legs out on a sofa and rest in her new quarters as well as she could in Cambridge. But when Vanessa ignored her protests and refused her requests, Virginia began to suspect that her exile was a matter of convenience for her sister. After an irritating late October in the "brown paper" of her aunt's Cambridge cottage, Virginia did get to enjoy a nearly two-week trial stay in their brightly decorated new Bloomsbury home. During that exciting time, the Stephen brothers and sisters entertained Thoby's friend Leonard Woolf, who was about to leave for a colonial administrative post in Ceylon, for nearly seven years. Some eight years later, Virginia Stephen and Leonard Woolf would marry, but in 1904 she hardly noticed him.

Virginia now was not thinking of men or marriage but instead was delighting in the radicalism of the Stephens' new life. The dark, heavily draped, ornately wallpapered Stephen home at 22 Hyde Park Gate was now a relic of the past. Vanessa had sold most of their carved, overstuffed Victorian furniture to Harrods, and in its place she had bought functional couches, tables, and chairs; chosen cheerful chintz fabrics for the upholstery, bedspreads, and curtains; and installed a plain red carpet to contrast with the freshly painted white walls. The house at 46 Gordon Square, Bloomsbury, was large and open enough to provide private space for each of the Stephen children and their very different interests.

Just as the Stephen children had changed their physical surroundings from ornate to plain, hidden to open, so they changed their emotional surroundings as well. They had grown up in a welter of family connections: In addition to their parents, the four Stephen children had had two half sisters, two half brothers, one unmarried aunt, four sets of married aunts and uncles, twenty-three first cousins, two "aunts" who were not blood kin, one living grandfather, an invalid grandmother, and an ever-changing retinue of nurses, governesses, and maids. By 1904, even though all the Stephen children were in their twenties, their living in Gordon Square without any elders to supervise them so shocked their more conventional relatives that the Stephens effectively severed both their physical and their emotional ties with South Kensington.

Virginia was elated with the freedom and independence their new Bloomsbury home offered, but Vanessa soon sent her away again. She arranged with the talented and lovable Madge Vaughan, wife of their schoolmaster cousin Will, to invite Virginia to visit in Yorkshire, a half-day from London by train. Vanessa's rationale was that Virginia's stimulating two weeks in London had so nearly caused "another serious breakdown" that she needed a quiet holiday in Yorkshire to calm her nerves.

Virginia was depressed by this second forced exile and angry to find Madge turning into a conventional headmaster's wife. She found the inescapable company of masters and schoolboys deadly, the constant ringing of church bells jarring. She had few books to read and no one to talk with her about them. Furthermore, Vanessa had instructed Madge not to let Virginia walk far alone and never into busy streets or shops, but Virginia knew only that Madge placed limits and prohibitions on her whenever she went out. Although she paid calls with Madge in the village and managed two longer walks alone, the frozen craggy moors and blustery weather usually kept Virginia confined to the schoolmaster's house and garden.

Meanwhile, Vanessa filled her letters to Virginia with descriptions of her decorating projects, the art exhibits and parties she attended in London, and visits from Clive Bell (whom Vanessa would later marry) and from a fellow art student (who was staying in Virginia's room, no doubt to Virginia's irritation). Vanessa also packed her letters with inquiries and admonishments about Virginia's health and warnings that Virginia would not be able to lead a normal life unless she ate and rested properly and allowed her nerves to mend. But reading about Vanessa's gadding about in glamorous London infuriated Virginia, exiled in the frigid north and then again packed off to Cambridge. Virginia suspected that Vanessa was colluding with her doctors, but the angrier she became, the more she proved, Vanessa wrote her, that she was still not well.

Rather than preventing another breakdown, these exiles could have precipitated one, for each of the conditions that would destabilize Virginia Woolf throughout her life was present: She felt guilty about her relief at her father's death; she felt trapped in dull domesticity yet was forced to remain in boring households supposedly for her health's sake; and because Vanessa kept shipping her away, she feared she had lost her sister's love.

Although the "pigheaded" family doctor permitted Virginia to write a contribution to a biography of her father, he otherwise expected her, while she was in Yorkshire, to lay down her pen, while at the same time, in London, Vanessa gleefully took up her brush. Virginia had felt the "desire of the pen" in her blood since early childhood. For her, not writing was tantamount to not living. When she could not write, she felt that she was "sinking down, down" into nothingness.

Rather than let these waves of despair ravage her, however, for most of her life Virginia became her own self-therapist. She would laugh about feeling her brain like a pear to see if it were ripe. What she really was doing was testing her ability to write, for more than anything else, her literary talents could rescue her from the abysses into which she sometimes sank. Therefore, in 1904, she rehabilitated herself by giving in to the longing deep in the "richest strata" of her mind: She defied her doctor's orders and used what was supposed to be restful boredom to compose a multitude of writing assignments, including her first essays to be published. More than rest, nourishment, medicine, and even love alone, Virginia's writing sustained her, as it would for the rest of her life.

Early on, the compulsion to write had defined Virginia in opposition to Vanessa, who felt an equally intense compulsion to paint. Accordingly, this book is about a sibling rivalry that also became a rivalry between these siblings' arts. How each of these women, born into upper-middle-class Victorian England, managed to follow her talents despite the repressions and biases of their time is in itself an extraordinary story. And how Virginia's own ambitions became so entangled with Vanessa's that Vanessa's successes seemed to become Virginia's failures, and vice versa, is part of the curious story I tell here. Yet even though Virginia defined the art of the pen in opposition to the art of the brush, to her considerable irritation she had to admit that modern painting had much to teach modern writing. Sadly, no matter how much she wrote, flattered, supported, and entertained her sister, after 1910 Virginia felt she could never secure Vanessa's love again.

Until her fifty-ninth year, the "insatiable desire to write something before I die" sustained her. If the need to write was the warp in the fabric of Virginia Woolf's being, the need to be loved was the woof, without which that fabric would unravel. This biography tells the story of the peculiar intertwining of art and affection in Virginia Woolf's life.

Baton Rouge, La. P. R.
April 1996

Acknowledgments

This book is dedicated first to my husband, John Irwin Fischer, whose example of meticulous archival research has inspired and whose love has sustained my research and writing. With John Fischer, the phoenix riddle does indeed, as John Donne wrote, have "more wit."

My Scottish cousin Amanda was often with us in London at the inception of this book. In 1988 when she and her husband, Matthew Bannister, were on the southern coast of Spain celebrating their new positions with ITV and BBC Radio, Amanda swam too far into the Mediterranean, where an undertow seized her. As Kristina Olsen sings, "I cried you a waterfall."

I owe special debts to three scholars besides John Fischer, one I know slightly, another I have met once, and the other, never. Elizabeth Heine, editor of the E. M. Forster Papers and Virginia Woolf's *The Voyage Out*, read this text in manuscript, offering me her editing skills and knowledge of Bloomsbury. In our single meeting, Elizabeth Richardson introduced me to her vast collection of Bloomsbury visual materials. Thereafter, she carefully read a version of this typescript and gave me the benefit of her infallible graphic memory, her extensive expertise regarding Bloomsbury, and her remarkable copyediting skills. John Bicknell, whom I have never met, most generously lent me his typescripts of Leslie Stephen's letters to Julia Stephen. He has kept up a lively conversation with me by mail, answering my many questions and, with John Fischer, Elizabeth Heine, and Elizabeth Richardson, proving that a tradition of scholarly sharing still exists.

During the years of researching and writing this book, Elizabeth Inglis of the University of Sussex, Sally Brown of the British Library, and Ann Wilson of the Strachey Trust have been remarkably helpful to me. Frederic Spotts, editor of the *Letters of Leonard Woolf*, answered my queries and kindly allowed me to reproduce two photographs that he was the first to publish. Dan O'Donnell graciously volunteered to read an earlier version of this manuscript. I am grateful to all of them.

Thanks for invaluable assistance also go to Helen Bickerstaff and Joy Eldridge of the University of Sussex; Jacqueline Cox, archivist of the Modern Collection at King's College, Cambridge; Michael Halls, her predecessor; Karen Kukil, assistant curator of Rare Books at Smith College; Christopher Fletcher, Sally Brown's assistant at the British Library; Philip Milito, technical assistant; Lisa Browar, acting curator; and Frances O. Mattson, former curator of the Berg Collection at the New York Public Library; Jill Mansfield, curator, Tate Gallery Archive; Louise Ray, her predecessor; and Judy Collins, who cataloged the Tate's Charleston Papers; Susan Halpert of the Houghton Library and Michael T. Dumas of the Harvard Theatre Collection, Harvard University; Russell Maylone, curator of Special Collections, Northwestern University; and Cathy Henderson, research librarian, and Thomas Staley, director, Harry Ransom Humanities Research Center, University of Texas. My thanks also go to Aidan Weston-Lewis, assistant keeper of Italian and Spanish art at the National Gallery of Scotland; Carlotta Gelmetti of the Tate Gallery; Paul Cox of the National Portrait Gallery; Herbert Goodman of Louisiana State University's School of Art; and Mark Kleiner of LSU's Graphic Services.

I want to thank as well my various correspondents during the years of composition: April Armstrong, Joseph Blotner, Susan Dick, Annabel Farjeon, Diane Gillespie, John Graham, Barbara Strachey Halpern, James Haule, Anne Meade Heine, Peter Jacobs, Robert Langenfeld, Leila Luedeking, Veronica Makowsky, Regina Marler, Allan Parkas, Christopher Reed, Philip Rieff, S. P. Rosenbaum, Angus Ross, Bill Ross, Brenda Silver,

Frances Spalding, Janice Stein, Alastair Upton, Stanley Weintraub, Deborah Wilson, and Cecil Woolf. And I thank all the students over the years whose interest in Virginia Woolf has inspired me at least as much as mine has them.

Lisa Ross, my agent, and Joseph Spieler of the Spieler Agency have been terrific sources of support. My appreciation also goes to the competent staff at Oxford University Press, especially Susan Chang and Elizabeth Maguire, her predecessor; Elda Rotor, editorial assistant; Irene Pavitt, editor; and Margaret B. Yamashita, copy editor.

I also want to acknowledge my children, Reid Broughton and Hannah Fischer, who, despite jokes about the dim prospects of my ever finishing the revisions of Chapter 1, have been truly supportive.

I began my research with grants from the National Endowment for the Humanities and the Council for the International Exchange of Scholars. Over the years, I also held an NEH Travel to Collections grant, a summer research stipend from LSU's Council on Research, and a Manship Grant from LSU's College of Arts and Sciences. My thanks to each of these organizations.

I sincerely thank the following persons for their generous permission to quote from unpublished letters to me and interviews with me: Igor Anrep, Anne Olivier Bell, Quentin Bell, John Bicknell, the late Pamela Fry Diamand, Diane Gillespie, Barbara Strachey Halpern, Elizabeth Heine, Regina Marler, Nigel Nicolson, the late Trekkie Parsons, Elizabeth Richardson, Bill Ross, and Richard Shone. For permission to quote from unpublished materials for which they hold the copyright, I thank the Society of Authors as the literary representative of the Estate of Virginia Woolf, © 1996 Quentin Bell and Angelica Garnett; Quentin Bell for Vanessa Bell's letters and Julian Bell's papers; Angelica Garnett for Vanessa Bell's letters and memoirs; Annabel Cole and Betty Taper for the Roger Fry Papers; Trekkie Parsons for the Monk's House Papers; the Society of Authors, as agents for the Strachey Trust, for Lytton Strachey's letters; Sidney Wilkinson for her late husband's note on a lecture by Roger Fry; the Trustees of King's College, Cambridge; and the C. D. McCormick Library of Special Collections, Northwestern University Library, for permission to quote from the papers of Maynard Keynes.

I want to thank the owners of the copyrights for permission to reproduce illustrations: Angelica Garnett for artwork and photographs by Vanessa Bell and Quentin Bell for Stephen family photographs; the Estate of Duncan Grant, © 1978 courtesy of Henrietta Garnett; Annabel Cole and Betty Tabor for artwork by Roger Fry; Trekkie Ritchie Parsons for photographs by Leonard Woolf; Nigel Nicolson for the photograph of his mother; Barbara Strachey Halpern for her photographs of both Leonard and Virginia Woolf; Photo Researchers, Inc., and Gisèle Freund for her photographs of the Woolfs; Olive Kennedy for her husband's drawings; Alastair Upton and the Charleston Trust for photographs of Charleston interiors; and the Duke of Sutherland for Raphael's *Bridgewater Madonna*. In addition to the museums and libraries whose archives have provided me with various illustrations, Sandra Lummis Fine Art, Tony Bradshaw and the Bloomsbury Workshop, and Anthony d'Offay Gallery, London, have been especially helpful.

Contents

Collections and Abbreviations of Unpublished Sources

This list of unpublished collections, their abbreviations, and the abbreviations for the materials they contain indicates where they may be found. Since the text identifies the materials, the abbreviations for collections are used only when the materials exist outside the expected location or when (as with Virginia Woolf's reading notebooks) there are two "expected" locations. Otherwise, the materials in these collections are cited by date (day, month, year).

I use the initials VB and VW even before Vanessa Stephen married Clive Bell and Virginia Stephen married Leonard Woolf. The initials of the writer and addressee also are cited (for example, RF/VB for an unpublished letter from Roger Fry to Vanessa Bell). If no other location is listed, the reader may assume that the letter can be found in the source given in the following list. The note indicates whether the letter is in a different collection. For example, under Berg, I explain that the correspondence from Vanessa Bell to Virginia Woolf can be found in the Berg Collection at the New York Public Library. But Vanessa's last letter to Virginia is in the Monk's House Papers, as the note accompanying its citation indicates.

Selected Letters of Leslie Stephen, edited by John W. Bicknell, was published too late for me to cite references to it by page number. Hence I handled these letters in the same manner as I did unpublished letters.

Whereas Diane Gillespie abbreviates Virginia Woolf's mother's initials as JDS, as she signed herself (see JDS, 215), I follow Bicknell's example and abbreviate it JPS, for Julia Prinsep Stephen. Abbreviations of individuals' names may be confirmed by checking the text itself.

Throughout the book, I have followed as closely as possible Anne Olivier Bell's editing principles for reproducing sometimes peculiar punctuation and abbreviations. For example, Virginia and most of her friends characteristically wrote "dont" for "don't" and "&" for "and"; were careless about apostrophes, including those for possessives; and were profligate with colons and semicolons. Although I cite the published edition of Vanessa Bell's letters, I follow the manuscript when the editor did not. For example, *&* is substituted here for the *and* in the *Letters*. I sometimes clarify abbreviations by adding missing letters in brackets, but I do not attempt to make any original writing conform to late-twentieth-century practice.

Berg — Henry W. and Albert A. Berg Collection, New York Public Library, Astor, Lenox and Tilden Foundations: Virginia Woolf papers, including her unpublished early diaries; letters from Leslie Stephen to Julia Duckworth Stephen; letters from Vanessa Stephen Bell to Virginia Stephen Woolf; Stella Duckworth's diary and letters, notes for, and manuscript and typescript versions of *Roger Fry* and *Between the Acts*; and Virginia Woolf's reading notebooks

BL — British Library: *Hyde Park Gate News* (abbreviated HPGN), late typescript of "A Sketch of the Past," early versions of *Mrs. Dalloway*, and all letters to and from the Strachey family unless otherwise noted

C,AD1 — Charleston, Addition manuscript 1: letters from Maria Jackson to Julia Prinsep Stephen, deposited by Quentin Bell at the University of Sussex

CP — Charleston Papers, Tate Gallery, London (with photocopies at King's College, Cambridge, and the University of Sussex): all letters to and from Vanessa Bell and Clive Bell, including those from Roger Fry; and letters from and to Julian Bell

FP — Roger Fry Papers, King's College, Cambridge: all Fry materials, including drafts and final versions of articles and lectures, clippings about exhibits and other press notices, and correspondence, except for some letters to Virginia Woolf in MHP and letters to the Bells in CP

LWP — Leonard Woolf Papers, University of Sussex (all letters to Leonard and Virginia Woolf, unless otherwise noted, are at the University of Sussex): all Leonard Woolf material, especially correspondence to and from Leonard Woolf (except when identified as among the MHP or CP), and his diaries (abbreviated LWD)

MHP — Monk's House Papers, University of Sussex: letters to and from Virginia Woolf (and when so marked, Leonard Woolf), manuscript and typescript notes for and versions of *Roger Fry*, and reading notebooks. Subdivisions within MHP are noted with a sequence of letters and numbers

MRF — Vanessa Bell's "Memories of Roger Fry": collection of Angelica Garnett, soon to be published in a collection entitled *Sketches in Pen and Ink: Six Memoirs by Vanessa Bell*, ed. Angelica Garnett and Lia Giachero (London: Chatto & Windus)

Abbreviations of Published Sources

References to Woolf's novels and major nonfiction works are to the readily available Harcourt Brace Jovanovich paperback editions. The abbreviations are BA, *Between the Acts;* FL, *Flush;* FW, *Freshwater;* JR, *Jacob's Room;* MD, *Mrs. Dalloway;* ND, *Night and Day;* O, *Orlando;* RF, *Roger Fry;* ROOO, A *Room of One's Own;* TG, *Three Guineas;* TL, *To the Lighthouse;* VO; *The Voyage Out;* W, *The Waves;* and Y, *The Years*. For the major periodicals in which both Woolf and Fry published articles, I provide the original publication and date and the most accessible reprint in a modern collection. My reason for citing the original reference is to show how interwoven in both time and place the Woolf and Fry essays and reviews were.

Bell, 1 and 2	Quentin Bell, *Virginia Woolf: A Biography*, 2 vols. (New York: Harcourt Brace Jovanovich, 1972)
CE	Leonard Woolf, ed., *Collected Essays*, 4 vols. (New York: Harcourt, Brace & World, 1953–1967)
CS	*Congenial Spirits: The Selected Letters of Virginia Woolf,*

	ed. Joanne Trautmann Banks (London: Hogarth Press, 1989)
CSF	*The Complete Shorter Fiction of Virginia Woolf*, ed. Susan Dick (London: Hogarth Press, 1989)
D	*The Diary of Virginia Woolf*, 5 vols., ed. Anne Olivier Bell, assisted for 3 vols. by Andrew McNeillie (New York: Harcourt Brace Jovanovich, 1977–1984)
JDS	*Julia Duckworth Stephen: Stories for Children, Essays for Adults*, ed. Diana F. Gillespie and Elizabeth Steele (Syracuse, N.Y.: Syracuse University Press, 1987)
LWA	Leonard Woolf, *An Autobiography*, 2 vols. (Oxford: Oxford University Press, 1980)
LWL	*The Letters of Leonard Woolf*, ed. Frederic Spotts (San Diego, Calif.: Harcourt Brace Jovanovich, 1989)
MB	Virginia Woolf, *Moments of Being*, 2nd ed., ed. Jeanne Schulkind (London: Hogarth Press, 1985)
MFS	"Some New Woolf Letters," *Modern Fiction Studies* 30 (Summer 1984): 175–201
MOE	Virginia Woolf, *The Moment and Other Essays* (London: Hogarth Press, 1947)
PA	*A Passionate Apprentice: The Early Journals of Virginia Woolf*, ed. Mitchell A. Leaska (London: Hogarth Press, 1990)
RFL	*The Letters of Roger Fry*, 2 vols., ed. Denys Sutton (London: Chatto & Windus, 1972)
VBL	*Selected Letters of Vanessa Bell*, ed. Regina Marler (New York: Pantheon Books, 1993)
VD	Roger Fry, *Vision and Design* (1920; Oxford: Oxford University Press, 1981)
VWE	*The Essays of Virginia Woolf*, ed. Andrew McNeillie (London: Hogarth Press, 1986–1994)
VWL	*The Letters of Virginia Woolf*, 6 vols., ed. Nigel Nicholson and assisted by Joanne Trautmann (London: Hogarth Press, 1975–1980)

Art and Affection

The imagination is to reality as the shadow to the body that casts it and as poetry is to painting. . . . For painting does not talk; but reveals herself as she is, ending in reality; and Poetry ends in words in which she eloquently sings her own praises.

Moreover, the works of poets are . . . often not understood and require many explanations, and commentators very rarely know what was in the poet's mind. . . . But the work of the painter is immediately understood by its beholders.

LEONARDO DA VINCI, *Paragone*

Is it not possible that some writer will come along and do in words what those men [Postimpressionist painters] have done in paint?

VIRGINIA WOOLF, "Books and Persons"

Is that the origin of art he asked himself: . . . making yourself immune by making an image?

VIRGINIA WOOLF, EARLY VERSION OF *The Years*.

1
A Most Precocious Little Girl

1882–1895

Virginia Woolf began life with the ideal Victorian parents, an esteemed father and a sainted mother. Her father, Leslie Stephen, the first editor of the *Dictionary of National Biography*, was, after Matthew Arnold's death, the foremost man of letters in England. Her mother, Julia Prinsep Jackson Stephen, was revered for her beauty and wit, her self-sacrifice in nursing the ill, and her bravery in facing early widowhood. Posing as the Virgin Mary for Edward Burne-Jones's famous painting *The Annunciation*, Julia appeared to be the Victorian male's ideal woman: beautiful, virtuous, and accommodating. She was also enormously capable. Although when Leslie Stephen began to court her, the young widow was melodramatic enough to maintain that she longed only for the happiness of the convent, he overcame her leanings toward chastity.[1] Her motive for marrying him might have seemed almost as extreme as the convent, however, for he was himself a widower, the father of a retarded child, and fourteen years older than Julia. As Virginia wrote, "Perhaps there was pity in her love" (MB, 91).

There was also respect in her love, for by 1877, when he began courting Julia in earnest, Leslie was an important literary figure. His ancestors had entered the middle class only in the last century, but they had quickly become prominent statesmen and polemicists. Like his friend and relation by marriage, William Wilberforce, Leslie's grandfather had devoted his prodigious energies to evangelical religion and the emancipation of slaves. As a member of Parliament, he fought for the eradication of the slave trade, and, as a pamphleteer exposing the ways in which French and American traders were slipping through the British blockade, he inadvertently provoked the War of 1812 between England and the United States.

Leslie's father, James Stephen, was an exemplary mid-Victorian gentleman, a civil servant in the Colonial Office who was as religious and as dedicated to good works as his father had been. Despite being knighted for his work protecting Africans and trying to reform colonialism, Sir James Stephen remained a melancholy, puritanical man. Lady Jane Stephen (née Venn) was a sensible woman from a similar family of dedicated public servants who also had connections to the Wilberforces. Her cheerfulness helped mollify Sir James's depression. Although Virginia did not know her Stephen ancestors, she inherited her Stephen great-grandfather's powerful way with words, her grandmother's sense of humor, and her grandfather's tendency toward depression.

Virginia's father, Leslie Stephen, was the middle child of three. His older brother, James Fitzjames Stephen, became a prominent barrister and journalist and fathered seven children, cousins all considerably older than Virginia. Leslie's younger sister, Caroline Emelia, became a dedicated Quaker who never married and was known by Virginia as the Nun. Leslie was a sensitive son who inherited his mother's wit and a fair share of his father's gloomy outlook on life; as a youngster, he possessed little of his older brother Fitzjames's robustness. While he was at Trinity Hall, Cambridge, however, Leslie overcame his reputation for delicacy by becoming an oarsman on the college's team. He was a tall lanky young man who filled out his form not only by crewing but also by walking great distances and becoming an accomplished mountain climber.

After his graduation, Leslie was made a fellow of Trinity Hall. In those days, a man had to take holy orders in order to become a fellow of (or teacher in) a college. Accordingly, Leslie was ordained in 1855 and took priest's orders in 1859, but the evangelical faith that had guided the reformist impulses of his grandfather and father became an empty husk for Leslie. By 1862, his skepticism regarding such miracles as Noah's flood led him to leave the church and hence his academic profession. He then became a journalist, a campaigner for the Union in the American Civil War, and eventually a man of letters. Like his brother Fitzjames, Leslie entered the upper-middle-class world of intellectuals and statesmen.

In 1867, Leslie married Minny Thackeray, the younger daughter of the famous novelist William Makepeace Thackeray. The following year, Minnie had a miscarriage but then gave birth to a daughter, Laura, in 1870. The next year, Leslie became the editor of the premier English magazine the *Cornhill*. The Stephens were a happy pair, despite growing suspicions that Laura was not a mentally sound child. In 1875, Minny was pregnant again, and that November they were visited by a friend, Julia Prinsep Jackson Duckworth. Later the same night, on Leslie's forty-third birthday, Minny went into convulsions and died. Julia Duckworth, herself recently widowed, consoled Leslie in his despair.

Julia also came from upper-middle-class English society, but her family had more aristocratic and celebrated connections than Leslie's did and more mediocre ones as well. While Sir James Stephen campaigned to reform Britain's colonial enterprise, Julia's father, Dr. John Jackson, enjoyed what he could get out of it. After he married Maria—one of the Pattle sisters famed for their beauty—he traveled to Calcutta, where he set up a lucrative medical practice. He and Maria, who was fourteen years younger than he, had three daughters, all born in India: Adeline, Mary, and Julia. Fearing contamination from India's heat and disease, Maria Jackson soon sent the two older girls back to England to stay with her sister Sarah. Then when Julia was two, Maria returned to England with her. Jackson remained in India. Growing up in England without their father, Julia and her sisters were virtually adopted by their Pattle aunts, especially Sarah.

Maria Jackson was a friend of Coventry Patmore, who wrote the quintessential Victorian paean to self-sacrificing womanhood, "The Angel in the House." But Maria herself seems to have preferred that others sacrifice to her. She had few interests beyond beauty (chiefly her daughters') and illnesses (chiefly her own). Her husband was, according to Leslie Stephen, entirely conventional and utterly uninteresting.

But even though Julia's parents were rather dull, several of her aunts were not. The eccentric, extravagant, and beautiful Virginia Pattle became Countess Somers. Sarah Pattle married Henry Thoby Prinsep, a member of the council of the East India Company. Their home, called Little Holland House, was far from little. A quaint rambling place with a thatched porch, gabled roofs, and additions on several levels in different architectural styles, it could house the Prinseps' four children and innumerable guests. Although it was on the edge of suburban London, Little Holland House enjoyed a seemingly country setting of lawn, garden, and croquet court. The Prinseps made it a gathering place for writers, statesmen, and especially artists. They turned part of their upstairs into a studio for the eminent Victorian painter G. F. Watts, and later they built a studio for Watt's sculpture on the grounds of Little Holland House. The Prinseps also took in Edward Burne-Jones when he was ill, unknown, and impoverished (Hill, *Julia Margaret Cameron*, 57, 61). Julia Margaret Cameron was the least beautiful but the most eccentric and talented of the Pattle sisters. In 1863, when her daughter and son-in-law gave her a camera, Julia Margaret immediately transformed herself into an avid photographer of the famous men and beautiful women who frequented Little Holland House and her own home, Dimbola Lodge, at Freshwater Bay.

The various Pre-Raphaelite painters who frequented Little Holland House saw the Prinseps' young niece Julia Jackson as a work of art herself. Watts made a chalk drawing of her head when she was a child and two paintings of her as a young woman. Julia's cousin Val Prinsep drew her profile when she was quite young, and before she was eleven, someone took a photograph of her wearing a cape and artistically posed against a large tree (Figure 1).[2] When Julia was about sixteen, the sculptor Thomas Woolner asked Maria Jackson for permission to sculpt busts of Julia and her sister Mary, but Maria refused, lest such attention destroy her daughters' simplicity or unselfconsciousness. But having been drawn, painted, and photographed since she was small, Julia was already highly self-conscious. As both the Prinsep drawing and Figure 1 show, at an early age Julia knew that a profile or near-profile view was her best pose.

Burne-Jones often sought models like Julia for his many paintings of beautiful women from classical, biblical, or medieval narratives, and her profile became a source for the "Burne-Jones type." Pre-Raphaelite art, in turn, imitated life, for in dress and manner Julia adopted a pose seemingly borrowed from painting. Her wistful beauty made Maria Jackson fear that any man who saw Julia, even in a railway carriage, would fall in love with her, and at least two artists did. Both the painter William Holman Hunt and the sculptor Thomas Woolner proposed marriage to the young Julia, and after she turned them down, they each married women who looked startlingly like her. Thus Julia became not only an artist's model but also a model for those same artists' choices of spouses.

After Jackson returned from India, the family continued to be frequent visitors to Little Holland House. They themselves lived in several locations, including the graceful and elegant Saxonbury near Tunbridge Wells.[3] In 1856, Julia's oldest sister, Adeline, married Henry Halford Vaughan, a former historian who was serving as a clerk of assize on the South Wales Circuit and took Addy (as she was known) to live in a castle in Wales. In 1862, Julia's middle sister, Mary, married Herbert Fisher, then tutor to the Prince of Wales. They went to live on the Brighton coast.

Maria Jackson's fears of India and illness were exaggerated, as her sufferings from rheumatism also seem to have been, but Julia was entirely sympathetic to her ailing mother (Figure 2), who, by the time of this photograph, seems to have lost most traces of her good looks. By 1862, when she was sixteen, Julia became Maria's principal caretaker. They traveled to Venice that year to meet the newlyweds, Mary and Herbert Fisher, and to enjoy the warmer climate. There and elsewhere on their travels, they met a twenty-nine-year-old vacationing barrister named Herbert Duckworth. After having rejected proposals from eccentric, driven artists, Julia found the handsome, wealthy, conventional, and not very hard-working barrister appealing. They married five years later and were by all reports very happy. Two children, George and Stella, arrived in as many years. Then in 1870, while visiting Adeline and Halford Vaughan at Upton Castle with his once again pregnant wife, Herbert Duckworth reached up to pick a fig off a tall tree and burst an undiagnosed abscess. He died within a day.

Six weeks later, Julia gave birth to her third child, Gerald Duckworth. The twenty-four-year-old widow with three very small children spent much of her time visiting her aunt Julia Margaret Cameron at Freshwater, on the Isle of Wight. There her aunt resumed photographing the younger Julia. In somber black dresses, she appears as a tragic subject whose sadness her aunt fully exploited in dramatic photographs. Refusing to remarry, young Julia spent her time tending after her two sons and daughter and her increasingly enfeebled mother.

The senseless death of her husband left Julia unable to believe in divine providence, and so she began to read the agnostic publications of Leslie Stephen and found comfort in his freethinking skepticism. Steeling herself against despair, Julia created meaning in her life by doing good deeds for others. Thus she felt great sympathy for the newly widowed Leslie, and in the late spring of 1876 she helped him, his retarded daughter Laura, and his sister-in-law Anny Thackeray, who agreed to keep house for him, move next door to her on the street then called Hyde Park Gate

South. Then Julia and her children left London to spend the summer in Brighton with the Fishers and perhaps her parents.

That year, Leslie established his scholarly credentials with his *History of English Thought in the Eighteenth Century*, and he also published "An Agnostic's Apology," which further confirmed the faithlessness that he and Julia shared. In early 1877, he went mountain climbing in the Alps for the first time. Rejuvenated by the exercise, he also felt cured of his melancholy and able to put his thirteen months of mourning behind him. Against Leslie's wishes, however, Anny Thackeray was planning to marry and so leave him without a housekeeper. Becoming embroiled in the debate, Julia argued that Anny had a right to marry whom she pleased, and so Leslie dropped his objections and lost his household manager. Leslie's sister, Caroline Emelia, tried to help out, but she was too pliable for Leslie and he too demanding for her. And so he was left alone with Laura. Living side by side on a fashionable London street were the beautiful widow with her three young children and the famous writer with his afflicted daughter. The adults shared the experience of arbitrary death, a faith in faithlessness, and great emotional needs, a situation ripe for romance. Leslie had long admired Julia's poignant beauty (Figure 3), and by 1877, he (predictably) had fallen in love with her.

Julia thought herself too uneducated for this learned man and wished instead to remain only his platonic friend. Nonetheless, he wrote to her daily and made various overtures, but at first Julia replied that it would be wrong for her to marry. Then she asked Leslie to make up her mind for her. He refused. Julia feared that she could not "come to life again," but after a year of resistance, she agreed "to be a good wife to you." Between their engagement and marriage, however, Julia left Leslie to nurse her uncle Thoby Prinsep on his deathbed, thus establishing a pattern of leaving Leslie to care for members of her family that would continue throughout their marriage.

They married on 26 March 1878. For their honeymoon, Julia's aunt Virginia, Countess Somers, and Lord Somers lent them Eastnor Castle, a nineteenth-century version of a medieval baronial hall, near Ledbury, in Herefordshire. Afterward, they settled in London with their four children: Laura Stephen (aged seven) and George (aged ten), Stella (aged eight), and Gerald Duckworth (aged seven). The combined families lived in Julia's house, which was soon renumbered 22 Hyde Park Gate. Julia was thirty-two; Leslie, forty-six. Four infant Stephens were born in rapid succession (Vanessa in 1879, Thoby in 1880, Virginia in 1882, and Adrian in

1883), raising the tally to eight children from three different marriages living at 22 Hyde Park Gate, London.

Maria Jackson, increasingly incapacitated by rheumatism, gout, and the morphine treatments to which she turned frequently, continued to summon Julia to her side whenever it pleased her to do so. Leslie suspected that his mother-in-law was not dangerously ill but that her pathetic weakness was "fagging" Julia. Worried about the strain on her and tired of Julia's absences from him and the children, Leslie repeatedly urged her to bring her mother home to them. But his wishess did not prevail until late in 1891 when Maria became, in Virginia's terms, "the invalid of 22 Hyde Park Gate" (Stephen, *Mausoleum Book*, 43, 73; LS/JPS, 4 December 1879; LS/James Russell Lowell, 21 January 1880; HPGN, 7 March 1892). The house then held ten adults, including seven maids and nurses, and eight children.

The house stood tall and dark behind a masonry wall, interrupted by spiked iron rails and a gate on a cul-de-sac just off Kensington Gardens. Behind it was a rather large back garden, walled off from its neighbors. One in a series of elegant row houses, it was a fairly formidable address.[4] The Stephens, however, damaged their house's stateliness by adding, to its original five, two extra stories of "atrocious design" (Bell, 1:22). The children's nurseries and Leslie Stephen's study were in the added top stories, the servants' quarters in the basement, and Maria Jackson's sickroom somewhere in between. A curtain of Virginia creeper outside and heavy Victorian furniture, dark paint and draperies, and poor lighting inside all heightened the sense of enclosure.

But if 22 Hyde Park Gate confined the Stephen children, Talland House, in St. Ives, liberated them. This home, to which the whole family and innumerable guests migrated every summer after Virginia's birth, rested on a craggy hill above the rest of the fishing village of St. Ives in Cornwall. From its French doors and elongated windows overlooking the bay, the children could watch the sun set to the west beyond Ireland and follow in their imaginations the beams of the Godrevy lighthouse. They themselves would sail with their father to the lighthouse; play cricket, tennis, and bowls in the garden; swim in the sea; and tramp over the countryside. With doors and windows flung open, breezes and fragrances from the surrounding gardens meandered in and out of Talland House, as did the Stephen children, at will.

Virginia's history was intricately tied to those two houses and to the other three little Stephens. Vanessa, the eldest Stephen offspring, was born on 30 May 1879, her half sister Stella's tenth birthday.[5] Although the Stephens hired an official nurse, Stella took her birthday "gift" as her own special charge. Called "Bunch," for "Honeybunch," by her grandmother, Stella adopted a little-mother role because Julia was weakened first by undergoing a difficult childbirth and then by nursing her mother through a bout of rheumatic fever from which Maria Jackson never recovered enough to walk again. A month before the birth of the second Stephen child—although she was sometimes worried that Bunch would "transfer her devotion to the new baby"—Maria Jackson assured Julia that Bunch's "maternal reach is large eno' I think for both" (Stephen, *Mausoleum Book*, 72; MJ/JPS, 12 August 1880).

Thoby Stephen not only was within his mother's and Bunch's maternal reach but also engaged Vanessa's. Although only sixteen months older, little Vanessa assumed a mothering role toward the new baby. Once, for example, Leslie wrote to Julia that baby Thoby kissed Vanessa's legs while she was sitting on Leslie's knee. She called him "funny little boy" and pulled him onto her lap. With stable parenting from Julia and Bunch and (especially when Julia was with her mother in Brighton) their father, these two Stephen children became an emotionally tied pair, nicknamed collectively by Leslie as Vato, Tova, Nessaby, or his Ragamice.

Leslie was impressed by the force of character that led Vanessa at twenty months to contradict her half brother Gerald in a surprisingly argumentative spirit. Leslie played games with her, showed her photograph proudly to his friends, and drew pictures for, with, and of her. He was delighted to write to Julia about Vanessa's reciting a version of "Froggy Went A-Courting" when she was less than two years old. Projecting from her own experiences of an absent mother, little Vanessa identified a picture of a boy writing as a "boy writing to Mamma!" Leslie had to coax her to send a kiss to Julia, for she had "odd little misgivings in her little mind as to doing anything demonstrative." Hardly six months old, Thoby amazed Leslie with his cheerfulness. He was always "squeaking" with joy and bouncing about so much that Leslie concluded he must be "in training as an acrobat." Leslie wrote to Julia he had never seen "such a picture of satisfaction. The creature is evidently going to be an optimist."[6]

In April 1881, when Thoby was seven months old, Julia left home to care for her sister Adeline Vaughan, who was seriously ill in the drafty and rat-infested Upton Castle in Pembroke, which Leslie considered a barely habitable old ruin (Stephen, *Mausoleum Book*, 67). Julia weaned Thoby rather abruptly (his parting gesture was a bite). While Julia was away,

Bunch and their father cared for the little ones at 22 Hyde Park Gate; the Duckworth boys stayed first with their grandparents in Brighton and then with their stepfather in London. Maria Jackson regretted not having Vanessa, Thoby, and Bunch come with the Duckworths to stay with her but exulted in Bunch's happiness in "seeing V & ToTo again" in London.[7]

Adeline's husband, Halford Vaughan, did not hire a professional nurse, despite nagging from his mother-in-law, advice from his father-in-law, and clear evidence of Julia's exhaustion. Despairing over her inability to alleviate Addy's suffering, Julia nevertheless shouldered sickroom duties in hopes of at least comforting her sister. Several times a day, the mail brought letters of worry and regret from their mother, suggesting, for example, that Julia hang up enema bags after washing them, put Addy on top of four mattresses, and "pray never be without all your Petticoats [as] one gets so chilled towards morning." Maria cautioned, "You can not go on night and day. Pray my own remember yr. breasts I am so afraid they may be painful [from weaning Thoby] & make you feel ill and take care not to take cold in them."

Even though her mother thought she was suffering from "descent of the womb," Adeline died of thyroid disease in mid-April. The circumstances must have been gruesome, as Maria instructed Julia to be certain that Addy was "*really really* dead—by *unmistakable* signs"—and to guard the dead body against rats (MJ/JPS, 13 April 1881; Stephen, *Mausoleum Book*, 70, 60; MJ/JPS, 17, 18, 15 [Good Friday], and 16 April 1881). (For a commentary on Maria Jackson's letters, see Appendix A.)

Julia stayed on, Leslie feared, precisely because caring for the bereaved Vaughans was "painful" and therefore valuable according to Julia's moral economy. She was so melancholy about life she expected that cheerful little Thoby would "never be so happy again." Leslie vigorously protested her pessimism, but to no avail. At last Julia returned home, on 22 April 1881, after at least eleven days of this harrowing and exhausting ordeal. Beforehand, Leslie wrote to her that he looked forward to her being "in your nest again, close to your poor old husband & with the little ones near to us,—I am not very sorry that Toto will be in another room, for you will want a good bit of sleep & you shall have it in my arms." Julia stayed home for a week but then went to Brighton to visit her mother, who found her "worn" after her deathbed watch. Maria concluded—too late—"You ought not to have been allowed to undergo all the physical as well as mental strain" (LS/JPS, 18 and 20 April 1881; MJ/JPS, 30 April 1881).

With four older children and then Vanessa and Thoby, Julia and

Leslie had not intended to have more children, but after Julia's emotionally draining time with the dying Addy, perhaps immediately on her return—with Thoby now sleeping in the nursery and Leslie holding her in his arms—Julia and Leslie conceived their third child, Adeline Virginia. Late in her life, Virginia wrote that her parents "did what they could to prevent me" (MB, 127). Perhaps only a euphemism for contraception, the phrase "prevent me" nevertheless points to Virginia's residual suspicion that her existence was unwanted and highly provisional.

A new pregnancy began to threaten further Julia's nearly exhausted strength, so in June 1881 Vanessa and Thoby were sent, with Stella and Gerald, to visit their grandparents in Brighton. After their return, Julia traveled alone there several times during the summer, leaving the little ones in London with Leslie and, presumably, Bunch. In October, Julia was again with her mother in hopes of developing what Leslie called "strength for the next brat." He was determined that this be their last child and urged Julia not to contradict or laugh at him: "Dearest, whatever the difficulty there must not be a successor to Chad." As the birth of the third Stephen child, "Chad," approached that winter, plans were made to send Thoby alone to the Jacksons, but Maria claimed that her "dear Bunch" would be unhappy without Vanessa too (LS/JPS, 28 July and 17 and 20 October 1881; MJ/JPS, 22 January 1882).

Despite Leslie's objections to sending Vanessa away, when a girl was born, on 25 January 1882, both her siblings were sent to the Jacksons' grand home in Brighton where they enjoyed the attentions of Gerald and Stella Duckworth, and their hypochondriac grandmother. There was a "family dissension" regarding the name—with Maria hinting, on black-bordered note paper, "As for her name I am sure I hope she is going to do better than have her hideous old Granny's"—but the expected "Chad" was named Adeline (after Julia's deceased sister, aunt, and grandmother) and Virginia (after Julia's aunt, Virginia, Countess Somers) (LS/James Russell Lowell, 28 January 1882; MJ/JPS, 12 February 1882).

Leslie wrote to his friend James Russell Lowell (the American poet and minister to England) that he had been anxious about Julia because all her worries were a "bad preparation for this trial." Nevertheless, mother and baby had made a "good start," and Lowell was asked to be Virginia's secular godfather. A "Mrs. G" was hired for a month to care for baby Virginia. Two weeks after Virginia's birth, Maria asked, "Is little Virginia like my Vanessa?" She wondered whether her eyes were as large and round or "orblike" as Vanessa's. She asked if Julia were giving "the little one bottles as you ought to do? yr nursing ought really to be only nominal." Maria passed on her doctor's theory that Julia was suffering from what Maria called "vacancy" or "the look of weakness," which indicated nervous exhaustion from nursing babies too long. Scolded by her mother, an exhausted Julia stopped nursing Virginia. Julia feared that Leslie would disapprove of her "bottling" Virginia, but he insisted that he did approve.[8]

Around the time of Virginia's birth, the publisher of the *Cornhill* removed Leslie Stephen as editor (subscriptions had drastically dropped) and invited him to edit instead a monumental dictionary of all the important English figures throughout history.[9] Also in 1882, Leslie leased Talland House in St. Ives, partly as a retreat from the daunting task he had assumed. Mrs. G stayed on with the family until Virginia was five months old and apparently did not accompany the Stephens on their first trip to Talland House, summer migrations that Maria suspected would be a fresh burden and further drain Julia's strength. Meanwhile, Maria reported that Stella already had begun "to take an interest in the Daisy Chain"—that is, someday marrying and having her own children.[10]

The next winter, when Virginia was just a year old and Julia and the children were staying with the Jacksons in Brighton, Leslie instructed her to "kiss my ragamice & Ginia" and affirmed "there will be no more of that breed." On the anniversary of her birth night, while sending a kiss to "my darling Ginnums," he wrote of his relief that this year was not a repeat of the last when Julia was in labor (LS/JPS, 23 and 24 January 1883). What he did not know was that Julia was probably already pregnant with her seventh and last child, as Virginia was later to explain, "in spite of precautions" (MB, 127).

Combining great beauty and charm with the skills to organize the enormous Stephen household, Julia seemed to Virginia "the very centre of that great Cathedral space which was childhood. . . . My first memory is of her lap; the scratch of some beads on her dress comes back to me as I pressed my cheek against it" (MB, 81). In another memory, which also seemed "first" and "most important," Virginia recalled feeling as if she were "lying in a grape and seeing through a film of semi-transparent yellow." This impression was all-enveloping or "globular": "The buzz, the croon, the smell, all seemed to press voluptuously against some membrane" (MB, 64–66). Virginia's impression of ecstasy and rapture inside a membranelike enclosure seems to have been located at Talland House. It hearkened back to the infant's early sense of fusion with her mother and the world, but for Virginia that fusion did not last long enough. From an infant's perspective, however, it never does, as the psychoanalyst Melanie Klein explained in London in the 1920s.[11] Weaned early and abruptly and then losing her substitute mother at five months, baby Virginia longed for her mother.

Her energies drained by watching over sister's deathbed, attending her mother's sickbed, and giving birth to her sixth child, Julia seems to have taken better care of herself during her next pregnancy, although perhaps at baby Virginia's expense. Leslie assured Julia, when she was again with the Jacksons in the fall of 1883, "Ginia most affectionate. She sat on my knee to look at Bewick & every now and then said Kiss and put her cheek against mine." Virginia's affection for her father may have compensated for her mother's absence, a separation she would feel more completely with the birth of the Stephens' last child seventeen days after Leslie wrote that letter to Julia. Julia felt "some special affinity of temperament"–although he also was unplanned–with the new baby, Adrian. Whereas she had apologized to her husband for weaning Virginia early, this time, after six months Leslie was urging her to "turn l'Adrian adrift."[12] Later Virginia realized:

> I see now that [my mother] was living on such an extended surface that she had not time, nor strength, to concentrate, except for a moment if one were ill or in some child's crisis, upon me, or upon anyone—unless it were Adrian. Him she cherished separately; she called him "My Joy."

Late in her life Virginia wondered, "Can I remember ever being alone with [my mother] for more than a few minutes?" (MB, 83).

When Julia's energy began to return, she devoted her time to caring for the sick and needy, making tiring rounds by bus to hospitals and workhouses. Leslie compared her to a flying fish, jumping from one obligation to another with "a dozen people tiring you to pieces & Adrian fretting & so forth." When Julia was away, "Little Ginia" became an especially "accomplished flirt." When Leslie once said that he must leave her, she "squeezed her little self tightly up against me & then gazed up with her bright eyes through her shock of hair & said 'dont go, Papa!' She looked full of mischief all the time. I never saw such a little rogue." He hoped that the Ragamice were beginning to make her "more one of the little party" as they together tried putting her to bed. Julia's frequent trips to her parents' elegant house on the Brighton seacoast, often with Adrian and sometimes with more of the children, left Leslie lonely and impatient. He longed for domestic stability: "When you & the little ones are all round me, I feel like an animal in his burrow" (LS/JPS, 3, 13, and 14 April 1884 and 8 February 1885).

As John Bicknell explains, since Leslie "often worked at home as much as he did at the British Museum or at Waterloo Place [the *Dictionary* office], almost daily he spent time with the growing brood, especially when Julia was away nursing her mother or some other relative."[13] Stephen published the first volume of his *Dictionary of National Biography* in late 1884 and continued to organize and edit others' contributions while he himself was writing the lives of most of the major English authors on such a tight schedule that another volume appeared each quarter. He relieved the tension with annual winter mountain-climbing expeditions to Switzerland.

These duties and absences made him an erratic surrogate mother, but there was no other lasting substitute except the children's half sister, Stella. Maria's letters mention not only Mrs. G but also Lisa, Suzette, Mrs. W, Mrs. V, and Nurse as caretakers for the Stephen children.[14]

Vanessa and Thoby, Vanessa recalled, "had had an intimate friendship before [Virginia] came on the scene." Virginia's arrival soon set up a rivalry between the sisters that Vanessa interpreted as "simply the result of two little females and a male" (LWP, 2:6a). Vanessa's first surviving letter, signed "Vanessa Stephen Butterfly," tells her father of their privileged life visiting their grandmother at Brighton, riding donkeys and getting "airballs." Her account of the fate of their balloons presages later interactions: "First Ginia's was busted, then Thoby's was busted and then Ginia busted mine" (VBL, 5).

When the children were sent back to their father while Julia stayed on at Brighton, Leslie wrote to her about Ginia's "very gravely" reporting on her visit. Virginia's hot temper, exhibited when she tried to bite or scratch Thoby, had already earned her the label "naughty." Thus when she explained to her father that in a Punch and Judy show, a substitute puppet had to be found because the old Punch was "so naughty," the joke seems to be that she was projecting onto "old Punch" the title she herself so often earned. When Leslie wrote to Julia, "I am sorry that my 'Ginia is out of sorts, dear little soul," Virginia was either ill or grumpy over being away again at Brighton, albeit this time with Julia (LS/JPS, 22 July and 16 November 1885).

If the first two Stephen children formed a girl–boy pair, the second set might have been expected to follow their example. But Virginia had no interest in pairing with her little brother, as she later sometimes regretted

(D, 2:242). Instead, she tried to enter into or spoil the twosome her elder siblings had established. And she looked on the younger interloper as a rival for her mother's affections. Virginia's unhappiness is suggested in another of Maria Jackson's innumerable letters to Julia: "Have you a double perambulator for them [Virginia and Adrian]. How pretty they must look. God bless you my own heart. Poor little Ginia." In the dynamics of the nursery, Thoby and Vanessa had each other, but "poor little Ginia" had no one because first Julia's energies were exhausted and then Adrian absorbed them. Maria thought of giving the little girl a locket with a lock of her mother's hair in it, as if to compensate (MJ/JPS, 10 May 1884).

Despite her profound pessimism and her frequent absences, Julia was almost universally lauded as the perfect woman and ideal mother. James Russell Lowell said he considered Julia a saint, even if she was one of his "own canonization," and the children suspected that he was in love with her (Bell, 1:25). Leslie Stephen envisioned Julia as both a disembodied goddess and a perfect mother. Analysts now tell us that the mother is experienced not as a saint but as an enabler who, as she becomes more separate, encourages the baby gradually to establish its own independent self. But if the mother is lost or her interests are suddenly deflected, the child's development into an autonomous self is jeopardized.[15] With seven children of her own, a retarded stepdaughter, and countless nursing commitments to her mother and almost anyone in need, Julia's nurturing skills were spread thin. She and Bunch had enabled Vanessa to develop such a secure sense of herself that she could mother her little brother, but Julia was too worn down to do the same for Virginia.

Virginia reacted to Julia's defection to Adrian by trying also to remain her mother's baby. She would not speak. Her orblike eyes observed alertly, but her pouting lips refused to say what she saw or felt. Vanessa remembered her drumming on the table, demanding breakfast without words. Then, as if conceding that Adrian could remain their mother's baby and she could grow up, little Virginia adopted a different strategy. When she was about two, she began to speak as a suddenly articulate self.

When Leslie scolded her for scratching Thoby with her fingernails, for example, the two-and-a-half-year-old looked "very thoughtful for some time and then says, Papa, why have we got nails?" Leslie was clearly delighted with this "bit of infantile teleology" (Bell, 1:22, 23–24, note). Stella wrote to their mother a less endearing example that shows Virginia using words instead of fingernails:

> This morning Ginia wanted Thoby to give her something which he had but he wouldn't, so she went up to him and gave him a hug and said "Please Thoby give it, Darling Sweetheart Boy" but Thoby still said "No I won't." Then she went up to him and tried to bite him and said "Nasty, Pigswash horrid, disgusting boy" and afterwards he gave it her.[16]

Speech had become Virginia's deadliest weapon, Vanessa remembered, soon aimed less at Thoby than at Vanessa, the little mother who babied Thoby and not Virginia and who was, as her parents were not, fair game. Virginia laughed at Vanessa's truthfulness and matter-of-factness, qualities she later associated with painterly simplicity. She "persecuted" her older sister "with horrid titles" (MB, 31).[17] Indeed, her verbal assaults may have further isolated Virginia from both the offended Vanessa and the disapproving Julia, but nonetheless, through speech, the toddler Virginia came into possession of her self.

Although many myths about Julia Stephen have survived the century since her death, concrete evidence is scarce.[18] The only letter to Virginia that seems to remain is not a comforting missive: Julia wrote about a woman who accidentally killed her child when she "waltzed [the baby] round the room till all the breath was out of its body. . . do mind you never waltz children round the rooms" (Berg, n.d.). Despite that attempt at wit, the letter seems a thoughtless one to send to one's own child without loving reassurances. Perhaps Julia was trying to warn her daughter that dependence on one's mother carried its own dangers. But Virginia so adored Julia that the more distant she was, the more clinging Virginia became. Much later—and using the present tense—Virginia remembered waiting "in agony peeping surreptitiously behind the blind for her to come down the street, when she has been out late the lamps are lit and I am sure that she has been run over" (MB, 84).

Among the surviving documents are a number of stories and essays that Julia wrote when Virginia was a little girl that dispel several myths.[19] First, Julia was ambitious: She turned her altruism into a practical publication, *Notes for Sick Rooms*, and, with Leslie's encouragement, tried unsuccessfully to publish her stories and essays. She was a creature of her class, however, and so did not extend her sympathies to the servant class when its claims threatened her privileges. Julia also was highly moralistic, intending her stories to be as edifying as her essays. But she was inconsistent, arguing that women should be respected as professional nurses and as independent thinkers yet should not be free to rethink traditional gender roles. Julia clearly was not introspective, as she seems to have been as oblivious of the contradictions in her positions on women and servants as she apparently was of the irony of posing as the Virgin Mary for the painter Burne-Jones when she was a committed agnostic.

Julia's stories are filled with private references to the Stephen children. In several stories, the boy Harry seems to be Thoby, and in "Wandering Pigs," when the pigs land their boat on a beach and seek refuge in a little cave, the "lovely Vanessa butterflies dashed backwards and forwards" (JDS, 145). The story "Cat's Meat" is Julia's more generalized answer to any of her children who dared object to her missions of mercy to the poor and the sick. The children in the story run away from home to the suburbs in the hope that, as they later tell their mother, "if we were really poor children, we *should* see you because you are always with the poor" (JDS, 187). The children are found and are returned to their upper-class home, and the story ends. Julia's moral was that the children should appreciate their comfortable existence and not resent their parents' absences. She did not believe that parents like herself should be more attentive to their own children. Julia apparently was trying to be humorous when she wrote that a little boy and girl "had quite settled they should be married when she knew how to darn" (JDS, 85), but the line is less funny when one realizes that the stories actually endorse the notion that girls need not learn much beyond darning.

Another of Julia's stories openly features the three eldest Stephen children, even identifying Virginia by name. In "The Monkey on the Moor," set at Talland House and in the Cornish countryside, there are two older children, Annie and Harry, and "little Ginia, the youngest" (JDS, 47), who does not scratch or bite but is still considered very naughty at the seashore: "Instead of running up and down in the hot white sand to dry her feet as Annie and Harry were doing, this wicked little girl popped her shoes and socks into a hole and covered them well up and then ran off to the sea again." Later, Annie and Harry are allowed to go on an excursion to town to buy new shoes, but Ginia has been "so mischievous she must stay at home" (JDS, 47). Feeling rejected, Ginia becomes sick, but her illness does not produce its apparently intended results, for at the end of the story, she has been "quite forgotten" (JDS, 63). In Julia's treatment of the naughty Ginia in "The Monkey on the Moor," I suspect a psychological source for Virginia's propensity to be ill.

Julia's assumption that "to serve is the fulfillment of woman's highest nature" (Annan, *Leslie Stephen*, 120) endeared her to many eminent Victorian men, but one contemporary female witness remembered that "there was something awful about Mrs. Stephen's wide-apart eyes and the perfection of her chiseled brow and nose and lips." Julia was awe inspiring, but that "something awful" may have also been Julia's severity toward young women. This seventeen-year-old witness with literary sensibilities had anticipated visiting Talland House, "seeing the great author and perhaps having a glimpse of his library. Judge of my disenchantment when we [her mother and herself] were separated at the drawing-room door, and Stella Duckworth was instructed to take me to nursery tea!" (Swanwick, *I Have Been Young*, 107). Clearly, Julia's sense that young women, even precocious literary ones, were to visit in the nursery, not the library, was thought demeaning and out-of-the ordinary, at least by that adolescent girl.

The letters to Julia suggest her habit of indulging her sons but not her daughters. When George Duckworth failed to win a prize, Leslie observed that a week's reading might have helped his chances.[20] In 1884, the novelist George Meredith wrote to Julia that he wanted to see Thoby "before his father has taught him that he must act the superior, and you have schooled the little maids to accept the fact." Meredith went on to tell Julia that gender roles were "largely (I expect you dissent) a matter of training" (*Letters of George Meredith*, 2:360). Julia's stories preach that a girl "must always do what she was told and never think of what she liked" (JDS, 123). Thus her insistence that little girls must learn their roles implies tacit agreement with Meredith, even though she would train them in passivity, he in assertiveness.

In the days of her widowhood, Julia had easily schooled Stella Duckworth to be passive. Leslie admitted that Julia considered Stella to be essentially a part of herself (Stephen, *Mausoleum Book*, 59). She called Stella, no doubt under the guise of affection, "old cow." Virginia recalled that Julia "was stern to [Stella Duckworth]. All her devotion was given to George [Duckworth] who was like his father; and her care was for Gerald, born posthumously and very delicate" (MB, 96). Julia sent her Duckworth sons to Eton and Cambridge, whereas Stella had very little formal education.

As Virginia reflected, "Stella she treated severely. . . . A pale silent child I imagine her; sensitive; modest; uncomplaining; adoring her mother, thinking only how she could help her, and without any ambition or even character of her own" (MB, 96). When Stella kept a diary, she had so little sense of herself that she recorded mostly the activities of the other members of the family (Stella Duckworth's Diary, 1896 [Berg]).

Whereas the self-effacing Stella was accommodating, the little Stephen girls were not. Indeed, Julia intended her stories to teach her Stephen daughters to be more like her Duckworth daughter.[21] Instead, Vanessa was mothering but never self-effacing, and Virginia was neither. Vanessa expressed herself with paint, Virginia with words, making her needs and opinions known by using speech like a weapon.

Julia reserved her nursing energies for adults. Since adolescence, she had cared for her mother. In 1873, she nursed her Prinsep aunt and uncle, as she did again in 1878. In 1881, she cared for the dying Addy in Pembrokeshire. In 1884, she presided over a friend's deathbed. And by 1885, she began to worry about the strain the great dictionary was inflicting on Leslie's health. He continued to protest that "I am not an invalid" and logically argued that a man who climbed mountains or "tramped" from London to Cambridge and back for recreation could not be ill.[22] Nevertheless, the enormous editing and writing chores, as well as dealing with complaints about errata, plagiarism, and unfairness, were fraying Leslie's nerves. He cursed the dictionary:

> That damned Dictionary is about my bed and spies out all my ways, as the psalmist says. . . . The damned thing goes on like a diabolical piece of machinery, always gaping for more copy, and I fancy at times that I shall be dragged into it, and crushed out into slips. (quoted in Annan, *Leslie Stephen*, 111)

Leslie was unconventional enough to say such things at dinner parties, distressing the ladies but amusing his friends with "outbursts of invective against some obtuse contributor." Julia was not amused, however, as she saw that this work was destroying his peace of mind (Annan, *Leslie Stephen*, 111). Caring for him despite his protests, she frequently sent the children to stay with their grandparents in Brighton.

Also living in Brighton were Julia's sister Mary Fisher; her husband, Herbert; and their eleven children. Often enough, the five motherless and now fatherless Vaughan children stayed with their grandparents and cousins in Brighton. Except for the youngest Vaughan cousin, Emma, whom she called Toad and considered her best friend for a time, Virginia had little use for her sixteen cousins on her mother's side of the family. Most of the aunts and uncles and cousins, on both the Jackson and

Stephen sides, were too dull, pious, and commonplace for the lively Stephen siblings. Furthermore, the freethinking Leslie encouraged their contempt, at least for their Jackson relatives. His attitude toward her sisters and brothers-in-law created some friction with Julia, who was always willing to leave her husband and children to tend the sickbeds of her mother and family.

When Virginia was four, Julia went to Paris over Easter with her eldest son, George Duckworth, leaving Leslie and the children in Brighton. First the family stayed in lodgings where the nurses were, Maria reported, "so happy." Then Leslie returned to London on Easter Sunday or Monday, and the children moved to Maria's grand home on Brunswick Terrace overlooking the Brighton coast. Meanwhile, Maria busied herself by sending travel tips to Julia. For example, she counseled Julia to take a berth on the returning ferry in the gentleman's cabin, which was "always safer as [gentlemen] are not sick." Julia returned to Brighton in the middle of the week and remained there with her mother and the children. This led Leslie to claim: "You are not much nearer to me at Brighton than at Paris" (LS/JPS, 26 April 1886; MJ/JPS, 27 April 1886; LS/JPS, 30 April 1886).

The Brighton beach was composed of hard pebbles; the surf in the English Channel was dull; and pony rides were confined to the well-manicured parade between the house and the sea road. Even though both Talland House and Brunswick Terrace overlooked water, they were physically and psychologically near opposites. Whereas Talland House was overgrown with unkempt vines (Figure 4), Brunswick Terrace was a model of Regency elegance in white stone with fluted pilasters outside and a multistoried entranceway inside. Talland House perched above a wild coast, and Brunswick Terrace sat back from but almost level with the calm Channel waters. Talland House was unique and isolated, but Brunswick Terrace was a row of town houses that varied from one another only in such minute architectural details as cornices, pilasters, and half capitals. Talland House had its own lawn, garden, and greenhouse, whereas Brunswick Terrace abutted directly on pavement stones. Talland House was open to the country, and Brunswick Terrace was at the center of fashionable Brighton (technically, in the part called Hove). Talland House sat above a simple fisherman's village, and Brunswick Terrace looked down the coastline to the Brighton Pavilion, an absurd (and doomed) example of Victorian ornament and decoration. And at Talland House the children played cricket and other games outdoors (Figure 5), whereas at Brunswick Terrace they seem only to have taken well-supervised pony rides. Because Virginia was sent there so often without either parent or with just one parent and because her mother was so often at Brunswick Terrace without her children, the ordered elegance there became, for Virginia, associated with separation from her parents.

By the fall of 1886, Leslie Stephen had simplified what he termed his "burrow," by putting the retarded Laura into foster care and reserving his fatherly energies for his four younger children, particularly Virginia. One summer, Leslie hiked back from St. Ives along the southern coast of England in about five days to meet Julia and the Stephen children at Brighton. Writing to Julia every night, Leslie worried about all the family but consoled himself: "I can always call up their little faces, especially Virginia's."[23]

Again in 1887, Julia spent a prolonged time in Brighton, this time nursing her seriously ill father. She left the children with Leslie. Then after her husband died in early April, as Leslie explained, Maria Jackson's "nerves broke down," and so Julia stayed on to care for her. Leslie protested that she was "sacrificing yourself by staying away from me & the children for weeks. Your mother knows it, you know it but you dont think it right to speak of it." Leslie thought that Maria was "unintentionally cruel" to her daughter and that Julia was "calmly submitting to be tormented." He urged her gently to tell her mother so. But she remained with her mother at Brighton, making Leslie, he confessed, "rather impatient" and their children, especially Virginia and Adrian, rather foresaken (LS/JPS, 5 and 6 April 1887; Stephen, *Mausoleum Book*, 71).

Leslie sent Julia engaging news of the children's activities, especially Virginia's verbal precocity: "Ginia tells me a 'story' every night—it does not change much but she seems to enjoy it." She also wrote a letter "in a most lovely hand." Virginia, whom Leslie now called Br'er Fox, certainly for her cleverness and perhaps also for her scheming ways, seems to have compensated for her mother's absence by becoming an accomplished, or at least a ready, storyteller. Her father believed that she would become a novelist, although he was expecting her to rival only a Miss Veley. One evening, Virginia announced that she would make a speech: "She stood in the window & declaimed a long rigmarole about a crow & a book till her hearers coughed her down. She would have gone on till now." While passing on such jolly stories, Leslie repeatedly urged Julia to return to her family, bringing her mother with her if necessary. Instead, Julia extended her visits (LS/JPS, 15, 12, and 17 April 1887; Stephen, *Mausoleum Book*, 71). Nervous exhaustion and hypersensitivity began to jeopardize Leslie's own health, almost, it seems, as a means to reclaim Julia from her mother.

In the summer of 1887, the Stephens brought the widowed Maria Jackson with them to St. Ives, further cementing a connection between the family's being together at Talland house and separated at Brunswick Terrace. Virginia's distaste for Brunswick Terrace may have been the reason for her first house-finding venture, at the age of six. Leslie had speculated that Maria might "sacrifice herself to the furniture" by staying on in the expensive and oppressive grandeur of Brunswick Terrace. In February 1888, the Stephen children were once again visiting Maria, the Fishers, and the Vaughan family, who spent their holidays there. Virginia was staying at Powys Grove with the "invalids," probably the Vaughans. When her grandmother came to visit in a horse-drawn "fly" or carriage, Virginia was "very radiant" over a house in Powys Grove that she had found for her Granny. To humor her, Maria went with Virginia to see the house. Although it was much more suitable for a widowed invalid than the multistoried Brunswick Terrace was, it was too plebeian for Maria's tastes, and she joked about the excursion to Julia.[24]

Maria speculated that Leslie was suffering from indigestion, a "horrid faintness," and an "over-worked brain." Nevertheless, she continued to call Julia to her bedside, away from Leslie and the children. In the spring of 1888, when all the Stephen children caught whooping cough, Julia cared for them and then left to take care of Maria. By June, Leslie called Julia "my poor battered darling" and urged her to return to him. She left Brighton but did not return home immediately, for Maria lamented that she was too infirm to stay at "HPG with the little ones whilst [Julia was] away." Instead, the little ones, recovering from their whooping cough, were sent to stay with her in Bath. On 18 June, Maria reflected, "It is a mercy the children have got over their w. cough." She was invited to join the family at St. Ives again that summer but declined as "gouty people get worse by the sea" (MJ/JPS, 13 February 1888; LS/JPS, 1 June 1887; MJ/JPS, 9, 3, 18, and 12 June 1888).

In her "Notes on Virginia's Childhood," Vanessa remembered that late spring when

> in the end emerged four little skeletons & were sent to Bath for a change. The rest of us quickly recovered, but it seemed to me that Virginia was different. She was never again a plump and rosy child & I believe had actually entered into some new layer of consciousness rather abruptly, & was suddenly aware of all sorts of questions & possibilities hitherto closed to her. (LWP, 2:6a)

Her father had called her "that plump little partridge Ginia" and considered Virginia "the healthiest of the lot," but after the whooping cough she changed physically and, Vanessa suggested, psychologically. Maria wrote to Julia, "As for Ginia's [cough] it must be a spurious kind I think she goes on coughing but no whoop or interruption of appetite" (LS/JPS, 14 November 1885; MJ/JPS, 18 June 1888). Virginia seems to have hoped that at six years old, after three periods away from her mother that spring, remaining ill—even with a "spurious" cough—offered her the best chance to reclaim her mother's affection. But like little Ginia in "The Monkey on the Moor," she learned that even illness could not guarantee her mother's presence or attention. In a very early letter she imagined, "I AM A LITTLE BOY AND ADRIAN IS A GIRL" (CS, 2), perhaps surmising that as a boy she might displace Adrian in her mother's affections.

There was a man who appeared from time to time in Hyde Park Gate and exposed himself to girls. Both Vanessa and Virginia had seen him (Bell, 1:35), but such behavior was not reserved for the streets:

> There was a slab outside the dining room door[25] for standing dishes upon. Once when I was very small Gerald Duckworth lifted me onto this, and as I sat there he began to explore my body. I can remember the feel of his hand going under my clothes; going firmly and steadily lower and lower. I remember how I hoped that he would stop; how I stiffened and wriggled as his hand approached my private parts. But it did not stop. His hand explored my private parts too. I remember resenting, disliking it. (MB, 69)[26]

The event happened "once," perhaps when she was four or five and Gerald was sixteen or seventeen. Indeed, the vividness of the recollection seems to result from that singularity.

Although this memory in no way justifies claims that Virginia was a victim of continuing sexual abuse, it may explain why Virginia grew up to fear male sexuality and her own body.[27] In 1939, when she first (as far as we know)[28] put that experience into words, Virginia also remembered a playful fight with Thoby: "Just as I raised my fist to hit him, I felt: why hurt another person? I dropped my hand instantly, and stood there, and let him beat me. . . . It was as if I became aware of something terrible; and of my own powerlessness" (MB, 71). Both these recollections brought on helpless despair, for "they seemed dominant; myself passive" (MB, 78, 71–72).

Virginia told no one of her terrifying experience with Gerald and may have completely suppressed it for some fifty years. But like the experience of whooping cough, it marked her.

Virginia's literary gifts distinguished her from her siblings and began to cause some friction. For Christmas in 1888, James Russell Lowell sent the Duckworth and Stephen children £1 each. Thoby composed a poem to thank him, and Virginia (Lowell's goddaughter) transcribed the poem. Leslie wrote to Lowell that "Thoby was complimented on his first verses: but explained that he should prefer shooting tigers, as a profession, to poetry" (LS/JRL, 30 December 1888). Virginia, not yet seven, probably felt irritated that Thoby could invent a poem, that she had to copy his work, and that he was contemptuous of the literary profession. Under her parents' tutelage, she had learned to write early, but Julia and Leslie were not always such successful teachers. Their absences made the lessons erratic, and their impatience sometimes made them explosive. Only Thoby learned much math from Leslie. And Virginia might have felt that only Adrian got the full teaching attention of a weary Julia (Figure 6).

When Julia was at home in London, Maria Jackson feared that nursing Leslie—who by the late 1880s had suffered several nervous collapses—would make Julia an invalid. And when Julia was with Maria, Leslie feared that nursing Maria would accomplish the same end. Observing the competition between grandmother and father over who needed Julia more and who was harder on her health, the children must have felt uneasy. Virginia, with the new depth of consciousness that Vanessa said she had developed, probably was alarmed.

In January 1889, Julia and Leslie left the children with Maria to go to Grindelwald in the Swiss Alps. Despite her worries about Julia's catching cold if she went out in the night air, Maria hoped that both Stephens could rest, Julia could gain weight, and Leslie's "nervous energy [could] be repaired." In a photograph of them at that time, Leslie, with no gloves or overcoat (but with good boots), looks determined to show himself healthy, but Julia, turning her profile to the camera and wearing furs and an inadequate hat, appears cold and self-absorbed (Figure 7). The Stephens stayed on the Continent until mid-February, missing Virginia's seventh birthday.[29]

When Leslie made his other mountain-climbing pilgrimages to the Alps, Julia usually took the children to Brighton, further cementing the association between being at Brighton and being separated from at least one parent. During Leslie's 1891 mountain climb, Julia did stay at home with the children, but when he returned, she fled to care for her worsening mother in Brighton. As Julia's absences continued, Virginia seems to have become more and more attached to her father, as he was to her. When the dying Maria Jackson was moved to 22 Hyde Park Gate, her needs still diverted Julia's attention away from the children. Not surprisingly, therefore, when Maria died there in 1892, they do not seem to have much lamented the loss of the "invalid of Hyde Park Gate."

It seems that one needed to be ill to attract Julia's attention. By this time, Leslie had given up protesting. The dictionary haunted his dreams, as did fears of failure or bankruptcy. Julia would soothe and reassure him during his troubled nights, but her solicitousness exhausted her energy and only encouraged his increasingly petulant behavior. Adrian Stephen recalled that in the summer of 1893, because Stella had mumps, Julia stayed behind to care for her when the Stephens set out for their holiday in St. Ives. Leslie traveled alone with Vanessa, Virginia, and Adrian on the train. (Thoby, now in boarding school, met them at the end of his school term; the Duckworth young men, now both in their twenties, might have joined the family briefly and then gone their own ways.) The Stephens carried a basket lunch, but their lemonade spilled over their sandwiches. Vanessa, who suffered from motion sickness, was excused from eating them, and apparently Leslie was excused because he was the father. But Adrian remembered that he and Virginia were "forced" to eat the soggy sandwiches, which he found "perfectly disgusting!" (AS/VB, 1 October 1945 [Berg]). Virginia might well have concluded that Stella, being sick, had the better time of it.

From St. Ives that summer of 1893, when Virginia was eleven, Leslie sent Julia descriptions of enjoying long hikes with the children and of their running up and down rocks with the "inevitable nets" for butterflies. He repeatedly urged Julia and Stella to join them, even though Stella had not quite recovered:

> There will be no difficulty whatever in keeping the children away [from Stella's contagious mumps]—the real difficulty is in ever getting hold of them: They are always about in the garden—bug hunting or up in the tree or playing cricket. Stella will be able to look out of her window & see them & the sea & I am sure it will be good for her.

Not all their activities were so safe. After Emma Vaughan joined them, the children went climbing along the rocky coast and "left Emma stuck somewhere. She was brought back in a boat, for wh. she paid 3 pence."[30]

Nor was life always harmonious. When Adrian did rather well at billiards, "to the scandal of the girls," Leslie noted that the "children seem to resent the success of smaller children," perhaps as a "way of protesting against the excessive praises given the little ones." Vanessa, Emma, and Virginia joked (until Leslie forbade them to) that Adrian's successes were mere "flukes." As Leslie half-surmised, Virginia insulted Adrian as payback for Julia's devotion to him (Figure 8). But Virginia, too, received excessive praise, at least from Leslie. He predicted to Julia that whereas Stella's vocation might be doing good deeds, writing articles would be more "Ginia's line unless she marries somebody at 17." He also pointed out that Vanessa painted, and he "discussed George II with 'Ginia. She takes in a great deal & will really be an author in time; though I cannot make up my mind in what line. History will be a good thing for her to take up as I can give her some hints" (LS/JPS, 2 August 1893).

Two recently discovered letters show different sides of the young Virginia: The much earlier letter, written from London to her mother, retells a tale she had heard from a member of Julia's family about an old man who got his legs "caute in the weels" of a fast train. The lugubrious narrative ends, "He called out for somebody to cut off his legs but nobody came he was burnt up. Good bye your Loving Virginia" (CS, 3). The later letter to Julia tells of visiting, along with her siblings, the painter G. F. Watts and catching butterflies. It also includes a joke about Leslie's unconventionality: "Has father done anything shocking yet, and has he found any flowers." This undated letter also tattles on Adrian, who "picked up an old whip with which he means to beat Shag [their dog]." The morbidity and fearfulness of the first letter are balanced by the liveliness of the second. But the second also discloses a writerly fear: "I expect that the others are writing the same things as I" (CS, 3–4).

Given his literary career and his longevity, Leslie Stephen's influence on Virginia has been easy to trace.[31] Virginia's first account speaks warmly of his instilling in her a love for books and inspiring her critical faculty by asking the children to evaluate the merits of the books he read to them (VWE, 1:127–30). Although Virginia's 1907 "Reminiscences" hardly mentions Leslie, her late memoir "A Sketch of the Past" is decidedly severe. Yet even there she praises him: "Just as a dog takes a bite of grass, I take a bite of him medicinally, and there often steals in, not a filial, but a reader's affection for him: for his courage, his simplicity, for his strength and nonchalance, and neglect of appearances" (MB, 115–16). Probably the medicinal dose was of Leslie's simple but noble notions about art. He judged a work for its "truth," which for him usually meant the values it promulgated and the decency of its author. That is, he assessed literature largely by ethical rather than artistic standards.[32]

Virginia recognized that although strong, her father's was "not a subtle mind" but "an impatient, limited mind." For example, his work on Jonathan Swift is "assertive by temper" rather than reasoned. As scholarship, it is "imperfectly informed" and even unfair, for Stephen simply dismissed Swift's beliefs as "prejudices."[33] Stephen's lack of subtlety meant that he was selectively unconventional (MB, 113, 115). For example, during the American Civil War he had visited the United States in support of emancipation, but like most Victorian inheritors of class and privilege, neither he nor Julia saw any analogies between their dependence on servants and the southern plantation owners' dependence on slaves.[34]

Leslie also failed to see any inconsistencies between being enough of a freethinker to reject Christianity but not enough to question patriarchal values. (In fairness, of course, such biases are easy to identify a century later.) He entrusted Julia with the financial management of their household, but his fears of insolvency brought on moods during which he would accuse her of nearly sending them to the poorhouse. Even though he adored Julia, "it never struck my father [Virginia believed] that there was any harm in being ill to live with" (MB, 109). James Russell Lowell considered Leslie the "most lovable of men," but to Julia and Stella he was becoming an emotional bully.

On Virginia's ninth birthday, Leslie had written to Julia from the Alps that Virginia "is certainly very like me, I feel though I cannot say how but so much more life in her than I had at her age. I can see her little face" (LS/JPS, 25 January 1891). As Virginia's verbal talent continued to develop, they were often "in league together. There was something we had in common" (MB, 111). They both were hardworking, ambitious, intense, hypersensitive, obsessively literary, and brilliant. But Virginia became a creative genius, whereas Leslie only wanted to be one.

A less sanguine affinity further aligned father and daughter, for lurking in both was the specter of mental instability. Julia felt Leslie's moods to be unstable and theorized that only Leslie's "absolute release" from the *Dictionary of National Biography* would "assure his becoming a sound man again."[35] With Leslie, being of unsound mind seems to have meant being unpredictably explosive or depressed, imagining himself so put upon that he had failed to develop his genius. Virginia's contrariness typed her, in her father's terms, as being full of "life" and, in her mother's, as "the naughty little Ginia" of the "Monkey on the Moor." In her siblings' terms, she was an ornery "Billy Goat." Vanessa remembered that she and Thoby liked to tease Virginia until, "purple with rage," she created an "atmosphere of tense thundery gloom."[36] Julia's severity with Virginia (MB, 82) may have been an attempt to prevent her moods from developing into an unstable state like her father's.

Often finding her rages counterproductive, Virginia also learned to use her tongue as charm rather than weapon. She was five when she told her father a story every night and made a long speech standing on the windowsill. At the same age, Virginia wrote her grandmother a letter "so beautifully written it ought to be framed." Leslie sent James Russell Lowell a letter that Virginia had written when she was six, with the proud testimony "This is a spontaneous production of Miss Stephen" (VWL, 1:1).

By the time she was seven, Virginia had found a way to use words with her grandmother against her mother. Maria scolded Julia, "Do not subject y[ou]rself to either of the conditions which Virginia says you live in!" Maria's letters suggest that the "conditions" the seven-year-old reported on probably were sleeplessness, thanks to caring for her husband or children, and overwork, thanks to her active social and charitable life.[37]

If tattling on her mother's fatigue had gotten her grandmother's attention, at about the same age Virginia learned how to enthrall her siblings who had resented the "excessive praises" given to her. In the night nursery, under heavy covers, with a fire flickering on the hearth, she told stories that entranced the other Stephen children. Meanwhile, Vanessa, who had recited poetry before she was two and delighted her father with her argumentative skills before she was three, was becoming the silent sister.

When she was nine, Virginia ventured into a more permanent medium. With Thoby, she began a weekly newspaper called the *Hyde Park Gate News* (usually two long sheets marked into columns and handwritten on both sides). Produced from Virginia's ninth to her thirteenth year, the paper offered family news and mocked standard journalistic features.[38] There is an "Easy Alphabet for Infants" with such rhymes as

G for Goliath
So Great and so strong.
H for 8th Henry
Who to his wives did great wrong.
. . .

W for Watts
A great painter is he.
X for Xerces [sic] who died B.C.

An 1891 "NOTICE" invited "any young people who whish [*sic*] for assistance" to write to the authors. Soon thereafter, an advice-to-the-lovelorn column appeared, along with model love letters "to show young people the right way to express what is in their hearts." There were scholarly notes, as on the "singularly unfitting" name Guinea Pig; jokes; a series of extracts from a fictional diary; and a serialized story (HPGN, 7 December, 30 November, and 7 December 1891, and 19 December 1892). Virginia's self-proclaimed precocity, her contempt for male authority figures, and her wicked tongue surface even in her nine-year-old phrasing. For example, when one "Sir Fred Pollock and his better half" arrived for tea, Virginia refused to "say much about them as they were not very interesting." On the same day, a certain doctor "was most unceremoniously observed by a most precocious little girl to greatly resemble a bull frog!" (HPGN, 7 December 1891).

Clearly, Virginia was the genius behind the production (one penciled reference to her as "Editor" is crossed out and "Author" is substituted), although Thoby and Vanessa did contribute to the paper as well.[39] Nevertheless, if Julia's story of "little Ginia" suggested an alliance of the older children against Virginia, the *Hyde Park Gate News* might have been Virginia's revenge, at least on Vanessa. For example, she mockingly remarked that the birthday present of a gold necklace "perhaps wasted it's [*sic*] splendour on" Vanessa. Furthermore, as it had no illustrations, the paper denied Vanessa an outlet for her own talents.[40]

The *Hyde Park Gate News* suggests why Vanessa had become the silent sister. On her tenth birthday, Virginia became "the happy possessor of a beautiful ink stand the gift of her grandmother. . . . She had also a blotter a drawing-book a box with writing implements inside." Maria Jackson bought presents to suit each Stephen granddaughter's tastes: writing implements for Virginia and painting supplies for Vanessa. After Maria's death, when Vanessa received, perhaps from her parents, an atlas on Homer, albeit a pictorial one, Virginia regarded it as a gift "not perhaps solely adapted to Miss Vanessa's tastes." Nor were the visual arts to Miss Virginia's tastes: "Our correspondent is not an art connoisseur" (HPGN, 25 January and 23 May 1892). Thus the paper shows Virginia claiming writing as her province. Willing to be a scribe but not a writer for the *Hyde Park Gate News*, Vanessa channeled her talents into painting. But when the Stephen siblings gathered around Vanessa at her easel for a family photograph (Figure 9), Virginia held her own emblem, a book, in her lap to signal their disparate interests.

The *Hyde Park Gate News* shows Virginia (or her scribe) not always in command of spelling conventions, as in "An Article on Chekiness" or the 1891 "Cristmas Number." But she was mistress of the mock-heroic tone. Even though the Duckworth brothers make only rare appearances in the paper, Virginia recorded that the "Mater familias of the Stephen family" was anxious over "her second son's maladies." She described her own great loss: The "youngest female residing at 22 Hyde Park Gate is plunged in the very deepest mourning by the loss of her beautiful boat the Fairy"; then an "Oh astounding event!!!"—the recovery of the *Fairy*—elated her. With a haircut, "the Editor now looks so like a cockatoo that she is ridiculed on all sides." Their Irish terrier's "appellation is Shag a derivation from Shaggg [*sic*] as he is long-haired and numerous haired." When Thoby arrived at St. Ives and was reunited with his mother, Virginia's tone indicates envy: She drew "the grey veil of silence over the joyous scene that ensued as it is too tender to be described." Virginia created drama out of the flare-up of a lamp and then undercut that drama at Adrian's expense. She even turned illness into a joke: "What is the difference between a camera and the whooping cough? One makes facsimiles and the other makes sick families." Virginia's droll retelling of family events and imitations of regular newspaper features delighted her readers: "Curious. The circulation increases weekly" (HPGN, 11 January 1892, 21 December and 30 November 1891, 15 February 1892, 14 December 1891, 11 July, 1 August, and 8 February 1892, 6 April 1891, and 7 December 1891).

The *Hyde Park Gate News* testifies to the Stephen children's vitality and good times. They could be unkind, as in their April Fool's joke on "her Ladyship the Lady of the Lake" (their handicapped half sister, Laura) when they told her she had a smudge first on one side of her face and then on the other and laughed to see her trying to brush it off. On April Fool's Day, 1892, when Virginia was ten, the children sent Julia a "false epistle," and "many other fooleries passed between the infants that day too many to relate." Later that year, an early release from Hyde Park Gate to Talland House was a "heavenly prospect to the minds of the juveniles." The *Hyde Park Gate News* tells of Leslie there as a delightful father. He taught the children the names of neighborhood plants while taking the family and their visitors (young people and eminent Victorian men of letters) on long walks, once sighting a seal and once saving the life of a bird caught in a net, but when the walks were too strenuous, Virginia and Adrian had to stay behind with Julia and Laura.[41]

Virginia was highly conscious of clichés, as her quotation marks show. She described Julia as "a 'Good Angel' to the poor of St. Ives, now trying to get enough 'Filthy Lucre' to start a nurse in the town." The paper continued its narrative of exciting activities at St. Ives. There the exuberant children took rides on hay carts, and Virginia and Thoby went to the lighthouse with two local youths and a boatman. No doubt Adrian was jealous, for a few weeks later Leslie took Virginia and Adrian on a sail to the lighthouse that "ended happily by seeing a sea pig or porpoise." Despite Julia's worries about rain, he took the children to the St. Ives regatta and hired a boat to "take them to the scene of action." Back in London, the paper reports that Leslie took Virginia, as "special correspondent," to a gathering "to rescue Carlyle's house from the tooth of time" (HPGN, 3 October, 18 July, 12 September, and 15 August 1892, and 25 February 1895).

Julia Stephen seems not to have participated in such activities, in either St. Ives or London. The paper reports only one excursion on which Virginia was alone with her mother. Just before her tenth birthday, "to the great delight of Miss Virginia Stephen [she] went with her mother to the Police Court on Saturday to bear testimony" against "the big dog." Virginia was both witness and reporter: "Miss Virginia Stephen was called up and she stated that the dog had run at her and bitten her cloak besides knocking her up against the wall." Julia testified as well, and together they restored safety to the Hyde Park Gate neighborhood, with the verdict that the dog was to be either controlled or killed (HPGN, 18 January 1892). The trial of the "big dog" established that her mother cared about safety, for herself and others, and also discipline, of the dog and others.

Despite the offenses of the "big dog," Virginia developed no fear of dogs or any creatures. She was delighted when a stray dog, perhaps ironically named Beauty, "adopted" the Stephens. Because "he is not renowned for cleanliness," Julia hired a boy to take Beauty to the London Dog's Home. But the home "did not admit dogs from private persons. So the boy turned him loose to wander." The loss, instigated by their mother, was traumatic: "This grief came like a thunder clap on the heads of the infants of H.P.G. on that memorable morning of the 2nd of March [1892]." At St.

Ives, it was necessary to get an "Iris [*sic*] terrier" to rid Talland House of rats. Waiting for the dog, Leslie "could not withstand the anxiety and excitement which reigned omnipotent," but to Julia the dog was "not for pleasure but for business." When it was "rather constantly seriously ill (under the side-board)," the dog became "a beast to be dreaded rather than liked by Mrs. Stephen." When the children were given a butterfly box, Julia also disapproved, for she was afraid of having caterpillars in the house (HPGN, 22 February, 7 March, 11 July, 4 July, 5 December, and 10 October 1892).

The *Hyde Park Gate News* reveals a Julia who was very stern about household regulations but very permissive toward her sons. In 1892, Stella gave to Julia, for her birthday, photographs of Adrian, clearly because he was her favorite child. When Thoby was at the top of his class, Virginia wrote, without apparent bitterness, "It brought a little misty moisture into the eyes of the parents." On George Duckworth's twenty-fourth birthday, Julia took him "to see her beloved Thoby," who had gone off to Evelyns School at about the age of twelve. Then Virginia wrote in such a saccharine vein that she seems ironic: "It must indeed be sad for the Mother to see her sons growing older and older and then to watch them leave the sweet world of child-hood behind and enter into the great world of manhood." When Gerald Duckworth returned from Eton, "our author was much touched to see tears in the maternal eyes." The household was in some suspense about whether or not Thoby would win a school prize. "Miss Vanessa Stephen said philosophically that he couldn't get it," but Julia, against Leslie's judgment, believed he would, and in a teary-eyed scene she was proved correct (HPGN, 8 and 29 February, 7 and 21 March, and 19 December 1892). In the spring of 1893, Julia went to Eton, presumably to use her influence on Thoby's behalf, and then she went there when he had his examinations that summer (LS/JPS, 10 April and 11 July 1893). Nevertheless, Thoby did not win a scholarship to Eton. As Virginia wrote, he was "not precocious" (MB, 125).

Julia appears in these pages as an angel to the poor and sick, an administrator of discipline and cleanliness, and a doting mother to her sons. The paper describes no scenes of affection between Julia and her daughters. The sentimental depictions of ties between mother and sons were calculated to win Julia's approval, but they also inadvertently convey Virginia's envy.

Analysts who study the significance of art theorize that at some deep level, aesthetic pleasure can recall the all-too-brief union with one's mother.[42] According to this view, Virginia tried to create art not to recall but literally to secure a union, because she wrote to claim Julia's attention. When her mother appreciated one of Virginia's amusing, precocious features in the *Hyde Park Gate News*, Virginia later described the experience as feeling "like being a violin and being played upon" (MB, 95). As long as she could use words wittily and well, little Ginia could expect that people, including her mother, would value and love her.

But the paper must have caused as much friction as delight. Despite the repeated references to the "younger Miss Stephen," Vanessa is rarely mentioned. References to "Miss Virginia Stephen's poetical work," "the quick eyes of our writer," and herself as "special correspondent" probably irritated her siblings; certainly they confirmed Vanessa's propensity to leave writing to her sister. Virginia called Adrian "the baby (but Alas! he is now no longer one)." On his ninth birthday,

> Mrs. Stephen with the maternal tenderness which . . . appears in everything she does wishes it was only his fifth as she says one is much nicer when one is young. Master Gerald Duckworth who is not quite so gentle [and would be twenty-two two days after Adrian's birthday] says that though Master Adrian is nine in years he is five in intellect. (HPGN, 30 November 1891, 25 April 1892, 25 February 1895, and 25 January and 31 October 1892)

Adrian had reason to resent such reports and determined in the next month to begin a rival paper, which Virginia, correctly assessing parental loyalties, knew would "not be underrated by Mrs. Stephen nor overrated by Mr. Stephen." After that announcement, the paper failed to appear: "Mr. Adrian Stephen's little 'squinty' paper was supposed to come out on Thursday but as we feared he is not blessed with the spirit of punctuality." Leslie Stephen attempted to avert a family feud by advising Adrian to join Virginia's "most respectable paper." Virginia offered obligatory encouragement but could not resist adding, "not that we are in want of writers." Adrian's inadequacies, rather than cooperation, averted a feud. The rival editor of the "Cork Screw Gazzette" was apparently "suffering from overwork," and so he gave up in disgrace his bid for journalistic fame.[43] Virginia's satire no doubt jeopardized the approval she had purchased, for Julia could withhold her endorsement when "little Ginia" made fun of little "Master Adrian."

Although Leslie had to work to avert a family feud, he probably found amusing Virginia's reactions to Adrian's journalistic pretensions, for Leslie also found Adrian "infantile for his years, though he is certainly intelligent." Leslie knew Julia's bias, and he feared that Adrian might be "moony" like his father, but then he apologized for the comparison, "though I dont mean to compare myself with him in other respects. You might be hurt," apparently by a suggestion that Adrian was in any way less than perfect (LS/JPS, 29 January 1894).

Just as Julia favored Adrian, so Leslie favored Virginia. Her account of their visit on behalf of Carlyle's house must have gratified him: "Mr. Leslie Stephen thanked the Lord Mayor very briefly for his kindness in permitting the meeting to take place in his Mansion House, upon which, the Lord Mayor said he had never been present at such an intellectual feast, and then he went away." Virginia noted that Leslie enjoyed having his discourses written up (HPGN, 5 February 1895, and 17 October 1892). Certainly Leslie was delighted with the following coverage of his accession to the presidency of the London Library:

> Mr. Leslie Stephen whose immense literary powers are well known is now the President of the London Library which as Lord Tennyson was before him and Carlyle was before Tennyson is justly esteemed a great honour. . . . The greater part of Mrs. Stephen's joy lies in the fact that Mr. Gladstone is only vice-president. She is not at all of a "crowy" nature but we can forgive any woman for triumphing when her husband gets above Mr. Gladstone. We think that the London Library has made a very good choice in putting Mr. Stephen before Mr. Gladstone as although Mr. Gladstone may be a first-rate politician he cannot beat Mr. Stephen in writing.

That encomium to masculine power gratified Leslie's pride. Writing to Charles Eliot Norton two weeks or so later, he borrowed from little Ginia's account: "As it is, [Gladstone] is my Vice, or one of them, and I am one of the few people who can be called his superior"[44]

Such houseguests as Henry James, George Meredith, Oliver Wendell Holmes, and James Russell Lowell took more objective pleasure in little Ginia's precocious journalism (Stella Duckworth's diary, 1896 [Berg]). By Christmas 1892, Meredith confessed that "my heart is fast going to Virginia" (*Letters of George Meredith*, 2:457). Leslie recognized her brilliance largely as a reflection of his, and later he proclaimed with satisfaction: "Virginia [is] becoming as literary as her papa" (LS/Charles Eliot Norton, quoted in Stephen, *Mausoleum Book*, xxviii). He expected her to become a journalist or to write history or ladies' novels. This most precocious little girl began her career by sending an early story, no doubt handwritten in her childish scrawl, to the popular journal *Tit-Bits* (Bell, 1:31). The inevitable rejection did not deter the fledging writer, nor did her siblings' resentment of her way with words, for Virginia already felt what she would call the "desire of the pen in the blood" (VWL, 1:169).

2
Dreadful Fixes

1895–1903

Whereas Leslie considered Virginia his career descendant and Virginia practiced the profession of writing in the *Hyde Part Gate News*, Julia regarded female professions as a contradiction in terms. Before their marriage, Leslie had written to Julia about his "chiefly" held belief that "women ought to be as well educated as men." But such views depressed Julia, as she could not "believe with such views on female education you can care for me." She claimed to be "utterly uneducated" and inadequate except in love. When he saw how his views upset her, Leslie assured Julia that a "noble character" was much more important than that a woman "should learn anything whatever." Consequently, as Noel Annan writes, "his daughters were brought up to suit Julia's ideas rather than his."[1]

When Leslie wrote this discussion of what Virginia Woolf later called "professions for women," he was thinking of the then seven-year-old Laura. Even when the handicapped Laura did odd things like throw scissors in the fire (Bell, 1:35) or amuse herself "by beating me," her father still expected her to learn a "bit of Greek & Latin & arithmetic." Such a disparity between her abilities and her father's expectations could only have exacerbated Laura's dysfunctionality.[2] Leslie Stephen's blindness to her disabilities suggests (as he acknowledged in another context) his "cowardice," a tendency "to shut [his] eyes" (LS/JP[S], 18 and 19 July 1877). However much Leslie denied Laura's infirmity, he did not deny a woman's need of an education and a profession. But Julia did.

As Virginia's talents blossomed, she probably began to feel professionally undermined by Julia, whose idea of female education outside the home seems to have consisted largely of sending Virginia to a "'graceful deportment class'" (HPGN, 5 December 1892). But it would have been difficult for Virginia to voice resentment of her adored mother and, as she had called Julia without irony in 1892, "reverend younger parent."[3]

Once, however, when the two sisters were bathing together, with typical inquisitiveness Virginia asked Vanessa which parent she preferred. Vanessa named her mother, and then Virginia claimed her father. Vanessa remembered that the event began "an age of much freer speech between us. If one could criticize one's parents what or whom could one not criticize?" (LWP, 2:6a). The moment permitted Vanessa to criticize Leslie, and Virginia, Julia. For her, Julia no longer seemed the sainted mother but, instead, the representative of restrictions, proprieties, "the consciousness of other groups impinging upon ourselves; public opinion; what other people say and think."[4]

But resentment of either "ideal" parent could not be articulated outside the nursery, and the result was repression and an obsession that, for Virginia, lasted until 1927: "How deep they drove themselves into me, the things it was impossible to say aloud" (MB, 108). A family photograph (Figure 10) shows Adrian absorbed with his mother, Leslie rather morose, and Julia (looking old beyond her years) rather sour. The Stephen children are dressed more casually than their half brothers or their mother. While the Duckworth brothers and Thoby and Vanessa seem jolly, Virginia is staring pensively, perhaps thinking of something that could not be voiced.

One of Julia's letters to George Duckworth hints at her coercive powers: The Stephen children were invited out to tea, and "Adrian wanted to go more than anyone & Ginia was rather alarmed so Nessa agreed to go they enjoyed it very much."[5] One can imagine Virginia, shy and "rather alarmed" among strangers, angry with both her brother, who got his way, and her mother, who honored Adrian's rather than Virginia's wishes and assumed that all three children enjoyed themselves. As she reached adolescence, the ambitious, supremely talented Virginia must have wondered how she might escape from her mother's pressure to "always do what she was told and never think of what she liked."[6]

We can tell by the handwriting that the 1895 volume of the *Hyde Park Gate News* had become exclusively the thirteen-year-old Virginia's. By then, the paper typically carried only two features: a family news column and a philosophical column written in the first person. One column about a man who gave up friends and fortune to live in the slums and help the inhabitants is a wonderful satire of upper-class prejudices against reformism. Another describes a dream that the author was God. Still another reveals Virginia's heightened sense of alienation: "But I believe after all, that human beings would find it very difficult to exist together if they knew each other." Really knowing each other would have meant knowing each other's innermost thoughts, a prospect that both intrigued and repelled Virginia. The idea of taking "possession of other people's minds, for a short time," offered power and insight. That insight, however, might reveal "that my best friend hated me," perhaps in reaction to Virginia's habit of satirizing even her "best friend."

In January 1895, while Julia was away on an unspecified "melancholy" visit, Leslie reported that the children were in "uproarious spirits." While "they were alone" with him, they went to the zoo and to talk with the old friend who had become a more successful editor of the *Cornhill* than Leslie had ever been.[7]

In early February, Leslie went on a mountain-climbing jaunt in Wales. On 4 March 1895, the *Hyde Park Gate News* noted that the "fiend" influenza had visited Mrs. Leslie Stephen a fortnight before. Then the 11 March issue records her near recovery and her renewed energy at "generally superintending the household as if she had never been ill in her life." Certainly Julia played the role of one who was never sick, but perhaps, too, the correspondent was in denial. In the issues of 18 March, 1 April, and 8 April, Virginia did not mention Julia's health. Instead, in the 8 April issue she wrote about their spring travel plans. The Duckworth children were going abroad, and "the rest of the family may stay for some days at Great Tangley Manor and High Ashes." That issue of the *Hyde Park Gate News* was the last of this high-spirited—although lately more serious and reflective—journalistic enterprise (LS/JPS, 2 and 3 January and 1, 2, and 3 February 1895; HPGN, 8 April 1895).

Despite her lingering sore throat, Julia took the Stephen children to visit at High Ashes, the elaborate country home of Sir Roland Vaughan Williams and his wife, one of Vanessa's godmothers. Leslie briefly visited nearby with Mr. and Mrs. Wickham Flowers at Great Tangley Manor, perhaps to see the imitation medieval library that the Flowerses had added when they redecorated the manor the previous year. On 15 April, Leslie walked over to see his family at High Ashes, which he called "that astonishing place." He was appalled to find that the "fiend" had returned. Julia's sore throat was worsening, and Virginia, too, was sick. Returning home to Hyde Park Gate, Leslie wrote to Julia that he was "still rather haunted by your looks—you were so tired & weak."

If her looks on that one-day visit so haunted him, her children must have been even more distressed to watch Julia weakening while caring for Virginia. Because High Ashes was so cold and drafty, Leslie feared that Julia had been "shivering ever since" he left her. He also was "vexed with all the trouble you have had. Poor darling Ginia; it is maddening!" He was referring to Julia's troubling over Virginia's health and no doubt taxing her own strength. Words like *vexed*, *trouble*, and *maddening*, however, also suggest that taking care of Virginia may have weakened Julia. Leslie urged Julia to stay at home more so that there would not be "so many letters to write." He begged her to return: "My own, it is horrid to feel that you are ill away from me." She returned, but too late. There would be no more letters between Leslie and Julia.[8]

After Julia and the Stephen children returned home, Virginia recovered, but Julia's health continued to worsen. She retired to her bed, and Leslie summoned her Duckworth children home from Europe. Soon Aunt Mary Fisher took the household under her control (MHP/A5c). No doubt she heightened the maudlin atmosphere at 22 Hyde Park Gate. Virginia must have known that her own illness had troubled Julia and have feared that she had exhausted Julia's strength, but she suppressed all mention of the condition that had made Leslie write about "poor darling Ginia" that "it is maddening!" On the evening of 4 May, Virginia tiptoed into the darkened sickroom to kiss her mother. As she "crept" out of the room, Julia whispered: "'Hold yourself straight, my little Goat'" (MB, 84). Her last words to her youngest daughter might have been an attempt to wean or toughen her emotionally, but they seem only to have numbed her (Caramagno, *Flight of the Mind*, 120).

Early the next morning, Julia Prinsep Stephen died in the great ornate bed where her Stephen children had been conceived and born. As George Duckworth led the children in to say good-bye, Leslie "staggered from the bedroom." Virginia stretched out her arms to her father, "but he brushed past me, crying out something I could not catch; distraught." For

Virginia, "everything had come to an end" (MB, 91, 84). Although Julia had looked haggard and aged for years, she was only forty-nine.

For Virginia, the normal adolescent process of gradually distancing herself from her mother had been abruptly and tragically terminated. In the memoir she wrote more than twelve years later, she at last spoke of the "dull sense of gloom" that followed her mother's death and did "unpardonable mischief by substituting for the shape of a true and most vivid mother, nothing better than an unlovable phantom" (MB, 45). Then in her 1939 memoir, "A Sketch of the Past," Virginia revealed more about her immediate reaction to Julia's death. Holding herself emotionally straight, "I said to myself as I have often done at moments of crisis since, 'I feel nothing whatever.'" When she last saw her mother before burial, "her face looked immeasurably distant, hollow and stern. When I kissed her, it was like kissing cold iron" (MB, 92).

Having discerned the discrepancy between Julia's image as the perfect mother and her too-absent self, Virginia for the moment felt that she had lost nothing and hence had no reason to grieve.[9] Her inability to grieve also led to even more appalling conclusions: If she had troubled and maddened her mother with her own illness less than three weeks before Julia's death, had she not caused that death? If she had rebelled against her mother's strictures, had she not willed her mother's death?[10] Virginia felt first anesthetized and then guilty, an unfortunate pattern repeated all her life at other "moments of crisis" (MB, 92).

Talland House, with its hodgepodge of cast-off furniture; openness to sea, garden, and countryside; and ever changing assortment of guests, had tempered Julia's insistence on the proprieties. But after her death, there were no more holidays at Talland House. At 22 Hyde Park Gate, Julia's contributions to "natural life and gaiety" were now overshadowed by a sense of her conventionality (MB, 94). She became an "unlovable phantom" because her death erased all her complexities. The only acceptable way to remember her was as a lost angel. Draped in black, Aunt Mary Fisher visited and wept for Julia's husband and motherless children. A photograph of the family in mourning (Figure 11) shows Virginia looking numbed and rather lost.

Sixteen days after Julia's death, Leslie began writing a private autobiography, describing himself and Julia and their life together until "that fatal morning of the 5th of May," when he saw "my beloved angel sinking quietly into the arms of death" (Stephen, *Mausoleum Book*, 96–97). The handwritten volume, ending with a series of obituaries, was so lugubrious that his children dubbed it *The Mausoleum Book*. Rather than blame himself for some part in Julia's fatigue and ill health, Stephen reassured himself that at least he lacked the "fault of blindness to a wife's devotion." But he was blind to the way his overweening grief denied his children their own grief and preempted Stella Duckworth's life. Leslie henceforth called her "my darling Stella" (Stephen, *Mausoleum Book*, 98, 70, 104).

Stella now assumed Julia's household responsibilities, managing the servants, ordering the food, buying underclothes, paying bills. Virginia remembered Leslie as "one of the most dependent of men. Not intellectually. But emotionally. Emotionally he fell then upon Stella. . . . He couldn't do anything for himself" (MHP/A5c). The playful father who had always found some way of amusing his children was lost to them, his earlier petulance now having become emotional tyranny (VWE, 1:127). Out of self-pity, he demanded comfort from Stella, but "what comfort could she give?"[11] Leslie Stephen became for Virginia "the alternately loved and hated father" (MB, 116), but just as resentment of her mother had had to be suppressed, her "denunciation" of this formerly lovable man had to be kept private in the bereaved Stephen household's "oriental gloom."[12]

In an unpublished version of her 1939 memoir, Virginia remembered that "my mother's death fell into the very middle of that amorphous time" between childhood and adulthood. Virginia's age made the effect of the tragedy "much more broken. The whole thing was strained. This brought on, naturally, my first 'breakdown.' It was found that I had a pulse that raced. It beat so quick I could hardly bear it. No lessons, no excitement: open air, simple life."[13] Suddenly the specter hiding behind her early temper emerged. For the first time, what we now understand as Virginia's manic-depressive syndrome manifested itself, and the "most precocious little girl" seemed "mad."

The concept of "madness" sounds outdated, but firsthand testimony and modern neuroscience suggest that the term is not inappropriate. The novelist William Styron revealed ravages of his own depression whose symptoms—despite differences in time, place, and gender—are almost identical to Virginia's. Styron explains that "depression, in its extreme form, is madness." This biologically based illness is "*periodic*: It comes and goes, and when it is gone, individuals are not sick or insane (unlike neurotics, whose unconscious conflicts seep into and determine even 'normal' behavior)."[14] Manic-depression might best be thought of as the propensity to exaggerate real emotions; hence, sadness becomes despair, and happiness, wild exuberance. (Today, manic-depression often can be controlled with lithium. For further information, see Appendix B.)

In the handwritten version of her late memoir, Virginia remembered that after her mother's death, she became

> terrified of people—used to turn red if spoken to. Used to sit up in my room raging—at father, at George . . . ? [*sic*] and read and read and read. But I never wrote. For two years I never wrote . . . [*sic*] The desire left me: which I have had all my life, with that two years break.

The blank space and question mark in this manuscript suggest that in 1939, Virginia wanted to list a third person against whom she raged. The only person whom Virginia would have been reluctant to name was, I suspect, her deceased mother, who had spoiled and indulged her husband and sons, sacrificing her own health to them and others, and who had abandoned her daughters to a situation in which they were expected to make the same sort of self-sacrifice.[15]

The passage also connects this rage with a two-year hiatus in Virginia's compulsion to write. In her 1895 depression, Virginia retraced the pattern of her infancy.[16] She raged against her mother for abandoning her and suppressed those language skills that had signaled the infant's separateness from her mother.

For two years, Virginia lived "in a state of physical distress." Whereas Vanessa emerged as a sensuous, even rather sultry young woman (Figure 12), Virginia felt "the excitement and the depression of the bodily state." She wondered how to separate the body and the mind. The bodily changes that Virginia experienced between the ages of thirteen and fifteen simultaneously excited, frightened, and depressed her. Guilt over her unspeakable rage at her mother and her initial failure to grieve lacerated Virginia's consciousness. Soon grief and a sense of being at the mercy of her body further contributed to her prolonged disturbance. Because she was unable to write, Virginia's self in this disturbed period is largely lost to us and almost lost to herself as well. One photograph (Figure 13), however, shows her as prettier and less thin than she had been at the time of her mother's death, but she still looks dazed and perhaps angry. The painful memory of losing her desire to write may explain why, in 1939, after typing "for two years I never wrote," she followed that sentence with three spaced periods.[17]

Besides Virginia's late recollections, another source of information about this period survives in the diary that Stella Duckworth kept during 1896. Stella avoided the first person and came close to commentary only with such conventional observations as "in aft[ernoon] to old Masters thought them even more beautiful than before. then to workhouse." She usually took Virginia along on her frequent visits to the poor women's workhouse.

We can assume that Virginia was still suffering from depression sixteen months after Julia's death, because on 13 October 1896, Stella wrote: "Took Ginia to [Dr.] Seton & he says she must do less lessons & be very careful not to excite herself—her pulse is 146 [?]. Father in a great state. We are to see Seton again in a week."[18] The next week, Seton instructed that Virginia "must give up lessons entirely till January & must be out 4 hours a day." In November, as a diversion, George Duckworth and his aunt Minna Duckworth (one of the "aunts" who was no kin at all to the Stephens) took Virginia and Vanessa for a week's holiday in northern France. Traveling with George and Aunt Minna, who was a "rich, fat old lady and entirely commonplace" (Bell, 1:58), hardly improved Virginia's state of mind.

Virginia was taken to see Seton twice more in 1896 (Stella Duckworth's Diary, 1896 [Berg]). Seton's assumption was that intellectual activities put too much stress on a pubescent girl's nervous system. Too tortured by guilt, grief, and rage to express her creative self and now too coerced by authority to assert her intellectual self, Virginia's depression must have seemed irremediable.

She began to cure herself, however, according to a pattern that would sustain her—as it does many trauma survivors and creative artists—for the rest of her life:[19] Defying her doctor's orders, Virginia reconstituted herself by writing. Indeed, after only sixteen entries, her 1897 diary surpassed in length her (now lost) 1896 diary, and she could call herself "Wonderful creature!"[20] For example, Virginia's January 1897 diary records not only what she was reading but also such events as seeing baby crocodiles ("most delightful critturs") at the zoo, bicycling, walking, and going to the theater, to "Animatographs" (movies), and to the National Gallery (PA, 8, 9). Virginia also began a (now lost) work of the imagination entitled "Eternal Miss Jan." (Miss Jan was one of her own nicknames.) She read prodigiously and planned to "reread all the books father has lent me."[21] The emotional shock waves created by Julia's death subsided. Virginia's brilliant writing self was again intact, and the family resumed their normal activities.

Leslie Stephen also emerged sufficiently from his self-absorption to walk regularly with her in Kensington Gardens. He escorted her to the London Library and again to Carlyle's house. Although Leslie had had some scruples about young ladies reading books like Balzac's *Cousine Bette* (LS/JPS, 4 February 1893), he now gave Virginia free access to everything in his library. Active and healthy and (above all) writing again, Virginia declared 1897 the "first really lived year" of her life (PA, 134).

Not only had Stella Duckworth inherited Julia's roles in caring for the Stephen–Duckworth household and placating her stepfather, but she also continued to be a Julia-like angel of mercy.[22] Then in July 1896, Stella became engaged to Jack Hills (a fact she did not record in her diary). Their happiness afforded Virginia "my first vision—so intense, so exciting, so rapturous was it that the word vision applies—my first vision then of love between man and woman" (MB, 105). But the "engagement lasted all the months from July till April." This "clumsy, cruel, unnecessary trial for them both" (MB, 106) was caused, as Leslie Stephen himself admitted, by his "selfish pangs" over losing Stella's support (Stephen, *Mausoleum Book*, 103). Later, Virginia was more explicit: "He was jealous clearly," but he had "entrenched himself away from all truth" and could not admit it.[23]

In February 1897, Stella proposed that Virginia accompany her to the southern coast of England, where Jack was going on holiday. Virginia found it "impossible to be alone" with the engaged couple, whom she called "those two creatures," but Stella would not go without her. Vanessa's company would improve the trip, but being urged to go against her own wishes "bewildered" Virginia and left her in a "dreadful fix" (PA, 27). The prospect of a holiday by the English Channel in February, especially with an engaged couple who had time only for each other, could not have been enticing; yet Virginia's aversion to going has provoked speculation about its less obvious causes.[24] I find that Virginia's own words sufficiently explain her behavior. She felt herself in a "dreadful fix": "If I do not go, Stella will not, and Jack particularly wishes her to." It was a "*quandary* as Vanessa calls it" (PA, 27).

Despite her protests, a week later Virginia, Vanessa, and Stella traveled to Bognor, where they stayed in a house and Jack stayed in a hotel. Unfortunately, Virginia carried only "2 vols. of Scott," which, given her boredom, promised to have "a very quick ending" (PA, 32). Except for her Scott, the trip was boring, the countryside ugly, and the weather awful. When Leslie Stephen joined the vacationers for two days, he announced that he had never seen "such an ugly country and such bad weather in my life," and Virginia agreed. Stella proposed that they stay for another week, but Virginia's temper erupted again. After nearly six days away from her "comfortable arm chair, paper and books," she "absolutely refused" to spend "another week of drizzle in that muddy misty flat utterly stupid Bognor (the name suits it)" (PA, 32–35).

If being ill had been a means of commanding her mother's affection, being literary would ensure her father's. After the dreadful fix of having to go to Bognor, these two paradigms warred against each other. Although her health was shaky, Virginia began doing Greek exercises "by way of beginning lessons" on 15 February, and on 27 February Dr. Seton allowed her to study Latin. Vanessa had begun lessons the year before at Sir Arthur Cope's art school in South Kensington, and now Vanessa went to the studio every day for a week "to make up for Bognor," but Virginia's literary privileges were parceled out sparingly (VB/Thoby Stephen, 14 February 1892). In fact, it was the loss of her literary life that was most apt to damage Virginia's health. But no doctor recognized how much being ill and being literary were mutually exclusive options for her.

Being sick did, however, rouse Stella's attention. She had the unrewarding task of monitoring Virginia's health, seeing that she took beef tea ("most disgusting stuff" [PA, 16]), and ensuring that she did not tax her brain with too much reading and writing. Stella took Virginia along on her typically female missions of shopping and making charitable visits, activities that Victorian doctors never understood were far more exhausting than reading or studying. The second time in 1897 that Virginia recorded going with Stella to visit the workhouse, the two young women "were inspected by the old ladies—Most of them were blind, so that they had to feel us over to find out how much we had grown" (PA, 56). For Virginia, such a physical experience associated the female, the fleshly, and the unpleasant with Stella. Even though she was not, like "little Ginia," "quite forgotten," Virginia then would have preferred to be. Supposed to go on another trip with Stella, "with the stubbornness of a mule and the ardour of a marmoset (my new title) I *refused*" (PA, 61).

In March 1897, Vanessa wrote to Thoby about a street panic that Virginia was developing: "As Ginia and Father were coming back from their early walk in the Gardens they saw a bicycle accident." Although Leslie saw only part of it, Virginia witnessed the whole episode of a woman on a bicycle running into a cart. Vanessa told Thoby, "The poor old Goat was in a dreadful state as you may think and now she wants me to give up riding altogether, which of course I shan't do. It's very unlucky that it should always be the Goat who sees accidents" and who panicks.[25]

Virginia must have thought that Stella's marriage on 10 April 1897 promised freedom to reclaim her literary self, especially after being "tryed [*sic*] on" (PA, 37) for outfits and caught up in the "hurly burly" (PA, 68) of wedding preparations. The ceremony itself seems to have been rather a nightmare for Virginia, yet at the end of the day she joyously exclaimed, "Mr and Mrs Hills!" and pressed petals from wedding bouquets between the pages of her diary (PA, 68).

After Stella's marriage, Leslie summed up the condition of his two daughters (Laura no longer figured in his calculations):

> Virginia has been out of sorts, nervous and overgrown too; I hope that a rest will bring her round. Nessa has been hard at work at her drawing class and is, I hope, getting on well. 'Ginia is devouring books, almost faster than I like. Well, the young ones are satisfactory. (Stephen, *Mausoleum Book*, 103)

But Leslie's grief over Julia's death and his irritation at Stella's defection increasingly isolated him from his children.

Taken away from reading Pepys's diary to shop and run errands, Virginia's street panic returned. Indeed, she was too upset to tolerate the noise of London streets, "in a most fiendish state of uproar." Trying to pack enough books for a family holiday in Brighton (already a place of bad associations), Virginia became "very furious and tantrumical" after her experiences on the streets of London.

At Brighton, the Stephens rented a home from a vicar's widow. There "all the books are poetry—except a few bad novels . . . and a sprinkling of terribly religious works—sermons and missionary reports. . . . Everything rather gruesome." Virginia mostly kept herself occupied finishing the first two volumes of Macaulay's *History of England*. She and Vanessa avoided their Fisher cousins as much as possible, although Virginia did enjoy a search with Emmeline Fisher to see whether Aunt Mary had saved any of Maria Jackson's mawkish letters.[26] The wind and rain were fierce; holidayers were "all penned up in their hotels"; and Virginia concluded that "truly such a dismal place was never seen." Brighton was "disgraceful," and the wind off the water was "diabolical." When the weather relented slightly, the families did bicycle together and once took their bicycles on a train to Lewes, where they enjoyed a "most beautiful ride" in the East Sussex countryside, which Virginia would come to love. Back in Brighton, Virginia was irritable enough to break her umbrella in half. London should have been a relief, but they arrived home only to receive disquieting news: Stella and Jack had returned from Italy, but she had "a bad chill on her innards." The family felt "miserable" and depressed (PA, 69–77).

Stella and Jack moved into a house just three doors away from number 22 (chosen to placate Leslie over "losing" Stella) (VB/Thoby Stephen, 14 February 1897). Now Virginia observed more street horrors: On 8 May, she "managed to discover a man in the course of being squashed by an omnibus." She became so obsessed by such events that on 9 May, Dr. Seton prescribed work in the back garden. On 11 May, the Stephen children filled a prescription for Virginia and bought gardening boots so that she could be "healthfully employed out of doors—as a lover of nature" (PA, 84). On 12 May, she had "the pleasure of seeing a cart horse fall down." On 13 May, after a stampede, "a glimpse out of the door—to which the young ladies all crowded to get a better view—showed one horse on the ground and a second prancing madly about it" (PA, 83, 85). Even though she was still reading her "beloved Macaulay" on 14 May, she was forbidden "any lessons this term" (VWL, 1:7–8). The common denominator in all the horrors she saw on the streets is one creature's helplessness while others objectively or pleasurably observed it at a distance. Each, I believe, recalls Julia's death and Virginia's numbness.

Virginia was confined in the gloomy house, made to cultivate its enclosed back garden, or sent down the street to sit with the pregnant Stella, who also was ill with peritonitis. Virginia was more resentful of her exile than sympathetic to Stella. Once she even used their mother's supposedly laughing nickname, "that old cow," for Stella and became "extremely gruff and unpleasant" (PA, 83, 86). Virginia was irritable and anxious enough to try to calm herself by sleeping several times with Vanessa, but her greatest comfort lay in books (PA, 78, 79). She especially resented being forbidden to study and read and write when Vanessa was free to paint.

Early entries in Virginia's 1897 diary begin simply, "Nessa went to her drawing." But by May and June, Virginia referred to Vanessa's drawing and her studio with a blank that implied "damned," as in "her most —— (to be filled in as desired) studio," "her —— drawing," and "Nessa bicycled off to her —— studio" (PA, 91, 95, 101). After Stella's marriage, Virginia and Vanessa were "promoted" to separate quarters. The old day nursery became Vanessa's "studio," where she painted while Virginia read Victorian novels to her. Virginia took over the old night nursery, where she had entertained her sister and brothers with fantastic stories. Virginia's long room was divided into waking and sleeping ends that "fought each other" (MB, 123). The opposition between waking and sleeping seems to have represented to Virginia the differences between mind and body, herself and Vanessa, writing and painting.[27]

On 22 June 1897, Vanessa, Thoby, and Virginia rode in a carriage to view from St. Thomas's Hospital the parade celebrating "the great Diamond Jubilee Day," the sixtieth anniversary of Queen Victoria's reign. Virginia observed carefully and with some kindness: "The Queen was lying back in her carriage, & the Pss. of Wales had to tell her to look up & bow. Then she smiled & nodded her poor tired head, & the whole thing moved on" (PA, 103, 105).

Virginia was less kind to Stella. When Stella announced that Virginia "should have to" accompany her on an excursion, Virginia, with "great vehemence," declared her expectation to be "*impossible*." Now she "growl[ed] at everything" (PA, 99). Eventually, having to answer the "invariable" queries from concerned relatives about Stella's health caused Virginia to hate "poor Stella & her diseases" (PA, 105). Her fury was compounded by knowing that one of her Fisher cousins considered her to be a burden on Stella: "'So bad for Stella to have Ginia always with her!'" (PA, 112). Virginia's anger reveals why, years earlier, Julia could have gotten a laugh from Vanessa and Thoby by talking about "naughty" little Ginia and why Leslie was amused by little Ginia's telling him about the "naughty" old Punch. Now "naughty" Virginia resented the expectations Stella imposed on her and the sympathy Stella elicited from others for being ill.

Stepsister, doctor, and the ghost of her mother all had nonliterary expectations for the precocious young Virginia. The strain between meeting their demands and asserting her own literary desires put Virginia in a disturbed dreadful fix. As a result, on 13 July, Dr. Seton ordered Virginia to bed, and Stella (although still sick herself) sat by Virginia's bedside. On 14 July, Virginia "had the fidgets very badly." Even though Virginia had resented having to sit with her, Stella "sat with me till 11:30—stroking me till [the fidgets] went" (PA, 114).

But Stella was in more need of care than Virginia. Her illness got worse, and on 17 July she had surgery. But on 19 July 1897, she died.[28] Once again, caring for Virginia seemed to have precipitated a death, and once again, Virginia's resentment seemed to lead to a tragedy.

Virginia recorded Stella's death in her diary, noting, "It is impossible to write of" (PA, 115). Like the "pleasure" of seeing victims on the street,

Virginia had the questionable pleasure of getting rid of the "old cow." If her uncharitable wishes somehow had contributed to the illness and death of her unselfish half sister, she did have reason to feel guilty: "I have many regrets: not then, but now I have them; chiefly with regard to Stella." This regret "with regard to Stella" was a late recognition of guilt that she could "never make . . . up" (MHP/A5c). But in 1897, Virginia denied that guilt. Finding it "impossible" to write, she suppressed the self that had judged Stella so severely.

Virginia could acknowledge neither guilt nor terror:

> My mother's death had been a latent sorrow—at thirteen one could not master it, envisage it, deal with it. But Stella's death two years later fell on a different substance; a mind stuff and being stuff that was extraordinarily unprotected, unformed, unshielded, apprehensive, receptive, anticipatory. . . . But beneath the surface of this particular mind and body lay sunk the other death . . . the blow, the second blow of death, struck on me; tremulous, filmy eyed as I was, with my wings still creased, sitting there on the edge of my broken chrysalis. (MB, 124)

This second death was so compounded by the first as almost to crush the vulnerable young butterfly about to emerge from its chrysalis.

Years later, Violet Dickinson confided to Vanessa what she had learned from the gossip in 1897: Stella "was malformed & marriage finished her off. I only knew this 20 years later. Her nurse nursed an Aunt of mine!" She explained that "everyone was told [Stella] had appendicitis . . . but as far as I remember the Nurse said she had inversion I suppose of her uterus, something was concave that ought to have been convex; or vice-versa. . . . I think Jack must have been a tiring lover. . . . Stella an unselfish saint."[29] However imprecise Violet's understanding of female anatomy, her letter establishes that gossip associated Stella's death with sex and marriage. Virginia also probably knew that her father's first wife had died during her third pregnancy.[30] And she could assume that bearing seven children had weakened Julia. Furthermore, Stella's death—coming so soon after marrying and becoming pregnant—may have reinforced the idea that sex is dangerous. In turn, such an idea must have made the fifteen-year-old Virginia—in her "unshielded, apprehensive, receptive, anticipatory" state—worried for herself and more immediately for her sister, who was growing into a gorgeous and marriageable young woman. In the months after Stella's death, Virginia left many blank pages in her diary. On the edge of her broken chrysalis, she could not express her sadness or guilt or fear. Once again, her emotional distress threatened her writing self and hence her very being.

Leslie was worried enough about Virginia to write to Julia's old friend Mary August (Mrs. Humphrey) Ward that Virginia had said how much she "wanted a *woman* to confide in." He admitted that he had "perhaps been a little selfish in shutting myself up of late...but I must try when we return to make a rather more cheerful home for the girls" (LS/MAW, 24 July 1897).

Three days later, Dr. Seton "came & saw me & Adrian & reported very flourishingly on us" (PA, 117). But his report is less consequential than Virginia's willingness to record it. By 9 August 1897, tired of diary-writing, Virginia had begun a "great work" (PA, 121). She was also recovering from the "fidgets" that followed Stella's death and from her guilt, panic, and fear. Through both her diary and her imaginative writing, Virginia reclaimed her stability, as she was to do all her life.

Virginia's quick recovery was rewarded by the opportunity to study at the University of London. But even as she recorded in her diary, "I go to King's College, Nessa to her studio," she admitted, "Life is a hard business—one needs a rhinirocerous [*sic*] skin—& that one has not got!" (PA, 132). Without such a tough skin, Virginia felt vulnerable and cowardly, but she urged herself to be courageous: "Nessa preaches that our destinies lie in ourselves, & the sermon ought to be taken home by us" (PA, 134).

Those destinies, however, were jeopardized by a number of factors, especially by Victorian "strictures on [female] behaviour" and by the flesh itself (MHP/A5c). With Virginia, helpless defenselessness was associated with sex. Even though she had learned from the women in the workhouse and the dressmaker how distasteful being fondled could be and from the deaths of Julia and Stella how dangerous sex could be, her suppressed reason for fearing sex was probably the residual memory of Gerald Duckworth's hand under her dress. When this memory finally surfaced, Virginia offered a rather skewed interpretation, that her "instinctive" reaction to Gerald's misdeed showed that "certain parts of the body . . . must not be touched" (MB, 69). Thus she perpetuated the Victorian idea that sex is always dirty, and she failed to see that the evil of that situation lay in Gerald's abuse of both sex and power.

All her life, Virginia could easily feel "ecstasies and raptures spontaneously and intensely and without any shame or the least sense of guilt, so long as they were disconnected with my own body." Gerald's violation explained her fear of the body and her "looking-glass shame," the "feeling of guilt" she experienced when looking at her reflection or being looked at (MB, 68–69). Violet Dickinson remembered that when the Stephen girls were young, everyone spoke of them as "the most beautiful babies." Accordingly, the girls "wore large hats pulled down over their eyes so that people might not look at them." Violet also recalled "Virginia [being] so nervous at crossing the road & so furtive at being looked at" (VD/VB, 8 May 1941, and 16 June 1942).

In response to these early memories of being vulnerable to others' stares, Virginia later wrote that she had added another: Looking at a flower bed at St. Ives,

> "That is the whole," I said. I was looking at a plant with a spread of leaves; and it seemed suddenly plain that the flower itself was a part of the earth; that a ring enclosed what was the flower; and that was the real flower; part earth; part flower.

That perception, which she knew would be "very useful to me later" (MB, 71), recaptured early maternal fusion and so could counter the alienation in "looking-glass shame."

The summer after Stella's death, Jack Hills sought female consolation in his grief. At first, both Stephen girls "bore the brunt of [Jack's] anguish" (MB, 141), but Vanessa's sympathy began to flower into a romantic attachment. Meanwhile, as Leslie aged, George Duckworth became the titular "man of the house" and took it upon himself to introduce Vanessa properly into society, as Julia would have wished. Vanessa accepted George's gifts; allowed herself to be escorted to various balls, operas, and theatrical events; and put up with his flirtatious manner even while she was becoming very fond of Jack Hills. Meanwhile, Thoby, too, was maturing. But "as for sex, he passed from childhood to boyhood, from boyhood to manhood under our eyes, in our presence, without saying a single word that could have been taken for a sign of what he was feeling" (MB, 139). However young womanhood and manhood affected Vanessa and Thoby, neither confided in Virginia.

Virginia followed the example of her father and most Victorians in suppressing her anxiety. Leslie Stephen had convinced himself that Laura was merely a stubborn—not a retarded—child. Of Julia's death, "not the very slightest foreboding of the reality had even crossed my mind." After Stella's death, Stephen recorded that the remaining family had returned from holiday and that "Ginia, I hope, is improving, though still nervous."[31] Leslie had a well-developed capacity for denying signs of trouble, as did most Victorians of his class who would have said "stubborn" rather than "retarded" or "nervous" rather than "mentally unstable."

Virginia later analyzed the Victorian tendency to conceal motives and wondered, "What is the psychological effect of lying?" (MHP/A5c). She had tried to break the pattern by asking Vanessa which parent she preferred. But after these disasters, the Stephen children retreated behind the cover of denial, suppressing all feelings about their mother and half sister, even their memories: "We never spoke of them."[32]

The tragic deaths of Julia and Stella had proved that life was indeed a "hard business," and they further prompted Virginia to conclude that life was harder on women than men. Men who were guilty of insensitivity, misbehavior, abuse, and professional bungling seemed impervious to the consequences of their misdeeds, whereas women were worn down even to death. Virginia had dreaded the fixes in which Julia's sense of a woman's role, Gerald's abuse, the doctors' restrictions, and the Victorian habit of denial had placed her. As a young woman, though, she must have felt especially vulnerable, for these tragedies suggested that a masculine power structure would always trap women.

Leslie now was putting Vanessa in such a fix, expecting her to keep the household accounts, as Julia and then Stella had done. His habit of protesting Julia's supposed extravagance had been aggravated by grief, loneliness, and paranoia. Now he groaned and complained bitterly. His accusations both frightened and infuriated Virginia, especially because "not a word of what I felt—that unbounded contempt for him and of pity for Nessa—could be expressed" (MB, 144). As a defense, the sisters "formed together a close conspiracy. In that world of many men, coming and going, in that big house of innumerable rooms, we formed our private nucleus" (MB, 143). This conspiracy still does not seem to have included discussions of any intimate matters, but for a time it ended their rivalry. The two shared their efforts at their separate arts and hid them from the rest of the family, now all-male.

Vanessa was so shy about her accomplishments that when she won a prize for drawing, she told only Virginia, expecting her sister's powers of language to shape the news for the men (MB, 30). Virginia empathized and ceased making remarks about Vanessa's "—— drawing." Their female alliance granted Virginia what she needed: Vanessa's motherly attentions and an environment in which the two supremely talented young women could work in harmony on their arts. Threatening that harmony for Virginia, however, was the possibility that a man would capture Vanessa's affections and end their close relationship.

As Vanessa proved herself an increasingly accomplished painter, Virginia proved herself a dedicated student, continuing her studies at King's College and beginning Greek and Latin lessons with Walter Pater's sister Clara. Virginia wrote a series of wonderfully spirited and comic letters, and by the summer of 1899, she began keeping her diary again. In her holiday diary, Virginia made a correlative for the unspeakableness of the situations in which mother, half brother, and father had placed her. The seventeen-year-old wrote in the tiny, cramped, barely legible hand that had become a family joke (MB, 122). Then to make her work even more inaccessible, she pasted her essays onto the pages of Isaac Watts's *Logick: or the Right Use of Reason*. Virginia's mutilation of the *Logick* became a "right use" of reason because it ensured her privacy and made her journal virtually inaccessible.[33] Thus, even when it "suppose[s] a reader" (PA, 144), this journal signals Virginia's retreat from others' expectations about her writing and her conduct (MB, 80).

In her journal, Virginia practiced the skills that she had begun with the *Hyde Park Gate News*: mastering events by writing about them. For example, after she and Adrian and their cousin Emma Vaughan were dumped from their punt into a pond, Virginia transformed the unpleasant event into a narrative. Their supposed death in a "carpet of duckweed—the green shroud alas of three young lives!" (PA, 151) became "Terrible Tragedy in a Duckpond." This "Tragedy" was successful and comic enough to emerge from the pages of *Logick*. Virginia copied out and expanded her original narrative (PA, 150–52) to send to Emma, adding to the journalistic account of the drowning "A Note of Correction and Addition" by "One of the Drowned." She instructed Emma to "read my work carefully—not missing my peculiar words—and then tell me your criticisms and humble thanks. Really I am an angel—no other word describes my character so well."[34]

Despite Vanessa's complaining about her "eternal writing" (PA, 142), Virginia had discovered that "I love writing for the sake of writing" (PA, 139). She knew now that "there is no end to writing, & each time I hope that I may make better stuff of it" (PA, 150). As a fledgling artist, Virginia pondered the function of art. She initially voiced the platonic assumption that "we are a world of imitations[;] all the Arts . . . imitate as far as they can" and "try & reproduce," but she also explored distinctions between verbal and visual imitation (PA, 143). Later she wondered whether art did more than imitate, and if so, she argued with Vanessa about which art best succeeded in surpassing nature. Because there is no English word to identify the unending debate between the sisters about which was the superior art, I use, after Leonardo da Vinci, the Italian term *paragone* to signify the rivalry central to Virginia's understanding of herself and her art. (For more on Leonardo and *paragone*, see Appendix C.)

Virginia assigned herself a painterly subject—a sunset—to describe in words. She admitted that an unclouded day's sunset is unapproachable and that only a poet "can express in words or paint the human significance & pathos of the suns unclouded rain of light." At the end of her ornate description of a real sunset, Virginia confidently proclaimed, "Well may an Artist despair!" for in this early instance of *paragone*, the pen had bested the brush (PA, 155–56). However, she also admitted that her description lacked "human significance & pathos"; thus she proposed that describing nature, like painting it, was a limited art. And when she reverted to drawing, instead of describing, the winged look of the sunset's clouds, she acknowledged that the visual image could convey some things better than verbal images could (Figure 14). In this way, the seventeen-year-old author addressed the problem of representing not only a "real" sunset but also the "reality" of human experience. Here she encountered the aesthetic dilemma on which Leonard's theory about painting and poetry had foundered: Should the artist imitate and hence try to reproduce nature or invent and hence try to improve nature?

As it always was for Virginia, the theoretical debate between the arts had its roots in her private emotions. In 1900, when George Duckworth took Vanessa to Paris, Virginia urged him not to let Vanessa become so enraptured with the Parisian art world that she would not return. Appealing to George's sense of propriety and mastery, Virginia instructed him not to expose Vanessa to "too many improper studios" lest the "volcanoes" of Vanessa's "artist's temperament" erupt (VWL, 1:31). Vanessa's delight in Paris made Virginia conclude that "pictures have got into her brain" (VWL, 1:32). As when she had railed against Vanessa's "—— drawing," Virginia transferred her sense of loss and resentment from Vanessa to pictures, making painting itself her rival.

By the turn of the century, the playful good times recorded in the *Hyde Park Gate News* were distant memories. The father who had taken the children sailing, butterfly catching, and book borrowing had turned inward. He was now deaf, irascible, morose, and ill. Despite honorary degrees from Edinburgh, Cambridge, Oxford, and Harvard; despite what would be ten volumes of collected essays (DNB, 413); despite the imposing volumes of the *Dictionary*; despite receiving the KCB (knight commander of the Order of the Bath), "Sir Leslie" was plagued by a sense of failure and loss. His complaints made 22 Hyde Park Gate seem like a gloomy dungeon. Although the younger Stephens found holidays—even with their father—an escape from the atmosphere at home, their most consistent escape was their studies.

Thoby went to school in Bristol and in 1899 to Cambridge; Adrian attended day school in London and later followed Thoby to Cambridge. In 1901, Vanessa won admission to the Royal Academy, although Virginia cattily suggested that, despite Vanessa's talent, the paucity of pupils made her acceptance less of an accomplishment (VWL, 1:43). Vanessa now was studying under John Singer Sargent, who, Virginia reported, seemed to like her work, but giving a gibe even as she offered a compliment, Virginia claimed that "artists are so inarticulate" it was hard to be certain (VWL, 1:60).

In late 1901, Leslie wrote in his *Mausoleum Book*: "Nessa—I am ashamed that I forgot to note this—got into the Academy School last summer and has been working there steadily since our return to town. She sent the drawing for the Academy to a competition at her old studio, where it got the medal" (110). Leslie's regret at forgetting to note Vanessa's success shows that he was not so unappreciative of Vanessa's achievements as she later claimed.

Although Virginia seems to have enjoyed her lessons at King's College in Greek and history, she did not sit for exams (VWL, 1:12). In early 1902, Janet Case, a rigorous and admirable teacher, replaced Clara Pater as her Greek instructor (PA, 181–84). Virginia continued to raid her father's library and the London Library for English literature and history and the works of ancient Greece, and whenever she could, she retreated to her room to digest book after book.

Unlike Stella, Vanessa bore without flinching Leslie Stephen's routine accusations that her housekeeping was sending them to the poorhouse. Virginia may have feared that Leslie eventually would wear down Vanessa's strength, just as (all too obviously to everyone but Leslie himself) he and other dependents had worn down Julia. But Vanessa had immunized herself against Leslie's accusations by severing her emotional bond with him. Unlike Stella and Virginia, she refused to become dependent on him or to let him become dependent on her. Confronted with Vanessa's coldness, Leslie turned to his younger daughter for comfort and found this bookish daughter "most fascinating" (Bell, 1:84–85). Virginia took on the duties of reading to, writing for, and consoling Leslie in his grief, loneliness, and sense of failure. But doing so put her in another fix, although her anger at his insatiable demands for attention was tempered: "In me, though not in [Vanessa], rage alternated with love" (MB, 108).

As the young editor of the *Hyde Park Gate News*, Virginia had seen the comedy in marital negotiations, but as a young woman she was baffled by them. She now was tall and strikingly attractive, but awkward and shy. She found dancing difficult and social chitchat impossible. In 1900 and 1901, she and Vanessa stayed with their cousin Katherine Stephen, vice-principal of Newnham (a women's college), and attended the May Balls at Trinity College, Cambridge. There they met the friends that Thoby's stories and her own imagination had mythologized: Clive Bell, Lytton Strachey, Saxon Sydney-Turner, and Leonard Woolf. These introductions seem to have made little impression on Virginia, but Leonard Woolf never forgot. He recalled the sisters in their white dresses and large hats, carrying parasols against the sun, so beautiful it was "almost impossible for a man not to fall in love with them . . . [but] they seemed to be so formidably aloof and reserved that it was rather like falling in love with Rembrandt's picture of his wife, Velasquez's picture of an Infanta, or the lovely temple of Segesta" (LWA, 1:119). With Virginia at least, this aloofness was a self-protective carapace. She did not like "being looked at" and was frightened of male attention. As she later told her husband, Vanessa had a natural curiosity about and intuitive understanding of sex. But Virginia could not discuss her sexual anxieties with Thoby or apparently with Vanessa either (MB, 139).

With women, however, Virginia was neither aloof nor reserved, at least in her letters. Her friendships with her cousin Emma Vaughan and Madge Symonds Vaughan (daughter of the writer John Addington Symonds and now the wife of Emma's brother Will) deepened, and she was developing a new friendship with the aristocratic but "unreserved" Violet Dickinson. Virginia's letters to these women showed her to be no angel (despite her claim to the contrary), except in her extraordinary talent with words.

When normal sibling rivalry felt to Virginia like being victimized or bullied, she retaliated in words. In her first novel, Virginia's protagonist, an only child, escapes being "bullied and laughed at by siblings" (VO, 34). Years later, another of her characters speculates that "people who have been bullied when they are young, find ways of protecting themselves," as Virginia had done: "Is that the origin of art he asked himself: . . . making yourself immune by making an image?"[35]

Virginia also endeared herself to others through her writing. She made a first try at biography in 1902 with an admiring description of Violet (Bell, 1:82–83). Affectionately addressing Thoby as "Milord" and "Your Highness," Virginia also entertained and flattered him while expressing her longing to discuss literature with him. In this way, she used her pleasingly manic powers of verbal invention to secure her friends' and her brother's affections. Indeed, her effusiveness toward Thoby seems to have been almost calculated to disrupt the intimate friendship that he and Vanessa had established before Virginia's birth.

Nevertheless, the sisters managed to overcome their rivalry with an endeavor that, unlike the *Hyde Park Gate News*, united their interests and skills. In 1902, both sisters took lessons in bookbinding. In this way, Vanessa honored Virginia's interest in books, and Virginia used Vanessa's visual skills to serve her verbal interests. The materials of binding, "leather—linen—silk—parchment—vellum—japanese paper etc. etc. etc.," each carried its own physical appeal, and Virginia proved herself, as she bragged to Thoby (VWL, 1:56, 52), to be an efficient and practical binder.[36] Virginia also proved herself a competent draftsman, making a number of bookplates.[37]

Despite such satisfactory resolutions of the debate between their arts, Vanessa's 1902 picture-viewing excursion to Rome and Florence with George Duckworth suggested to Virginia again that Vanessa would be seduced, not by George, but by pictures. Since no letters from Vanessa survive (and Virginia kept a lifetime of them), her boast that Vanessa sent her long love letters rings rather hollow (VWL, 1:50). Desperate for affection, Virginia half-jokingly complained that she got no attention: "Nessa gets it all. She has so many intimates" (VWL, 1:55).

About this time, when Thoby's Cambridge friend Leonard Woolf came to visit him in the gloomy, death-haunted house, he found that he had to shout into an ear trumpet in order for Leslie to hear him, but Leslie's groans could be heard by everyone, including the men he insulted (Noble, *Recollections of Virginia Woolf*, 247). Virginia suspected the ailing Leslie of being "a fraud, only an invalid for the sake of his ladies." Harking back to her childhood, Virginia wished that she, too, could pose as "an invalid and have ladies" (VWL, 1:69). Instead, she feared herself unlovable, "compared with Nessa . . . vain and egotistical, irritable" (MH/A5c). Not only had she lost her mother and half sister and not only was she, like various sympathetic ladies, expected to give rather than receive attention from her father, but Virginia also feared she might lose her sister to other "intimates."

Virginia escaped from such worries through reading and writing. Her 1903 diary is an indexed assortment of occasional essays, all legible exercises in practicing the art that someday would earn public acclamation and secure her self-identity. One deletion she made in an essay, "Country Reading," shows Virginia questioning how much art could *re*present the world. When discussing the "great books," she first equated them with "the books that create something like life" but then crossed out that equation (PA, 205, 413). If literature did not imitate or attempt to reproduce "something like life," she was uncertain what it did do, but clearly by 1903, she was searching for an aesthetic that prioritized imagination over likeness to life.

In an essay entitled "The Serpentine," Virginia offered a rather melodramatic commentary on a "nameless woman" who drowned herself in the Serpentine, an artificial lake stretching across Hyde Park and Kensington Gardens. Virginia used her imagination to envision the woman's desperate condition, acknowledging that she could not get the words of the woman's suicide note "out of my head": "'No father, no mother, no work'" (PA, 212).

While Virginia was questioning fiction's ability to reproduce experience "faithfully" (PA, 152), Vanessa was making comparable readjustments in her theories about painting. At the Royal Academy, Sargent was teaching her to use thick paint, to "try to get the right tone at once," and to enliven her gray palette (VBL, 11). When Vanessa and George Duckworth visited the eminent Victorian painter G. F. Watts, Vanessa laughed to her friend Margery Snowden about Watts's antiquated theory that "no art can be great unless its beautiful any more than music could be great if it were all discords." He protested against Impressionism by priding himself on painting "cleanly," as one could not do with thick paint. When George played up to the old man by saying he was "ashamed" to look at the paintings (no doubt nudes) at the Royal Academy, Watts prudishly responded by telling Vanessa she need not paint from a model.[38] Vanessa may have laughed to her sister, as to Margery, about Watts's out-of-date notions of art, but Virginia seems to have remained loyal to Julia's friend, who had painted their parents' portraits (VWL, 1:174) and whom the *Hyde Park Gate News* used to identify with the letter "W for Watts / A great painter is he."

Gerald Duckworth, the younger brother who had traumatized

Virginia's youth, seems largely to have faded out of her life. The brighter of the two brothers, he had set up a publishing firm, and George Duckworth was pursuing his social and political ambitions. George was courting various titled ladies and was now the unpaid secretary (no doubt expecting to be rewarded with a prestigious post) to the Tory politician Austen Chamberlain.

Meanwhile, Vanessa's friendship with Jack Hills promised to become a serious courtship. Because it then was illegal for a widower to marry his deceased wife's sister, their romance promised to create a scandal, especially for that upholder of the proprieties, George Duckworth. Leslie would not intervene, perhaps because he was too ill but also because he respected Vanessa's wishes, so George enlisted Virginia's help to prevent a marriage. But when Virginia followed George's instructions and cautioned Vanessa, her sister bitterly replied, "So you take their side too." Virginia remembered, "Confusedly, I wobbled at once from George's side to her side" (MB, 142, 143, 142).

Virginia herself thus nearly ended the close conspiracy she felt the two had formed against the men who controlled so much of their lives. By siding with the forces of propriety, she risked losing Vanessa's loyalty. She also risked losing Vanessa to adult sexuality. In 1903, she remarked, "Jack back tonight, so beloved Nessa has her cisterns full again—not a wholly apt metaphor."[39] The awkwardness of Virginia's metaphor suggests her failure to understand "beloved" Vanessa's response to being seriously courted by a man.

Virginia's naïveté only made her own problems worse. When Vanessa refused go out with him, George Duckworth set about socializing his tremulous younger half sister. Virginia's instincts warned her against George and high society, but he invoked memories of her dead mother and half sister and claimed that he was only following their wishes. Later she came to recognize—in a phrase that echoes Julia's instructions for little girls—how he trapped her by insisting, "We should go where he wished us to go, and do as he wished us to do."[40] Her instinctive resistance was undermined, for "duty and emotion muddied the stream. And over that turbulent whirlpool the ghosts of Stella and Mother presided. How could we do battle with all of them?" (MB, 156). Those female "ghosts" encouraged her to accede to George's notion of upper-class female behavior, even though at nineteen or twenty, Virginia found dances, social conversation, and George's attentions inexplicable, boring, and frightening.

While claiming that loyalty to Julia made him the Stephen girls' protector, George violated his brother-protector role. Especially after the dances, he would "fling himself on my [Virginia's] bed, cuddling and kissing and otherwise embracing me in order, as he told Dr Savage later, to comfort me for the fatal illness of my Father—who was dying three or four storeys lower down of cancer" (MHP/A16; Bell, 1:96). Although George made a ludicrous parody of his mother's wishes[41] to have her socialized, Virginia was almost helpless to resist him, either when he entered her bedroom to kiss her goodnight or when he coerced her into attending fashionable parties so that he might test her social skills (MB, 156). Later she described Victorian society as a "ruthless machine. A girl had no chance against its fangs. No other desires—say to paint, or to write—could be taken seriously" (MB, 157). Furthermore, "Thoby & A[drian were] no help" against her sense of entrapment.[42]

Of course, "all the old ladies of Kensington and Belgravia" raved about how blessed the Stephen girls were to have such a wonderful half brother (MB, 168). Years later, Virginia punctured that image in a "fearfully brilliant" Memoir Club paper (D, 2:77). An unpublished portion of that 1920 or 1921 memoir first detailed George's generosity: "It was always George who made the peace, said the tactful th[in]g, broke bad news, braved my fathers irritations, read aloud to us when we had the whooping cough, remembered the birthdays of Aunts, sent turtle soup to the afflicted, attended funerals, gave the servants umbrellas at Xmas." He was often found on his knees "adoring" Julia, and this "exuberance of emotion" was felt to be to his credit.[43] Lest George's "exuberance of emotion" credit him and detract from her final judgment and lest "cuddling and kissing" seem fairly innocent, Virginia deleted those passages from her memoir's final version. She ended, "Yes, the old ladies of Kensington and Belgravia never knew that George Duckworth was not only father and mother, brother and sister to those poor Stephen girls; he was their lover also" (MB, 177).

Clearly, George was sexually frustrated, but he was hardly the Stephen sisters' sexual "lover."[44] In 1918, Vanessa told their cousin Fredegond Shove about George's distasteful behavior. Fredegond wondered, "Do you think that it was conscious (the sexual impulse in him?)." She concluded that George was "not quite conscious" of the sexual nature of his behavior.[45] In 1940, trying to find reasons for George's excesses, Virginia noted that one reason—in addition to desires to dominate and compete with Jack—was "as became obvious later, some sexual urge" (MB, 154). Only later did Virginia, with Vanessa and Fredegond, realize the "obvious" fact that George's attentions were sexual.

A letter from Vanessa to Virginia describes George's actual behavior. In 1904, when she and Adrian visited George's fiancée with him, "George embraced me & fondled me in front of the company—but that was only to be expected."[46] Clearly, "embraced" and "fondled" meant a distasteful and inappropriate show of affection, an "exuberance of emotion" but not sexual intercourse in "front of the company." At that time, the term *lover* meant "suitor," as when Virginia described Jack as Stella's "lover" even when she had refused him (MB, 99). Indeed, the *Hyde Park Gate News* and Virginia's diaries during the period of George's offensive behavior present an approving portrait of "Georgie" that is hard to square with her subsequent attacks.

Not until later did Virginia realize how much George's masking of sexual urges as brotherly affection had confused her adolescent sensibilities. She explained:

> Everything was drowned in kisses. He lived in the thickest emotional haze, and as his passions increased and his desires became more vehement—he lived, Jack Hills assured me, in complete chastity until his marriage—one felt like an unfortunate minnow shut up in the same tank with an unwieldy and turbulent whale. (MB, 169)

George's kisses and hugs violated neither his "complete chastity" nor his half sisters'. As Janet Malcolm put it, George "couldn't keep his hands off Vanessa and Virginia while affecting to comfort them," but he did not rape them ("A House of One's Own," 58).

What George did do was traumatic enough, and feeling like a helpless minnow in a tank with a whale, Virginia lived in fear of what he might do. His abuses of power further exacerbated the legacy from George's brother, Gerald, of associating male sexuality with female helplessness. The two brothers also trapped the sisters between Leslie's "large conception," perhaps of female nobility, and George's "little one. So we caught it both ways."[47]

Vanessa and Virginia did not discuss with each other until later the impropriety of George's embraces. Nor did they protest. As Quentin Bell explains, if Vanessa and Virginia had complained, the world "would either have called it a wicked lie or made an appalling scandal in which the sufferers, as always in Victorian scandals of this kind, would have been the victims" (letter to me, 1 April 1988). And to whom would Vanessa and Virginia have complained? With Thoby, Virginia could "recall no confidences, no compliments; no kisses; no self analysis between me and him" (MB, 139). Violet Dickinson was too much the typical maiden lady for the Stephen girls to have confided in her. Indeed, there is no evidence that they talked with anyone about how helpless they felt in regard to George's inappropriate behavior. If Virginia had suspected but been unable to voice the discrepancies between the ideal and the real Leslie and Julia, then the discrepancy between the publicly solicitous and adoring and the secretly exploitative and lustful George must have been even more incomprehensible and inexpressible.

Virginia had one means of escape from her emotional entrapment: She could write. If 1897 had been her first "really *lived* year," it was so, at least in part, because it was her first *recorded* year. Although her 1899 diary may have been too curiously assembled to be read, it did address a hypothetical reader and hence envision a self that would write for the public. By resorting to a few visual instead of verbal images, Virginia's words reasserted the pen's hegemony over the brush. Whereas her 1899 diary was secretive and mute, her 1903 diary showed the twenty-one-year-old actually practicing writing essays for public consumption. If speech were a weapon to turn on others, writing was a means for subduing a hostile world, escaping the dreadful fixes in which it placed her, and winning others' affections.

3
Put Some Affection In

1903–1909

When the *Hyde Park Gate News* had pleased Julia, Virginia experienced such intense delight that henceforth writing seemed an entrée to motherly affection. Virginia's letters to Violet Dickinson, an old friend of Stella's and seventeen years Virginia's senior, replicated that pattern, as Virginia exaggerated the intensity of their relationship and her own dependence.[1] As Violet's little "Sparroy" (not "Goat"), Virginia assumed the role of child in their "romantic friendship" (VWL, 1:90, 75).[2] "Sparroy" could beg for rewards for her sparkling letters: "Put some affection in to your letters—like sugar in otherwise rather bitter tea" (VWL, 1:91). Violet's affectionate praise for Virginia's "brilliant" letters established, Vanessa claimed, that Violet's taste was "peculiar" (VBL, 18). Although once she claimed to fear that Vanessa had "filched" Violet from her, Virginia for the most part was assured that her dazzling way with words had purchased her friend's loyalty (VWL, 1:83).

By 1903, Leslie Stephen had survived one operation and several near-fatal crises and hence could not believe, despite his chronic "haemorrhage"

(VWL, 1:91), that he had incurable intestinal cancer. Vanessa knew better, but she persevered in the Stephen family habit of denial. She wrote to Margery Snowden that she wondered whether she "ought to tell . . . the Goat & Thoby" that death was near, but the truth was, in any case, impossible to hide. Virginia also knew that her father was dying and felt keenly the agony of the process. She remembered how by making "a fetish" of Leslie's health, Julia had worn "herself out and died at forty-nine; while he lived on, and found it very difficult, so healthy was he, to die of cancer at the age of seventy-two."[3] On her twenty-second birthday, Leslie gave Virginia a beautiful ring and told her she was a "very good daughter!" She confided to Violet, "It amazes me how much I get out of my Father, still" (VWL, 1:123).

The emotional pressure to comfort the ailing Leslie was intense, but rather than weep over him, Virginia shouted literary talk to him. Thus she strengthened the alliance she had formed years before when she chose him as her favorite parent. Later Vanessa wondered to Virginia whether as parents they also would commit "all the old abuses & that I shan't have any idea what my children are like or what they want to do?" (VBL, 54). Vanessa may have felt unappreciated because Leslie did take more interest in Virginia's bookish achievements than in her painterly ones, but his early letters show his delight in drawing with little Vanessa, his pleasure in drawing her, and his great fondness for her.[4] After the turn of the century, however, Leslie's accusations convinced Vanessa that he had always been opposed to her.

Virginia believed that Leslie "wants to die," but the wait was so prolonged that she vowed "to ruin my constitution before I get to his age, so as to die quicker!" (VWL, 1:125). In the long process, the "relations Swarm. I liken them to all kinds of parasitic animals etc etc" (VWL, 1:108). Virginia detested their hypocrisy (VWL, 1:97) and the duty foisted on her of writing innumerable "virtuous letters" (VWL, 1:124). She was furious that despite all her bulletins on Leslie's decline, her aunt Mary Fisher scolded her for not doing her duty by her relations. Leslie Stephen dictated to Virginia the last paragraph in his *Mausoleum Book*, dated 14 November 1903, asking her to record: "It comforts me to think that you are all so fond of each other that when I am gone you will be the better able to do without me" (112).[5] He died on 22 February 1904.

The Stephen children felt themselves sufficiently able to do without him to go on a family trip a few days after his death to Manorbier on the Pembrokeshire coast in Wales. Even more rugged than their beloved

Cornish coast, this countryside offered the thrice-bereaved Stephens a release from the emotional and physical constrictions of 22 Hyde Park Gate. They walked along the cliffs and pondered their future. Virginia could not forget her father and wondered "how we go on as we do, as merry as grigs all day long" (VWL, 1:135).

She must have had a depressive mood swing, for she wrote an article at Manorbier "to prove to myself that there was nothing wrong with me" (VWL, 1:154). She later recalled that as a young woman she had been determined to write "a book—a book—But what book?" The need to write predated a vision of what to write about and how to do so. Then, "walking the down on the edge of the sea" at Manorbier, the "vision came to me more clearly" (D, 2:197). This vision appeared "on the edge of the sea," and Virginia did pedestrian writing to prove that there was "nothing wrong" with her. But we can surmise (from her sense of the vision's strangeness and from later references) that the vision revealed the visible universe—not as an object to be analyzed, proving her powers of description, as in a "Chapter on Sunsets," but as a subject with which her self could mystically fuse.

In the nine years between her mother's and her father's deaths, Virginia's love for her father had become colored with anger and resentment. In 1940, she described the period between Stella's and her father's deaths as a time when "Nessa and I were fully exposed without protection to the full blast of that strange character" (MB, 107). But at Manorbier and afterward, that "strange character" began to seem more sympathetic. Virginia wrote to Violet again and again about her regrets: She wished she could have "made up" with her father for her anger with him (VWL, 1:136). She feared, "I never did enough for him all those years. . . . If he had only lived we could have been so happy" (VWL, 1:130). She regretted, "If one could only tell him how one cared, as I dreamt I did last night" (VWL, 1:133).

After Manorbier, 22 Hyde Park Gate, with its memories of three lost lives and its accessibility to such solicitous and sentimental relatives as Aunt Mary Fisher and "Aunt" Minna Duckworth, was stifling to "us four" (VWL, 1:137). At Easter time in 1904, the Stephens fled London for Italy in the company of the Duckworth brothers. Vanessa was clearly relieved that Leslie was dead. According to her son,

> Vanessa had got what she wanted—her freedom. . . . Her happiness in being delivered from the care and the ill-temper of her

> father was shockingly evident. She was clearly and unequivocally delighted, and Virginia, emotionally strained, exhausted and exasperated by the long months of Sir Leslie's last illness, still guilty and still inconsolable, found this more than she could bear. (Bell, 1:89)

Perhaps angry with herself for being "as merry as grigs" just after their father's death, Virginia was definitely angry with Vanessa (VWL, 1:135). She may have known that Vanessa had more than once dreamed of killing Leslie (Spalding, *Vanessa Bell*, 40). Certainly she knew that Vanessa did not grieve. But if Virginia was horrified by Vanessa's relief at Leslie's death, she also was distressed to recognize her own complicity, for she too had secretly wished to be rid of Leslie Stephen. Thus even though their reactions to Leslie's death had seemingly ended their sisterly conspiracy against him, their common guilt restored it.

In Venice, Vanessa wrote to Margery Snowden that the Stephens looked at pictures all morning and at churches, which also were full of pictures, all afternoon. She decided that "Tintoretto is the greatest" (VBL, 13), and Thoby agreed: "Certainly one has never seen a picture till one comes here. Tintoret's supremacy is completely beyond a doubt." He also named Tintoretto the "Shakespeare of painting" (TS/Clive Bell, ? and ? April 1904).

Virginia picked up his line: "Still the pictures *are* pictures: till you have seen Tintoretto you don't know what paint can do." She was trying to share Vanessa's enthusiasm for paintings, frescoes, and altarpieces by Tintoretto, Titian, Bellini, Giotto, Botticelli, Filippo Lippi, and others, but she found the effort depressing and confining. She wrote to Emma Vaughan that "Venice is a place to die in beautifully: but to live [in] I never felt more depressed—that is exaggerated, but still it does shut one in and make one feel like a Bird in a Cage after a time" (VWL, 1:138).

In a letter to Clive Bell, Thoby pompously dismissed Clive's aesthetic opinions: "Until a man has been there [Venice] he has no more right to speak of painting than a man who has read neither Sophocles nor Shakespeare to criticize literature."[6] If Thoby pontificated in person as he did in his few surviving letters, he must have added to Virginia's distress.

In Florence, the Stephens were joined by Violet Dickinson, to Vanessa's relief, "as I dont think the Goat & I should manage very well alone" (VBL, 15). Violet's appearance should have relieved Virginia's sense of alienation, but Violet too seemed to have defected to the visual arts: She knew an art critic who agreed to show the young women even more of what paint could do; thus Florence became for Virginia "even worse than Venice" (VB/MS, 21 April 1904; interview with Quentin Bell, 26 July 1993).

Regarding the trip as a virtual celebration of her father's death, unable to speak Italian, unable to see on her own those sights connected with literature, and being forced instead to visit picture galleries and churches all day long put Virginia in another dreadful fix. She felt caged; she was irritable and "tempersome"; and her depression threatened another breakdown (PA, 222). Then, to conclude this painterly holiday, the Stephens (now without Violet or the Duckworths) stopped off in Paris, where Clive Bell planned to take them to Rodin's studio. They talked of "Art, Sculpture and Music" with Clive and an English painter. But even Vanessa found the painter difficult, so she "left Thoby & Bell & the Goat to talk" with him, no doubt to Virginia's considerable irritation (VBL, 17).

While contemplating the expedition to Rodin's studio, Virginia apologized to Violet for her bad temper: "You had much to stand: I wish I could repay all the bad times with good times." The only way to shake off such a mood was writing. She knew "I must work," but traveling and viewing art galleries made writing impossible (VWL, 1:140). Furthermore, Thoby had arranged with Clive for accommodations away from his sisters and brother.[7] Presumably he was enjoying Parisian nightlife, possibly with Clive, and wished to hide his indulgences from his siblings. But his absence must have aroused both Virginia's curiosity and her apprehensions.

The deaths of her mother, half sister, and father had heightened Virginia's sense of helplessness, as loss usually affects vulnerable, especially young, people. Later she described her anxiety as "not unnaturally the result" of so many deaths, traditions, and complications (MB, 183). Virginia also coped with conflicts between traditional expectations about a young lady's proper deportment and her own writerly ambitions. "Not unnaturally," her anxieties escalated, and by the end of the trip, she was in a fix that developed into acute nervous anxiety.

When Virginia returned from Italy and Paris to 22 Hyde Park Gate, already physically exhausted, the grief and guilt she had felt before the trip and the resentment she felt on the trip were at a peak. The morose house where her mother had died and where she had watched and reported on her father's decline until his death must have seemed haunted by ghosts. On the day after returning, back in the room that held so many memories and tensions, Virginia "first heard those horrible voices" (MB, 123).

On that May day in 1904, delusions about the external world settled into Virginia's mind, and paranoid hallucinations tormented her. She seems to have experienced what modern science calls *derailment*, defined as "loose associations" or "a flight of ideas"—that is, "spontaneous speech in which the ideas slip off the track onto another one that is clearly but obliquely related, or onto one that is completely unrelated."[8] For the first time, Virginia suffered a breakdown that landed her fully in the possession of her own demons. (For a discussion of manic-depression, see Appendix B.) Afterward, recuperating at Violet's home in Burnham Wood, Welwyn, Virginia lay in bed, refusing to eat and "thinking that the birds were singing Greek choruses and that King Edward was using the foulest possible language among Ozzie Dickinson's azaleas" (MB, 184). Although such illusions could be construed as referring to masculine power and its abuse, Virginia regarded Vanessa as the abuser.[9]

Fasting gave her, Virginia decided, the power to see into others' minds. And what she saw seemed to confirm the fears she had expressed in the *Hyde Park Gate News* some nine years ago, that her "best friend hated" her. Accordingly, Virginia refused even to see Vanessa. Thoby dealt with this embarrassing turn of events by lying to Clive Bell about "my sister having started Scarlet Fever in the Country and I being quarantined here for a week . . . the devil of a nuisance." The more truthful Vanessa simply explained that her sister was ill in the country and that she had stayed with Virginia and then returned to London.[10] Vanessa went home because Virginia now imagined her a tyrannical legislator, a heartless fiend who felt neither grief nor guilt over their father's death, and a mere painter who dared abuse literary sensibilities.[11] But having transformed Vanessa into a villain, Virginia now had no family ally. Finally, she made a feeble and relatively harmless suicide attempt by flinging herself out of one of the windows in Violet's spacious home.[12]

Between her thirteenth and her twenty-second year, Virginia lost her mother; her half sister, who was also her surrogate mother; and her father. The first two deaths and her helpless guilt unsettled her stability, and the third unhinged it. But in all these difficult times, the combination of "complications" for Virginia seems not to have been just loves and deaths but also deeply opposed emotions, a sense of being torn in half (MB, 183). In 1895, 1897, and 1904, she could neither voice nor escape her emotions, so conflicted were they between grief and relief, loss and guilt.

In 1904, Virginia was further torn between sisterly love and artistic resentment. Loss of her father, guilt over what had seemed a secret complicity with Vanessa against him, and helplessness to resist Vanessa's painterly itinerary on the Continent seem to have been the causes of Virginia's hallucinatory condition.[13] Loss, guilt, and helplessness formed a hazardous constellation of feelings against which Virginia's healthy ego needed to guard itself throughout her mature life.

Virginia knew "I can write, and one of these days I mean to produce a good book," and this knowledge saved her (VWL, 1:144). After a summer of enforced, supervised rest, in September 1904 she was well enough to leave Violet's supervision and take a holiday with her siblings and one nurse at the Manor House, Teversal, Nottinghamshire, courtesy of George Duckworth.

That month, with "immense pomp," George married Lady Margaret Herbert, whose father owned the Manor House, among other properties. Vanessa was a bridesmaid, but Virginia did not even attend (Bell, 1:96). The voices that told her "to do all kinds of wild things" had subsided, and she now recognized that overeating had not caused the voices, so there were no more "disgusting scenes over food." Her animosity to Vanessa also dissipated, as did her guilt and the grievous sense that Vanessa disliked her (VWL, 1:142, 143).

Adrian Stephen resumed his studies at Cambridge, and in October the other Stephen children returned from Teversal to London. While staying with family friends, Vanessa and Thoby supervised the Stephens' dramatic move from Hyde Park Gate to Gordon Square. Perhaps still under a nurse's supervision, Virginia wrote no letters and played no physical part in the Stephens' psychological and geographical break with all that the "old ladies of Kensington and Belgravia" represented.

Whereas 22 Hyde Park Gate had been dark, crowded, draped, and partitioned, 46 Gordon Square, Bloomsbury, was bright, airy, and open. Vanessa handled the arrangements and finances. She was astonished that Harrods paid £48 for the family's antiques, which she had thought "worthless" (VB/VW, 21 November 1904). She invested in fresh paint and new furniture; lined the hall with five of the best photographs of Julia Stephen taken by their great-aunt Julia Margaret Cameron, finding them "very beautiful all together"; and hung Watts's portrait of their mother, but not that of their father.[14] She also hung Cameron photographs of eminent Victorian men. These pictures and the mantelpiece from 22 Hyde Park Gate were almost the only reminders in Gordon Square of the Stephens' past (VBL, 32).

Just three blocks behind the British Museum, Gordon Square was on the periphery of the University of London. The Georgian row houses around the square were handsome but unpretentious and, at the time, offered a Bohemian and distinctly déclassé address. There the Stephens could live comfortably on income inherited from their parents and from Stella. Although 46 Gordon Square lacked some of the grander architectural touches of its neighbors—like fan windows and ornate balustrades—after Hyde Park Gate, its simplicity and the rather wild garden in the square, with its tall plane trees and paths between flower beds and tennis court, were distinct assets. Gerald Duckworth now preferred to live on his own, and George, by marrying Lady Margaret Herbert, had acquired other, truly elegant properties. The Stephen children were at last alone, and Vanessa was rude enough to outraged relatives like Aunt Mary Fisher to ensure that they would be left alone. (Later Vanessa referred to her aunt, Julia's sister, as a "relic" [VB/VW, 25 June 1916].)

By the middle of October, 46 Gordon Square was ready for the Stephens to move in. Vanessa and Dr. George Savage, however, thought that London endangered Virginia's health and so sent her to their Quaker aunt, Caroline Emelia Stephen, in Cambridge. At first, Virginia felt better away from the "whirlpool" of house decorating and arranging (VWL, 1:145). But she must have felt left out when she received a black-bordered postcard from Vanessa that triumphantly proclaimed, "Here I really am under our own roof!" and insisted that "London would do you harm now" (VB/VW, 22 October 1904).

In September, Thoby had written to Clive Bell about the responsibility he had assumed for collecting materials for Frederic Maitland's biography of Leslie Stephen (TS/CB, 8 September 1904). In Cambridge that October, Virginia visited Maitland, and he invited her, rather than Thoby, to contribute to his book. Savage agreed to the exercise of this filial duty, since, as Vanessa told Virginia, "You understood Father better than anyone else did" (VBL, 19). In a gesture of Victorian scruple unimaginable for a late-twentieth-century biographer, Maitland declared her parents' letters to be "so private" that he would not "look at them himself"; thus he assigned Virginia the task of copying out extracts (VWL, 1:148).

Although Virginia wanted to write something herself, she accepted the assignment (VWL, 1:145). But she was sometimes irritated to have her work interfered with by "Freds wish," and she was outraged by advice from the "authoritative" Jack Hills, widower of Stella, who "never knew nor understood Father, and has no more sense of what a book ought to be than the fat cow in the field opposite."[15] In Cambridge, Virginia was excruciatingly bored. Her visits with the Maitlands and Adrian helped, but reading her parents' letters provided the greatest distraction.

Vanessa explained to her "Beloved monkey," "Beloved William," "My own William," or "Beloved Monkey & Wombat" that Dr. Savage believed Virginia's early October "fortnight in London had brought [her] very near another serious breakdown," for she was "quite clearly not yet strong enough to stand London." Furthermore, as a friend of Leslie Stephen, Savage proclaimed, "I will have no fees." Vanessa used that "extraordinary kindness" to pressure Virginia into following Savage's advice, but her message must have put Virginia in another bind: If she refused to follow Savage's advice, she would appear not only recalcitrant and unwell but ungrateful too. Vanessa protested, "Dont go and imagine that I *want* you to stay away," which was exactly what Virginia now did imagine (VBL, 19, 20). She claimed Vanessa said that no one cared where she was, "which made me angry, but then she has a genius for stating unpleasant truths in her matter of fact voice!" (VWL, 1:147).

Although Vanessa did not save Virginia's letters from Cambridge, in late October she did reply to one: "Your letter this morning made me scream with its account of the Quaker household & the meeting." But then she undercut that compliment by telling Virginia, "You sound somewhat rabid," and by proclaiming her own verbal inadequacy: "You say so much about the faculty of expression that I am quite afraid of putting my feeble sentences before your critical eye." After shelving the family's calf leather–bound books in their new study, Vanessa did acknowledge that "though you may hook your learned nose at me in disdain," even she harbored affection for old books (VBL, 21). Vanessa continued to fill her letters to Virginia with stories of social and painterly activities in London and advice to eat more and follow Dr. Savage's instructions (VB/VW, 24, 30, and 31 October 1904).

Virginia repeatedly protested about being kept away from London, where she needed only "a large room to myself, with books and nothing else" (VWL, 1:147). Before her wish to be released from Cambridge could be granted, however, Vanessa asked Madge Vaughan whether it would "be possible for you to have Virginia to stay with you for about a fortnight" in Yorkshire, after a limited stay in London. Vanessa assured Madge that Virginia had recovered from her breakdown, except for sleeplessness. Vanessa's only stipulation was that Virginia "ought not to walk very far or for a long time alone" (VB/MV, 28 October 1904). Virginia's right to a brief stay in London seems to have been contingent on her agreeing to go to Yorkshire afterward, where, Vanessa promised, she could "play with the children, or be a hermit. It all sounds as if it might be what you would like" (VBL, 22).

Even though it was not at all what Virginia wanted, she agreed. And so in November 1904, she was allowed nearly two weeks at the exciting Gordon Square address, which seemed "the most beautiful, the most exciting, the most romantic place in the world. . . . Everything was going to be new; everything was going to be different. Everything was on trial" (MB, 184–85). Virginia's stock of mythic houses had increased by one. She later wrote, "46 Gordon Square could never have meant what it did had not 22 Hyde Park Gate preceded it" (MB, 182). At Gordon Square, "the Watts–Venetian tradition of red plush and black paint [from 22 Hyde Park Gate] had been reversed; we had entered the Sargent–Furse era; white and green chintzes were everywhere; and instead of Morris wall-papers with their intricate patterns we decorated our walls with washes of plain distemper" (MB, 185).

As Virginia's trip to Yorkshire approached, Vanessa assured Madge that Virginia now could walk alone; the only danger was that her mind might "become too active" (VBL, 24). Vanessa admitted to Madge that given Virginia's illness, "it was impossible not to interfere a good deal in physical ways because [Virginia] had rather morbid ideas on the subject of health & would have allowed herself to get ill if one had not prevented her." Vanessa said she did not "enjoy myself having to be always on the lookout for her." Because she knew that her interference bred "unhappiness" and resentment, she hoped that the exile would encourage Virginia to become independent (VBL, 26).

At the conclusion of Virginia's stay in London, apparently on 17 November, the Stephens held a farewell dinner there for Leonard Woolf. Thoby invited Lytton Strachey: "Woolf is coming, I hope you will be able both to see him & inspect this home which is just becoming habitable." To Clive Bell, Thoby gave the impression that only he entertained Leonard: "Poor old Woolf sails for Ceylon tomorrow—he dined with me tonight."[16]

Leonard Woolf, too, had known sorrow early. His father, a prominent barrister, had made "vast sums of money" as a queen's counsel but had "spent it all on living." He had enjoyed "too large a house, too many children, too many servants, and then he died quite suddenly" when Leonard was only eleven, leaving his widow and large family in seriously reduced circumstances (interview with Leonard Woolf, BBC, 1967). Mrs. Woolf had to sell her spacious home and move to relatively cramped quarters in

Putney, a suburb of London. Outwardly undaunted but inwardly crushed by these events, Leonard, the third oldest child and second son, attended St. Paul's (preparatory school) and Trinity College, Cambridge, on scholarships.[17]

At Cambridge, Leonard was a member of the supposedly secret elite intellectual society, the Apostles. With them and other serious-minded young men, Leonard spent his time and energy discussing literature, art, and G. E. Moore's abstract theories about "the good." Given that such talk was more absorbing than his courses, Leonard scored too poorly on the tripos exams to become a fellow at Cambridge. Given his family's reduced finances, therefore, he did not study law but instead took the exam for the Colonial Service.

Despite Thoby's ignoring his sisters when he wrote to Clive about entertaining "poor old Woolf," Leonard was very aware of their presence. They were not so remote as they had been at Cambridge when he had likened them to Rembrandt or Velázquez paintings, but they now would be even farther away from him. In Ceylon, Leonard was about to begin, just before his twenty-fourth birthday, a lifetime career as a colonial administrator. The next day, he left, taking with him a fox terrier and seventy volumes of Voltaire. Virginia did not refer to the farewell dinner for Leonard. She was distracted and furious because after only a few days at Gordon Square, she was, on the day after the dinner, about to be shipped away to Yorkshire.

Virginia repaid Vanessa for her incarceration in Yorkshire by sending a satiric piece to Violet (and not Vanessa). Because Violet "howled over the Life of Caroline Emelia," Vanessa begged to read the "Life" of their aunt herself (VBL, 27) and scolded Virginia for not eating enough.[18] Virginia complained that Yorkshire did nothing to cure her insomnia and longed for her "perfectly quiet room at home, where I need never talk or be disturbed" (VWL, 1:159).

Virginia found stifling Madge's marriage to Will Vaughan, whom Madge had met when they both were visiting his Stephen cousins at 22 Hyde Park Gate. Will, in turn, soon "disapproved of the way the Stephen girls conducted their lives" in Bloomsbury.[19] Familiar with Virginia's letter-writing style, Madge rightly suspected that Virginia might be laughing at her behind her back, although the truthful, sensible Vanessa tried to rationalize that Virginia never said a "really unkind thing" and that "everyone who has brains must be critical when they are young" (VBL, 25).

Virginia's impassioned protests against being exiled again did not persuade Vanessa, who told her how Madge exclaimed that Virginia had come "as a boon and a blessing to them (like the pen)."[20] Virginia was not mollified. She saw her life as "a constant fight against Doctors follies" and resented Vanessa's deference to the "pigheaded" Savage (VWL, 1:159, 153). Vanessa was "distressed" at Virginia's talk of returning soon, for Dr. Savage would "much rather you stayed away." Vanessa argued that she should stay with Madge until at least 1 December, for her letters sounded "distraught." After much pleading, Vanessa finally agreed to let Virginia return on 28 or 29 November, "for a few days only" (VB/VW, 22 and 24 November 1904).

Virginia spent four or five nights in London but then was sent back to Cambridge for another week. In this third exile, she resumed her combined biographical and filial duty. Although Vanessa approved Virginia's extracts of their parents' letters, she was embarrassed by their tenderness: "I only feel on reading the expressions of affection & such bits that I want no one else to see them & would rather not read them myself in spite of their beautiful expression." Instead, Vanessa wanted the original letters burned. Her attitude toward emotional displays seems to have been set before Virginia was born, when Leslie wrote to Julia that Vanessa had "odd little misgivings in her little mind as to doing anything demonstrative."[21]

As the Preface to this book suggests, finding herself exiled in Cambridge and Yorkshire was pivotal for Virginia, for henceforth she doubted that Vanessa really loved her, and so thereafter she wrote not just to secure motherly affection but also to gain public regard. In Cambridge, Virginia had vented her irritation in a private satire. In Yorkshire, she proved her self-sufficiency by writing for publication for the first time since as a child she had sent a story to *Tit-Bits*. She wrote one book review and visited the bleak Brontë parsonage at Haworth and wrote an essay about it.

Violet had shown some of Virginia's writing to the editor of the women's supplement of the religious paper the *Guardian*, where her review was published on 14 December 1904 and the Haworth article on 21 December.[22] Despite editorial tampering, these publications fulfilled Virginia's "desire of the pen in my blood," so that she had "hard work not to write" (VWL, 1:169). She composed an article on Boswell for Leslie Stephen's former journal, the *Cornhill*. But when the *Cornhill* bluntly rejected her and Violet's friend Richard Haldane offered criticism (albeit mild) of the Haworth article, Virginia's writing confidence was nearly destroyed (Bell, 1:94).

Years later, Violet asked Vanessa whether she remembered "the Goat blowing me up sky high because Haldane didnt praise her article on the Brontë family enough?" (VD/VB, 16 June 1942). Virginia's written response to Violet did not explode but compensated for the feeling of rejection with a threat based on the curious opposition she posited between the arts of the pen and the brush: "If Haldane is severe, I shall give up literature and take to art." Her reasoning was that "pictures are easier to understand than subtle literature, so I think I shall become an artist to the public, and keep my writing to myself" (VWL, 1:170).

Virginia was voicing the quandary she had illustrated by writing the 1899 diary within the pages of *Logick:* She needed to write for the public in order to assert her identity, but she could hardly bear to be exposed to public reaction. She was serious enough about becoming an "artist to the public" to send Violet several drawings she had made, copies of drawings by Blake and Rossetti and a head of Hypnos whose wing is reminiscent of her early drawing of a sunset. Diane Gillespie argues that Virginia, like Blake and Rossetti, could have been creative in both the visual and verbal arts.[23] I see these drawings more as a self-defense, to prove that even her visual efforts "dont turn out so bad" (VWL, 1:173). They also established, however, that Vanessa's simpler art was easier to master than Virginia's more subtle one. Then Vanessa's comparable ploy backfired, however: She sent an article she had written on Watts to the *Saturday Review*, but its rejection suggested that she was no master of Virginia's art (VWL, 1:178 and note).

When Fred Maitland asked Virginia to do more than copy extracts—that is, to write about Sir Leslie's old age—she was gratified at last (VWL, 1:172). She balanced her portrait of Leslie with both a positive account of his literary influence and a negative account of his domestic tyranny. But she feared her work would be subsumed under Maitland's notion of what a life should be, as indeed it was. Maitland quoted Virginia's warm portrait of Leslie's bookishness, but he summarized Virginia's account of some of her father's weaknesses, indeed, so softening them they became virtues (VWE, 1:127–30). For example,

> Stephen would have liked to be both father and mother, and was grieved when he was told, as he had to be told, that his anxious and self-sacrificing solicitude was doing harm. But, like the eminently reasonable man that he was, he took the proffered advice, and then all went well.

Maitland quoted one friend who endorsed the Julia-as-angel myth: "'Apart from his gifts, the highest to be said of [Stephen] is that he was the worthy husband of such a woman.' That is high; but it can be said." Maitland also repeated Dr. A. W. Ward's insistence that Leslie Stephen never wrote "'either a meaningless or an intentionally unfair word, never spoke a vapid or an unkind one.'"[24] Clearly, Maitland intended to keep hidden the man behind the esteemed image, and Virginia must have been furious that he bowdlerized much of her writing. Nevertheless, she could be proud that she had shaped her divided experience of her father into words and thereby subdued the conflicts within her.

At last back in London, Virginia's manic delight struck Thoby's friend Lytton Strachey as "wonderful—quite witty." She was "full of things to say, & absolutely out of rapport with reality." Strachey continued this letter to Leonard Woolf with an outsider's acknowledgment of the Stephen family's instability. He expressed sympathy for "the poor Vanessa [who] has to keep her three mad brothers & sister in control. She looks wan & sad" (LWL, 75). Although we know little about Thoby, clearly Lytton and Leonard had seen behavior that justified labeling him as mad. Apparently, all the Stephen siblings showed symptoms of depression (Caramagno, *Flight of the Mind*, 97–113). Adrian experienced several breakdowns, like Virginia's but milder. Both Adrian and Vanessa suffered prolonged periods of depressed lethargy, but in 1904 Vanessa stood out as the Stephen with the energy to manage her new household and her brothers and sister as well.

Soon Vanessa could stop worrying, however, for in early 1905 Virginia proclaimed the end of her "horrible long illness" (PA, 222). Keeping an evocative and informative diary between Christmas 1904 and late May 1905, Virginia again amused herself by writing, despite being "ordered not to write for my brains health" (PA, 230). Albeit entertaining, writing was serious business. She read the *Poetics*, "which will fit me for a reviewer!" and she praised Aristotle for "laying down so simply & surely the rudiments both of literature & of criticism" (PA, 240, 241). Around New Years Day 1905, vowing "to keep myself in pocket money at least this year by my writing!" Virginia took a major step toward constituting herself as a financially independent woman (VWL, 1:172).

She began teaching classes for working women. Then Bruce Richmond of the *Times Literary Supplement*, who had asked her father to review for him, now approached Virginia, and she agreed "with joy" (Bell, 1:104). Richmond promised to send her more books to review, "& thus my work gets established" (PA, 234). Having learned a sense of audience from both her lectures and her reviews, she chose to expand her lectures on Greek myths and Greek history to cater to women's curiosity about "whether there were fleas in the beds at Venice. I shall have to invent some" (PA,

225). In addition, she experienced writing to deadline, having some of her reviews cut, and having one review rejected because of her point of view and another because it was not "academic" enough. When a title was "changed, half is cut out, words are put in & altered, & this hotch potch signed Virginia Stephen," she found her author's blood boiling, but she learned to endure such editorial prerogatives (PA, 243).

On 1 March 1905, the four Stephen children held a "garrulous and successful party" at Gordon Square and launched what would become the social and artistic alliance later known by its location as the Bloomsbury Group (PA, 245). The next week, Clive Bell came by, and they "talked the nature of good till almost one!" (PA, 249). Thoby's friends—including Bell, Desmond MacCarthy, Saxon Sydney-Turner, and Lytton Strachey and his cousin the painter Duncan Grant—began regularly dropping by on Thursday evenings to talk among themselves in front of the Stephen sisters about such abstract questions as the nature of good and the necessity of beauty in art.[25] These young men carried over from Cambridge (where several had been members of the Apostles) such theoretical interests and commitments to absolutely open discussions.

At these unchaperoned gatherings, soon the young women, perhaps to the conventional Thoby's surprise, began to talk as well and found their remarks accepted or rejected objectively. After the emotionally dishonest and sexually threatening environment at Hyde Park Gate, these candid, intellectual evenings, when Virginia might be complimented for the clever way she argued her case, were "reassuring" (MB, 191, 192). They allowed her to emerge from her cocoon of shyness and display her dazzling cleverness in words.

The evenings must have been too literary for Vanessa, however, for the following fall she started her Friday Club in order to discuss painting. Virginia wrote to Violet that the young Bloomsbury men were not "robust enough to feel very much. Oh women are my line and not these inanimate creatures" (VWL, 1:208). Later Virginia realized that this atmosphere was "narrow; circumscribed" and that the ease of conversation resulted partly from the homosexuality of many of the young men (VWL, 4:180). For Virginia, with so many reasons to fear male sexuality and power, this unthreatening time among young men responsive mostly to her argumentative eloquence was restorative, and she remembered those early Bloomsbury years as a "kind of Elizabethan renaissance" (VW/Julian Bell, 21 May 1936, MFS). Of course, "Bloomsbury rapidly lost the monastic character it had had" (MB, 195), and soon the eminently respectable and charming Hilton Young joined the Thursday evenings and began to pay restrained gentlemanly attention to the younger Miss Stephen (Bell, 1:131).

At the end of March 1905, Virginia and her brother Adrian traveled together to Portugal and Spain. This was Virginia's first trip abroad since her ill-fated trip to Italy in 1904. Despite seasickness, monotony, and problems with lodgings and money, it was "a good journey," probably because Virginia included literary and historical sites in their itinerary (PA, 267). In Lisbon, they "let loose a caged bird that was singing by Fieldings tomb—a pious act!" (PA, 262). Virginia's journal displays a professional's eye for both journalistic material and a public, noting, for example, that one of her strained metaphors "may be read 2 ways" (PA, 263).

The spring of 1905 was a harmonious time. While Vanessa was anxiously awaiting her art professor's criticism, Virginia "sat in the Studio, & tried to comfort—& did not work whatever" (PA, 256). Without resentment, Virginia spent a "morning devoted to art!" and shared in the "general rejoicing" over the public exhibition of Vanessa's portrait of their friend Nelly (Lady Robert) Cecil (PA, 268). She laughed at Vanessa's Friday Club as a committee that "sat for two hours on very few eggs" and debated the merits of "Whistler and French impressionists" versus the "stalwart British."[26] Despite her teasing tone, Virginia would comfort, attend exhibits, and even participate in artistic debates if doing so would secure Vanessa's affections.

This harmony seems related to Virginia's own writing success. Her diary, she pointed out, was destined to "a premature grave" because "writing an extra page every day, when I write so many of necessity bores me, & the story is dull" (PA, 273). Not surprisingly, therefore, she stopped making entries at the end of May but did list in it all her journal articles, estimated the number of words, and made an incomplete guess at how many shillings and pounds she might earn.

Clive Bell, who moved easily between the literary Thursdays and the painterly Fridays, was now paying considerable attention to Vanessa. He had been mythologized by Thoby as "a sort of mixture between Shelley and a sporting country squire" (MB, 187). As the son of wealthy but philistine parents, Clive led a life of masculine indulgence. As a young Cambridge aesthete, he cultivated and critiqued the arts of writing and painting. In no need of earning a living, he planned to write about art, especially painting. Virginia felt that her sister "might marry 20 Clives and still be the most delightful creature in the world" (VWL, 1:297). She laughed at the affected melancholy in *Euphrosyne*, a privately printed collection of poems by Clive

Bell, Lytton Strachey, Walter Lamb, Saxon Sydney-Turner, Leonard Woolf, and others. Given such an example of literary pretentiousness, she asked Violet for reassurance about her own talent: "Do you feel convinced I can write?" (VWL, 1:202).

In August 1905, for the first time in eleven years, the four Stephen children journeyed to Cornwall, hoping that they would "find our past preserved" (PA, 281). The first night, they walked to Talland House, now leased to strangers, where they peered through the hedge and "hung there like ghosts" (PA, 282). Their past, it seemed, had been guarded and treasured by the villagers, who revived the image of Julia as a noble and sainted mother (PA, 285). Sailing in the bay, the Stephens saw a "sudden exclamation of porpoises; & not far distant we saw a shiny black fin performing what looked like a series of marine cartwheels" (PA, 283). (That image of the fin rising out of the water—benign if it belongs to a porpoise but ominous if a shark—became a keystone of Virginia's literary imagination.) Virginia spent her afternoons tramping alone along the Cornish coast, and her evenings writing.

The Stephens also enjoyed visiting the country people they had known eleven years before. Virginia collected local material, finding in their washerwoman's life story a "rather psychological study for a novelist" (PA, 287) and in the St. Ives regatta, a sight for a "French impressionist picture" (PA, 291). At the end of the trip, "certain circumstances"—perhaps Clive Bell's marriage proposal to Vanessa—interrupted Virginia's diary. But Vanessa rejected Clive, probably because, as she wrote to Margery Snowden, "I only wish we could always go on like this, but after all we may for a long time yet. I dread every day to hear that Thoby is in love" (VB/MS, ? ? 1905 or 1906). Feeling that there was "something of our own preserved here" at St. Ives, Virginia, like Vanessa, feared that the comradely band of Stephens could not be preserved much longer (PA, 299).

Virginia's travel diaries show her practicing careful and evocative descriptions, especially "descriptions without adjectives" (VWL, 1:206). One such description caused Thoby to tell "Nessa, who told me, that he thought I might be a bit of a genius" (MB, 130–31). This label was the one that had eluded her father (MB, 110), so having it applied to herself in the "divine art" of writing was especially gratifying (VWL, 1:209). (She would have been less pleased to know that Thoby's letters to Clive Bell hardly acknowledged her existence.) But Virginia was no Mozart or Pope (who first published at twenty-one). Rather, her art of writing could blossom only after she had long practiced the craft of writing.

During the next year, Virginia studied history, taught her classes, and wrote and wrote (PA, 274–80). Vanessa, too, did "a lot of work," which Virginia thought her best, and managed the Friday Club with unsentimental good judgment (VWL, 1:209, 213). She even allowed the reactionary Thoby to lecture on the "Decadence of Modern Art" (VWL, 1:225). Virginia's reading was designed to fill in the gaps in her education. When she visited in Yorkshire in the spring of 1906, she stayed in a boarding house where she read, among many self-imposed assignments, Landor's "Pericles and Aspasia for the first time" (VWL, 1:223). After hiking around the bleak countryside, she tried to describe the "great melancholy moors" but had to acknowledge the descriptive inadequacies of language: "Words! words! You will find nothing to match the picture" (PA, 305). Still, words were her tools, and she consoled herself that George Eliot was "near 40 I think" when she wrote her first novel. With Vanessa still being courted by Clive Bell, Virginia's "vague and dream like world, without love, or heart, or passion, or sex" remained "the world I really care about" (VWL, 1:227).

In the fall of 1906, the two Stephen sisters, along with Violet Dickinson, met Thoby and Adrian in Greece. Virginia wrote a satire of her brothers' attempts to speak Greek with the natives, which Thoby and Adrian probably found less amusing than she did.[27] When practicing writing scenic descriptions in her diary, Virginia confronted the rivalry between the arts as she had in 1899. She feared that she—like Leonardo's poets who offered only the "shadow of the body"—could only "pile words," which were an empty "pretense" beside the reality (PA, 319). Language was so physically limited that she would "not try to reproduce here the whole" (PA, 321). Often "there never was a scene less easy to fit with words" (PA, 333). Now recognizing the limitations of her medium, Virginia no longer threatened to abandon it for painting or to pretend that language could emulate paint. Rather, she could see the emptiness in the analogy of the "mute poem" to the "speaking picture" attributed by Plutarch to Simonides: "When you speak of 'the colour' of the Parthenon you are simply conforming to the exigencies of language; a painter using his craft to speak by, confesses the same limitations" (PA, 323). She reflected that the color of words and the speech of pictures are mere analogies based on Horace's doctrine of *ut pictura poesis* (as with pictures, so with poetry). It was tempting, however, to use such analogies: "If statues & marble are solid to the touch, so, simply, are words resonant to the ear" (PA, 331).

Virginia's 1906 travel journal is a highly impersonal exercise book.

Only a sentence about "reading that you do in a sick room" suggests that one of the party was ill (PA, 340). There were "circumstances again, at which a discreet diary can only hint, gloomily" (PA, 346), but Virginia barely hinted. She fondly recalled Yorkshire moors, London squares, and warm fires at home, for England now seemed "clean & sane, & serious" beside the "inhospitable east" (PA, 345–46). Virginia did not record that Vanessa had suffered a physical collapse, perhaps—as Lytton Strachey suggested—from the strain of keeping her siblings "in control."[28] Vanessa herself explained her collapse to Clive Bell as the result of getting "very much tired by nursing Virginia two years ago" (VBL, 44).

Thoby wrote to Clive another manufactured excuse, saying that Vanessa had appendicitis and so could not leave Greece for a week. After arriving with Adrian at Gordon Square on 25 October, Thoby planned to travel to Cambridge for the weekend (TS/CB, 10 and 25 October 1906). He did not make it. By the time the rest of the party returned to England, Vanessa, Violet, and Thoby all were seriously ill. Vanessa recovered; Violet languished; but Thoby became delirious. Virginia, Adrian, and Clive were allowed to see the invalided Stephens in their separate quarters in Gordon Square, but they could do little. On 17 November, Clive reported that Thoby, who had been in great pain, was "round the corner." But then on Tuesday, 20 November 1906, Thoby died of what had been just the week before diagnosed as typhoid fever (CB/Saxon Sydney-Turner, 17 and 20 November 1906).

Virginia Stephen's young adulthood had been shaken by a succession of tragic deaths when she was thirteen, fifteen, twenty-two, and now twenty-four. After Thoby's death, Virginia grieved but seemed to suffer no emotional crisis. The disparity between her extreme distress after the first three deaths and her calm after the fourth cannot be explained by indifference to Thoby. As her early letters testify, Virginia adored and loved him as she did not Stella. Furthermore, Leslie's death at seventy-one had none of the irrationality of the other three deaths, yet Virginia's reaction to it was the most extreme. Her equanimity after Thoby's death could not have been resignation, for Virginia remained all her life distressed by death and terrified of loss.

The only denominator common to the first three but not to the fourth death that would explain the disparity in Virginia's reactions is guilt. She had rebelled against her mother in troubling ways; she had resented Stella's illnesses and the attention they garnered; and she had raged against her father. Then each died. But she had only adored Thoby. For his death, therefore, she needed to feel no guilt, and hence she suffered no emotional crisis.

In fact, rather than precipitate a relapse for Violet, who also had typhoid, Virginia wrote to her for nearly a month cheering words about Thoby's supposed recovery. On the day of his death, she did not mention Thoby but ended her letter: "Goodnight and God—have I a right to a God? send you sleep. Wall nuzzles in and wants love" (VWL, 1:248). (Wallaby was another of her animal personas in her relationship with Violet.) On 22 November, she told her, "Thoby is well as possible. We aren't anxious" (VWL, 1:249). Three days later (and five days after Thoby's death), she wrote:

> Thoby is going on splendidly. He is very cross with his nurses, because they wont give him mutton chops and beer; and he asks why he cant go for a ride with Bell, and look for wild geese. Then nurse says "wont tame ones do" at which we laugh. . . . Thoby has been reading reviews of the Life [of Leslie Stephen], and wants to know if you are up to that? (VWL, 1:250)

On 28 November, Virginia admitted to Violet that Vanessa was "exhausted by years of unselfish labour on behalf of a sister" so that now she "neither writes, reads, nor in any way toils or spins." Although sleep was difficult, Thoby was "all right, and is having whey, and chicken broth—chicken pounded to dust" (VWL, 1:252, 253). Virginia let surface some guilt over Vanessa's "unselfish labour" on her behalf, but no truth about Thoby. For four weeks, she kept up this fiction, which not only protected Violet but diverted her own shock.

Wondering to Virginia whether Violet "knows anything," Vanessa showed herself ignorant of Virginia's hoax (VB/VW, 13 December 1906). It is unlikely that Adrian felt any need to protect Violet from bad news. Thus it seems most likely that despite having suffered serious mental distress after three previous deaths, Virginia set herself the task of maintaining the fiction of Thoby's recovery in order not to give Violet a traumatic shock. That Virginia Stephen maintained the fiction so convincingly (and with no apparent damage to her own psyche) testifies to the power of her imagination and the strength of her denial. Also, as Mitchell Leaska points out, by 1906 Virginia had established her identity as a writer; hence she was no longer helpless (PA, xxxvii). Among the threatening constellation of loss, guilt, and helplessness, in 1906 only loss troubled her, and loss alone did not trigger her manic-depression.

Ironically, when Violet was "up to" reading reviews, she at last discovered Virginia's deception. In a review of Maitland's biography of Leslie Stephen, Violet read, "This book appeared almost on the very day of the untimely death of Sir Leslie Stephen's eldest son, Mr. Thoby Stephen, at the age of 25" (VWL, 1:266, note). Violet must have been horrified, not only by Thoby's death, but also by Virginia's deceit. Virginia defended her subterfuge as if she had had no choice and then remarked, "You must think that Nessa is *radiantly happy* and Thoby was splendid to the end" (VWL, 1:266). That is, Vanessa was happy because she had become engaged to Clive Bell two days after Thoby's death. Virginia's linkage seems to have reflected her sense that Vanessa's engagement was a betrayal of both Thoby and Virginia. This betrayal, moreover, jeopardized the harmony the two sisters had achieved in the past year and a half.

After years of the usually thankless guardianship of Virginia's health, finding her own health at risk and her favorite brother dead, Vanessa abandoned her caretaker role and let herself be loved. Her emotional defection must have seemed as sudden as Julia's had to the baby Virginia when another infant displaced her. Saxon Sydney-Turner wrote to Leonard Woolf in Ceylon that Vanessa now would see only Clive, thus making "the other two feel rather more lonely." Leonard had schooled himself since the death of his father to expect little from life, and Thoby's death only confirmed his pessimism.[29] Virginia asked Saxon to see whether Leonard would like Thoby's copy of Milton. Thinking perhaps of Milton's reaction in "Lycidas" to a friend's death, Leonard was gratified to be given the Milton to help him remember Thoby in this "accursed" life (LWL, 122).

Despite feeling abandoned and betrayed, when Vanessa visited her future in-laws three weeks after Thoby's death, Virginia wrote Clive an extremely generous and gracious letter (VWL, 1:268–69). In reply, Vanessa wrote to Virginia that she and Clive talked "7 or 8 hours a day" (VBL, 46); it was "all too wonderful"; and she did not mention Thoby. But Virginia came to think such denial was callous. In a late December letter to Violet, she described Vanessa's behavior as "strange and intolerable sometimes. When I think of father and Thoby and then see that funny little creature [Clive] twitching his pink skin and jerking out his little spasm of laughter I wonder what odd freak there is in Nessa's eyesight" (VWL, 1:273). With Clive, Vanessa was dishonoring the memory of father and brother, leaving Virginia, and taking over 46 Gordon Square.

Virginia's letter to Clive clearly expected the newlyweds to find a new house, but instead she and Adrian had to find one. Externally, with its Georgian fan windows, 29 Fitzroy Square (where George Bernard Shaw had lived before the turn of the century) was grander than 46 Gordon Square, but this new neighborhood across the Tottenham Court Road was even less respectable (VWL, 283–84). As Duncan Grant explained, Fitzroy was "a derelict square. The houses of the great had gradually decayed and were taken as offices, lodgings, nursing homes and small artisans' workshops" (Noble, *Recollections of Virginia Woolf*, 25). Internally, 29 Fitzroy Square lacked the style that Vanessa's presence had given to 46 Gordon Square. Virginia marked the house with her own taste by hanging Watts's portrait of Leslie Stephen, but little else made 29 Fitzroy Square her home. Even the transferred Thursday evenings that Virginia and Adrian hosted did not redeem this unprepossessing dwelling.

Displaced, Virginia expected never again to see Vanessa alone because Clive was now "a new part of her, which I must learn to accept," although becoming reconciled to loss seemed impossible (VWL, 1:276, 285). Six months after the marriage, Virginia reflected on "what one calls Nessa; but it means husband and [expected] baby, and of sister there is less than there used to be" (VWL, 1:307). With Vanessa's affections now channeled toward Clive, Virginia endured another separation. Virginia was so shaken by this loss that Vanessa felt she must scold her: "Are you being sensible or shall I soon have to nurse you through a nervous breakdown?" (VBL, 57).

Over the years, Vanessa had often needled Virginia about writing, claiming that Virginia's taste for old books was "disreputable." At a party when a man was astonished to find that Vanessa did not read the *Times Literary Supplement*, Vanessa had to explain at length "that I was I & you were you & that I didn't indulge in literature" (VW/Mildred Massingberd, ca. 1901, BL; VB/VW, 13 April 1906). Similarly, Virginia should not "indulge" in art criticism. Before her engagement to Clive Bell, Vanessa had told him, "You will be glad to hear that when I took my sister to see the Watts exhibition her last scruples of loyalty to him disappeared, & even she admitted that there was nothing to be said!"[30] In Vanessa's view, "even" Virginia, with her deficient aesthetic judgment, had to admit that Watts was now passé. During the courtship, Virginia had complained that Clive and Vanessa isolated themselves, supposedly talking and laughing about art. Virginia did not believe them and concluded, "Well, you may mix art with many things" (VWL, 1:256). Thus her first reaction to being displaced by Clive was to graft her resentment of him onto a revived resentment of painting. Not for thirty more years would she again sacrifice her writing to Vanessa's painting.

Virginia's second reaction was to try substitution: If she had lost Vanessa, she might secure Violet's love by writing about her. Virginia proposed writing "the essence of truth" about Violet and insisted, "This Biography is no novel but a sober chronicle." Despite spoofing her ungainliness in such lines as "Miss Dickinson grew to be as tall as the tallest hollyhock in the garden before she was eight," the "life" was a love letter. Virginia typed it using a violet ribbon and bound the book in violet leather with a gilded title on its cover.[31]

Meanwhile, Clive and Vanessa had developed a "social ruthlessness. Clive let it be known that he thought Violet Dickinson and her brother second-rate" (Spalding, *Vanessa Bell*, 63). Perhaps influenced by them, Virginia loosened her ties. In August 1907, she first mentioned in a letter to Violet the ongoing pain of Thoby's death (VWL, 1:303), and then her letters decreased in number and intensity.

Vanessa lived only a few blocks away, but she was so absorbed by her marriage and prospective maternity she could have been in another country. Virginia filled the emptiness at the undistinguished Fitzroy Square house by beginning another substitution, a biography of her sister. But lest Virginia appropriate her life, Vanessa countered with a plan "to write my biography" herself, although she admitted, "I dont think you will find me a rival to you" (VBL, 51). Vanessa's own efforts came "to a full stop at page 13" six months after beginning. Irritated at her verbal failure and Virginia's fluency, Vanessa found it was "really horrible how you succeed in your profession" (VB/VW, 14 and 2 August 1907). Perhaps to placate Vanessa's anxiety, Virginia expanded her life of Vanessa, not into autobiography, but into "Reminiscences" of their childhood.

In this period, with Vanessa pregnant, Clive became less a "funny little creature" and more an ally. With Virginia, he talked not about painting but about writing, specifically her writing. She now conceptualized writing in sculptural terms: "solid blocks of sentences, carven and wrought from pure marble" (VWL, 1:300). Again, Virginia used language, as she had from her childhood, to secure affection. In late August 1907, after having tea with Henry James in Rye and claiming to be unsure whether Violet wanted to read about Rye or to see a picture, Virginia announced that although she could draw simple shapes, she instead had decided "to write out many small chapters that form in my head" (VWL, 1:307). She took the leap for which she had apprenticed in journals, letters, reviews, autobiography, biographies, and stories and began writing her first novel.[32] Clive became her literary confidant and intelligently critiqued her novel in progress, *Melymbrosia* (later to become *The Voyage Out*). He thought her writings "thrilling because they revealed—so I thought and thought rightly—in a person I cared for, genius" (*Old Friends*, 93). He was the audience she needed.

Such praise was thrilling, but another development carried both thrills and threats: As Vanessa's pregnancy advanced, Clive sought to expand his literary sponsorship of Virginia into a more intimate relationship. He began a campaign to awaken her sexual awareness. His first (known) letter challenged her maidenly notion of genius by describing *Les Liaisons dangereuses* as both a "most indecent work of genius" and "one of the greatest." Clive began 1908 with a rondo for Virginia's twenty-sixth birthday that alluded to her "emerald sleeping passions" that in "some wakeful dream" would break through deep-seated barriers.[33] Although Virginia craved affection, Clive's pursuit was more than she had bargained for.

Vanessa and Clive's first son, Julian, was born on 4 February 1908. Six days later, Clive was hinting at his emotional independence. He invited Lytton Strachey to dine with him, with the suggestive aside, "I am, of course, always 'in,' and always alone, unless Virginia chances to stay with me" (CB/LS, 10 February 1908). Unaware of Clive's innuendos to others, Virginia worked to finish "Nessa's life," addressing her "Reminiscences" to the young child. Because she planned to send Clive two chapters, Virginia explained the emotional and aesthetic importance to her of writing biographies: "I ask myself why write it at all? seeing I never shall recapture what you have, by your side this minute" (VWL, 1:325).

Spurred by her sense of the replacement value of writing biographies, Virginia wanted to write a "very subtle work on the proper writing of lives" (VWL, 1:325). Vanessa expected "Reminiscences" to be "probably your master-piece so far." Predictably, she liked "the part about Mother best." Apparently trying to awaken her sister's sexual awareness, Vanessa wrote to Virginia about a Mrs. Raven-Hill, who was so "wildly improper in conversation" that she discussed contraception and "the joys of married life" (VBL, 61, 62, 61). Despite Vanessa's pleasure in Virginia's portrait of Julia, the completed "Reminiscences" misfired. Even though she found it "extraordinarily interesting," Vanessa "felt plunged into the midst of all that awful underworld [their adolescence] of emotional scenes & irritations & difficulties again as I read—How did we ever get out of it?"[34]

Although Virginia enjoyed writing biographies, she was learning that it was a prickly undertaking. She had made considerable efforts to temper and conventionalize her life of Violet, but Vanessa still wondered "how you ever dared to show it to her" (VBL, 59), and Violet judged it a "little harsh" (VWL, 1:336). Finding that biography could go awry, between 1906 (when she satirized Thoby's and Adrian's attempt to speak Greek to the peasants) and 1909 (when she tried to satirize Clive), Virginia turned to mock biographies, such as "Phyllis and Rosamond":

> In this very curious age, when we are beginning to require pictures of people, their minds and their coats, a faithful outline, drawn with no skill but veracity, may possibly have some value. . . . And as such portraits as we have are almost invariably of the male sex, who strut more prominently across the stage, it seems worth while to take as model one of those many women who cluster in the shade. (CSF, 17)

However revolutionary in treating obscure women rather than prominent men, this statement suggests that Virginia Stephen's art was still shackled to the tenets of realism: faithfulness, veracity, portraiture. She was looking for a "vision" that could supersede realism and uncover "some real thing behind appearances" that she would make real and whole "by putting it into words" (MB, 72). As yet lacking such a vision, Virginia pretended simply to describe faithfully.

The letter to Clive quoted on page 87 not only suggests a connection between biography and the desire to "recapture" or fuse with Vanessa's presence but also posits an inverse connection between biography and the novel. Virginia told Clive that she "dreamt last night that I was showing father the manuscript of my novel; and he snorted, and dropped it on to a table, and I was very melancholy, and read it this morning, and thought it bad" (VWL, 1:325). Immediately after dreaming that her father disapproved of her novel, Virginia considered "the proper writing of lives." The dream about her father and her novel suggests that even though her first commitment was to the novel, she retreated to biography when she felt inadequate to write fiction. The pattern replicates the mental dynamics of 1904 at Manorbier when she followed her "vision" of an art mystically united with the visible world, with a factual account written to prove her sanity.

A month or so after her son Julian's birth, Vanessa wrote a generous letter to Virginia offering to take her on a holiday to St. Ives to repay her for financing Margery Snowden's holiday with Vanessa some years before (VB/VW, March 1908?). But then an unexpected development defeated such generosity: Vanessa's husband and sister become jealous of her son. Clive complained to Virginia that the baby screamed all day long (CB/VW, 13 April 1908). Virginia took his side, announcing that "a child is the very devil" (VWL, 1:328). Sure that "nobody could wish to comfort it, or pretend that it was a human being," Virginia concluded, "I doubt that I shall ever have a baby" (VWL, 1:331). Now the trip to St. Ives would include both Clive and the baby, with the Bells sleeping apart, because as Vanessa told Virginia, the baby kept Clive awake (VB/VW, 20 September 1908). Virginia's resentments echoed her experience in the Stephen nursery when Thoby had Vanessa, Adrian had Julia, and Virginia had no one. In 1908, however, she could "have" someone: her brother-in-law.

Virginia now bargained for Clive's affection, asking him to kiss his wife "most passionately, in all my private places—neck—, and arm, and eyeball, and tell her—what new thing is there to tell her? how fond I am of her husband?" (VWL, 1:325). That was the invitation Clive needed. Given their shared sense of now being outside the constricted orbit of Vanessa's affections, his letters to Virginia took on a newly bold tone. From the Bell home in Wiltshire, he wrote to her about the "bawdy talk" between men and women that Vanessa now enjoyed and insisted that Virginia was as "feminine as anything that wears a skirt." He even tried hinting that females might be the superior sex (CB/VW, 18 April 1908).

In two letters Clive mentioned the Raven-Hills to Virginia, who at least knew of Mrs. Raven-Hill's reputation for bawdiness; perhaps Clive thought she also knew that this woman was his former mistress.[35] His reference might also have served to signal that he had renewed his old affair and no longer, given Vanessa's absorption in their baby, felt bound by marital ties. But Virginia did not understand his allusion and was confused by the signals Clive was sending her. Jack Hills attempted to disburden her of her illusions. Later, Virginia remembered, "His address to me in Fitzroy Square after V[anessa]'s marriage on the sexual life of young men. Can they be honourable? I asked—when he said how all male talk was about women; how every young man had his whore" (D, 5:198).

Without accusing Clive, Jack's cynical picture of male sexual behavior seems to have been calculated to rid Virginia of her naïveté about men in general. He was "amused," presumably at her shock and her sense of the "honourable." Although she had resented his advice about writing on Leslie Stephen, Virginia did appreciate this conversation, remembering Jack, "of all our youthful directors [as] the most open minded, least repressive, [who] could best have fitted in with later developments" (D, 5:198). He "shocked me a little, wholesomely" (MB, 104). If Jack's revelations countered Virginia's repression, Clive's behavior challenged it. He had funneled his aesthetic energies toward Virginia and his sexual ones back to his former mistress.[36] Now he was pressuring his sister-in-law toward both an aesthetic and a sexual liaison.

While the Bells were in Wiltshire, Virginia traveled alone to St. Ives. She read, worked on a review, and tramped about in the cold spring winds, wondering how she might describe the color of the Atlantic Ocean. She was soon joined by Adrian and then by Vanessa, Clive, Julian, and his nurse. Virginia wrote to Lytton Strachey that the baby brought out the worst in all of them (VWL, 1:328). Vanessa's preoccupation with her child induced Clive to court Virginia's affections openly, much to Vanessa's distress. Moreover, Virginia so encouraged the flirtation that tensions escalated; indeed, they may have forced Virginia to leave early. In her haste, she forgot the book she was reviewing. Clive fetched it and fell as he was trying to hand it to her as the train pulled away. From London, Virginia apologized for Clive's injury, although her "subtle sense of what is due" finally made her admit the truth: "Nessa is the injured party." In return, Vanessa seemed to be punishing Virginia by not writing (VWL, 1:329).

Given that she knew she was injuring Vanessa, we must wonder why Virginia flirted with her sister's husband. On the simplest level, she was once again using her writing to gain affection, praise, and now helpful advice. On another, she was responding to Clive's infatuation, which validated her femininity and desirability. At that level, she adopted an innocent pose, hoping to reengage Vanessa's affections. Virginia admonished Clive for flattering her and maintained that they had committed no transgression. But Vanessa was always Virginia's emotional Achilles' heel. Virginia wanted to punish Vanessa as she had done at the age of (probably) three, by bursting Vanessa's balloon after her own had burst. In adolescence, Virginia had regularly written to Thoby, not Adrian, out of love but also, it seems, out of a desire to steal him from Vanessa. Now she was trying to punish Vanessa for her emotional defection by filching Clive from Vanessa.[37]

Clive admitted that Virginia had warned him against flattering her, but he reminded her that at St. Ives she had wondered whether they would "ever achieve the heights" in their talk (CB/VW, 5 May 1908). Teasingly, Virginia offered the opinion that "though we did not kiss—(I was willing and offered once—but let that be)—I think we 'achieved the heights' as you put it." She admitted that the factor that "restricted" her willingness was the sight of the Bells' domestic life (VWL, 1:329–30). Clive was not deterred but responded, mixing sexual and aesthetic praise. He told her that he had wished for nothing but to kiss her and cautioned that they should not leave unsaid and undone what "we know that we ought to say and do." He masked that proposition with the claim that his belief in her "perhaps almost equals your own."[38]

Against her better judgment, Virginia responded to this flattery. Having a male "worshiper" eased her sense of alienation at Fitzroy Square and buttressed her self-esteem. Furthermore, Clive's attentions offered a pleasant counter to Adrian and the friends who "accept my lowest estimate of myself." She encouraged more overtures by writing to Clive, "Nessa has all that I should like to have" (VWL, 1:334). He interpreted this to mean that she would like to have him, too. But Vanessa had other plans. She wrote to Virginia that having left her book, nightgown, and sponge bag at St. Ives, Virginia "had really better marry a practical man who will pick up your leavings" (VB/VW, 2 May 1908). Hilton Young was an obvious prospect. Virginia showed no interest in him, but Clive made marriage even more unattractive by labeling Hilton as Virginia's "future tyrant."[39]

In the summer of 1908, the Bell entourage again left London for a visit with the senior Bells in Wiltshire. There Vanessa found her father-in-law reading an article that Virginia had at last placed in her father's magazine, the *Cornhill*. Vanessa advised Virginia to marry, wondered whether Lytton Strachey had proposed, and suggested that "perhaps on the whole your genius requires all your attention" (VBL, 65, 66). Still, Vanessa wanted to know whether or not Virginia had had a proposal.[40] Virginia's sense of loss appeared in a dream about Vanessa's "fatal injury from an omnibus" (VWL, 1:341). When Vanessa assured her that both she and Clive admired Virginia for something other than genius, Clive wrote to Virginia the same day that she should "blush for the havoc you have made" (VB/VW, 7 August 1908; CB/VW, 7 August 1908).

Vanessa's jealousy then provoked her to ask to see Clive's letters to her sister, but he refused to let her to see these "too private" letters. Vanessa told Virginia that she would prefer Lytton to Hilton Young as a brother-in-law, but she thought Lytton more likely to fall in love with Adrian. Vanessa's fear and jealousy surfaced in a dream that she would lose her art to her family, whereas Virginia would lose a husband to her art (VBL, 67). Meanwhile, Clive gave bawdy poems written by Lytton Strachey to Virginia, and Vanessa did not like the implications (CB/LS, 22 August 1908). In the peculiar dynamic that conflated emotional and aesthetic issues, Vanessa's incoherent reaction was couched in terms of the painter's verbal inadequacy: "I feel painfully incompetent to write letters & [am] becoming more & more so as I see the growing strength of the exquisite literary critical atmosphere distilled by you & C`live with your wits alert to pounce upon convoluted sentences & want of rythm [*sic*]" (VB/VW, 27 August 1908).

Both Vanessa and, incredibly, Clive blamed Virginia for the discord in their relationship. Both were pressuring her to marry someone but discouraging her from marrying her most eligible suitor. To escape these emotional machinations, Virginia spent a holiday alone in August 1908 in Wells and Manorbier. Work on her novel cheered her up. She wrote delightful long letters to Vanessa designed to reclaim their sisterly closeness, but she and Clive still kept up a bantering correspondence. Virginia coyly claimed to be too shy to put her affection for him into words (VWL, 1:345). But she shared his irritation with Vanessa and the screaming baby, asserting that Vanessa saw no one but baby Julian.[41]

Chaffing under Vanessa's jealousy, Clive reopened safer aesthetic issues, praising Virginia's plans for a new form for her novel, although he warned her against making it too "splintered." He also discouraged any responsiveness to Hilton Young and a second potential suitor, Walter Lamb (CB/VW, 23 August 1908). Vanessa cautioned Virginia that if her letters were published without Vanessa's replies, "people will certainly think that we had a most amorous intercourse. They read more like love-letters than anything else. Certainly I have never received any from Clive that could compare with them." But Vanessa also encouraged "compliments & passion" (VBL, 71), even from someone whose life "sounds most chaste & maidenly compared to" the one Vanessa was leading.[42]

At Manorbier, near the craggy coast and away from all the emotional "havoc" that she and Clive had jointly created, Virginia read the bible of the Cambridge intellectual elite: G. E. Moore's *Principia Ethica.*[43] Every day, wearing heavy boots and a raincoat, she took a walk. Once she slipped at the cliff's edge and then regained her footing, having "no wish to perish" (VWL, 1:358). Although she wrote Emma Vaughan one wild, manic letter, she did not swing into manic exhilaration, no doubt because she was free from emotional stress and could exercise her critical and creative skills

(VWL, 1:358–60). For example, she could disagree with G. E. Moore and critique the style of her own travel notebook (VWL, 1:364).

Here, away from London, Virginia was able to finish about a hundred pages of her novel, an accomplishment of which she was proud. She used various memories of her Portuguese excursion (including her experience at Fielding's grave) in this tale of young Cynthia's voyage to an exotic Latin American setting. Cynthia (a Violet-like woman) is well enough educated to discuss Sir Thomas Browne's *Religio Medici*, but soon she is rechristened "Rachel" and loses her Violet-like and Virginia-like sophistication (Heine, "Virginia Woolf's Revisions," 410–11).

With the waves pounding the shore below, Virginia walked carefully along the cliffs, past the ruined castle, and fleshed out her 1904 "vision" with a formulation that united the verbal and the visual: "I think a great deal of my future, and settle what book I am to write—how I shall re-form the novel and capture multitudes of things at present fugitive, enclose the whole, and shape infinite strange shapes" (VWL, 1:356). Her notion of a holistic enclosure that transcended divisions replicates that first early "grape" memory and hints that the aesthetic experience involved a mysterious fusion (Bollas, *Shadow of the Object*, 16).

At the same time, Virginia wrote a piece of short fiction satirizing her current accomplishments. Purporting to be the "Memoirs of a Novelist," it is actually a commentary on a fictional Victorian woman novelist and her biographer. Virginia projected her own depression, "terrible self-consciousness," and romanticism onto the novelist, who "invented Arabian lovers and set them on the banks of the Orinoco . . . [where] the scenery was tropical, because one gets one's effects quicker there than in England" (CSF, 72, 73, 75–76). Despite the substitution of English for Arabian lovers, Virginia had chosen a similar tropical setting for *Melymbrosia*. However, as Phyllis Rose writes, when Virginia herself attempted such lush scenic descriptions, "her fluent pen drie[d] up" (*Woman of Letters*, 69). The confusion between the title, "Memoirs," and the content, a commentary, is indicative of Virginia's ambivalence toward articulating such fictional self-revelations. When the *Cornhill* rejected the piece, Virginia's own "terrible self-consciousness" intensified.

With their relations briefly normalized, Virginia, Vanessa, and Clive set out for Tuscany together in the fall of 1908. Their goodwill was nearly destroyed, however, when Virginia chose Clive as her next biographical subject. Her sketchy notes (PA, 383–84) for a "char of CB" suggest that Virginia planned her "character" to follow eighteenth-century models.[44]

But the sketch ended abruptly, apparently because Clive quarreled with her over her satiric account of his background and his egoistic discovery "that he belonged to the select race of people who are termed clever."[45]

While they quarreled in Siena over the ravages of Virginia's satiric pen, Lytton Strachey wrote to Virginia warning her about Clive. Saxon Sydney-Turner, too, tried to lecture Virginia about her flirtations. Comparing Virginia to a painting or a statue by Blake, Praxiteles, or Botticelli, Clive boasted to Lytton of the rivalry he had touched off, with Vanessa insisting that she was the more beautiful sister (CB/LS, 9 and 19 September 1908). Lytton was so outraged that he wrote to Leonard Woolf in Ceylon asking him whether he did not think it the "wildest romance? That that little canary-coloured creature we knew in [Trinity College, Cambridge] New Court should have achieved that?—The two most beautiful and wittiest women in England!" (LWL, 139, note).

In Italy, Virginia considered more theoretical aesthetic issues. She tried "to grasp some ideas about painting" but concluded that the idea in the artist's mind "has nothing to do with anything to be put into words." Noting that a painting could be unified by lines and color according to "some view of beauty in [the artist's] brain," Virginia puzzled over the Aristotelian sense that conflict in fiction could be compared with symmetry in painting (PA, 392–93). Then, playing pen against brush, she articulated a "vision" of her writing:

> I attain a different kind of beauty, achieve a symmetry by means of infinite discords, showing all the traces of the minds passage through the world; & achieve in the end, some kind of whole made of shivering fragments; to me this seems the natural process; the flight of the mind. Do they really reach the same thing? (PA, 393)

In an unguarded moment, Virginia had allowed painting to help her formulate her "vision" of an aesthetic whole.

Their trip ended with a stop in Paris, where the inevitable Bohemian art talk was a treat for Virginia, now a successful journalist and prospective novelist (VWL, 1:369). After their return, Virginia gave *Melymbrosia* to Clive to critique. He accepted the assignment as both personal and aesthetic flattery, writing to her from Gordon Square: "You only half understand, I think, why it is thrilling that you should care" (CB/VW, 4 October 1908). Then he assured her that "the wonderful thing that I looked for is there unmistakably" and that "this first novel will become a work that counts." Even though he voiced reservations about her linguistic excesses and immaturity, he saw "immense improvement on all your other descriptive writing. The first 3 pages are so beautiful as almost to reconcile me with your most feverish prose" (Bell, 1:207–8). We do not know Virginia's response, but she told Clive that Vanessa "said I never gave, but always took" (VWL, 1:376). Perhaps Vanessa's accusation was a protest against her husband's absorption in her sister's art.

Lytton's irritation with Clive resulted in his deepening friendship with Virginia and a pleasant trip with her and Adrian to her precious Cornwall. Back in London at Fitzroy Square, Virginia continued to write reviews and work on her novel, made friends with Lady Ottoline Morrell, and briefly slipped to the periphery of the Bells' circle. When they were in Wiltshire again at Christmastime, Virginia must have tried consoling Vanessa for maternity's drain on her, another strategy that backfired: Vanessa wrote to her on Christmas: "If you really knew what I was feeling you would not pity me for having Julian" (VB/VW, 25 December 1908).

Clive, rather than Vanessa, begged for her pity. He wrote to Virginia that he saw "nothing of Nessa. I do not even sleep with her; the baby takes up all her time" (CB/VW, 30 December 1908). While Vanessa was wrapped up in her maternal triumph, Virginia reveled in her writing triumphs. In 1908, the *Times Literary Supplement* published more than a dozen of her reviews, and the *Cornhill* had featured six of her essays. She rewrote seven chapters of her novel, and Clive, with a clear aesthetic conscience, told her they were "wonderful" (Bell, 1:209).

Virginia claimed not to "believe in wars and politics" (VWL, 1:325). But political issues, especially women's suffrage, were challenging that stance. Clive asked Lytton whether his sisters had been asked to join "the tramps for women." In early 1908, Virginia thought the campaign laughable. She "received a neat type-written notification which, somehow, struck her and us as essentially ridiculous" (CB/LS, 10 February 1908). But Marjorie Strachey and Lytton's other sisters did become involved in the suffrage movement. Vanessa later concluded that Marjorie was "rapidly going off her head about suffrage. I made her very angry too by saying that women were more hysterical than men" (VB/LS, 27 October 1909). However much the two Stephen sisters had rejected their mother's adversity to women's careers, they still followed her views on suffrage.

During 1908, the Bells and Stephens, with Saxon and Lytton, formed a Play-Reading Society, in which they read aloud rather risqué Restoration dramas. At the final meeting, the Stephen sisters' readings revealed their very different ideas of relations between women and men. Virginia read from Spenser's beautifully idealized marriage celebration, "Epithalamion," and Vanessa, according to Clive's notes, read a "long and improper passage from *Moll Flanders*."[46]

Finding herself unwell, Vanessa had reason also to feel "peevish & suspicious." In early February 1909, the Bloomsbury friends embroiled themselves in what Quentin Bell terms a very "dangerous adventure." They wrote a series of letters under false identities that "served only to embolden the participants. From behind his mask Clive felt able to renew his gallantries with unusual openness and ardour" (Bell, 1:142). The more bold Clive became, the more solicitous Lytton appeared, warning Virginia that her behavior made him "jump to the most extra-ordinary conclusions," presumably that she and Clive were actually having an affair. Still using assumed names, Virginia wondered whether his conclusions were not "(though I really won't admit it) a little uncomfortable" for Vanessa. Thus Virginia faulted Lytton more than herself and Clive for upsetting Vanessa (VWL, 1:382; Woolf and Strachey, *Virginia Woolf & Lytton Strachey*, 28–31). While this charade was progressing toward its inevitable blowup, Clive, in his better persona, wrote Virginia a long, careful critique of her novel, faulting only the first part for being too "didactic, not to say priggish" (Bell, 1:209).

Lytton's distress over Clive's attention to Virginia and her incautious responsiveness widened into an attempt to rescue her: The last letter under an assumed name is dated 14 February. On 17 February, Lytton (a homosexual) proposed marriage to Virginia, and she accepted. Then, as they quickly extricated themselves from this ill-conceived engagement, Lytton wrote to Leonard Woolf, explaining that a virgin could not understand him and that "she *is* her name." Thereupon Lytton developed an alternative plan, suggesting that Leonard marry Virginia, for he "would be great enough and you'll have the immense advantage of physical desire" (LWL, 147).

Far away in Ceylon, the idea of marrying a woman like Virginia struck Leonard as "the only way to happiness," a settled escape from the extremes of debauchery and guilt to which he had become accustomed. Leonard debated whether Virginia might have him and facetiously told Lytton to "wire to me if she accepts. I'll take the next boat home." Lytton claimed that he had told Vanessa to pass on Leonard's proposal to Virginia, as no doubt she jokingly did.[47] Seven months later, Leonard was still pondering how wonderful a "completed and consummated life" with Virginia actually would be; however, the "horrible preliminary complications" of marrying a virgin appalled him (LWL, 145, 150).

Whatever she knew of Leonard's facetious proposal, Virginia did not take it seriously, but Clive's attentions were another matter. When Virginia and Adrian traveled to Cornwall with the Bells, Vanessa was disgusted with Adrian's "languor" and conviction that "we had all combined to sit upon him in his youth" (VBL, 79, 80). The next day, when Virginia and Adrian left the Bells, Virginia began to show signs of stress, perhaps provoked by the guilt that Adrian's accusations aroused, perhaps by guilt over flirting with her brother-in-law, or perhaps by fear of the consequences (in several senses) of Clive's affections. Vanessa warned: "You know I shall have to come & nurse you if you are ill & how will you like that?" (VB/VW, 16 March 1909). That, of course, was exactly what Virginia would have liked.

Instead, she got Clive. He was glad to have her manuscript but sorry not to have "alas the authoress." He therefore wrote a poem dubbing her "the great contemporary novelist." He told her she was "exciting," that he wanted to kiss her but was too shy, that he wanted her to feel half the affection for him he did for her, and that because she had asked for an affectionate letter he did not regret saying such things (CB/VW, early April and 9 April 1909). The price that Virginia continued to pay for accepting Clive's attentions was risking losing Vanessa's.

In April 1909, Caroline Emelia Stephen died. While writing her aunt's obituary, Virginia reflected that "if one could only say what one thinks, some good might come of it" (VWL, 1:390). Probably Virginia felt guilty for being so dismissive of the "almost maternal" "Nun," who now made Virginia's career possible by leaving her a substantial legacy of £2,500.[48]

Suddenly rather comfortable financially, Virginia traveled with the Bells to Florence in late April. In her travel journal, she reflected that writing descriptions was both "dangerous & tempting" because rather than represent the thing in itself, "what one records is really the state of ones own mind" (PA, 396). (Later she recognized that one's own mind is unavoidable and is itself a suitable subject.)

The trip to Florence was not a success. Virginia and Clive's continuing flirtation caused scenes with Vanessa, who now found Virginia "tiresome" (Bell, 1:143). Virginia thus returned home early, and alone, only to receive a letter from Clive, still in Florence, telling her he loved her (CB/VW, 19 May 1909). Now Vanessa had a painting on exhibit at the avant-garde New English Art Club, and Virginia wrote to Violet that all of Vanessa's friends, perhaps meaning herself, were envious (VWL, 1:394). Whether or not she knew that her husband had declared his love for her sister, Vanessa projected her emotional dilemma onto a painting, *Iceland Poppies*, a "still life based on triplicates."[49]

In May 1909, Hilton Young offered the long-expected proposal.[50] Without hesitation, Virginia refused him (VWL, 1:424, note). In August, she and Adrian accompanied Saxon Sydney-Turner to the Wagner festival in Bayreuth. In Dresden, when the threesome "went to the pictures," Virginia was genuinely responsive to the Dutch paintings, especially a Vermeer, but she was defensive about discussing art with Vanessa and bored with listening to Saxon, which was "like reading a dictionary" (VWL, 1:410–11). Later she remembered pondering "the Lytton affair" and wisely counseling herself "never [to] pretend that the things you haven't got are not worth having" (D, 2:221). Although she might have snared Saxon, Virginia's boredom meant that Vanessa need not fear "at least for [her] chastity."[51] The Bayreuth experience eliminated Thoby's old friend as a possible suitor (Bell, 1:149–52).

Meanwhile, Virginia wished "(as usual) that earth would open her womb and let some new creature out. They are grown very stale" (VWL, 1:413). Her old love, Violet, now seemed too spinsterish to offer amatory thrills. But then, Dora Sanger, Vanessa gossiped to Lytton, reinforced all of Virginia's sexual fears by speaking of her "want of love" for Charlie Sanger, providing a "detailed account of their nuptial night," and pronouncing both sex and childbirth "degrading." Whereas Dora's confidences endorsed Virginia's spinsterhood, Vanessa's rebuked it. She told Virginia that she preferred bawdy talk with women in Wiltshire who express "unblushing animal delight" to the "vague & prurient insinuations & flirtations which abound here among the spinsters of the country" (VB/LS, 27 October 1909; VB/VW, 29 August 1909).

Vanessa was trying to shame Virginia's spinsterhood, counter Dora's notion that sex was degrading, and arouse Virginia's interest in other men, but in late 1909 Clive was again complaining that he and Virginia "talked of intimate things without being intimate." He urged her to share "ecstasy" with him (CB/VW, 27 October and 16 November 1909).

At Christmastime 1909, Virginia took an impulsive trip alone to Cornwall, and Clive wrote a poem "To V.S." His references to

> *the pregnant hours, the gay delights,*
> *The pain, the tears maybe, the ravished heights,*
> *The golden moments my cold lines commend.*[52]

must have outraged Vanessa, and she wrote to Clive, "The Goat said you wrote her very nice letters but what did you mean by her's [*sic*] being literary?" Vanessa resorted to her dismissive notion of the literary: "Did you find it very dull?" (VB/CB, 31 December 1909). But Clive did not find Virginia's "literary" letters dull, nor did she find his seduction poem boring.

Vanessa reacted with various strategies, including that Virginia "had better marry" and that Vanessa and Clive had better carry out an earlier plan to move to Paris. Clive found the idea of living in Paris appealing for aesthetic reasons (London having "no beauty of design") and for a personal reason, his "friendlessness" (VB/VW, 10 May 1909; CB/LS, 22 October 1909). Then perhaps because "the Goat now declares that she has made up her mind that if we go she will come too," the Bells canceled their moving plans. Instead of leaving Virginia, Vanessa closed the valves of her affection, as she had done years before with her father. She told Clive, for example, of having to "curb" Virginia's "attempts at an affectionate whispered conversation with me" (VB/CB, 29 November and 31 December 1909).

Whereas Virginia maintained that her relationship with Vanessa was "too intimate for letter writing" (VWL, 1:343), Vanessa found Virginia's letters, with their "Sapphist tendencies" (VBL, 84), too intimate for their relationship. The cold tone of Vanessa's letters now served as a corrective to Virginia's effusiveness. But the more distant Vanessa was, the more Virginia feared losing her; yet the more Virginia needed her, the more distant Vanessa became.

The terms of the sisters' relationship are suggested by the forms of their letters: Virginia's salutations to Vanessa were "Beloved" and "Dearest," but after the "affair," Vanessa's letters no longer began "Beloved Billy" or "Beloved Monkey & Wombat"; now they began "Dear Billy" or "Dear William" and were signed "Yrs, VB." The flirtation between Virginia and Clive and Clive's renewed affair with Mrs. Raven-Hills began the "transformation" of the Bells' marriage, as their second son writes, into a "union of friendship" (Bell, 1:169). The flirtation also sealed Vanessa's emotional alienation from Virginia, an alienation she expressed privately to others and publicly to Virginia by ridiculing her nonvisual tastes. Yet while Vanessa was protecting herself, her alienation and ridicule threatened the core of Virginia's being.

That the flirtation between Clive and Virginia remained sexually innocent does not excuse its inappropriateness and thoughtlessness. Seventeen years later, Virginia remembered it as an "affair with [both] Clive and Nessa" (VWL, 3:172). Clive flirted with her and Virginia flirted with him because both of them were trying to reengage Vanessa's attentions. But Virginia's plan failed. Having trespassed on her sister's emotional territory, Virginia was once again experiencing loss and guilt. At thirteen, she feared that if she could look into another's mind, she would find hatred for her. And at twenty-two, she interpreted Vanessa's behavior in Italy after Leslie's death as such hatred. At twenty-eight, however, she did not need to read her mind to sense Vanessa's distrust.

Like George Duckworth, Clive violated a brother-protector role. George had taken advantage of Virginia's helplessness, and Clive preyed on her ambitions and her competitiveness. But if Virginia had been helpless to resist George's attentions, she was not so with Clive. Even if she was partly ignorant of his sexual motivations, she certainly knew that Clive was her sister's husband. Yet that knowledge hardly affected her conduct. She craved affection from Clive just as she had from Violet. Clive was playing a game with her, masking flirtation with literary encouragement, and Virginia also was playing a game with Clive: She pretended not to understand his overtures in order to protect her virginity and to profit emotionally and aesthetically from his attentions. And indeed, this flirtation had some positive results. It allowed Virginia to "show off" for a man, a pleasure she later described as "one of the chief necessities of life" (MB, 194). It also gave her the benefit of Clive's astute advice about revising *Melymbrosia.* The flirtation was otherwise a disaster, whose repercussions lasted for the rest of Virginia's life. Decades later, Vanessa's daughter remembered her mother's wariness around Virginia, and "on Virginia's side a desperate plea for forgiveness" (Garnett, *Deceived with Kindness*, 28).

Virginia's need for affection and Vanessa's withdrawal were inextricably tied to their competition between words and paint. Vanessa could not "remember a time when Virginia did not mean to be a writer and I a painter. It was a lucky arrangement for it meant that we went our own ways & one source of jealousy at any rate was absent" (LWP, 2:6a). Although

Vanessa asserted that their differing interests prevented jealousies, her phrase "one source of jealousy at any rate" testifies to the considerable rivalry between the sisters. We have seen how that rivalry surfaced in the summer of 1897 when Stella was ill and Virginia had "the fidgets" and resented Vanessa's freedom to go to her studio. That rivalry subsided in easier times, and Virginia took pleasure in Vanessa's artistic triumphs, but when Vanessa distanced herself, Virginia again projected her resentment onto the visual arts.

Clive remembered that Virginia felt "a sort of jealousy, no doubt, that made her deprecate her friends pursuing the arts or professions which seemed in some way to put them in competition with herself." He illustrated his point with the story of Virginia's writing desk:

> Someone said in her presence that it must be very tiring for her sister, a painter, to stand long hours at the easel. Virginia, outraged, I suppose, by the insinuation that her sister's occupation was in any way more exacting than her own, went out at once and bought a tall desk at which she insisted on standing to write.

The desk seems to have been less a desk than a symbol that writing was as challenging as painting.[53] If Vanessa had an easel, Virginia could have a tall desk. Likewise, if Vanessa had Clive physically, Virginia could have him aesthetically.

Unable to accuse her sister of trying to steal her husband, Vanessa accused her art. Sometimes her criticisms of writers were general: "You writers, however, do not know the joy of experimenting in a new medium" (VBL, 81). At other times they were personal: "We spent this morning at the Accademia [in Florence], which of course you [in 1904] did not half see" (VBL, 84). Vanessa continued to connect painting with the body when she told her single sister about sketching a nude man (VB/VW, 8 May 1909). However unconsciously, she followed Leonardo's line, arguing that with writing, one could not "count upon the reader getting just the right impression, as you can in a painting, when it comes to describing the looks of things" (VBL, 87).

As the sisters became more emotionally opposed to each other, so did their arts. Vanessa complained and threatened: "You are always telling me how incapable of speech I am & then you are always expecting me to grow a tongue. . . . Shall I turn into a writer one of these days do you think?" (VB/VW, 25 June 1910). This threat, like Virginia's 1904 threat to "give up literature and take to art" if she were harshly reviewed, revived the notion of enmity between the arts (VWL, 1:170).

As we have seen, writing was Virginia's one sure way of surmounting— perhaps surviving—the hard business of life that had taken from her one emotional prop after another. Already in a fix emotionally because Clive offered unwanted—and Vanessa withheld wanted—affection, Virginia was about to be in a dreadful fix aesthetically, for painting was beginning to challenge writing's hegemony in England.

As Postimpressionist painting's champion and later as Vanessa's lover, completing the second step in transforming the Bells' marriage, Roger Fry further destabilized Virginia's artistic and emotional identity. Virginia reacted to Vanessa's defection to Roger and the new art by the pattern established long before, when she could not write but had to weed the garden or sit with Stella while Vanessa went to her "—— drawing." Embedded deep in her ego structure, this pattern (the *paragone*) transformed disaffection into rivalry between the pen and the brush.

4
I Fear That You Abuse Me

1910–1912

The year 1910 opened auspiciously for both Virginia Stephen and Roger Fry. Virginia declared her independence of her mother's notion of gender roles by volunteering to work, but not to write, for women's suffrage. She scribbled, typed, and retyped her lengthening novel but took time away from it to address envelopes for suffrage in an office that might have figured in an H. G. Wells novel (VWL, 1:422). Meanwhile, the home that Roger designed in Guildford—called Durbins and described by Vanessa as "perhaps one of Rogers most successful works"—was completed.[1] This landmark of domestic architecture was designed so that Roger could work, the two Fry children could play in its gardens, and his wife, Helen, could enjoy a nonstressful home environment.

Roger Fry was a generation older than Virginia and a generation younger than Leslie Stephen. Even though they were strict Quakers, his family, like the Stephens, belonged to an upper-middle-class English intellectual aristocracy. (His father was Lord Justice Sir Edward Fry.) At King's College, Cambridge, Roger had fulfilled his parents' ambitions for him by earning double firsts in the natural science tripos. He also fulfilled Leslie Stephen's dream for himself (and no doubt for his sons) by becoming a member of the secret society known as the Apostles.[2]

The members dined together at the weekly Apostles' meetings and presented hypothetical papers with outrageous titles. In the 1880s, Roger's papers had titles like "Shall We Sit on the Safety Valve?" "Shall We Wear Top Hats?" and "Ought We to Be Hermaphrodite?" One serious topic, despite the rhetoric, was "Are We Compelled by the True & Apostolic Faith to Regard the Standard of Beauty as Relative?" Another was "Shall We Obey?" (FP).

When he was an undergraduate, Fry sought a definable standard of beauty and a rationale for social and sexual disobedience. The issue of obedience had been significant for the young Quaker, who had arrived at King's disciplined and repressed into intellectual passivity and "total immunity from any understanding of sex" (RF, 29). King's, and especially the Apostles, transformed him: Although he dutifully and brilliantly fulfilled his curricular obligations, socially and intellectually he rebelled, abandoning a promising scientific career to become, to his parents' dismay, a painter.

Fry also shocked his parents by marrying in 1896 an exuberant and fascinating but penniless young artist. Helen Coombe was attractive, a talented painter, and daringly unconventional. For example, one evening on the Normany coast in France when they substituted wine for their usual cider, Helen made a rash vow. She promised that if the surf was calm in the morning "she wld. cast herself headlong into the sea . . . & so the dreadful deed was done amid the tumultuous applause of the populace assembled on the shore who declared her to be 'très intrépide'"[3] But Helen's intrepid behavior soon escalated into irrational outbreaks. She became increasingly erratic, unstable, and even dangerous, especially to the two Fry children, Julian and Pamela, born in 1901 and 1902.[4] Indeed, during the first decade of the century, she often was confined in institutions.

Meanwhile, although a far more successful art historian than painter, Roger had been turned down for the Slade Professorship of Fine Art at Cambridge. From 1906 to 1910, he was the European curator for the Metropolitan Museum of Art in New York City. He traveled around Europe, often with the banker J. P. Morgan, advising him and the museum about the purchase of fine art. Vanessa Bell had heard of his reputation "as the newest and most learned of young critics" but, "hating lectures," had avoided hearing him speak. Then, later, meeting Fry at Desmond Mac-

Carthy's home sometime in the early Bloomsbury period, Vanessa found it "almost unbelievable that one could really talk, chatter, express oneself to one of these dreaded members of the upper world."[5] To the Bloomsbury Group, Fry might have been a character from the elite world of Henry James's *The Golden Bowl*, for he seemed old, well connected (a member of the patrician Reform Club), prominent, straitlaced, and tragic.

In his commitments to work, skepticism, and personal integrity, Fry was rather like Leslie Stephen and became something of a father figure for Virginia Woolf.[6] Both Stephen and Fry believed in the ultimate moral good of art, but they were radically different in their notion of where to locate that goodness. Stephen had praised literature chiefly for inspiring a sense of beauty and inculcating right behavior, whereas Fry saw the morality of art in aesthetic experiences totally dissociated from everyday practicalities. Through those experiences, he believed, humankind becomes civilized. Almost instinctively, Virginia had questioned her father's aesthetic assumptions. Even as early as the *Hyde Park Gate News*, she had exposed the sentimentality inherent in moralizing. All her life, the naïve notion "that morality is essential to art" seemed questionable to her (VWL, 1:317). She valued beauty, but in presentation, not necessarily in subject matter.

Like the rest of his generation, Leslie Stephen had not doubted the concept of "truth" in art, but in the 1890s Roger Fry had begun to do so. He explained to his mother that "painting is not mere representation of natural objects" (RF/Mariabella Fry, 27 October 1893). His own painting, however, was then precise and highly representational, and his writing on the Old Masters valued content as much as form. As he became more and more appreciative of the painting of Cézanne and other modernists, however, Fry expanded his theories to encompass—and his painting to reflect—this new art. Thus the movement he labeled *Postimpressionism* enabled him to supersede what he began to see as old-fashioned notions of truth.

Virginia was of many minds regarding the concept of representation, as reflected in her early writings: In her stories and biographies, she favored faithfulness over nature; in her journals, she both endorsed and questioned it; in her letters, she imagined it; and in her novel, she was trying to create a vision that would supersede "likeness to life."

In the first decade of the twentieth century, Roger Fry developed a Postimpressionistic aesthetic that spoke to the very issues about which Virginia was ambivalent. In 1909, Fry wrote "An Essay in Aesthetics," a radical rejection of the principle of "likeness to life," dismissing all questions of morality, association, referentiality, and emotional impact in art. For him, art's morality lay in fostering the imaginative life, not in inculcating values or depicting goodness. Whereas Virginia sought some vision that prioritized the imagination, Roger made the essentialist argument that "the fullness and completeness of the imaginative life [that the artist] leads may correspond to an existence more real and more important than any that we know of in mortal life" (VD, 16). And whereas Virginia wanted to create a new, undreamed-of form, Roger argued that a true aesthetic appreciation was a response to the harmonious relationship of formal elements in an artwork, not to that work's relationship to the world beyond. These were arguments that might have inspired Virginia.[7]

The paths Virginia and Roger were traveling approached each other when in January 1910, the Bells spotted Roger Fry waiting by them on a railway platform. Vanessa summoned her courage and reintroduced herself. Traveling together in a first-class compartment on a train to Cambridge, the three immediately formed an alliance. Roger's involvement with the European cognoscenti and his knowledge of Bond Street art dealers opened new vistas for both Bells, as did his belief that new art movements would be the salvation of painting, criticism, and English taste.

On 19 January, Roger wrote Vanessa to invite her and her husband to visit at Durbins, and he offered to "start a discussion at your club." After a first abortive attempt, the Bells visited the Frys in February. Helen was incoherent and irrational, but Vanessa had seen comparable behavior in Virginia and, just the month before, had witnessed Adrian's "complete breakdown," which she feared would trigger a "second collapse" (VB/Lytton Strachey, 12 January 1910). With so much experience with bright people poised on the brink of breakdowns, Vanessa registered no surprise at Helen's condition and was pleased that Helen seemed to notice the early stage of her own second pregnancy. Vanessa "nearly" put her arms around Helen but found herself "too shy." She was impressed by Roger's efforts both to fulfill Helen's needs and to preserve an artistically and emotionally functional home life for the family.[8] As usual, Vanessa did not reveal her feelings, but as always, her response to Roger was conditioned by her aesthetic judgment.

While a new pregnancy and a new friend diverted Vanessa's energies, Virginia joined an expedition that would have horrified both her parents. On 10 February, Adrian Stephen, Duncan Grant, Virginia (in black face), and others pretended to be members of the entourage of the emperor of

Abyssinia and demanded a tour of the British ship HMS *Dreadnought*, of which the Stephens' first cousin William Fisher was flag commander. Even though Adrian was well over six feet tall, Fisher did not recognize even him (or perhaps he was merely keeping up appearances). At any rate, he treated the entourage with elaborate ceremony.[9]

The event did not remain a private joke for long, however, as the hoax's initiator released the story to the press, seeing it as a tweak of the nose at official pomposity. But his plan backfired, for the public was infuriated at this affront to the honor of the British navy. Threats of recrimination and increasing consciousness of the risks they had taken further undermined the group's sense of victory. Virginia's cousin stormed into 29 Fitzroy Square, claiming "that we ought to be whipped through the streets, did we realise that if we had been discovered we should have been stripped naked and thrown into the sea?" (Bell, 1:215). If the sailors discovered the hoax (as they certainly should have), they would have "revenged themselves by some violent practical joke," and if they had discovered that "one member of the entourage was female," it would have been "awkward," as Vanessa understated it (VB/Margery Snowden, 13 February 1910).

While recriminations from the British navy were still anticipated in the Stephen household, Roger Fry was facing fallout from J. P. Morgan, who had developed the habit of using Fry's expertise to increase his private holdings rather than the Metropolitan Museum's collection. The previous summer, for example, Fry had obtained a lower price for a splendid Italian primitive (from the collection of Belgium's King Leopold) on his word that he was buying it for the museum. Then, when Morgan bought the painting for himself (at the reduced price), Fry protested to at least one of the Metropolitan's trustees.[10] Thereupon Morgan, resenting such insubordination, pressured the board to fire Fry. In February 1910, he resigned.

Fry hoped to earn a living by lecturing and writing and accepting finder's commissions, and on 17 February he signed an agreement to clean and restore the nine huge Mantegna canvases depicting "The Triumphs of Caesar" at Hampton Court.[11] Then on 25 February 1910, he put aside his worries and obligations in favor of lecturing to Vanessa's Friday Club. From then on, he was a formative presence in both Vanessa's and Virginia's lives.

Virginia left no reaction to Fry's entry into her immediate world in February 1910, as she was showing signs of renewed distress. The reasons were many: She could no longer continue walking a sexual tightrope by flirting with Clive; she now thought that her novel failed to embody her vision; she must have been frightened by talk of "awkward" retaliation from the navy; she had destroyed her close relationship with Vanessa; and she had little sympathy for the art world that Roger had opened for Vanessa and Clive. Guilt, helplessness, and loss again gnawed at her confidence. Perhaps the suffrage work exacerbated her sense of loss, for it further marked her separation from her mother's values.

A week or so after Roger Fry lectured to the Friday Club, Virginia spent her last day working for suffrage and made the last revisions to a version of her novel she later rejected (VWL, 1:422). She was thought to be on the verge of another breakdown.

Despite Vanessa's having distanced herself after Virginia had taken up with Clive, in 1910 Virginia's unstable condition once again reclaimed Vanessa's attentions.[12] She and Clive organized for Virginia three holidays with them. In early March, they traveled to Cornwall and stayed briefly at the same hotel Virginia had visited alone the preceding Christmas. But this time, rain prevented the Cornish coast from working its restorative magic. Virginia claimed to Violet that the *Dreadnought* event "had no bad results for anyone" (VWL, 1:423), but when they returned from Cornwall to London she was confined to bed. The dates suggest that the hoax, along with the other factors mentioned, exacerbated her tensions and stress. Trying to avert a breakdown, the Bells took Virginia and Adrian to Studland Beach in Dorset for most of April. From the beach, Clive wrote to Saxon Sydney-Turner that Adrian "complains of chronic lassitude" but that "our invalid is much better." She was punctually "packed off to bed at 10." Their amusements were "skills or walks (taken by me alone), donkey-driving, bawdy, and patiens [solitaire]." Clive delighted in telling Saxon that someone had mistaken Virginia for his wife. He did not indicate whether Virginia took part in the bawdy talk, but he made an ambiguously suggestive reference to the "pretty general demand for one well-known medicament," perhaps himself. He urged Saxon to read Frances Cornford's "charming" poems and soon assured him that the patient was "restored to her good health" (CB/SS-T, 2 and 13 April 1910).

Back in London, Virginia had a visit from a relative of Hilton Young's who hoped that Virginia was wavering in her resolve not to marry Hilton (VWL, 1:424). Such pressure only made Virginia's moodiness worse. She told Clive that her head felt "numb," so he wrote to make her "less unhappy" (CB/VW, ? May 1910). But headaches, toothaches, sleeplessness, and alternating excitement and depression threatened Virginia with another decline into supposed "madness." The Bells invited Virginia to join them (with their child and domestic staff) for a holiday near Canterbury on 7 June. The day before she left, Virginia wrote a letter to Clive that must have arrived just as she did. Claiming to be "doubtful whether a woman of my defective taste should write a letter of affection," Virginia told him that she had failed to understand her attachment to him but assured him that she now would be more appreciative (VWL, 1:425). If Vanessa had seen that letter, she might have regretted inviting her sister to join them.

Meanwhile, it had become clear that Helen Fry would not recover, no matter how propitious her surroundings, and that her presence distressed and possibly threatened the children.[13] Virginia may or may not have known about Helen Fry, but Vanessa certainly did. The Stephen sisters had witnessed the dissolution of character before. Their first cousin J. K. Stephen had been a promising young author and Fellow at King's College, Cambridge; then suddenly he went mad and died.[14] Vanessa had witnessed a late stage of Helen Fry's transformation from a beautiful, witty, and talented woman into an apparent lunatic.[15] Recalling Virginia's irrational behavior in 1904 and aware of oddities in her 1910 behavior, Vanessa must have worried lest Helen's fate predict Virginia's. Thus Vanessa's emotions were torn between jealousy over Virginia's flirtation with Clive and concern for her sister's well-being.

After Vanessa had spent much of the spring responding to Virginia's needs, she left her with Clive at Canterbury on Tuesday, 21 June, to return to London and await the birth of her second child. Apparently Virginia was still behaving erratically. On the first night back, Vanessa wrote to Clive that she and Adrian had "a desultory talk & have not yet picked your sister [Virginia] to pieces much though of course a few incidents were related & commented on. 'Well, she is *too* extraordinary' etc." Although it was a source of concern, clearly Virginia's behavior was a source of fun as well to her siblings, who also "abused [their cousins] the Fishers & talked about family funerals that we would attend. We thought ourselves witty but perhaps we weren't" (VB/CB, 21 and 25 June 1910).

Having given as much of herself as she felt she could, Vanessa wanted someone—other than Clive—to take care of Virginia. On the day she wrote her first letter to Clive, she also wrote to Dr. George Savage about Virginia's behavior. During their consultation the next day, Savage recommended a rest cure, thereby curtailing Virginia's time alone with Clive. Vanessa wrote to Virginia explaining the necessity of this cure, for she would "never get quite well" without a "slightly modified" Weir Mitchell rest treatment in an institution that Savage recommended.[16] Institutionalization at this time did not involve psychological counseling; instead, the Mitchell routine proposed rest, overeating, and boredom as a cure for overwrought nerves. Whether or not this routine could cure Virginia, it certainly would take her out of the Bells' company for a time. Vanessa was "anxious that the Goat shouldn't think that I have in any way tried to make Savage suggest a rest cure" (VBL, 92). She must have felt guilty, for she wondered "whether the Goat will be furious or not." But her sympathies were with Clive; she was "rather sorry for you having to cheer her up for I suppose anyhow she will be depressed" (VB/CB, 22 and 24? June 1910).

Clive was left with the responsibility of "keeping the balance between depression and excitement" (which we now understand to be depression and mania), since Vanessa deplored "how utterly incapable the Goat is of taking any care of herself." Perhaps feeling betrayed again at Canterbury by Virginia's affection for Clive, Vanessa's principal concern seems to have been getting Virginia out of the way: "With any luck the baby ought to have arrived before she comes out & then I think we could get her away again in about a fortnight which ought not to do her much harm." (Presumably she meant the fortnight with them would not do much harm, but the sentence also might mean that Vanessa recognized that trying to get Virginia "away again" could be a source of harm.)

Vanessa wanted Clive to leave Virginia and join her in London on Saturday, 25 June, but she was afraid that Virginia would want to come with him. Clive stayed on with Virginia until Monday, 27 June. Apparently he did so, for once, not to flirt with her but to comfort her. When he returned to London, he wrote to Virginia about Vanessa's "talking bawdy" and noted, "Nessa does have sympathy for any one who has to live with Adrian."[17] That note was little comfort to Virginia, who soon had no one to live with but persons she termed lunatics (VWL, 1:430).

In this period, Clive wrote another consoling letter, this one to Fry, who had been rejected for Oxford University's position of Slade Professor of Fine Art; the elderly and safe Selwyn Image was appointed instead. In a letter written on 28 June, Roger thanked Clive for his sympathy and rationalized that perhaps Image "should have this consolation prize in his old age; it is all very English & perhaps rather admirable" (RFL, 1:335). With that ironic tribute to English values and with his personal and professional stability lost at the age of forty-three, Roger shed all expectations of establishment success.

As Roger was cutting his losses and preparing to begin a new life,

Virginia must have feared that hers was ending. She was depressed, albeit without delusions or hallucinations (Caramagno, *Flight of the Mind*, 307, 308). She faced opposing but equally unsatisfactory conclusions: If demons could possess her to the point that she really needed to be confined in a nursing home, wouldn't she lose her genius as well? But if she did not require hospitalization, wasn't her sister just trying to get rid of her? Virginia suspected that the latter explanation was more accurate, for her internment—like her 1904 exiles in Cambridge and Yorkshire—was suspiciously convenient for Vanessa, who was already planning how to "get [Virginia] away again" (VB/CB, 26? June 1910).

Virginia did not know about those plans, nor did she know the pleasure her siblings took in picking her apart, but she had her suspicions: "I fear you abuse me a good deal in private, and it is very galling to think of it. . . . [A]s I never abuse you, I feel it rather hard; and possibly I ought to consider a scheme for the future" (VWL, 1:429). Vanessa's denial was not reassuring: "How can you talk such nonsense as you seem to have been doing—& how can you indulge in invalid [?] & depressing wonders as to how much I abuse you in private, etc. etc." She also sarcastically admitted, "Of course my nature is so vicious that abuse is always attractive." In the same letter, Vanessa reversed the terms to which Virginia had resorted six years before, threatening to "turn into a writer" if Virginia continued ridiculing her inarticulateness (VB/VW, 25 June 1910).

Virginia had used her gifts with language to tease Vanessa in the *Hyde Park Gate News*, and Vanessa believed that she continued to do so. She tried to explain to Saxon Sydney-Turner "how Virginia since early youth has made it her business to create a character for me according to her own wishes & has now so succeeded in imposing it upon the world that these preposterous stories are supposed to be certainly true because so characteristic." She told Clive that Saxon's response essentially confirmed her point. Saxon admitted that Virginia "does do such things, but Saxon still thought that it would have been characteristic of you & me to behave rudely to anyone who became sentimental" (VB/CB, 25 June 1910). Each Stephen sister was partly a victim of the other's characterization of her, but Virginia certainly did not have to invent Vanessa's rudeness. Since Virginia's letters show mostly adoration (as she said, "I never abuse you") and Vanessa's show mostly irritation, Virginia seems to have been the principal loser in this exchange.[18]

For the first time almost literally caged in a rest home, Virginia must have felt truly the loser. Again, her worst fear was confirmed: Her sister had been the agent of her incarceration. At Gordon Square with her husband and two-year-old son, Vanessa was awaiting the birth of her second child, which she assumed would be a daughter. She announced that she would not visit Virginia in the rest home, lest her baby be "born in the train or a taxi." Because Clive was genuinely concerned for Virginia, because she delighted in his attentions, because Savage did not object, and perhaps also because Virginia had again come to regard her sister as an enemy, Vanessa rather cattily said that she knew that Virginia would not mind having Clive visit instead of her (VB/VW, 13 July 1910).

Once more, Virginia felt so threatened that for a month she wrote nothing, not even letters. Her excited, or manic, phase exhibited her brilliance, charmed Jean Thomas (the home's proprietor), and irritated Vanessa. According to Clive, Virginia transformed Miss Thomas's "drab and dreary" life into something "thrilling and precious" (*Old Friends*, 117). But such tales made Vanessa feel her "cynical sisterly hairs rising down my back" (VBL, 94). Virginia suspected her sister's hostility, and a depressive mood swing loomed. In her first letter from the home, she wondered why Miss Thomas did not pass on the letters from her "Dark Devil" of a sister, and she concluded "that some great conspiracy is going on behind my back" (VWL, 1:430). Although Vanessa had worried that Virginia might guess she had "tried to make Savage suggest a rest cure," Vanessa protested to Virginia, "What do you mean by this talk of a conspiracy?" She decided that Virginia had "deluded Miss Thomas" by pretending to be normal. Vanessa warned: "You dont seem to realize that it is a great concession to your peculiarities that you are allowed to have letters at all" (VB/VW, 13 and 29 July 1910).

Virginia was being punished in the ways that hurt her most. Vanessa had withdrawn her affection and seemed again to be an enemy. She had confined Virginia with "a troupe of lunies" (VBL, 94). Now she intimated that even the right to read and write letters could be withheld, as reading and writing books already were. Virginia's behavior with Clive, which replayed the sisters' rivalry over Thoby, had dissolved her close conspiracy with Vanessa, provoked her own guilt, and resulted in this helpless state. Her demons seemed poised to return, and Virginia threatened "to jump out of a window" if she had to stay longer. She also assured Vanessa, in an image that conveyed her own self-displacement, "I have been out in the garden for 2 hours; and feel quite normal. I feel my brains [*sic*], like a pear, to see if its ripe; it will be exquisite by September" (VWL, 1:431). Still awaiting the birth of a daughter, Vanessa tried to coax a reformation in Virginia's behavior: "You would have been touched Billy could you have heard me telling Clive yesterday how much difference it would make to me if I could feel that you would really look after your own health." She suspected that she had "good reason for doubting your capacities in that way" but hoped that Virginia was changing her behavior (VB/VW, 8 August 1910).

However humiliating, rest and removal from emotional havoc did benefit Virginia, as did the assurance that her witty ways with words could still charm. Released in August into the company of Jean Thomas, she spent a long walking holiday in her beloved Cornwall. In London, the Bells' second child finally arrived, not a girl, but a second son, first named Claudian and later Quentin.[19] Jean's monitoring of Virginia's health was no doubt useful, but the landscape itself probably helped most in Virginia's recovery. Walking miles every day over the open countryside, Virginia felt "great mastery over the world." She read Pope and recognized, after being unable to express her talents for months, "how odd—that one writes oneself!" She was free of lunatics and demons and was happy, although the thought of marrying an as-yet-unknown sympathetic man would have made her happier. She vowed to avoid emotional excitement and headaches and asked Clive whether he would "try not to irritate the beast, for amusement?" (VWL, 1:434).

In September, Virginia parted with Jean and joined Vanessa again at Studland Beach on England's southern coast, east of Cornwall. There she first met her month-old nephew. Even though this baby offered her no reason to revise her former low estimate of children, he was, as Vanessa wrote Clive, a "funny brat. Everyone, including the Goat & myself, is overcome by his almost comic likeness to the photograph of me as a very small baby." Despite the Cornwall holiday, Virginia's health was still in jeopardy. She was resting, drinking milk and avoiding coffee, but still suffering from headaches and insomnia. Vanessa told Clive that "the Goat, according to [Jean Thomas], really did behave very well in Cornwall." However, Virginia had told Jean about the renewed pressure on her to marry Hilton Young, and Vanessa wrote,

> Jean didnt know whether she was really worrying over it, but suspected that she only worked herself up about it at times—with which I agreed. Jean thinks that she is still in a state of mind when she very easily gets hold of some trifle, exaggerates it & cant shake it off. I think that is evidently what she did about your letter.[20]

Overtures from Hilton Young or Clive, with their inevitable sexual implications, could set off Virginia's depressive syndrome, as she obsessed, in classic depressive fashion, on "some trifle." Ways to elude the syndrome, however, other than enforced rest, escaped everyone.

In the fall of 1910, Virginia's relationship with Clive and Vanessa was becoming as emotionally charged as it had been the year before. Virginia claimed to Vanessa that she had asked Clive to help her avoid emotional scenes. But Vanessa thought the blame outrageously misplaced, as she wrote Clive: "She really has no idea that they [emotional scenes] may be in any part due to her as much as to you." She thought Virginia "rather touchy—she of course thinks *you* are." Vanessa suspected that Clive was "rather worried . . . at the prospect of the Goat's not coming to Studland till I do" and hence having no time alone with her. Virginia showed Vanessa a letter that Clive had written to her that "provoked rather an absurd amount of friction," as Virginia wanted Vanessa to admit that she preferred Virginia's letters to Clive's. Vanessa urged Clive to avoid a quarrel when he joined them and admitted, "The Goat's letters are distinctly more amusing when she imagines you to be one of her audience" (VB/CB, 8, 9, 7, and ? September 1910).

After such ominous preliminaries, Clive and Virginia for once cooperated in easing tensions while they were together at Studland. In a photo taken there (Figure 15), Clive looks positively contrite, and Virginia (elegantly but overelaborately dressed for the seacoast and a bit fatter after her rest cure) looks mellow. Neither is looking at the photographer, presumably Vanessa.

Meanwhile, Helen Fry's condition continued to decline, and Roger found his only consolation in art. With his establishment ambitions dashed, he was free to rechannel his energies into undermining and reeducating the English aesthetic sensibility. He spent 28 September at Hampton Court assessing the enormous assignment in late-fifteenth-century restoration he had accepted, and then he went to Paris to choose pictures for his upcoming exhibition of contemporary painting.[21]

Vanessa felt that London might be tempting "after the Goat has gone," but Paris was even more so. She was desperate to send Virginia to friends and her children to the senior Bells, and to join Clive and Roger in "that exciting atmosphere where people really seem to realize the existence of art," as her sister did not (VBL, 96). Visiting in Oxford, Virginia claimed to be "very well again," even though she really would not be well again until she returned to London, where she could review books and resume writing *Melymbrosia* (VWL, 1:436).

When she returned to the house on Fitzroy Square, however, Virginia found the preeminence of writing itself to be in jeopardy. Now that he was back from Paris, Roger was preparing to do battle against "a huge campaign of outraged British Philistinism" (RFL, 1:337) by exhibiting art far more alien to Virginia than that of the Old Masters in Italy had been in 1904. A brief autobiographical fragment suggests Virginia's sense of displacement: "Thursday evenings with their silences and their arguments were a thing of the past. Their place was taken by parties of a very different sort. The Post-Impressionist movement" (Berg). The fragment ends there, but it suggests that the environment that had nurtured Virginia Stephen's fledgling talents had suddenly become "a thing of the past" because of Postimpressionism. Vanessa later explained that due entirely to Roger Fry's planning, the Postimpressionist exhibition caused significant repercussions beyond the art world: "The writers were pricking up their ears and raising their voices lest too much attention be given to painting" (MRF). By "writers," she meant chiefly Virginia, who maintained more vehemently than ever before that artists were "rather brutes" and that literature and poetry are "much finer."[22]

Roger Fry's exhibition, entitled "Manet and the Post-Impressionists," ran from 8 November 1910 to 15 January 1911. Introducing to the conservative London art world the paintings of Manet, Matisse, Picasso, Cézanne, Derain, Gauguin, Van Gogh, and other moderns caused not sensation, but scandal. It also raised the very issues of what art could or should do that had perplexed Virginia Stephen. One journalist described the Postimpressionist painters as "uneducated persons, who never could draw or paint properly, but who have been deluded, either by natural insanity or by over-indulgence in absinthe, into the belief that they can" (unidentified clipping, FP). Whereas Leonardo da Vinci had argued that painting surpassed poetry in imitating nature, the journalist's assumption was that because the Postimpressionists did not imitate accurately, they could not draw or paint. Virginia probably would have agreed, but she knew better than to express such philistine opinions around Vanessa and Clive and Roger.

Fry fired back at these attacks by explaining the terms he had introduced the year before. He insisted that the artists had "already proved themselves accomplished masters in what is supposed to be the more difficult task of representation." They departed from realism because they were "in revolt against the photographic vision of the nineteenth century" ("Grafton Gallery—I," 332, 331). Not able to imagine any other vision, the press argued that Fry was trying to dupe the public into believing that Picasso, Cézanne, Matisse, and their ilk were serious artists. The exhibition made Fry into a controversial figure and such aesthetic issues into matters of public debate. Indeed, English art and culture have not been the same since. According to Vanessa, "That autumn of 1910 is to me a time when everything seemed springing to new life" (MRF). And Virginia believed, "In or about December, 1910, human character changed" (VWE, 3:421).

At the time, Virginia did not feel sanguine about this change, because all her life her sense of self had been founded on the assumption that literature was the greater art, painting the lesser. By that logic, if painting became the greater, literature would be the lesser. In late November 1910, she confessed: "Now that Clive is in the van[guard] of aesthetic opinion, I hear a great deal about pictures. I don't think them so good as books." She did believe that "one mustn't say that [Postimpressionist paintings] are like other pictures, only better, because that makes everyone angry" (VWL, 1:440). Since Roger Fry did argue that Postimpressionist pictures continued a classic tradition and thus were "like other pictures," Virginia's "everyone" seems to have meant Vanessa, still maintaining her long-established opinion that Virginia knew nothing about pictures or, indeed, about life, which Vanessa defined in decidedly risqué terms. When Sydney Waterlow dined with the Bells on 8 December 1910, for example, he was delighted that no one else except Virginia was present and that they exchanged "really intimate," definitely un-Victorian talk: "Vanessa very amusing on paederasty among their circle." Vanessa's bawdiness did not, however, humanize her: "I realised for the first time the difference between her & Virginia; Vanessa icy, cynical, artistic; Virginia much more emotional, & interested in life rather than beauty" (Stape, *Virginia Woolf*, 25).

In 1908, Virginia had proclaimed her intention to "re-form the novel" (VWL, 1:356), but Clive nonetheless had feared that she would compromise "with the conventional" (Bell, 1:207–12). Virginia now feared him prescient, for the version of *Melymbrosia* that she had finished in early 1910 did not reform but instead followed the conventions of literary realism. She wanted to invent a new form, to achieve a whole made of "shivering fragments," but she had not decided how to do so (PA, 393). While she had been pursuing a near dead end, the Cubist painters had settled on a visual method for achieving an aesthetic aim comparable to hers.

They broke subject matter into fragments that they then reassembled into new kinds of wholes according to their own artistic visions. As their advocate, Roger was "discussed perpetually" (VWL, 1:450). His prominence—or notoriety—even landed him an invitation, which he rejected, to be the director of the Tate Gallery.[23]

During this period, Virginia kept away from Clive. Although she resumed working in the suffrage movement, her work seemed to have been wasted when on 18 November 1910, after three hundred suffragettes marched on the House of Commons, nearly half were arrested. Virginia had a number of reasons to feel depressed, but she was writing again and hence did not despair. She was reviewing several books, and that December she had the pleasure of hearing E. M. Forster speak to the Friday Club on "The Feminine Note in Fiction." Virginia thought it was the best paper the club had heard, perhaps because Forster used the new feminine approach to writing she sought (Furbank, *Growth of the Novelist*, 193). Recognizing, however, that her novel did not yet embody that approach, Virginia did summon the emotional and aesthetic strength to reject (and later burn) the version of *Melymbrosia* completed earlier in 1910.[24] Having cut her losses, Virginia planned toward the end of November to "begin my work of imagination" all over again, after over two years of work (VWL, 1:438). But the prospect did not daunt her.

An unpromising Christmas alone with Adrian in a hotel in Lewes was redeemed because in a tiny Sussex village nearby, Virginia found a house to rent as her country place. She named it Little Talland House, evoking both the beloved coastal home of her childhood and the legend of her great-aunt's Little Holland House. Vanessa vowed to call it Virginia's Villa (VB/CB, ? February 1911). Perhaps Virginia remembered how her "Granny" had dismissed the semidetached "villa" that little Virginia had found for her in 1889. This time, Virginia, with her own money, could choose for herself. But her semidetached village house offered neither the freedom of Talland House nor the exotic glamour of Little Holland House.

Meanwhile, Helen Fry's doctor, Henry Head, wrote that Roger had "shown a devotion I have never seen equalled. Unfortunately, the disease has beaten us." Helen had to be institutionalized permanently. On Christmas, Roger wrote to his mother, "It is terrible to have to write happiness out of one's life after I had had it so intensely & for such a short time" (HH/RF, 14 November 1910; RF/Mariabella Fry, 25 December 1910). Roger found himself "dried up and hardened by the utter tragedy of my love." Sometimes he "could cry for the utter pity and wastefulness of things, but life is too urgent" (RFL, 1:336). After its auspicious beginning, the year 1910 had threatened Virginia with the loss of her self and had completed Roger's loss of his beloved, yet each ended the year with a sense that art would offer some sort of redemption.

Having once again proved through pedestrian writing that there was "nothing wrong" with her, by the beginning of 1911 Virginia was "very much excited by the thought of going back and writing the work of imagination." The experiment was to be so radical that she no longer called her book a novel; thus she was relieved that her Quaker aunt was no longer alive to be shocked by her break with convention (VWL, 1:448).

Caroline Emelia Stephen would have been shocked by another early 1911 event: After collecting fabrics, necklaces, and sandals so that his friends could come dressed as natives in a Gauguin painting, Roger organized a Postimpressionist ball. Dressed "as bare-shouldered barelegged Gauguin girls" (Bell, 1:170), Vanessa and Virginia outraged the proper ladies at the ball and further alienated themselves from South Kensington.[25] Vanessa's painting (Figure 16) conveys impropriety (the clothes are gaudy and skimpy) and also the isolation of the large female figure.

Clive must have taken Virginia's willingness to dress so scandalously as a sign of a new libertinism. Accordingly, with Vanessa again occupied with a baby and her painting and increasingly isolated from both sister and husband, Clive renewed his flirtation. Virginia avoided one meeting with him, much to his irritation, by keeping the head of the Working Women's College with her when he called. But she apologized and promised better arrangements the next time (VWL, 1:439). About this time, Clive told Molly MacCarthy, Desmond's wife, that Vanessa was "still rather an invalid, Virginia as often as not on the threshold of the mad-house" (CB/MM, ? ? 1911).

By reminding Virginia that she promised to come to him when she felt lonely, Clive managed to make her condition even more uneasy. He wrote to her about his arguments about Postimpressionism; "up-to-date" conversations with the Cornfords in Cambridge about "Post-impressionists, and Roger Fry, and art in France"; and Vanessa's gossip "about copulations." Referring to Vanessa's "paramour," Clive told Virginia that he and Vanessa now talked about "the possibility of changes, and whether we were happy. And with whom we were intimate" (CB/VW, 14? November 1910, and 12 January, ? March, and February/March 1911). The possibility that Vanessa and someone else (probably Roger) were intimate threatened

Virginia with both further alienation from Vanessa and further intimacy with Clive, especially when Clive warned Lytton Strachey away from both sisters (VWL, 1:49, note). It also promised to make Virginia aesthetically "out-of-date."

If Virginia was feeling out-of-date, Vanessa and Clive were feeling au courant. Roger gave them a letter of introduction to Gertrude Stein, the experimental American writer and wealthy collector of modern art now living in Paris. Roger also organized a trip to Turkey with the Bells and H. T. J. Norton.[26] With Clive leaving, Virginia could safely walk the tightrope that had created such emotional upheaval. She wrote a teasing letter saying that she had been "tempted" to invite him to visit her in Sussex, suggesting the countryside as an alternative to Turkey and her own company as an alternative to Roger's (VWL, 1:453). Clive was delighted with this apparent overture and asked to see her before departing, to assure himself of his "niche" in her heart (CB/VW, 3 April 1911).

After the group left, taking the Orient Express for the pleasure of traveling through the Balkans by daylight, Clive calculated his letters to Virginia to make her empathize with him. As if to advertise a new sexual awareness, Virginia wrote to Vanessa that she was now reading "Liaisons Dangereuses with great delight" (VWL, 1:458). Clive complained to her about Roger's caring for Vanessa, painting with her, and collecting Turkish art for her. In the midst of articulating his regret for being away from Virginia, Clive mentioned a "fainting fit" that Vanessa had experienced. He told Virginia that he "would give a good deal to have you here even to be in your bad books, which I shouldnt be when chastened & unhappy."[27] Despite reassuring telegrams, Virginia must have feared a repetition of the disastrous trip with Thoby some five years before; therefore, she set out for Turkey in April 1911. Virginia found Vanessa in Broussa, installed in hotel rooms filled with exotic shawls, rugs, and pots from the bazaar and innumerable sketches and paintings. She wrote to Violet that Vanessa was "surrounded by males" and that the trip was "the oddest parody" of their excursion five years earlier.

Vanessa had suffered both a mental and a physical breakdown, brought about by a miscarriage. Virginia had learned "most of the parts of the female inside by now, but that is useless knowledge in my trade, the british public being what it is." Norton drew "diagrams of the Ovaries on the table cloth, taking it for granted that a woman is the same shape as a kind of monkey."[28] Clearly these men knew and cared more about the female body than Virginia did. Furthermore, they seemed to know a great deal about Vanessa's particular body.[29] She was in Roger's care, and he was delighted "to be so desperately needed by someone for whom he wanted to do everything" (MRF).

Virginia also wanted to have Vanessa desperately need her and to do everything for Vanessa, but her help was not required:

> Hitherto [for Vanessa] Clive's passion for Virginia had been a source of severe, though concealed, pain. Now it became a matter of indifference. Now indeed Vanessa would have been only too glad to see Clive more completely obsessed by his sister-in-law, instead of which, such is the perversity of things, that affair seemed to be cooling off. In fact there was a moment—or so I suspect—when Vanessa feared that her much loved but agonizingly exasperating sister might set herself to charm Roger. (Bell, 1:169)

Even if Virginia had set herself to charm him, replaying old patterns begun with Thoby, Roger did not take Clive's place in an emotional ménage à trois. He had fallen completely (and permanently) in love with Vanessa. From Virginia's point of view, another advocate of the visual arts had stolen Vanessa's affection. Again, paint and the body of her sister were associated.

Whereas Clive's mentoring aspired to be both literary and sexual, Roger's was exclusively aesthetic. Therefore, even though Vanessa's illness provided an opportunity for Roger to display ingenuity at nursing, he was "not absorbed to the extent of forgetting the presence of someone [apparently Virginia herself] reading a book.[30] What book was it? What sort of merit had it?" (RF, 170). With Virginia, Roger's mission was discrediting the mimetic aesthetic; he "read books by the light of [Postimpressionism] too." He judged by Postimpressionist standards not only the poems of Frances Cornford but also those of Wordsworth, Shelley, and even Shakespeare. "Cézanne and Picasso had shown the way; writers should fling representation to the winds and follow suit" (RF, 172).

"Why," Roger asked Virginia, "was there no English novelist who took his art seriously? Why were they all engrossed in childish problems of photographic representation?" (RF, 164). Suggesting that the advent of photography had made mimesis passé in painting, he categorized representation as a mere "literary" aesthetic. A "literary" appreciation would value art for its associations or meaning or ethical import, not for itself.[31] Roger thereupon proposed that "literary" appreciation was as unfit for literature as for it was painting.

Although Virginia had completed a third of the newly unconventional *Melymbrosia* before leaving for Turkey, such talk undermined her confidence in what was probably the third version of her work of the imagination (VWL, 1:461). When she returned, she attempted further unconventionality by making the novel less "literary" and more visual. Similes became metaphors, and language became shapes or "blocks" (VO, 81). She even integrated some of Roger Fry's interest in Turkish art into her South American setting.[32] At the level of word and image, her revisions reveal the influence of Fry and the Postimpressionists.

After the trip to Turkey, Vanessa convalesced in the village of Guildford to paint with and be loved by Roger while his Quaker sister Joan presided at nearby Durbins.[33] By now Vanessa had a husband, two children, a lover, and the self-assurance to paint the way she wanted to, not imitating Picasso or anyone else. Virginia suffered by comparison. Sometime during this period, Duncan Grant, a frequent visitor at Fitzroy Square, painted a portrait of Virginia in the rough brush strokes typical of the Postimpressionists. In a black hat and dress, her skin very pale, Virginia looks withdrawn and much older than her years (Figure 17).[34] Another major depressive episode seemed imminent. She found that she "could not write, and all the devils came out—hairy black ones. To be 29 and unmarried—to be a failure—childless—insane too, no writer."[35] In this depression, failure was both artistic and sexual.

Perhaps trying the old game she had played with Thoby and then Clive, Virginia invited Roger to visit with her at Little Talland House. Then he invited her to Durbins. Virginia told Vanessa that she found Roger charming and the beauty of Durbins surprising (VBL, 99). With its giant hall two stories high and a minstrel gallery above connecting Roger's studio, small bedrooms, and a dining room, Durbins was the most unconventional modern house that Virginia, or her contemporaries, had ever seen.[36] Virginia's surprise at its beauty suggests a further enlargement of her aesthetic sensibilities, but her sense of Vanessa's new intimacy seems to have depressed her even more than Vanessa's marriage had.

Since her marriage, Vanessa had hinted to Virginia of her own "unblushing animal delight." She had attempted to arouse (and deflect from Clive) what Quentin Bell refers to as Virginia's "normal proclivities."[37] Since her affair with Roger, Vanessa had adopted a "loose, bold, almost sensual" persona. Her casual clothes, often with open necks, suggested "a woman eagerly accepting and indeed leading the age in new audacities in art and morals" (Bell and Garnett, *Vanessa Bell's Family Album*, 26).

Vanessa's risqué talk sometimes was designed to stir up jealousies. For example, she informed Roger that she was painting "indecent pictures" with Duncan Grant: "I suggest a series of copulations in strange attitudes & have offered to pose, will you join? I mean in the painting?" (VBL, 100). Although Virginia did not read that letter, Vanessa's brazen delight in bawdiness must have fixed in her mind the alignment of paint and the body. Uncomfortable with the notion of animal delights, Virginia at last told an outsider (her old tutor Janet Case) of George Duckworth's misbehavior a decade earlier (VWL, 1:472).

An ocean away in Ceylon, Leonard Woolf's stern personal integrity had brought order and responsibility to both the natives and the colonials. He was destined, it was said, for a very high post in that administration, but first he needed a break from the constant stress of his post. Therefore, after nearly seven years away, he returned to England in the summer of 1911 for a year's sabbatical.

On 28 June, Vanessa asked Saxon Sydney-Turner to bring Leonard to dinner on 3 July. After dinner, Virginia, Walter Lamb, and Duncan Grant joined the group. Leonard found that "almost the only things which had not changed were the furniture and the extraordinary beauty of the two Miss Stephens" (LWA, 13–14). Then on 8 July, Virginia invited "Mr. Wolf" for a weekend in Sussex, but he was away.

Roger's love letters to Vanessa from the summer of 1911 combine a painter's sensuous observation and a writer's verbal felicity. He describes, for example, "the queer silkiness of the palm of your hand," the "little waves of hair that ripple around your ears, and your torso that's hewn in such great planes & yet is so polished, your squareness beneath the arm pits." Given such physical beauty and the "shape of your soul," he was grateful to have a place in her heart. He proposed that a taxi company subsidize her, for "I'm always thinking it's worth while to take one to get a few minutes more with you." Remembering the pleasures of their "little married life," Roger sketched "the shape of your breast when you're lying down" (Figure 18) for Vanessa, because "it's one of the things you can only enjoy through me" (RF/VB, 1911?).

Such letters made Vanessa for once wish that her vision was literary, "so that I might see the words in which I could tell you what I feel." She assured him that she knew "the shape of all of you pretty well now" but "luckily" had so far drawn him only in her imagination (VB/RF, 6 July and

23 June 1911). She was stimulated and gratified but also apprehensive, particularly that Virginia turn their affair into gossip. Vanessa considered that "Virginia is equal to saying anything to anybody's face" and warned Roger, "*Do* be careful—this is only a general warning—what you ever tell Virginia." Vanessa thought Virginia could be trusted, but "evidently one cant run any risks with her. Isn't it awful? I'm thankful I've never told her anything about you" (VBL, 99, 103).

Even as Vanessa distrusted Virginia, her teasing inquiries about her marital prospects disconcerted Virginia. On 21 July, Virginia wrote to her sister that Walter Lamb had spoken with her about marriage but that he feared the "hornets nest" of intrigues Virginia lived in would seriously complicate any relation with her. She was almost apologetic about his tepid declaration of interest: "Oh how I'm damned by Roger! Refinement! and we in a Post Impressionist age" (VWL, 1:469–70). Potentially aligned with the refined and cautious Walter, Virginia felt "damned" when Postimpressionism had freed Vanessa and Roger for unknown liberties. As if to compensate, Virginia developed friendships with a group of young, free-spirited Cambridge friends with whom she had exchanged "frank and indecent" conversations earlier in the year (VWL, 1:450).

Still recuperating at Guildford, Vanessa wrote to Roger when he had to be away from her that "Virginia is here and so far we have been very peaceful. [Maynard] Keynes is staying on which helps to that end. He's nice and easy and bawdy." One topic of conversation was how much indecent talk Roger could tolerate (VB/RF, 9 August 1911). Vanessa wrote to Lytton that the "W & V [Walter Lamb and Virginia] affair has caused me some anxiety" (VB/LS, 9 August 1911), but she did not explain why. Vanessa told Roger, "I dont *really* mind your going to see Virginia if you like!" (VB/RF, 15 August 1911), and Virginia worried lest another visit to Guildford from her might be "too tiring" for Vanessa. Seeing herself and Vanessa as a unit, she regretted to Roger, "We are callous—in the way we lay burdens upon you!" (VWL, 1:474)

In mid-August, Virginia went to stay with Philip and Lady Ottoline Morrell in the country, and then she went to Cambridge, where she visited the poet Rupert Brooke at The Old Vicarage, just outside town. There she saw the Cornfords and Sydney Waterlow and found herself uninhibited enough to swim naked in the evening with Brooke. Indeed, she may have been disappointed that such a startlingly liberated act caused no sensation at all among the young people that she and Vanessa called the Neo-Pagans (Bell, 1:174).

By then, Vanessa thought that gossip about the "W. [probably Walter] affair" was "too old to be very amusing. She seems to be rather wrapped up in the Neo-Pagans" (VB/LS, 29 August 1911). Adrian Stephen remembered that Virginia was "appallingly dishonest in her personal relationships. She just could not resist playing up to people & pretending to be really interested in them." He specifically remembered the summer of 1911 when Vanessa was with Roger near Durbins, "and Virginia was carrying on with Walter Lamb & Sydney Waterlow and I think Hilton Young" (AS/VB, 20 October 1944 [Berg]). If Virginia did carry on with and play up to these men, her conduct was surely a manic reaction to being "29 and unmarried . . . a failure . . . no writer."

Trying to sort out the terms of Vanessa's existence, "thinking a good deal, at intervals, about marriage," Virginia supposed that marriage offered more "unity" than an affair or a liaison did. Her objection was that the "pace is so slow, when you are two people." She hypothesized, however, that such a slowdown should not bother a simple painter like Vanessa (VWL, 1:475). Vanessa resented such disparagements of herself and her art, but she also had long ago internalized the same radical distinction between writing and painting that Virginia felt.

The Bells planned a September holiday at Studland, but as Vanessa explained to Roger, Clive was "very much disgusted" to hear that Virginia planned to join them (VB/RF, 8 August 1911). Vanessa wanted to warn Virginia not to create difficulties, but Clive expected Virginia to "make mischief between you & him by making him think I am in love with you—oh dear—what complications there are—I think we must be careful with her & give her no possible excuse for being able to say any thing as she's really too dangerous." Vanessa also told Roger that Virginia's letters irritated Clive by repeatedly referring to Roger and Vanessa as a pair. She complained, "I cant think why she does it & must try to stop her" (VB/RF, 2 and 5 September 1911). Vanessa and Clive still believed that each should be free to "have what friendships we like," but Clive did not always "live up to his principles" (VB/CB, 13 September 1911). Virginia seems to have enjoyed subverting those principles.

In September, Virginia invited Leonard Woolf and Lytton Strachey's sister Marjorie to visit her at Little Talland House, which, especially after Durbins, she now thought of as a "hideous suburban villa" (VWL, 1:476). After that weekend, Virginia did join the Bells at Studland and proved herself more "amiable and amusing" than Vanessa had expected her to be. One photograph (Figure 19) shows Virginia dressed primly in a sweater-coat and gloves and looking young and rather impish. Virginia's only surviving letter from this time conveys no hint of tensions (VWL, 1:478).

Meanwhile, Roger had secured a commission to create murals of London scenes for the Borough Polytechnic Institute. There he and Duncan Grant and others painted scenes of "London on Holiday" in the dining room of the workmen's college and thereby put to the test Roger's theories that good art could treat even everyday subjects and should be available to the public. In Roger's playful *London Zoo* (Figure 20), an elephant steps out of the borders of the artwork, and in Duncan's *Bathing* (Figure 21), the waves wash out of the water and onto a diving pillar. Roger's painting raises questions about the function of frames, and Duncan's intricate mosaic of men's naked bodies and waves allows rhythm and pattern to transcend "reality."[38]

Satisfied with his mural, Roger left London and, with his children and Lytton Strachey, joined the group at Studland. However amiable Virginia had seemed, Lytton observed deep emotional difficulties in the Stephen–Bell–Fry ménage: Virginia was "very shrivelled up internally" (as in Figure 17); Roger, "like some medieval saint in attendance upon Vanessa"; and Clive, "a fearful study in decomposing psychology. . . . It all seems to be the result of Roger, who is also here, in love with Vanessa" (Holroyd, *Lytton Strachey*, 21, 23, 24). Whereas Roger's picture of the bay with a mother and child passing along the road is restoratively calm,[39] Vanessa's almost monolithic painting (Figure 22), with its isolated female figure turning her back on both children and adults, indicates emotional distress and withdrawal.

After returning to London, Virginia began seeing more and more of Leonard Woolf. Vanessa told Clive that "Roger and the Goat have been trying to outdo each other in their attentions," until "Woolf came to tea & we had an argument as to whether colour exists." That was the sort of argument Roger would have begun, Leonard in true Apostolic fashion pursued, and Vanessa "always took to" (VB/CB, 10 October 1911). Even though the argument may not have captured Virginia's interests, it did deflect the competition between Roger and Virginia on that afternoon, just as Leonard's presence did thereafter.

Given the alignment of Postimpressionism with all that was up-to-date in the arts, Virginia wanted to prove that she need not be damned by Roger or Vanessa or the new art. She herself wrote a review of a 1911 book entitled *The Post Impressionists*, in which she presents herself as an enlightened sympathizer of "the now historic exhibition" and endorses the author's "plea for an open mind" about such shocking paintings.[40] She finds the author's "best point" (borrowed from Fry) to be the "reconcilability" of Postimpressionism with traditional painting but regrets that few art critics have been flexible enough to alter their presuppositions and that "the lay opponents of Post-Impressionism" never would (VWE, 1:379–80).

Vanessa told Virginia that Roger was pleased with her review: "Its just what he wanted to get said." But then she undercut the compliment: "He thinks it a great advantage to have art criticism written by those who know nothing about it. For my part, I couldn't see any art criticism in it." Roger added an asterisk in Vanessa's letter and a comment at the bottom of the page, "*This isn't at all what I said but you can read between the lines and allow for Vanessa's artistic distortion" (VBL, 110–11). Presumably Roger had said he was pleased to have his aesthetics endorsed by those who were nonpainters, not by people "who know nothing about" art. Proselytizing for the importation of Postimpressionist methods into all the arts, Roger was encouraged that his French friend Charles Vildrac had evolved a form of "Post-Impressionist poetry" (RFL, 1:352). He was equally glad to see Virginia developing Postimpressionist prose.

Back in 1897, after Stella's death, the misery of a holiday in Brighton had been alleviated by a train trip ten miles north to the town of Lewes and a bicycle ride around the surrounding East Sussex countryside. Accordingly, in October 1911 when Virginia discovered the gracious Regency house Asheham (or Asham) out among the East Sussex downs she had come to love, she immediately began negotiations about leasing it with Vanessa and ending the lease on the more suburban Little Talland House.

In November, Virginia made another move. She and Adrian Stephen left Fitzroy Square and began a living experiment even less conventional than the Stephen children's unchaperoned 1904 arrangements. The two of them leased another Bloomsbury row house, 38 Brunswick Square, a few blocks from Gordon Square, to share with Maynard Keynes, Duncan Grant, and, later, Leonard Woolf.

Long a source of affection for Virginia, Violet Dickinson had become a rather stodgy figure who disapproved of the radical living arrangements at Brunswick Square (VWL, 2:86, 88), perhaps because Virginia had laughed about the convenience of Coram's Fields, the Foundling Hospital, for a young woman living in a household of men (VWL, 1:484).

During this period, Vanessa walked an emotional tightrope of her own making. She treasured her time alone at Gordon Square with Roger and once was irritated that he had invited Virginia to join them for dinner when "we might have been alone" (VB/RF, 6 November 1911). No longer joking about her "paramour," Clive now objected to Roger's sleeping at Gordon Square. Then, while her household was being moved to Brunswick Square, Virginia stayed at Gordon Square. With Clive and Virginia under the same roof again and Roger expelled, Vanessa considered taking Virginia into her confidence: "She could let us meet at Brunswick sometimes, & I could have you at Asheham without Clive's knowing. Also you could write to me undercover to her." Vanessa observed to Roger: "If this had happened 3 years ago when Clive was thinking only of Virginia it might all have been easy. Though I dont know. Its wonderful how you did divert him to me again." Now Clive was becoming "really jealous" of Roger, and Vanessa retained some loyalty to her husband. She wrote again that day, promising Roger that they would "manage somehow even if I have to tell Virginia" (VB/RF, 15 November 1911). But she did not do so, and the next week she tried to "stir your jealousy!" by telling Roger that she had taken her bath in Duncan Grant's presence. However, she was "afraid [Duncan] remained quite unmoved & I was really very decent" (VBL, 112).

In late 1911, Virginia turned down a marriage proposal from Sydney Waterlow (VWL, 1:483, note). As Leonard Woolf showed more and more interest in Virginia, the joking proposal of marriage he had sent from Ceylon two years before hovered like a disembodied presence between them, but Leonard vowed to say nothing until he was certain of Virginia's feelings for him. According to Virginia's later, dramatic rendition, in his Cambridge days Thoby had described to her an "astonishing fellow—a man who trembled perpetually all over." He was eccentric, a Jew; he trembled because "he was so violent, so savage; he so despised the whole human race. 'And after all,' Thoby said, 'it is a pretty feeble affair, isn't it?'" (MB, 188). Leonard was at the same time established (as a friend of Thoby's and an Apostle) and exotic (as a Jew, a ruler of the colonies, and a passionate trembling man). Virginia was both excited and frightened by his interest in her.

Meanwhile, Roger's care sustained Vanessa through ill health and emotional tensions. In several sketches and paintings, Roger celebrated her sexuality. Paired paintings (Figure 23) deemphasize her individual features but use color, light, and shadow to define the planes of her body. Roger also made a woodcut for his 1911 Christmas card with a Vanessa-like Virgin Mary, in the same way as Burne-Jones had used her mother as a model for the Virgin.

Taking up residence on the top floor of the Brunswick Square house on 4 December, Leonard worked on his first novel, *The Village in the Jungle*, in which he faced, and for the first time articulated, his empathy for the native Sinhalese and his distaste for European imperialism (LWA, 2:29–30). He finished the book within a year, and on the second floor of the same house, Virginia entered her fifth year of work on her first novel. Describing political reformation was simpler than accomplishing aesthetic reformation.

Leonard's diary entries show Virginia spending less and less time with Walter Lamb and more and more with Leonard; "TALK VIRG" is a regular entry. By December, Leonard was recording their meetings in Sinhalese. On Christmas Day 1911, the Bells held a luncheon party at 46 Gordon Square for Maynard Keynes, Duncan Grant, Leonard Woolf, and Virginia. Clive reacted negatively, Vanessa positively, to Leonard's preoccupation with Virginia. Early in the new year, the Bells reassembled the same guests, but without Keynes. Vanessa wrote to Roger that after dinner they had tried to describe the "meanest actions we had ever committed." Vanessa "couldn't think of any, not because I'm a saint, but because I never think of them again once they're committed." Duncan startled her by revealing himself to be of a similar constitution, ashamed of nothing (VB/RF, 7 January 1912; Spalding, *Vanessa Bell*, 119). Both Virginia and Leonard, though, were haunted by self-accusations of the sort not to be shared with others. By this time, Leonard's feelings were focused exclusively on Virginia, and she was at once flattered and apprehensive.

Vanessa had written to Clive her version of what Leonard Woolf's exotic experiences signified: "The colour [in Ceylon] is amazing & one's animal passions get very strong & one enjoys one's body to the full" (VBL, 108). If Vanessa publicly associated Ceylon, Leonard, and bodily pleasures, she would only have heightened Virginia's unease. Despite being a creature of the body, Leonard, like Virginia, was interested in words, not paint, and actions, not contemplation.[41] Virginia and Leonard had a long talk about his scheduled departure to Ceylon in the late spring, and then he left, probably on 9 January, for a planned visit with an old friend in Somerset.

Leonard had intended not to mention love or marriage to Virginia unless he was sure she loved him, but the agony of parting with her for even a few days brought him face-to-face with the prospect of parting from her forever. He felt he would go mad unless he spoke openly of his feelings, so on 10 January 1912 he sent a telegram from Somerset telling her, "I must see you." He arrived at Brunswick Square that afternoon and proposed marriage. Virginia said neither yes nor no, and Leonard took the evening train back to Somerset. His train was late, and Leonard had to walk miles in dense fog to his friend's home. Exhausted as he was, however, he could not sleep until he wrote to Virginia, telling her that he loved her not only for her beauty but for her mind and character as well. He assured her: "I will do absolutely whatever you want" (LWL, 168). On the next day, he wrote a calmer letter, offering a "damnably truthful" confession of his faults but also reminding her of how many interests and values they held in common (LWL, 169–70).

Virginia wrote back a noncommmittal letter saying that she "should like to go on as before; and that you should leave me free, and that I should be honest" (VWL, 1:488). She told only Vanessa about the proposal and asked her not to tell Clive. Vanessa retreated to the Isle of Wight, where Roger joined her. On 13 January, she wrote to Virginia, "Dont marry if you arent in love & do if you are . . . dont bother about the drawbacks such as Jewishness, etc. for they dont really matter if you are in love." She told Virginia that "Leonard is the only person I have ever seen whom I can imagine as the right husband for you," but she threatened to warn him "of your temper & general difficulty" (VB/VW, 13 January 1912). She also wrote to Leonard that he was the only man for Virginia (VBL, 113–14).

Virginia desperately wanted to consult Vanessa about Leonard's proposal and may have resented having to wait until Roger left. But being "rather ill & tired out & generally decrepit," Roger stayed on. Virginia joined them on the Isle of Wight on 16 January 1912 but shared only "public news" in front of Roger. In a letter to Clive the next day, Vanessa sketched Roger sketching her and outlined Virginia slumped forlornly in the background (Figure 24). This sketch is a perfect metaphor for the artists' self-absorption and Virginia's exclusion. The content of Vanessa's letter reflects the sisters' alienation. Since Virginia was unwilling to talk about her romance in Roger's presence, Vanessa suggested that Clive move into Virginia's quarters at Brunswick Square "& so have the pleasure of a thorough investigation of her papers" (VB/CB, 12 and 17 January 1912). But even Clive would not go that far; when Roger left that evening, Vanessa probed Virginia's feelings herself and found her sister in a state of considerable agitation.

The pressure on Virginia was intense. After Roger left, she broached the subject of his relationship with her sister, as Vanessa sarcastically explained to him: "Virginia told me last night that she suspected me of having a 'liaison' with you. She has been quick to suspect it hasnt she?"

(VB/RF, 18 January 1912). Vanessa exaggerated Virginia's naïveté, but such comments show how alienated she was from her sister. Virginia felt not only the painful loss of Vanessa to Roger but also the dreaded conditions of helplessness and guilt. Although Leonard was willing to resign from the Colonial Office, his resignation might oblige her to marry him, and if she did not, she would be guilty of ruining his career for nothing. But if she did, she would become the sexual partner of a passionate man. Her jokes about being pursued by a man or being molested testify to her fear of male sexuality and also to the delusional character of the manic syndrome (VWL, 1:489, 491). Under this pressure, in late January—a month after Adrian Stephen had been released from a nursing home—Virginia suffered a relapse. Again, she was sent away for a "rest cure."

Understanding that Virginia was high-strung and hypersensitive, Leonard followed instructions not to write to her at the nursing home. The period of rest was beneficial, and Virginia was soon released. In early February, she held a housewarming in Asheham, with Adrian, Marjorie Strachey, and Leonard as guests. Then on 9 February, the Bells had their own housewarming party at Asheham House, with Virginia, Adrian, Roger, Duncan Grant, and Leonard. On 13 February, Leonard took Virginia to meet his family in Putney, and three days later, Virginia returned for two weeks to Jean Thomas's nursing home at Twickenham. Finally, on 17 February, Vanessa wrote to Leonard that Virginia was not to receive any letters or to see anyone. While Vanessa gathered a group of males (Clive, Roger, Leonard, Adrian, and Duncan) for her own housewarming party (LWD), Virginia must have felt as displaced as Vanessa's sketch had shown her.

When she was released, Leonard assured Virginia that she need not worry "about any thing in the world." He seems to have sensed that pressure put her in some sort of fix; hence he did not mention his marriage proposal (LWL, 171–72). Soon Vanessa told Leonard that Virginia was ready to lead "an invalid life" back in Brunswick Square, "worrying a good deal about you just now & cant make up her mind what she feels" (VB/LW, 1 March 1912). Clive wrote to Molly MacCarthy that

> Virginia came out of her home looking very ill and wretched, and though she's been to see a psychologist whom I take to be a quack, she doesn't seem to get any better. Her novel languishes, I fear, so do her young men. Both Lytton & Woolf here last night.[42]

During this time, Vanessa painted an almost featureless oil portrait (Figure 25) of Virginia knitting that repeats the basic shapes of her upper body in an abstract background pattern of triangles. The painting conveys both Virginia's weariness and Vanessa's reawakened concern.

Even as she received passionate and tender love letters from Leonard, Virginia herself wrote to Molly MacCarthy that she felt she need not worry about "W" because he was staying on in England, "so the responsibility is lifted off me." But the claim that she was not obligated only confirms that she felt she was. She also declared, "Now I only ask for someone to make me vehement, and then I'll marry him!" (VWL, 1:492).

Virginia gave Leonard ample opportunity to arouse her vehemence. They saw each other every day in London, and he was also her guest at Asheham once in March, twice in April, and twice in May (LWD). Along with Leonard, the weekend in March included Vanessa, Roger, Adrian, Duncan, and Marjorie Strachey. Photographs of Virginia, Vanessa, and Leonard with various guests at Asheham all show great liveliness and easy harmony (Spater and Parsons, *Marriage of True Minds*, plates 50, 51, 56, 57, 59). One (Figure 26) of Virginia with Roger shows her in the same sweater-coat that she is wearing in Figure 19, but now the coat is open and Virginia, in her big dirty boots, appears not at all prim. On one of these weekends, both Vanessa and Roger painted Virginia's portrait: In Vanessa's, Virginia looks young and tremulous (Dunn, *Very Close Conspiracy*, plate 27), and in Roger's (Figure 27), she is beautiful and contemplative (perhaps sad), and her features are full, sensuous, and mature, suggesting Fry's serious and also masculine response to Virginia.

In April, Leonard began leveling and laying a terrace at Asheham (LWD), and his activity seems to have made Walter Lamb appear even more boring (VWL, 1:495). Meanwhile, Leonard wrote a confession disguised by the use of Greek names: "I am in love with Aspasia," the "most Olympian of the Olympians." In this projection of his feelings, Leonard had Pericles, a Syrian, wonder whether Aspasia had a heart. Leonard's diary regularly refers to Virginia as Aspasia; for example, on 10 April he wrote, "Went ASP who read her novel to me lunch w her." Leonard showed "Aspasia her character as I had written it. I forgot that she was reading it in the pleasure of watching her face & her hair" (LWD).

But Virginia took no pleasure in his Olympian characterization, borrowed from Landor, of her. Leonard dutifully recorded her response: "I dont think you have made me soft or lovable enough." To his credit, he reread the piece and agreed. He found Virginia a compound of romance and reality. But he, who some three years earlier had thought virginity a "ghastly complication" and probably remembered Lytton's comment that "she is her name," noted that Virginia had the mind and soul—as well as the body—of a virgin (LWL, 150; LWP).

Recognizing this dilemma, Virginia tried writing a romance: "Still at first, later they quivered in each other's arms."[43] Leonard found himself in a quandary, which intensified each day: He was scheduled to return to work on 20 May 1912, but he no longer wanted to represent imperialism in the British colony (LWL, 172; LWA, 2:33). He did, however, want to marry Virginia Stephen. But if he resigned and she turned him down, he would have simultaneously lost his vocation and his love. Nonetheless, he took the chance and resigned from the Colonial Service.

Virginia invited Ka Cox to join Leonard and her at Asheham in mid-April (VWL, 1:494–95). While Ka was with them, on Tuesday, 16 April, Virginia and Leonard attended with joking irreverence the funeral of Lord Gage, the hereditary owner of Firle Place. Ka was amused: "V & Woolf dressed in plumes & trappings attended & passed the time of day with His Grace of Norfolk & others." Thinking that the lovers wished to be alone, Ka consulted James Strachey, "As you know how things stand here—ought I to be off at once disregarding the Virgin's protestations? I'm puzzled." But instead of Ka's leaving, Leonard did.[44] Virginia was disappointed that the picture Ka had taken of Virginia and Leonard in the funeral procession turned out "a mere smudge" and that she and Leonard did not appear in the official commemorative postcard (VWL, 1:495). These notes show both the couple's delight in half-joining and half-parodying the gentry and the reluctance of "the Virgin" to be left alone with the "Woolf."

On Friday, 26 April, Leonard returned to Asheham, and this time Ka left (LWD). Leonard and Virginia worked on the terrace, leaving Virginia with aching bones and blistered palms and a new awareness that "the management of the earth is a great art" (VWL, 1:494). That night, Leonard "talked late w[ith Virginia] about what we should do if we married" (LWD). On Saturday, the couple went down to the Sussex coast. Sitting on the cliffs by the sea, "she was extraordinarily gentle I kissed her" (LWD). But on Sunday, after they "talked all morn" and then took the train to London (LWD), Leonard feared "something seemed to rise up" in Virginia against him (LWL, 172). Presumably, the "something" was fear of what that kiss could lead to. With Clive, Virginia had been able to exchange vows of affection, even embraces, to talk "of intimate things without being intimate" (CB/VW, 27 October 1909). But Leonard was her suitor, not her brother-in-law. Virginia's early fear of the body and sense that sex could be dangerous resurfaced.

On Monday, 29 April, Virginia went back to Asheham (LWD), and Leonard began writing her a testimony he did not finish until about three the next morning. This love letter is remarkable for its lack of pressure. Leonard assured her that she need not decide about marriage "until you've finished your novel." He also insisted his greatest happiness lay in "talking with you as I've sometimes felt it mind to mind together & soul to soul." He knew that she could "write something astonishingly good," and he "would rather do anything than harm you in the slightest possible way." However, his desire had "grown far more violent as my other feelings have grown stronger" (LWL, 172–74). Leonard was writing the first honestly passionate love letters Virginia had ever received, and they seem to have both stirred and terrified her.

On 1 May, Virginia set out to explain her ambivalence in a candid manner that the Apostolic Leonard should appreciate. She felt "angry sometimes at the strength of your desire." Virginia bluntly explained that although her affection and appreciation for Leonard had increased, she felt "no physical attraction" for him. However, they did share an ideal of a marriage that would be "a tremendous living thing, always alive, always hot, not dead and easy in parts as most marriages are. We ask a great deal of life, don't we? Perhaps we shall get it; then, how splendid!" (VWL, 1:496–97).

By 3 May, Virginia was back in London, where the couple attended the inquiry into the sinking on its maiden voyage of the supposedly impregnable ocean liner *Titanic*. On 5 May, Leonard sat with Virginia in the evening, but she was "not well," although she went to a concert with him on 7 May (LWD). Despite what she had said to Molly MacCarthy, Virginia did feel responsible, trapped.

Meanwhile, Vanessa was in Italy with both Clive and Roger. She wrote from Siena that she and Roger were painting together and warned Virginia: "Don't have another breakdown. I don't believe you *can* both work and fall in love." Apparently, Vanessa's intention was to warn Virginia against both work and ill health, because either could interfere with falling in love. From the Palace Hotel in Florence, she reflected on "the time when you were just beginning your madness & all our adventures since. Now I suppose you're embarking on another Billy. Is Leonard as charming as ever & are you falling in love with him? I hope so" (VB/VW, ? May 1912).

Reading Vanessa's reminder of her madness must have undermined Virginia's confidence as a sane adult, and reading Vanessa's generous encouragements to love while having to explain her own unresponsiveness to Leonard revived Virginia's sense of failure as a woman. Finally, rewriting *Melymbrosia* for the third time revived her feeling of "failure as a writer" (VWL, 1:499). Because of such threats to her identity, Virginia felt guilty about Leonard but helpless to change her situation. Vanessa heard that she was suffering attacks of near blindness and was in a "parlous condition" on the verge of breakdown (VB/VW, 28 May 1912).

Virginia's illness did not lead to a breakdown, however, largely because of Leonard's supportive love. Her harsh words had not turned him from or on her. Instead, she remained "for me the dearest & most beloved creature in the world." The couple assumed beast identities, with Virginia the more dominant Mandrill and Leonard the subservient Mongoose (LWL, 176). During May, they walked, dined, and went to the theater, concerts, Hampton Court, and (with Adrian) Asheham together.

Clive was jealous. His letters display an ugly anti-Semitic resentment of Leonard. To Molly MacCarthy, he predicted that Virginia's children would encounter "what none of us can help feeling for Jews—'Oh, he's quite a good fellow—he's a Jew you know'" (CB/MM, 14 May 1912). The couple's contentment was interrupted when Adrian showed them Clive's letters (LWD). Vanessa then had to intervene with Clive, presumably to stop him from acting so disgracefully (VBL, 117).

On 29 May, Leonard wrote in Sinhalese code in his diary (translated by Anne Olivier Bell): "Lunch w Virg Talking aft. She suddenly told me she loved me Went on river Maidenhead dined there" (LWD). Then Virginia agreed to marry him (LWA, 45–46). On 2 June, they told Vanessa about their engagement. She seems to have been "inexpressive" face-to-face but wrote to them later that happiness "does make the whole difference in one's life, doesn't it?" And she offered one of the very few testimonies of affection for Virginia that she made after 1910: "You do know, however, Billy, and Leonard too, that I do of course care for you" (VBL, 117–18).

Virginia's letters to her friends were defensive (she made a "confession" that she was marrying "a penniless Jew"), assertive ("I *insist* upon your liking him too"), and associative (she often mentioned Leonard's friendship with Thoby), and they were colored by proclamations that she was not worth him. She also told Violet that Leonard thought "my writing the best part of me," as perhaps Violet did too (VWL, 1:500–506). Her letters also were happy and hopeful.

But Clive's "rather bitter" letters continued to rankle. Vanessa told Roger that they "contained a good deal of abuse of V which naturally made both her & L very angry." Vanessa thought it "tiresome" for Adrian to have exposed Clive's abuse, especially because Leonard and Virginia asked her to caution Clive "not to speak ill of [Virginia] again." Vanessa did not see "what good can come of my telling him any such thing" and asked them "to let things be for a few days till we have all calmed down & can consider the point." Calming them was difficult, for "they have been working themselves up into a fury." Vanessa was angry that this tempest took her away from her painting, although "a lot of it has been very amusing" (VB/RF, ? June 1911).

Vanessa continued to write to Roger, who was visiting in France with the painter Henri Doucet, that Leonard and Virginia's engagement had created "something of a vortex" and that the engaged couple was "much more resented [by Clive] than you are." Clive "wont see much of the couple & I think resents me doing so" (VBL, 118). There was a "tremendous quarrel," and the "atmosphere is still rather strained here." Vanessa concluded, "I shant see more than I can help of them for the moment as after all they dont count me now & engaged couples arent the best of company" (VB/RF, ? and 8 June 1912).

These frictions affected Leonard only minimally, as he had merely contempt for Clive's dilettantism and amorality and no need for Vanessa's affection. But such unpleasantries must have confirmed Virginia's fears that she was being abused behind her back and proved that in choosing Leonard, she had lost both Clive's and Vanessa's affections.

In the midst of these entanglements, an artistic matter surfaced that had emotional repercussions. Clive had been asked to write a review of a painting exhibition. Despite his "abuse of V" and the disagreement with Leonard over it, Clive was willing to pass on the assignment to Leonard. Vanessa supposed that Leonard could "do it all right," particularly because she would "tell him about pictures" (VB/RF, ? May 1912).

The engaged couple made obligatory calls to friends and representatives of both families, including Madge Vaughan, Emma Vaughan, Violet Dickinson, Janet Case, Ottoline and Philip Morrell, Mary Fisher, Herbert Woolf, Flora Woolf, Margaret LLewelyn Davies, Gerald Duckworth, and George Duckworth at his elegant establishment, Dalingridge (LWD). In photographs taken there (Figure 28), Leonard is carefully dressed, and

Virginia is wearing a rather wrinkled cotton dress with many ruffles and flounces. In neither this picture nor the more frequently reproduced one (VWL, 2:4) are Virginia and Leonard touching.

Just as the Stephen children had simplified interior design and pared down their familial connections, so the sisters expected a thoroughly modern Leonard to sever ties with his family. As a first step, Virginia and Leonard offended his mother by not even inviting her to their wedding.[45] They saw the Russian ballet, acted in a "Suffrage play," and made several trips to Asheham. Visiting nearby, Ka Cox decided to "look up the Virgin & the Wolf [*sic*] on Saturday." Since Virginia, too, had misspelled Leonard's last name, the remark is interesting chiefly because it shows how pervasively Virginia was still known as the Virgin.[46] Despite tensions and obligations, Virginia managed to "finish" the third version of her novel,[47] and Leonard completed *The Village in the Jungle* (VWL, 1:507, 2:6).

The wedding of Virginia Stephen and Leonard Woolf was planned for 12 August, but as Vanessa wrote to Margery Snowden, it was "put forward two days to suit our plans." Despite Clive's nasty behavior, Leonard's fury with him, and Vanessa's irritation over the resulting upset, the Bells joined in giving the couple a wedding breakfast at 46 Gordon Square on 10 August 1912. Then that afternoon, Vanessa, Clive, and Roger left to see an exhibition of modern French painters in Cologne, where Roger secured some Cézannes for a new show at the Grafton Gallery. Perhaps Virginia was irritated by having her wedding plans disrupted to suit the painters, or perhaps she was bothered by her sister's "vague but deliberate" interruption of the wedding procedures, which Jane Dunn sees as Vanessa's "reclaiming her share of the emotional spotlight." At any rate, Vanessa wrote to Margery Snowden that "no one except the Goat herself seemed at all agitated" at the ceremony.[48]

The marriage did promise to be a tremendous intellectual boost for Virginia, for Leonard totally supported her writing and Virginia approved of *The Village in the Jungle* and his plans to become a political journalist. The principal threat to Leonard was that Virginia was not loving enough; to her it was that he, like others, would abandon or abuse her and that his achievements, like those of the painters, would somehow upstage hers. The Woolfs spent their first two nights of married life at Asheham, then went to the Plough Inn at Holford in the Quantock Hills, and finally set out for France, Spain, and Italy.

Vanessa told Leonard, "Roger thinks I have behaved monstrously" by not writing to the couple more often, but when she did write, her sexual inquiries were worse than silence. On August 19, she wrote to Virginia, "Are you really a promising pupil? I believe I'm very bad at it. Perhaps Leonard would like to give me a few lessons" (VBL, 125). Such queries could hardly have made Virginia feel at ease. Scarcely two weeks after their marriage, Leonard wrote to Vanessa, as she in turn told Roger, that although Virginia thought herself to be well, she would not eat. To Roger, Vanessa explained, "Evidently it was just the change at first which diverted her from worry." To Clive, she ridiculed the Woolfs' inability to appreciate Europe visually (VB/RF, 27 August 1912; VB/CB, 24 September 1912). To them, she wrote teasing sexual letters. She joked about Leonard's having to sleep with Virginia and asked him, "Does he ["Billy," the Goat, or Virginia] like manly strength & hardness? Also do tell me how you find him compared to all the others you have had?" (VBL, 127). Usually, Vanessa's pet names for Virginia were sexually ambiguous, but here she makes them specifically masculine, perhaps to suggest sexual incompatibility. She wrote to Virginia, "Tell Leonard that I very much enjoyed his letter & account of his night with the Ape for which I pity him sincerely. . . . He ought to buy a whip" (VB/VW, 2 or 9 September 1912). Such comments would hardly have eased Virginia's insecurities.

From Italy, Virginia wrote to Ka Cox that Leonard was writing "the first chapter of his new great work, which is about the suburbs" (VWL 2:6). Leonard's "new great" novel would become *The Wise Virgins*, published more than two years later. In the first two chapters, the Jewish Davises have moved to the suburbs, much as the Woolfs had moved to Putney in Leonard's youth. Mrs. Davis and her new neighbor Mrs. Garland discuss in tedious clichés such matters as their servant troubles. Although the elder Garland daughters detest Jews, a younger daughter, Gwen, finds herself attracted to Harry Davis. He is a painter, since, he sardonically explains, "Anyone can paint—that's the great discovery of modern times." Leonard caustically exposes the dullness of lives entrenched in customs and unwilling to think or question (*Wise Virgins*, 30, 27, 31).

Virginia continued her letter to Ka by naming the books she had read "since I lost my virginity." Then she wondered:

> Why do you think people make such a fuss about marriage and copulation? Why do some of our friends change upon losing chastity? Possibly my great age makes it less of a catastrophe; but certainly I find the climax immensely exaggerated. Except for a sustained good humour (Leonard shan't see this) due to the fact

> that every twinge of anger is at once visited upon my husband, I might still be Miss S. (VWL, 2:6–7)

The point of these lines is that Virginia maintained her good humor about "copulation" because Leonard interpreted her irritation or disinterest as a personal reproach; as far as her own desires were concerned, she would just as soon have remained a virgin.[49] Earlier she had felt guilty because she could not properly grieve. Now her defensiveness suggests she felt guilty because she could not fully love.

Soon Virginia's sexual guilt was compounded by a fear of artistic inadequacy. In a now lost letter to Vanessa, she must have spelled out what the painters would say about the landscape in Provence. Proudly announcing that she had sold a picture, Vanessa dismissed Virginia's attempt at painterly language: "How do you know what Roger and I would say of the scenery? Probably our remarks would be quite unexpected and original." Then Vanessa wrote to Leonard, "Tell the Ape his criticisms are not worth notice. I handed them on to Roger who wasn't even amused" (VB/VW, 23 August 1912; VB/LW, 29 August 1912). Virginia wrote to Duncan Grant in response, "My letters have to be carefully castrated so as to avoid descriptions of country and works of art" (VW/DG, 8 September 1912; MFS, 177). If Virginia's favorable review of the Postimpressionist exhibition had transcended artistic rivalry, Vanessa's dismissive remarks reanimated it.

Leonard continued writing his novel, perhaps about this time turning from the suburbs to Bloomsbury. There he introduces Harry Davis's reflections on his fellow painter, the beautiful and witty Camilla Lawrence, who possesses the "remoteness of a virgin." Attracted to her, Harry fears only that they both will forgo "fierceness of love, mental and bodily." Virginia probably knew that Leonard was writing about a potential romance between the Leonard-like Harry and the Virginia-like Camilla, for she seems to have begun planning her own romance based on her version of their courtship.[50] Leonard has Harry's Jewish sensibilities revolt against Camilla's world of "art and intellect" in which people mostly gossip nastily but "'never *do* anything'" (*Wise Virgins*, 37, 41, 52). Virginia might have been pleased at the exposé of her friends' bad habits were she not sometimes guilty of the same offenses.

Vanessa's letters to Virginia avoid mentioning Virginia's worries or restlessness or troubles with eating, although she did wonder whether Virginia had had "a miscarriage or something." Roger responded more kindly. He wired Leonard in Spain to invite him to return and become the secretary for his second Postimpressionist exhibition (LWA, 2:65). Leonard accepted. Roger also wrote to Duncan Grant asking him to create the official poster for the exhibition. Vanessa, who was with Roger, added her own postscript to that letter, inviting Duncan to Asheham and joking about the "young couple" who admitted that "they cant speak French!"[51] She laughed to Clive in the same vein: "I don't know what scenes of misunderstanding can have brought [Virginia] to that conclusion" (VB/CB, 22 August 1912).

Virginia had good reason to fear that her sister, brother, and brother-in-law were abusing her behind her back. She thought that Leonard would never do so, but she did not know what he was writing in his novel. Privately, as her letter to Ka Cox establishes, Leonard had shown his sexual disappointment. Formerly an unconditional lover, now Leonard seemed to be a judgmental one. Formerly not interested in the visual arts, now Leonard was about to be appropriated by Roger Fry and those same arts.[52] Again Virginia was threatened by both paint and the body. Her health, lovability, and aesthetic judgment all were in question, and her self-esteem—dependent on being the center of attention and entangled in the hegemony of the verbal arts—was again at risk.

5
Vile Imaginations

1912–1913

The Woolfs returned to London just in time for Leonard to assume his secretarial post during the second Postimpressionist exhibition, held from 5 October to 31 December 1912. Although it showed English painters (including Vanessa Bell, Roger Fry, and Duncan Grant) as well as Continental artists, this exhibition caused almost as much furor as the first had. Fry argued that the public's reaction proved that British philistinism was "as unwilling to learn by past experience as ever." He maintained that the Postimpressionists distorted appearances in order to express spiritual experiences. They did "not seek to imitate form, but to create form; not to imitate life, but to find an equivalent for life." Fry then defended the painters' classicism in terms of their independence of "literary" associations ("Grafton Gallery: An Apologia," 249, 250; VD, 166–70). His disparagement of the term *literary* might have offended Virginia, but his insistence that the artist creates form paralleled her search for an "undream't of form."

Desmond MacCarthy associated Fry's defense of Postimpressionism with Kant's notion of disinterested beauty.[1] But associations with the Old

Masters or Kant did not mollify the popular press, which described the exhibition as "laughable," a "nightmare of ugliness." Fry replied that "every new work of creative design is ugly until it becomes beautiful."[2] But he could not persuade a writer for the *Court Journal*, who labeled himself a "lover of what is true and beautiful, that such distortions, ill-judged hallucinations, or hideous nightmares on canvas bear the faintest resemblance to what is revered as Art."[3] Those were the very assumptions—that art must be true (or representational), beautiful (or pleasing), and revered (or uplifting)—that had aroused Virginia Stephen's skepticism and Roger Fry's reformism.

While aesthetic arguments continued in the press, Leonard and Virginia passed on their Brunswick Square quarters to Geoffrey Keynes (Maynard's younger brother) and withdrew to rooms in Clifford's Inn just off Fleet Street, where neither Bell was likely to drop in, although Clive tried at least once. He was "balked of my tête-à-tête with Virginia," however, by Leonard's being "invalided home from the Grafton Galleries, sick-poisoned" (CB/Molly MacCarthy, ? November 1912). Clearly, Clive felt that he could not visit with Leonard at home, and Leonard no doubt was as unhappy to find Clive visiting his wife as himself "sick-poisoned," especially when he knew that Clive had written so meanly about both of them. In his novel, Leonard was retaliating through Harry Davis's contemptuous judgment of the Clive-like Arthur.

Getting away from the Grafton Gallery to write was a relief for Leonard, since British philistines turned onto him their outrage at the paintings there. The authorial voice in Leonard's novel explains: "Only a person who had lived among the four great castes of India" can appreciate how much more oppressive the English class system is (*Wise Virgins*, 116). Later, Leonard recalled "how much nicer were the Tamil or Sinhalese villagers . . . than these smug, well dressed, ill-mannered, well-to-do Londoners" who ridiculed the new art (LWA, 2:66).

Despite having to endure smug and wealthy Londoners, there were consolations at the gallery, especially hearing Roger Fry trying to convert a "distinguished visitor" into a disciple of the new art (LWA, 1:69–70, 2:66). Fry tirelessly crusaded for a Postimpressionist transformation of literature. He would take E. M. Forster to exhibitions, insist that Forster voice his Ruskin-derived opinions, and then exclaim, "But Morgan, can you really think that?"[4]

No doubt Fry was gentler with Virginia, but Leonard's tenure with this second exhibition gave Fry ample opportunity to renew his campaign, begun in Turkey, to convince Virginia to follow the lead of the Postimpressionists in creating her own forms. Fry and the painters had given her a glimpse of a revisionary approach to art, but Vanessa asserted that Virginia's aesthetic opinions were not any more "worth notice" than Leonard's. Indeed, Virginia's husband had finished a novel, her sister was selling paintings, and she—whose very identity depended on writing—had been working for five years on a narrative line that took the heroine from innocence to experience to death. Virginia had finished a third version (and there may have been others) before her marriage. After all that work, Virginia could not abandon the novel, but after Matisse, Cézanne, and Picasso, she felt she could not publish it, either.

Earlier, Virginia had revised *Melymbrosia* at the level of word and image, in order to be more visual. Now she rescued herself by studying Cézanne in particular and importing techniques from the new painting into a complete rewriting, a wholesale visual reconstruction, of her novel.[5] Hibernating from Bloomsbury, stalked by the demon of ill health, she rewrote at a frenetic pace. As Elizabeth Heine explains,

> Fry's influence begins about the time of the first Post-Impressionist show in October 1910, and may have been strongest, in terms of the novel, in the winter of 1912–13, when Leonard Woolf acted as secretary for the second Post-Impressionist exhibition and the idea of the Omega Workshop was coming to fruition. . . . The work of 1912–13 was not merely a matter of polishing. The revision was profound, a re-seeing, a recreation. ("Virginia Woolf's Revisions," 424–25, 426)

Virginia rewrote as "one works with a wet brush over the whole, & joins parts separately composed & gone dry" (D, 2:323). But re-creating her novel as much as possible under the influence of the visual arts seems to have taken a psychological toll. Too much of her ego had depended on belittling the visual arts and insisting on Vanessa's verbal incompetence. Now Vanessa's insistence on Virginia's visual incompetence, even in minor queries about whether her Christmas present was "too aesthetic" for Virginia's taste (VB/VW, 23 December 1912), was more difficult for Virginia to brave, especially when she was borrowing from painting its imagery, repetitive patterns, and descriptive techniques for her novel.

As compensation, Virginia continued to disparage painters and painting, announcing that "artists are an abominable race." She went on to vent her irritation over the attention the exhibition was garnering: "The furious excitement of these people all the winter over their pieces of canvas coloured green and blue, is odious." That excitement did not end with the exhibition but was diverted into another channel: "Roger is now turning them upon chairs and tables: there's to be a shop and a warehouse next month" (VWL, 2:15).

That is, Fry was launching plans to begin, with Vanessa and Duncan Grant as subdirectors, the Omega Workshop, a cooperative scheme designed to produce aesthetically pleasing practical goods and to provide artists with a regular income. He housed the workshop in a handsome Robert Adam building with a large window overlooking Fitzroy Square.[6] Virginia described the Omega as a sort of crazed Postimpressionist attack on furniture (VWL, 2:16).

Events in London in late December 1912 suggest that for Virginia, feelings of aesthetic inadequacy became associated with fears of private inadequacy. In November, she had wondered whether Leonard was buying what may have been contraceptives (VWL, 2:12, note), but their sexual life still was not going well. If the fates of Leslie's two wives and of Stella had somehow associated childbirth with death, if the Duckworth brothers had convinced her that sex was nasty, and if Dora Sanger convinced her that it was disgusting, Leonard would have needed great tenderness and patience to make Virginia "vehement." At the time, many men and women were reading Marie Stopes's *Married Love* and *Enduring Passion*, and men were writing the author for practical information about lovemaking, particularly about arousing their wives.[7] Marie Stopes wrote Leonard a letter of condolence after Virginia's death (LWP), perhaps because she had met Virginia in 1939, but perhaps also because Leonard might have corresponded with Stopes about Virginia. Whether Leonard or Virginia read books like Stopes's, we cannot know. But the Woolfs did seek help.

Vanessa's bawdy talk and openness about her delight in sex made her a likely resource, and so the Woolfs went to Vanessa on 27 December 1912 to ask advice about overcoming Virginia's sexual unresponsiveness. Vanessa, apparently, was less interested in being helpful than in passing on the gossip to Clive and Roger:

> As I was in the middle of dinner in came the Woolves. . . . They seemed very happy, but are evidently both a little exercised in their minds on the subject of the Goat's coldness. I think I perhaps annoyed her but may have consoled him by saying that I

> thought she never had understood or sympathised with sexual passion in men. Apparently she still gets no pleasure at all from the act, which I think is curious. They were very anxious to know when I first had an orgasm. I couldn't remember. Do you? But no doubt I sympathised with such things if I didnt have them from the time I was 2. (VBL, 132)

Even though losing her virginity had not made Vanessa immediately orgasmic, she did not empathize with but, rather, laughed at Virginia's problems. Vanessa wrote to Fry, "We had astonishing revelations about Virginia last night! She is too incredible." And Clive renewed his abuse of the couple by passing on to Molly MacCarthy the news of Virginia's "conjugal dismays."[8]

Virginia's dreadful fix had become not only aesthetic but emotional as well. In the repressive Victorian era, many women spent their married lives as passive recipients of male sexual desire. After the turn of the century, theorists like Sigmund Freud, educators like Marie Stopes, and individuals like Vanessa Bell refuted notions of female sexual passivity. Nevertheless, Virginia seemed not to have understood sexual desire in men or women; perhaps she had not known that there was such a thing as a female orgasm. Yet now both her husband and her sister seemed to think her inadequate because she had been married for more than four months and had not yet had an orgasm.

Two days after the "Woolves" visited her, Vanessa asked Margery Snowden about what she "really thought of the Grafton—& the Goat—or was it all too bewildering?" The juxtaposition of the references to the Grafton—home of the exhibition—and "the Goat" points to an event involving Virginia at the exhibition where Leonard was at the time hanging more modern French pictures with Fry.[9] Perhaps Virginia projected onto these paintings her manic anger at having to confess to her sister her own sexual inadequacies. In any case, these new ideas about both female sexuality and art's nonreferentiality were destabilizing her world. Her headaches became intense, and she wrote no letters between 26 December 1912 and the second half of January 1913.

Leonard seems to have continued working on his own second novel during this time. Although Camilla Lawrence is supposed to be a painter and her sister Katharine a writer, Leonard let those masks slip when one character makes a comment about Camilla's failure to understand painting (*Wise Virgins*, 80, 99). Furthermore, the Clive character says of women like

Camilla: "They don't realise that we've got bodies. . . . [U]nless they are loose and vile they have no passions. What's noble in us is vile in them. . . . [T]hey simply don't know what desire is. What they want is to be desired—that's all." Katharine advises Camilla to marry but insists that no man should marry her, for "'I imagine a husband might not always be content merely to be in love with you. It isn't a normal male idea of the ideal wife'" (*Wise Virgins*, 96, 75–76). Leonard had heard from both Clive and Vanessa such discussion of Virginia's liabilities, but now his private experiences gave them particular pertinence.

When Katharine sardonically refers to Camilla's "flights" of imagination and her constant need for an audience, the connections with Virginia become even more transparent. Camilla's wish to have a happy marriage also sounds distinctly like Virginia's desire for a marriage that would be a "tremendous living thing" (*Wise Virgins*, 89, 85; VWL, 1:497). Virginia had warned Leonard about her lack of physical passion, and he was letting his sense of sexual loneliness surface, unbeknowst to Virginia, in the central sections of his book.

Because Vanessa was suffering from depression, Roger took her down to the restful Asheham. The distressed Woolfs joined them there in early 1913 for a melancholy stay in which sexual and psychological troubles were intensified for Leonard by the threat of malaria. Perhaps in reaction to Vanessa's attempt to "tell [Leonard] about pictures," Leonard defied his fever to write an article setting out his own views about art (VB/RF, ? June 1912). Virginia knew she must work, but on 4 January while Leonard wrote about art, Virginia "did not work" (LWD). As Vanessa's "usual rescuer," Roger helped her out of "the depths I had sunk to" (VB/Margery Snowden, 29 December 1912).

Roger may have done the same for the Woolfs before leaving for London. Then on several mornings, Leonard sat with both sisters or with Vanessa alone when Virginia could write in the mornings (LWD). While the sisters were so destabilized, the proofs of *The Village in the Jungle*, to be published in two months, arrived (LWL, 181). Vanessa reported to Clive that Leonard was busy "correcting his proofs which are beginning to come. He says his novel is very dull, but V[irginia] says it is amazingly good. They each have a great deal of admiration for the others works which must be a comfort" (VB/CB, 8 January 1913). But Virginia's genuine admiration for Leonard's remarkable firsthand account of human greed and corruption in a colonial setting intensified her anxiety over her own long-postponed novelistic career and, perhaps, over her husband's fictional foray into the suburbs.

Fry returned to Asheham on 10 January. The four—he, Vanessa, and the Woolfs—traveled to London by train together on 11 January, the women still in poor health. On 13 January 1913, when Virginia had had a "bad night," Leonard was worried and perplexed enough about her health to begin keeping a daily record of it. The next day, he met Fry and Henry James, among others, at the Grafton Gallery. Virginia later described the master novelist's encounter with Matisse and Picasso and praised Fry's exquisite delicacy in explaining to Henry James that "Cézanne and Flaubert were, in a manner of speaking, after the same thing" (RF, 180). On the actual day of the meeting, however, she worked only half the morning. Henry James seems to have dined with the Woolfs at the Cock Tavern, and then Virginia had another bad night (LWD). Perhaps, in addition to other problems, she may have feared that the master of modern prose, too, had defected to the visual arts.

However unsatisfactory, the Woolfs' sexual life seems to have continued. But given Virginia's poor health, Leonard worried about her becoming pregnant. On 16 January, Dr. Craig advised both Woolfs that it would be "gr. risk" for Virginia to bear children, and on 17 January, Jean Thomas concurred (LWP), even though she had apparently told Vanessa that Virginia could handle maternity (VBL, 134). Then on 23 January, Dr. Savage pronounced himself "in favour" of pregnancy. Leonard and Virginia, after meeting at the London Library and seeing a Friday Club exhibition together, consulted Dr. Hyslop, who "advised not for 1½ years" (LWD).

Prescribing a quiet life away from London, Vanessa assured Leonard that Virginia was "certainly better since you married," but now she thought childbirth would risk "another bad nervous breakdown—& I doubt if even a baby would be worth that." Given Virginia's precarious health and the fact that, in Vanessa's words, "one does plunge into a new & unknown state of affairs when one starts a baby," pregnancy was postponed. Vanessa urged Leonard to take "all precautions [against pregnancy] one can beforehand." On 2 February, Vanessa also explained to Virginia that having a baby when she was not strong could precipitate another breakdown.[10]

Perhaps the precaution Leonard resorted to was abstinence. If so, Virginia may have felt guilty, as she was introducing into her novel explicit protests over Rachel's sexual ignorance before her talks with Helen (a Vanessa figure) and her courtship with Terence (a Clive figure made more like Leonard in later revisions). Meanwhile, in his own novel, Leonard also explicitly blames Camilla's coldness on ignorance, for sex was "something she did not understand" (*Wise Virgins*, 154). In Virginia's novel, as Rachel comes to understand human passion, she seems to lose her fear of sex. Virginia may have felt that she, too, should conquer her fears.

Vanessa's own depression seems to have given her insight into and understanding of its threat to Virginia. In early January, for example, while Virginia was still ill, she asked Roger:

> Do you think I sometimes laugh at her too much? I dont think it matters, but really I am sometimes overcome by the finest qualities of her. When she chooses she can give one the most extraordinary sense of highness of point of view. I think she has in reality amazing courage & sanity about life. (VB/RF, 7 February 1913)

Unfortunately, Vanessa communicated her admiration for Virginia to Roger, not to Virginia. And she continued laughing at her.

Vanessa particularly resented the Woolfs' alienation from painters and painting. In one letter to Leonard, she mentions the "painters themselves of whom you've complained" (VBL, 127), and in another she refers to the Woolfs' distaste for "painters' company" (VB/LW, 20 January 1913). Beginning to be defensive himself, Leonard set out to adopt Roger's and Vanessa's theories to more traditional thinking about art, in an article on Postimpressionism, which Vanessa fiercely criticized. She said that Leonard failed to understand that people could derive strong emotions from form and color alone without thinking about what they represented (VBL, 133–34). Refuting Leonard's position on art seems to have occupied the painting advocates for about two weeks and only strengthened Leonard's distaste for their perspective. Vanessa explained to Virginia that "Roger's views of course are more mature than ours. He is at one pole & Clive at the other & I come somewhere in between in a rather shaky foothold—but none of us really agree[s] with Leonard. So your husband had better reconsider his position I think."[11]

This private controversy probably emerged from a public one. Fry's formalist artistic values initiated a debate in the pages of the *Nation* over the distinction he posited between an "ethical" and an "artistic" approach to art. He argued that an "ethical" writer like Bernard Shaw

> wants certain things done for the greater good of mankind. . . . But the artist qua artist does not want anything done; he has

> nothing to do either with action or opinion. . . . [H]e cannot consciously consider the effect of what he does on others any more than the scientific researcher can consider the effect of truth while he is seeking for it.

Shaw himself replied that not only was his work ethical but so was Dante's, Shelley's, and Fry's.[12] But Leonard accepted Roger's distinction, and Vanessa's dismissal of his aesthetic theory confirmed his predisposition to turn "from an aesthetic into a political animal" (LWA, 2:476). Whereas Leonard, like Shaw, believed in art's efficacy, Virginia, like Fry, believed in both art's significance and its disinterestedness.

Leonard was becoming a more political animal not only because of his contempt for disinterested art but also because of his dissatisfaction with philanthropy. In early 1913, Virginia's cousin Marny Vaughan (sister of Emma and Will) took Leonard to the slums of London to try to engage him in her charity work. He was appalled by the ineffectuality, the pettiness, and especially the condescension of the dispensers of charity: "Having refused to remain the benevolent ruler of Silindu and Hinnihami in Baddegama, I was not going to try to play the part of benevolent father to Mr and Mrs Smith in a Hoxton slum" (LWA, 2:70). Leonard's distaste for such charity work thereupon entered his new novel as Harry takes Camilla to see how people live in the suburbs; they attend a charity garden party to raise money "in aid of The Poor Dear Things" (*Wise Virgins*, 116).

Both families are embarrassed over Camilla's encounter with a crass "Mrs. Brown." Then, in a rather improbable scene at the garden party, Leonard places Gwen Garland on a bench among rhododendrons. There she overhears Camilla's assessment of suburbanites, and of the Garland girls in particular, as "uneducated, purposeless, sterile" with nothing to do in their lives except to wait for some man to marry them. Harry agrees but supposes that such people are happy in their boredom. He rightly expects that Camilla will blame men for the young women's predicament. Camilla herself does some soul searching, wondering: "What right have I to look down on those people?" She holds Harry's hand, but only, he realizes, as she might have held her sister's (*Wise Virgins*, 125–28, 136, 137, 147).

Desperately in love, Harry later confronts Camilla about his "'suspense—doubt—I can't stand it any longer. You must end it, one way or the other.'" Camilla replies, "'But I'm not in love with you, Harry,'" words that he takes as a final rejection, for it seems she cannot understand love or give of herself. After that encounter, Camilla barely figures in the novel (*Wise Virgins*, 152, 153, 154).

Leonard continued his novel, returning from the satire of Bloomsbury to a satire of suburbia. When the Davises and the Garlands take a seaside holiday together, Gwen tells Harry how much she has been hurt by overhearing his conversation with Camilla. Harry consoles her with kisses and then writes passionately to Camilla that the greatest thing in life would be to have his love returned, but the next best things would be simply to love her and see her (*Wise Virgins*, 189).

While the controversy over art's ethical obligations raged in the press, Gertrude Stein and Alice B. Toklas came to London and visited at Durbins. Fry impressed Stein because he had "discovered" Picasso independently of her and because he was trying to find an English publisher for her radical prose experiments. Vanessa wished that she had been invited to lunch with them and informed Roger that Duncan had "been making mischief" by telling her that Roger was disappointed in her painting. Vanessa's self-declared "nasty" letter had Roger turning to another woman, with herself left in an unfulfilling relationship with Duncan. She also mentioned a wild party at which "we should all get drunk and dance and kiss" as a way to gain rich patrons for the Omega (VBL, 135). Vanessa bragged to Virginia about Clive's and Duncan's meeting Stein. Such news, along with the testimony that "the air is teeming with discussion on Art" and the aggressive warning that "your husband had better reconsider his position," must have heightened Virginia's fear that both she and Leonard were out-of-date.[13] On the day she received that letter, Leonard noted that Virginia, still frantically revising, "wrote morn. Not v.w. sl.h. Not good n" (LWD).

Leonard's almost daily records—"not v.w.," "sl.h.," "g.n.," "f.g.n.," "b.n." (not very well, slight headache, good night, fairly good night, bad night)—may be terse, but they show that Virginia's health and her writing were uppermost in his mind. On 18 February, while Virginia worked on her novel, Leonard went to Gordon Square to confer with Vanessa about Virginia's health. That night, she had only a very slight headache and a fair night (LWD). Throughout January and February, Leonard recorded whether Virginia wrote; she did, almost every day in February. She was also, for the most part, fairly well (LWD). The previous spring, Leonard had urged Virginia to finish her novel before deciding whether to accept his marriage proposal. He assured her: "You are the best thing in life & to live it with you would make it ten thousand times more worth living. I shall never be content now with the second best" (LWL, 174). In 1913, Leonard stood by his words, protecting Virginia's health and writing to her day after day. Meanwhile, Leonard let his alter ego, Harry Davis, prove himself incapable of such unconditional love.

If Virginia felt guilty for not loving Leonard as physically as he (and Vanessa) thought appropriate, if she felt helpless over her own failure to write a novel in five years when he had written one in a year and was at work on another in a second year, if she felt inadequate because her artistic reformation did not match Picasso's or Gertrude Stein's, and if she had lost Vanessa's affection, Virginia had nonetheless gained Leonard's. Perhaps, however, she feared that his care was now the fruit of obligation rather than love.

Despite all these difficulties, Virginia transformed *Melymbrosia.* After five years, she took only five months to re-create it as *The Voyage Out*. This novel balances ominous hints of death, disease, and disaster, with hope for growth, maturity, and love. Although Virginia's illness threatened this frantic rewriting, it also enriched it. Rachel's dreams after Richard Dalloway kisses her and the nurse tries to quiet her and then plays cards become elaborate surrealistic passages inspired by Virginia's fears of male sexual aggression and her anger at female dissociation. They also borrow from her manic hallucinations in 1904, and the surrealistic passages from Rachel's feverish point of view as walls curve, the outside world retreats, nights continue endlessly, and her body seems no longer hers, are brilliantly realized.

Virginia's illnesses seem to have nurtured this re-creation in a less obvious but perhaps more significant way. Restored health and harmony with her world were all the more precious after being out of sync with it and even with her own self. Along with the artists' examples and Roger Fry's proselytizing, a fresh sense of being at one with her world inspired Virginia's profoundly visual reconstruction. In *Melymbrosia*, she had written: "Since religion has gone out of fashion, and the soul is called the brain, these enormous spaces of silence in which our deeds and words are but as points of rock in an ocean, are discreetly ignored; the novelist respects but does not attempt to render them" (68). But in *The Voyage Out*, Virginia herself rendered them.

The earlier novel had been mostly dialogue with little description, especially in the early parts, whereas the new book alters the proportions. Almost all the long descriptive passages are either elaborations of earlier short passages or new additions.[14] As her characters find "new forms of beauty" (VO, 90) in the South American setting, so Virginia created new forms. Furthermore, these descriptions often borrow from the Postimpressionists' reductions of figures to basic featureless shapes (as in Figures 16, 22, and 25). For example, Helen's figure "shared the general effect of size and lack of detail" (VO, 95), or "the movements and the voices seemed to . . . combine themselves into a pattern before his eyes; he was content to sit silently watching the pattern build itself up" (VO, 374). By using language to actualize undifferentiated shapes and "enormous spaces of silence," Virginia overcame the rift in her own consciousness and triumphantly forged an aesthetic resolution to the *paragone*.

Rachel herself has an aesthetic experience in which an ordinary tree

> appeared so strange that it might have been the only tree in the world. Dark was the trunk in the middle, and the branches sprang here and there, leaving jagged intervals of light between them as distinctly as if it had but that second risen from the ground. Having seen a sight that would last her for a lifetime, and for a lifetime would preserve that second, the tree once more sank into the ordinary ranks of trees. (VO, 174)

This experience illustrates the aesthetic moment as "a suspended moment when self and object" merge as the child and the mother once did (Bollas, *Shadow of the Object*, 31).

In *The Voyage Out*, Virginia conquered the demons that had plagued her after three deaths and that had been threatening ever since her "affair" with Clive had alienated Vanessa. Although this novel presents a number of possible explanations for Rachel's death (poorly cooked vegetables, bad water, a disease caught on the excursion up the Amazon, poor medical care, a malevolent universe), it does not endorse any of them. Instead, it confronts the irrational, denies to the characters (and the critics) a "reason why such things happen" (VO, 359), and refuses any semblance of blame. Terence and Rachel are helpless before her irrational disease, but Terence need not feel guilt for Rachel's death. Virginia herself very much needed that lesson in guiltlessness.

After a quiet weekend at Asheham, Virginia finished *The Voyage Out*, and on 6 and 7 March, Leonard read and approved it. Then on 9 March, he delivered the result of Virginia's last five years of writing to Gerald Duckworth's publishing firm and recorded "v f w g n" (Virginia fairly well good night) (LWD). Leonard must have taken the book to Duckworth knowing that Virginia's chances of being turned down were slim. Apparently, she had not shared with anyone her memory of Gerald Duckworth's exploring her private parts. Not until 1939 did she write about it, in her "Sketch of the Past," which she then kept private, and then in a letter in 1941 (VWL, 6:460). Unable to tell her husband or perhaps even to recover the memory of the "unthought known"[15] cause of her distaste, she was in a fix because Leonard had unwittingly placed her once again at the mercy of the half brother who had sexually traumatized her when she was a child.

Later, Virginia avoided at least the first phase of the author's helplessness by being, along with Leonard, her own publisher. She explained that it was "the greatest mercy" to write with "no editors, or publishers" overseeing her work: "I dont like writing for my half brother George [Gerald]" (VWL, 2:168). Her conflation of Gerald and George may indicate how much she equated them as emblems of male oppression playing on female helplessness.

On the same day that he placed her novel in Gerald Duckworth's hands, Leonard took Virginia with him by train to Liverpool (LWD). This trip was one of several he made to study the Cooperative movement in the Midlands and the north. By 1913, Leonard had experienced what Frederic Spotts calls "a personal political revolution" and had emerged "simultaneously a feminist and a socialist" (LWL, 371). This political revolution, with its feminist and socialist overtones, led to the Cooperative movement, an alternative economic system in which profits were shared with the (usually female) workers who produced the goods. Leonard went to the north to learn about baking, tailoring, shoe repairing, hosiery making, and cabinet and chair making, as well as efforts to secure better pay for women (LWD). Virginia was impressed by the potential power of a reformer who could energize "about 6,000 helpless women" to protest working conditions (VWL, 2:19).

Back in London and about to visit with Gertrude Stein, Clive facetiously compared her devotion to art with the Woolfs' devotion to improving the lot of the working classes: They

> have gone, or are going, to the great industrial centres of the north to study co-operation and found girl's clubs. They do all things well, except one apparently. And talk and write about Socialism and Life and Books which is far more real, of course, than writing and talking about art. (CB/Molly MacCarthy, Tuesday, ? April 1913)

Clive's repeated references to the Woolfs' sexual difficulties and his dismissive tone regarding good works were no doubt shared with other Bloomsbury members—behind Virginia's back. From Scotland, she wrote a cheerful letter to Duncan Grant, telling him about "seeing machines and factories" and hearing from Vanessa's friend Margery that when Roger lectured on Postimpressionism in Leeds, his audience thought him "mad as a hatter." Each of her letters from this trip mentions Leonard's *The Village in the Jungle*. To Duncan, however, Virginia sounded an ominous note harking back to her old jealousies and fears; Leonard got "a letter praising his novel at every town—which you must admit is pretty depressing for me" (VW/DG, 16 March 1913; MFS, 179).

Virginia would justifiably have been "very furious and tantrumical" (PA, 69) had she known that (thanks first to Vanessa and then to Clive) her sexual troubles were public knowledge. She had suspected Vanessa's "abuse," and now she felt growing hostility from both Bells. For example, when Clive was to meet Molly at the Cock, a tavern where the Woolfs often dined, he assured her that "the Woolves won't be there." Clive had met Vanessa there in a "fine old rage: they [the Woolfs] had just left."

Virginia had voiced a number of the suspicions that had haunted her since the last numbers of the *Hyde Part Gate News*. Despite Vanessa's claim that she could not remember the meannesses she committed, Virginia's accusations prompted her to ask Clive whether she was "an atrocious egoist, and intolerable to live with? Are all Stephens egoists?" Clive reported that every subject that Vanessa mentioned was made a topic for personal antagonism by Virginia. Leonard, too, offended Vanessa by his imperious manner, which Clive characterized as "I told you so," "I foresaw," "I took steps," and "I am life's master." Clive predicted that the Bells (with Adrian) and the Woolfs would drift apart and concluded that "the young couple are down in the [Bloomsbury] world at present, the feeling seems to be that Virginia + Woolf = 2/but it's none of my doing" (CB/MM, March 1913).

Virginia did not know that Clive had proclaimed the Woolfs to be a lone twosome whose popularity had fallen in Bloomsbury and that both he and Vanessa had spread tales about their private life, but she did feel Vanessa's alienation. That evening at the tavern, Virginia seems to have adopted not only Leonard's irritation with the Stephen egoism but also his accusatory method. This strategy, of course, backfired when Clive rejected both the straitlaced Leonard and the offended Virginia.

Vanessa wrote to Clive that Virginia was angry with Adrian, and he was eager "to hear all about my quarrel with the Goat." Her letter reveals how the Stephen family's habit operated: "We spent much of the time discussing the W[oolf]s." Adrian—who had yet to distinguish himself in any way—maintained that Leonard had a "superficial mind" (VB/CB, 31 March 1913). Whereas Vanessa and Adrian took pleasure in attacking the Woolfs behind their backs, at least Virginia (however ineffectually) attacked Vanessa face-to-face. Neither Vanessa nor Adrian seems to have recognized that Virginia's accusations expressed her deep sense of loss.

With her new visualization of her novel a risk, with her half brother holding her book (and hence her being) in limbo, and with her sister's affection lost to her, Leonard's love became Virginia's ballast. When her "little beast" had to visit his family, Marjorie Strachey stayed with Virginia, and they discussed many things, including the possibility that her book had been rejected or lost (VWL, 2:21). After Leonard returned, Virginia displayed a cheery persona even when she told Violet that she was postponing pregnancy until her health improved and that while Leonard was in the middle of writing his second novel, she expected to have her first rejected. Earlier, she had associated failure to marry with failure to write; now she aligned maternal and aesthetic failures. Virginia wrote to Violet that Leonard was in the middle of his new novel and that she expected to have hers rejected, "which may not be in all ways a bad thing" (VWL, 2:23 and note). Leonard saw the danger in such thinking, for he telephoned Gerald Duckworth that day and found out that "they accepted it" (LWD). Probably without knowing about his call, Virginia went to Gerald the next day and received the good news herself. She did not, however, mention the acceptance in her letters until late May, when the proofs arrived (VWL, 2:28).

Vanessa continued to translate this emotional quarrel into aesthetic terms, insisting that when Henri Doucet visited Roger, Leonard and Virginia "would have mocked at . . . the artistic atmosphere" (VB/VW, ? May 1913). Virginia defensively claimed—at the very moment she was checking proofs of *The Voyage Out* (which explicitly showed Fry's influence)—"personally I dont feel in Roger Fry the inspiration of Morris, but no doubt I'm wrong" (VWL, 2:28). Vanessa manifested the "drifting apart" that Clive had predicted, by writing to Leonard that she owed him a letter "& I'm afraid of getting into trouble if I dont pay what I owe to the Jews" (VB/LW, 25 May 1913). She also seems to have teased Virginia about her fears, for Jean Thomas wrote to Violet Dickinson that when Virginia got her proofs back, she "thought everyone would jeer at her. Then they did the wrong thing & teased her about it and she got desperate" (Bell, 2:16). Vanessa took the tack of describing Leonard to Virginia as "a successful novelist" (VBL, 139), thereby exacerbating Virginia's fear that the label was his and not hers.

Nor, it seemed, was Virginia as good a feminist as Leonard, who later wrote that "feminism [should be] the belief or policy of all sensible men." Although Margaret Llewelyn Davies, "who could compel a steam roller to waltz," convinced Virginia to support influential friends calling for liberal divorce laws (VWL, 2:30), Leonard was becoming a more active ally with Margaret, working for suffrage and the Women's Cooperative Guild. Exposure to the slums of London and the ineffectiveness of traditional charities had convinced Leonard of "some destructive disease in the social organism" that could be cured only by "social revolution." This exposure completed his "journey to the Left" and converted him from a "liberal into a socialist" (LWA, 2:69, 70). Leonard was invited by Sidney Webb, chairman of the *New Statesman*, to write for that leftist journal and to draft papers for the socialist Fabian Society (Spater and Parsons, *Marriage of True Minds*, 82).

After correcting proofs of *The Voyage Out*, Virginia wrote very few letters. Leonard continued to record her health and activities, including being thrown from a mare while riding in Sussex on 1 July. In midsummer, she fainted or, using Vanessa's word, "collapsed." Her condition was serious enough to require rest. On 25 July, Virginia was once again caged with the lunatics. Vanessa speculated to Roger: "Please be very careful not to say a word to anyone about her worrying over what people will think of her novel which seems really to be the entire cause of her breakdown" (VB/VW, 17 August 1913; VB/RF, summer 1913). Like the fear that her best friend hated her, the fear that the public would despise her novel traumatized Virginia.[16] The publication of *The Voyage Out* was delayed, presumably by Leonard in consultation with Gerald.

While Virginia was back in the nursing home at Twickenham in the summer of 1913, Leonard visited almost daily and wrote a series of love letters designed to absolve her of any worry. He assured her that although she would need to remain in the home for more than two weeks, they were not really separated: "When I'm away from you, you simply always are before my eyes & dance dance in my thoughts—you can't imagine how you obsess me & how I long to see you" (LWL, 186). Leonard also told her that his love was absolute, that he was the powerless partner before her, and that there was "no reason in the world why you should reproach yourself with anything or think that you have done anything to be laughed at" (LWL, 188). Against her self-reproach for unknown offenses and her fear of unknown abuse, Leonard's love was reassuring, but even his love could compound Virginia's guilt. When she thanked him for being "absolutely perfect to me," she added, "Its all my fault. . . . Goodbye, darling mongoose—I do want you and I believe in spite of my vile imaginations the other day that I love you and that you love me. Yr. M" (VWL, 2:34). Whether or not "wanting" carried any sexual connotation, Virginia's intensifier "do" shows that she needed to assert her want against doubt, perhaps both his and hers. Her "vile imaginations" exhibit the way that manic-depressives magnify entire theories (what Leonard termed "worries") based on small events or even nonevents.

While Virginia was being overfed and kept literally in the dark at the rest home, Leonard finished *The Wise Virgins* by acknowledging what it would have meant for him to lose Virginia: After Harry has written to Camilla of his hopeless love, Gwen steals into his hotel room one night, and he makes love with this "foolish" virgin. As Harry explains to Camilla on his farewell visit, he and Gwen "ruined each other, and now we're to marry—in two weeks." The book ends as the couple starts on their honeymoon and Harry can only console himself that "he had known Camilla; he had loved her" (*Wise Virgins*, 219, 197, 236, 247). Unlike Leonard, Harry accepted second best.

Leonard certainly had intended this novel as a satire of both suburban and intellectual milieus. And although he did reveal some private frustrations, I think Frederic Spotts is correct in considering this book an "homage to Virginia, whom he presents as existing on a plane above other women. It was also his monument of gratitude to her for saving him from the sort of conventional marriage that was the denouement of the novel" (LWL, 158). Leonard seems to have worried principally about what the Woolf family would think of his harsh portrayal of the Davises, so he sent the manuscript to his sister Bella for her approval.

In response, Bella wrote a nine-page critique, dated 12 August 1913, strongly disapproving of the book because of the pain it would give their family and neighbors. She also urged Leonard to give his protagonist "more of himself," more of his affectionate, gentle character (LWP). Even though Leonard's brother Philip found the novel depressing, he considered it too good to let family reactions interfere with its publication (LWL, 195, note). Over the next several months, wrangles over whether to publish this book added to, as we shall see, Leonard's distressed life.

Virginia was released from the rest home on 11 August, and she and Leonard went to Asheham. Even though Vanessa had "harrowing dreams" about Virginia and told her to "be good & sensible or I shall come & scold you," she kept her distance. She admitted, as she told Roger, that she ought to do more for her sister, but "after all one cant do much with married people." That is, Virginia's marriage excused Vanessa to leave Virginia's care to Leonard and to seal off her emotions further. Indeed, Vanessa's letters to Clive during this period convey less sympathy than irritation over having to share Asheham. Virginia's internment had delayed their holiday there; hence Vanessa complained, "It will be sharp work getting them out by the 25th." After Leonard agreed to leave on 25 August, Vanessa then changed her plans to arrive on Saturday, 23 August 1913.[17]

Vanessa's "sharp work" succeeded, for on Friday, 22 August, Leonard and Virginia left Asheham. They must have been angry at being so pressured and offended by Vanessa's insensitivity, and they went to London and consulted with Dr. Savage that afternoon. Savage advised Leonard to take Virginia on a holiday, but "Roger persuaded him to get first a second opinion from Head." Dr. Head told Leonard that a holiday would be "a great risk," but he felt that changing plans might upset Virginia more than the stress of traveling. That night, when the Woolfs dined with Roger (LWD), all three must have sensed the impending crisis.

On Saturday, 23 August, before she left for Asheham, Vanessa was called to help. She later described Virginia's behavior in terms we now can identify as characteristic of a classic depressive syndrome. Vanessa and (possibly) Roger

> saw the Ws. off at Paddington. It was all rather wretched, as she seems to me to be in just the same sort of state, only not so bad, as she was in the very first time. She worries constantly & one gets rid of one worry only to find that another crops up in a few minutes. Then she also has definite illusions about other people.[18]

Virginia seems to have elaborated the accusations she made at the Cock Tavern into paranoid "illusions" or "vile imaginations" about others, which Leonard tried to diffuse. He had long known that Virginia was hypersensitive and high-strung, but her racing mind, jumping from one worry to another, was a species of derailed, manic behavior for which he was utterly unprepared. He tried to cope by using reasonable discussions and, more effectively, love and assurances. Vanessa concluded, "I dont think Woolf can go on for long alone, the strain of always looking after her is so great" (VB/CB, August 1913; Muggeridge, interview with Leonard Woolf). Then she wrote Leonard an exuberant letter about taking pictures of her naked children on the terrace at Asheham. In her next letter to Virginia, Vanessa insisted that she, Leonard, and Dr. Craig—all "sensible" people—"know that you are ill."[19]

Leonard's diaries give details of Virginia's troubles, although not of his own. Several entries (originally in Sinhalese code) from Holford (where they again were staying at the Plough Inn) are illustrative:

> Cheerful, then worried. One moment very depressed. Rested 2–4:15. Thought people laughed at her. . . . [She] persists in saying that she is all right and will not eat. . . . Ate bad tea talked to her about breakdown on walk & she saw more clearly & ate good dinner cheerful & calm even. Slept badly, took veronal three o'clock & slept till 7:30. . . . Slight worry till 7 but no illusions. . . . (Delusions persisted I think all day.) Great difficulty over food all day.[20]

Clearly, Virginia's worries, illusions, and delusions were that she was being laughed at or abused and that her loved ones had turned against her. Vanessa probably had teased her about her novel; she took pleasure in picking her sister to pieces; she did acknowledge that she laughed at Virginia perhaps too much; but mostly she was oblivious to her own meannesses. And when she was in a depressive state, Virginia's sense of abuse escalated into paranoid proportions.

Vanessa's response to this crisis was not empathy but another "lecture." She warned Virginia that she was in the "same sort of badly nourished state that you were in that first time [1904] at Welwyn" when she was "incapable" of using her brain and insisted that "we were all mistaken [and] that there was nothing the matter." With Vanessa rationally instructing, "Dont make things difficult for Leonard," who was "far more sensible than you are" (VBL, 141–42), Virginia must have felt alienated from Vanessa, guilty for troubling Leonard, and helpless to resist the demons that could again make her incapable of writing.[21]

Ka Cox, who had become a real friend, joined the Woolfs on 2 September to help Leonard look after Virginia. Virginia was "afraid" to return to London lest she be incarcerated again in the nursing home (LWD). Because she adamantly insisted that there was nothing wrong with her, Leonard proposed an agreement:

> I suggested that we should return to London at once, go to another doctor—any doctor whom she should choose; she should put her case to him and I would put mine; if he said that she was not ill, I would accept his verdict and would not worry her again about eating or resting or going to a nursing home; but if he said she was ill, then she would accept his verdict and undergo what treatment he might prescribe. (LWA, 2:110)

The three returned to London on 8 September, planning to stay briefly at Brunswick Square. In the morning, Virginia and Leonard saw Dr. Wright, who "told V she was ill" (LWD). Remembering that Roger Fry had described Henry Head as a doctor sympathetic to Helen Fry's problems and, in Leonard's terms, an "intelligent man" as well as an "intelligent doctor" (LWA, 2:111), Virginia chose to see him again, a testimony, as Mitchell Leaska writes, to "her faith in Roger" (Afterword to *Pointz Hall*, 455). As soon as he saw her, Dr. Head prescribed another rest cure. In accordance with her agreement with Leonard, Virginia had only one other option if she did not wish to be "caged" again: Finding the hidden case of drugs, she took 100 grains of veronal.[22] The date was 9 September 1913.

Ka Cox found Virginia unconscious and barely alive. Then Leonard and Geoffrey Keynes raced in a taxi to St. Bartholomew's Hospital, where Keynes was an intern, "waving aside the policemen (in the then absence of traffic lights) shouting 'Doctor! Doctor!' if they tried to stop us." They commandeered a stomach pump, returned to Brunswick Square, and Drs. Keynes and Head spent the night "washing out" Virginia's stomach (Keynes, *Gates of Memory*, 166).

The precocious child-author of the *Hyde Park Gate News*, who had become an ambitious author with plans to reform the novel single-handedly, had become a wretched creature so haunted by demons that death seemed preferable to life. To Keynes and Head, she was an inanimate being whose organs had to be purged and reactivated. To Leonard, she was a nearly lost beloved whose death would have effectively ended his own life

(LWL, 205). To Clive, she was a probable casualty: "One begins to wonder whether she will ever get really sound again" (CB/Molly MacCarthy, 25 September 1913). And to Vanessa, who soon left town, Virginia seems to have become a threat.

Leonard slept for a few hours, and then Vanessa woke him early the next morning. After a night of hovering near death, Virginia was, Dr. Head pronounced, "practically out of danger." Neither Brunswick Square nor Clifford's Inn offered an adequate setting for Virginia's convalescence, but Leonard had gotten Virginia's message: She would die rather than go to a rest home. Vanessa thought of George Duckworth, whose snobbish "vices," as she wrote later in the day, had "really blossomed into perfection" with his aristocratic marriage; hence "all one can do now is to get what one can from him compatible with self-respect—& that's a good deal sometimes!" (VB/LW, 10 September 1913).

While Virginia slept, Leonard went to confer with George, who offered to the Woolfs rooms in the large, elegant, and well-staffed Dalingridge Place (Bell, 2:16–17). Vanessa sent the news to Maynard Keynes that Virginia "took veronal & medinal while alone for a few minutes at tea time." Without comment, she continued, "I think I shall go back to Asheham this evening" (VB/JMK, 10 September 1913). Vanessa herself periodically suffered "crippling bouts of lethargy," suggesting severe depression, so perhaps she went away to protect herself (Garnett, *Deceived with Kindness*, 32). Leonard put Vanessa on the 5:20 train from Victoria Station to Lewes, from where she could be driven to Asheham. Later, he recorded that Virginia was "unconscious all day" (LWD).

On 11 September, George Duckworth visited, and Vanessa, safe at Asheham, wrote to Leonard that once Virginia had a "developing consciousness" he had to "remove everything that could be dangerous. Scizzors. Knives. etc. & of course drugs." She thought that Virginia might try to assault a nurse and offered to come to see Leonard "anyhow if not her. We could meet outside the house" (VBL, 142–43). Presumably, she realized that her presence would upset Virginia's recovery, although she, again, may have been protecting herself as much as Virginia.

By 12 September, Virginia was "fully conscious." On the next day, she was "cheerful" (LWD), but her condition remained decidedly precarious.[23] Roger Fry volunteered to deal with the commissioners who were investigating to determine whether Virginia needed to be institutionized for her own good. From Asheham, Vanessa assured Leonard that Roger "managed [the commissioners] quite easily alone. He says that if they see that you are doing all you can & have good advice they are anxious not to interfere." Because of Leonard's care, Dr. Head was, as Vanessa reminded Leonard, "strongly against any kind of asylum" (VB/LW, 13 September 1913).

On 15 September, Vanessa came up from the country for the day. She and Leonard went to the Omega, where Wyndham Lewis and others were arguing with Roger. Together Leonard and Vanessa conferred with Drs. Belfrage and Savage. After Vanessa talked with Virginia in the afternoon, Leonard put her back on a train. He sat with Virginia after dinner when she was "cheerful" (LWD).

On 17 September 1913,[24] Leonard, Virginia, and a nurse were driven, apparently in George's car, to Dalingridge in Sussex. Virginia spent a "bad night" there. The next day, she was "depressed" and experienced "much worry." She slept very badly (LWD). Although the spacious accommodations for herself, Leonard, and now Ka, as well as the nurses, must have been welcome, at the same time being again the recipient of George's beneficences must have been distressing for Virginia.

Leonard's cryptic jottings in his diary were devoted entirely to Virginia's condition. A typical entry reads: "V very excited & worried. Gr trouble w food. Bad night." Vanessa told Roger that Virginia would not eat and that the nurses "have had to call in Leonard which of course is just what one didn't want," presumably because he should not be perceived as the enforcer of discipline. Furthermore, even though Ka Cox was helping out, Vanessa speculated that Ka "cant stay for ever. It's difficult to see who can go to them" (VB/RF, 23 September 1913).

On 24 September, Leonard bicycled from Dalingridge to Asheham, where he found Vanessa and Duncan Grant happily engrossed in painting panels for the Omega's Ideal Home Exhibition room. Vanessa reported to Roger: "I had a very gloomy visit from poor Leonard yesterday. She has been sleeping very badly." Leonard was the only person who could get Virginia to eat, which she only did "from affection & the danger is that that may fail." Virginia was "decidedly worse since the veronal episode" (VB/RF, 25 September 1913). On that same Thursday, Virgina was "v excited all day [. She took] 2 hrs over each meal. Did not sleep at all" (LWD).

Ka had to leave Dalingridge on 26 September, and the next day Leonard's activist friend Margaret Llewelyn Davies came to help out. As Vanessa warned Leonard, "You'll need patience for months I expect." She concluded, "All you told me made it quite clear that this is just the same as before," when Virginia had turned against her. She hoped to go with

Leonard to see Dr. Craig, to tell him more "about your effect on her" (VB/LW, 26 September 1913). In his diary, Leonard continued to record Virginia's worries, excitement, jumpiness, sleeplessness, difficulties with eating, and eruptions of violence. On 27 September, Vanessa received a letter from Leonard telling her "about the violence." She replied that she was asking Dr. Craig "to see [Dr.] Belfrage & arrange to come to see you as soon as possible." Despite being "obliged to lie down today," Vanessa offered to come if Leonard telegrammed. She hoped that Virginia's violent resistance to food was merely a side effect of "the sleeping draught" and would not continue.[25]

Ka reported to Janet Case that Virginia was eating better but that the "other symptoms are alarming." She felt it "so splendid when people like you and Margaret Davies love & understand," having "never seen anything so wonderful" as Leonard's loving care for Virginia. When she was back in London, Ka telephoned Leonard every day to ask about Virginia's health. Ka, whom Janet Case called a reliable "rock," returned to Dalingridge several times (KC/JC, 29 September 1913; JC/LW, 1 October 1913). While Leonard and Ka and Janet and even Margaret Llewelyn Davies (not a close friend) were giving so much of their time and energy to restoring Virginia's life and sanity, Vanessa's absorption in her own shifting emotional complications with Clive, Roger, and Duncan seemed callous to Ka. Back in London, she wrote, "I saw Nessa for a minute yesterday. Their amazing entanglements are very absurd, it seemed to me! They really struck me as so infinitely unimportant!" She then advised Vanessa herself to bring Dr. Craig to see Virginia (KC/LW, 30 September and 1 October 1913).

Perhaps feeling guilty after Ka had criticized her, Vanessa reminded Leonard that according to Ka, Dr. Head thought Virginia's difficulties were due to "nerve exhaustion" from not sleeping and that she herself, too, suffered from "nerve exhaustion." She also suggested that Virginia "must be very strong physically from the way she stood the veronal." But Vanessa's offer to visit is revealing: "I dont expect she would be too much agitated by me if you didnt tell her long beforehand & I will come when you like" (VB/LW, 29 September 1913). Apparently when Virginia knew that her sister was coming to visit, she could work herself up into an agitated state full of accusations, fears, and guilt. On 30 September, Leonard noted that Virginia had been very worried in the morning and "confessed" to him something that she had magnified into a source of great guilt (LWD), but he did not record what that was. Ka's chastisement provoked Vanessa herself to come to Dalingridge (for the first time) the next afternoon, 2 October, bringing both Drs. Craig and Belfrage with her (LWD).[26]

On the day after her visit, Vanessa invited Roger Fry and his friend Goldsworthy Lowes Dickinson to Asheham on 12 October. In what seems to be projection, she told Roger that Ka "seems to have been getting very tired with the Woolves. The nurses arent much good at the feeding & they are going to get others." Vanessa was still preoccupied with painting the panels for the Ideal Home Exhibition with Duncan, although she did mention new domestic possibilities. She told Roger, "If I have a baby, which on the whole I incline to more & more, you'll have a good deal more than you know what to do with I daresay."[27]

About the time he received this letter, Roger was attacked by Wyndham Lewis and three other artists, who claimed that Roger had stolen for the Omega the commission for the Ideal Home Exhibition, which had been intended for them. Roger simply ignored the attack and carried out his plans to paint with Henri Doucet in France. Then, when the disaffected artists sent out a circular to all Omega shareholders accusing Roger of fraud, Vanessa came up to London to assess the situation and work to restore the good names of Roger and the Omega.[28]

On 13 October, Vanessa wrote to Roger that Lewis had told Clive he created the fuss only because "he had his way to make etc!" Therefore, "Lewis was very disappointed that you had not rushed back from France at once! What they would really like would be an action for libel" (VB/RF, 13 October 1913). For the rest of October, Vanessa's letters focus on this contretemps and on learning how to make pottery.

Leonard visited at Asheham on 19 October (LWD), and Vanessa wrote to Roger that evening about both the wrangle with Lewis and news of Virginia, who "seems to be going on quite well though theres still occasionally great difficulty about food" (VB/RF, 19 October 1913). Otherwise, Virginia's condition seems to have slipped from Vanessa's consciousness (for example, see VBL, 150–51). Vanessa plunged energetically into Omega activities, so depressive lethargy was not what kept her away from Virginia; probably self-protective instincts did. On 10 November, Leonard went up to London and saw Ka and Vanessa. Apparently at Leonard's insistence, Vanessa came to Dalingridge that night and, after his visit, returned with Dr. Belfrage to London the next day, when Virginia was "very restless & excited all day" (LWD).

While she was recovering, Virginia needed to leave Dalingridge for her own peaceful country place. Accordingly, Vanessa vacated Asheham, returning to 46 Gordon Square, which she still shared with Clive, and on 13 November, Leonard took Virginia with two nurses to Asheham. While Ka stayed with them for a few days, Vanessa helped by advertising for another nurse. Except for two afternoons (15 September and 2 October) and one overnight visit (10 November), she seems not to have come near Virginia since the day after the suicide attempt. Vanessa's two small children needed her attention, of course, but since she often left them with their nurses while traveling, they could not be her excuse for avoiding Virginia. Leonard must have asked Vanessa to visit more often, but she rationalized her absence, telling him that she "really did do more harm than good at Dalingridge." Her presence "agitated" Virginia, who probably delved up those accusations voiced earlier at the Cock Tavern in a disturbed manic pattern, only to feel guilt in a depressive pattern. As Ka wrote to James Strachey, "Vs been rather bad this week." To Leonard, Vanessa defended her absence by explaining that her doctor advised against her seeing Virginia.[29] Virginia had demonized Vanessa, and Vanessa's withdrawal seemed to justify the demonizing.[30]

With Virginia's psychological survival still in question, Leonard slept poorly and wrote only a few journal articles. But he had a second completed novel to dispose of somehow, and so he left it up to others to decide what to do with it. In November 1913, Bella Woolf wrote again about *The Wise Virgins*, worried that Leonard had only changed names rather than eliminating offending depictions. Bella understood that Ka Cox thought her objections were foolish and that Virginia thought this a better novel than *The Village in the Jungle* (LWL, 195). Leonard's mother claimed that he had ridiculed the Woolf family and friends, urged him to alter these characterizations, and threatened a "serious break" if he did not. She observed that the Lawrence (Stephen) sisters' height and beauty were not really to their credit (LWL, 196–97), whereas Bella found the sisters "beautiful shadowy beings whose omniscience we must rather take for granted."[31] She and her mother thought that the book was too flattering to the Stephens and too insulting to the Woolfs. Exasperated, Leonard referred to Lytton Strachey the question of publishing *The Wise Virgins* and thus risking alienating his family.

Virginia's state of mind during her recuperation must be determined mostly from secondary sources, as she did not write, except to Leonard when he was away. When Leonard went to London to remove their belongings from their rooms in Clifford's Inn (now occupied by Ka), Virginia wrote him a revealing letter whose themes were love and guilt over "wasting" his life. She insisted, "Dearest Mongoose, I wish you would believe how much I am grateful and repentant. You have made me so happy" (VWL, 2:35).

Even though Leonard assured her again and again that she had nothing to regret, he sensed in all her breakdowns "some strange, irrational sense of guilt." Virginia invented her own explanations, that these breakdowns were "due to her own fault—laziness, inanition, gluttony."[32] Nevertheless, as Leonard wrote to Lytton, "her courage, as it has been all through, is amazing in facing it all" (LWL, 191). Virginia's courage was especially remarkable in the face of the inexplicability of her illness. In normal times, she must have wondered where her "vile imaginations" and fears came from.

Vanessa wondered whether housekeeping would be a "wholesome distraction" for Virginia or would create "more worries." She helped by trying to find the Woolfs a cook, by sending Virginia needlework patterns, and by advising Leonard to get Virginia somehow to spend large parts of her days "dawdling." Although it made her "rather melancholy" to give up their shared residency of the lovely Asheham, Vanessa conceded that her sacrifice was necessary for Virginia's recovery. In a rare moment of self-awareness, on Boxing Day (December 26) 1913, she remarked that she hoped her letters to Virginia "did no harm" (VB/LW, 10 and 26 December 1913).

Vanessa's habit of ridicule, which certainly did harm Virginia, was not limited to her contemporaries. Ka reported that Vanessa and Adrian had gone to the Woolfs' former London apartment to get Leslie Stephen's *Mausoleum Book.* Ka assumed that they wanted "to get it typed," with an eye to publishing it (KC/LW, ? December 1913). Actually, however, they used the lugubrious book to cheer themselves up by laughing at their father's mawkishness and by making comedy out of the various tragedies that had stalked the Stephen family. The sentimentality and denial of *The Mausoleum Book* were put into relief by another piece of autobiographical writing. Toward the end of 1913, Leonard gave Vanessa part of his satiric and self-critical *Wise Virgins.*

On the last day of 1913, which Leonard spent in London, Virginia wrote him a letter that he would have received, perhaps at Gordon Square, before returning to Sussex. Saying that she loved him more and more and found him both beautiful and indispensable, Virginia also offered him, in beast terms, what seems to be a sexual invitation: The Mandrill "wishes me to inform you delicately that her flanks and rump are now in finest plumage, and [she] invites you to an exhibition" (VWL, 2:35).[33] Leonard returned to Asheham to record that Virginia had spent a quiet day and did spend a "Good n" with him there (LWD).

After a year of "vile imaginations" so horrid that death had once seemed preferable to life, Virginia ended 1913 in peace and comparative health. Her great work of the imagination that was to reform the novel, however, had been read only by Leonard, the reader for Duckworth, and maybe Gerald Duckworth himself. But her husband had finished a second novel, and Virginia's health was still not stable enough for her to begin writing her version of their courtship in her own second novel.

6
Seas of Horror

1914–1915

Early in the new year, Ka Cox came to stay with Virginia, and Leonard took a break to visit Lytton Strachey. Leonard had sent Lytton *The Wise Virgins* nearly a month before; now Lytton gave it his imprimatur but advised a six-month wait (LWL, 197, note). Meanwhile, Vanessa had read the part (presumably the Bloomsbury section) Leonard had given to her. She warned Leonard about upsetting people "if they think that criticism of them is intended" but concluded that "feelings, after all, arent very important" (VB/LW, 14 January 1914). Apparently she was thinking not of herself or Virginia but of Clive. The usually truthful Vanessa hedged, however, by not telling Clive about Arthur, the character based on him.

Vanessa told Leonard that Clive thought it was "absurd of anyone to object," leaving Clive with the opinion that only Leonard's family might have reason to object (VB/LW, 14 January 1914). On the same day that Vanessa wrote that letter, she also wrote to Duncan Grant (and did not tell Leonard) that Adrian "amused us very much about Woolf's novel. He is furious about it." Adrian intended to send Leonard a bill for £70, presumably for damages (VBL, 154). While Leonard, unaware of this controversy, tried to make up his mind whether or not to publish, Virginia strove to secure her stability.

Vanessa's way of dealing with adults who needed her was to "curb" their show of fondness. This policy wounded both Virginia and Roger, the two adults who most needed her love. In June 1913, Roger had already begun to fear that he was losing Vanessa. He had sent her a paper entitled "Why I am happy & why I am unhappy." He was happy because he loved and was unhappy (like Virginia) because "our loves are not equal" (CP). By 1914, as Vanessa came to need Roger less, his adoration became more and more a burden. No longer did she consider bearing his child; his vitality and articulateness exhausted her.

Meanwhile, Duncan Grant had been "making mischief" between Roger and Vanessa. Duncan was not threatening precisely because he was not literary, but his egoistic, amoral absorption in himself and his art was refreshing to Vanessa (Garnett, *Deceived with Kindness*, 103, 35). Now seeing the two of them as kindred spirits, she set about breaking off her affair with Roger. As she told Clive, she tried to cheer him "up about life & art without being too kind" (VB/CB, 14 May 1914). She coldly explained to Roger that if seeing her when she no longer loved him gave him pain, the rational solution was not to see her. Vanessa had now cut her ties with all her 22 Hyde Park Gate connections except Virginia and Adrian. Speaking of their aunt Mary Fisher, she had written to Virginia, "It isn't worthwhile seeing these old hags."[1] Perhaps Virginia now worried that Vanessa also thought it was no longer worthwhile to see her.

Rest, isolation from emotional upset, the devotion of Leonard and friends, and her own courage were enabling Virginia's recovery. In early January 1914, Ka again stayed with her at Asheham while Leonard went to London. When she got back to London, Ka reported to Vanessa, as Vanessa relayed to Leonard, that Virginia was "ever so much better" and that "it's only at intervals one would know there was anything the matter." Vanessa left for Paris and on her return told Leonard how sorry she was that he had so much expense to face.[2]

After not seeing Virginia since the November visit at Dalingridge, Vanessa was coerced into a visit that, she told Clive, "I rather dread." (Had she seen them, such letters would have convinced Virginia that she had not invented her sister's enmity.) Vanessa contrived to make the dreaded visit as brief as possible, coming late on Thursday, 29 January, when, as she wrote to Leonard, she had to return to London in time for a Friday evening dinner party.[3] Vanessa arrived at Asheham laden with enough Omega crafts to keep the conversation away from explosive emotional topics (VWL, 2:39), but art was nonetheless an emotional topic for Virginia. Leonard recorded that with Vanessa there, Virginia had "Not v g n" (LWD).

Leonard still had to decide the fate of his novel but could get nothing like a consensus from friends and relations. Janet Case, who stayed with Virginia while he was away on two or three overnight trips to London, was mildly critical, but Leonard's publisher thought that it offered confirmation of Leonard's powers as a writer (LWL, 198, 199, note). Exhausted and irritated, Leonard vowed never to "write another book after these damned Virgins" (LWL, 197), argued with his publisher about objectionable passages and then altered them, and resubmitted the book. In February 1914, *The Wise Virgins* was accepted.

According to Virginia, "the question is what L's. family will think" (VWL, 2:39). But there is no evidence, other than Bella's assumption that Virginia had approved the book, that Virginia had read it. If she had, she might have remembered the portrayal of the Olympian Aspasia and worried. At last, Virginia could write a few brief letters to people other than Leonard, but she did not write to Vanessa. Except for the affectionate ones to Leonard, her letters are mainly expressions of information or thanks, but they do show her very much aware of the literary successes of Leonard, Lytton, and Clive and the artistic activities of Vanessa. Although Virginia does not mention her own writing, she volunteered to do typing for Lytton and Violet.[4] Vanessa had warned Virginia about becoming "incapable" of using her brain again. As she recovered from her 1913 illness and suicide attempt, the fear that she had ruined her mind for writing nagged at Virginia.

Because Leonard was exhausted from constant worry and care for Virginia, in late February, Janet Case proposed that she and Ka Cox visit and give him a much needed fortnight away. Leonard accepted their offer and went to visit Lytton Strachey and then his sister. One of Janet's letters to Leonard suggests Vanessa's destabilizing influence: When Vanessa came to see Virginia one afternoon, Janet went out for a while. Without alarming Leonard, Janet told him that she left Virginia looking "very well & happy looking," but that when she returned, something was "not quite so good" (JC/LW, 25 February and 10 March 1914).

Aware that caring for her had strained Leonard's health, Virginia adopted a nurturing role in her letters, telling him that if he could have seen her sadness when he left, he "would have had no doubts about my affection" (VWL, 2:40). Her tender letters almost made Leonard cry from being "depressed & wanting you so much today." He responded, "You cant realize how utterly you would end my life for me if you had taken that sleeping mixture successfully or if you ever dismissed me" (LWL, 205). Virginia promised, "My pet, you would never doubt my caring for you if you saw me wanting to kiss you, and nuzzle you in my arms. After all, we shall have a happy life together now, wont we?" (VWL, 2:44). Her letter sent him "to bed as happy as I can be not having the great Brute in person" (LWL, 207). She replied, "I love your little ribby body, my pet" (VWL, 2:45). She told Janet guiltily, like a naughty child, "In fact I think if only I can behave now, he will soon be quite right" (VWL, 2:45).

One rather odd reference in these love letters may give a clue to the basis for the Woolfs' new harmony. Leonard suggested that Virginia read a book that included a reference to Leslie Stephen and his brother, Fitzjames. Leonard told her that the author described them as men who were "always criticising" but never creating (LWL, 204).[5]

Leonard read Freud in May and wrote "the first article in a non-medical journal on Freud's theories" (LWL, 274).[6] He may already have known something about Freud's theories of the negative effects of psychological agendas imported from childhood. He himself had ceased writing gossipy criticisms as he had toughened into a colonial administrator (LWL, xiii). And in *The Wise Virgins*, Leonard exposes the destructive results of both intellectual backbiting and Harry's judgmental ways. Perhaps he was encouraging Virginia to recognize the damaging effects (on abused and abuser as well) of the Stephen family habit of always criticizing others.

Ka's letters to Leonard, like Janet's, offer insight into Virginia's condition while he was away: Although Virginia was able to engage in such activities as croquet games on the lawn at Asheham, Ka knew that they still had to be "very careful for a bit," for the nurse suspected that Virginia was trying to run away. Ka considered Leonard a "fine creature" and was sure "any other husband would have done something either selfish or stupid." She also told him that Virginia was being better about food and had confessed to Ka that "it was the only discipline she'd ever had in her life" (KC/LW, 8 March 1914).

Force and discipline, of course, locate an authority outside the self and threaten one's autonomy. Given Virginia's intellectual brilliance, such impositions from outside were especially humiliating for her. One letter from Janet to Leonard shows how much Virginia resented any authority:

Virginia "really doesn't mind us in ourselves only the feeling that we may be being put above for her—which we aren't" (JC/LW, 17 March 1914). During Virginia's rehabilitation, Leonard necessarily had to make decisions for her, but he was eager to limit his paternalistic authority in exchange for Virginia's accepting more responsibility for her health (LWL, 199).

Leonard must have spoken to Vanessa about this plan in January, for she hoped the new "system" would not give him "too many alarms." The doctor agreed that it was time to rely on Virginia's self-control, but he stressed the "enormous importance of ordering her life in the most careful and thorough fashion." Meals were to be punctual, and Virginia was to be in bed for ten out of twenty-four hours (VB/LW, 14 January 1914; Belfrage/LW, 5 February 1914). No doubt she was grateful for even a limited system that permitted her to begin monitoring her own health. Dr. Craig stressed that Virginia could not be left alone, but, Leonard pointed out, "it naturally depresses her to feel that I get people to come here when I want to go off" (LWL, 209).

In 1913, the publisher Chatto & Windus had approached Roger Fry and asked him to write a book on art. Because he was busy with founding the Omega, Roger generously passed on the offer to Clive. Bell's *Art*, published in 1914, was largely an oversimplification of Fry's work. While Leonard was away, Ka and Virginia read it. Ka thought it could be compressed into an article (KC/LW, 13 March 1914), and Virginia told Leonard that she read it "very laboriously, so as to be able to argue with you" (VWL, 2:40). Roger wrote a restrained review of this "most readable of abstract treatises," faulting Clive for arguing in a circle. He also made an analogy that must have intrigued Virginia: "I feel confident that great poetry arouses aesthetic emotions of a similar kind to painting and architecture." He found the human emotions even in *King Lear* to be "accessory, and not the fundamental and essential qualities of these works."[7] Virginia felt sorry for "poor old Roger" in his efforts to have Clive taken seriously (VWL, 11, 41). Leonard said that Clive's book was "condemned" by himself and Lytton (LWL, 203), and they argued "about books & Vanessa & Clive & Roger & Adrian," not about "Mistress Mandril" (LWL, 205).

Virginia's own reactions to *Art* suggest how shaky her aesthetic self-confidence was in these days of returning health. Rather than argue with Leonard's condemnation of it, she modified her first favorable opinion. She told Clive that she disagreed about "a great many things," but she did not specify them. Then she undermined her intelligent observation about

Clive's tendency to generalize history by falling back on a myth of aesthetic inferiority (VWL, 2:46).

Back on 30 September 1913, Leonard had noted that Virginia was very worried and "confessed" to him, but he did not record what she confessed. In March 1914, on a sad evening when Vanessa either was with them or was expected, Virginia revealed to Ka the cause of her troubles, and Ka told Leonard.[8] Virginia shared "mostly worries about people despising you & her. She got into the Clive Vanessa business & imagined sins of her own." Ka then described to Leonard what Virginia's day had been with Vanessa there: She had been "fidgety" in the morning, with "a hundred [worries] an hour I notice which shows that you're near the end of it my dear." Presumably Ka meant that by expressing her multiple worries, Virginia could purge herself of them. By the evening, Ka could assure Leonard that Virginia was "happy now" (KC/LW, 18 March 1914). Ka's letter isolates the central provocations for Virginia's bouts with depression.

Virginia did not have to imagine "sins of her own" regarding Clive and Vanessa. Furthermore, she did not know that Clive, while claiming that it was "none of my doing," was spreading the story that Bloomsbury was not sympathetic to the Woolfs, and he predicted a "drifting apart." Virginia did not read Vanessa's letters to Clive, Roger, or Adrian.[9] And she did not know that Vanessa wrote about curbing her attentions, or picking her to pieces, or dreading visiting her. But Virginia could perceive that Clive, Vanessa, and others disliked both her and Leonard.

If Vanessa had indeed teased her the year before when the proofs of *The Voyage Out* arrived, Virginia would have some reason to think that the world jeered at her and that Vanessa did indeed despise her. But in the six months since Virginia's suicide attempt, Vanessa had hardly seen her sister. She did stay with Virginia and Ka at Asheham on 18 March and expected to be "rather irritated" on another visit, but Vanessa rarely stayed alone with Virginia except for an afternoon (VBL, 159, 162). Moreover, when she did visit, Virginia's stability became precarious.

Although for the Woolfs, emotional issues had superseded aesthetic ones, back in London the painters were still very much in the public eye. The Omega Workshop had survived the attack from Wyndham Lewis and was successfully popularizing and domesticating Postimpressionism. Its crafts sold rapidly, and commissions to decorate homes and public rooms were flowing in. As one of the Omega directors and defenders, Vanessa was deeply involved in the project. When Leonard traveled up to London, he went to art exhibitions or to the Omega (LWD) and brought back news of these activities so different from Virginia's guarded convalescence.

On 6 April, both Leonard and Virginia traveled to London, staying with Janet Case. They took Vanessa along on a visit to Dr. Craig the next day. Although he gave "a cheerful account of Virginia," her condition was still precarious. Vanessa reported the news to Roger:

> He said she wasn't in a state where she could possibly be certified & that she couldnt therefore be made to go into a home or have nurses. He thought there was a certain amount of risk of suicide but that it must be run. They are going to Cornwall tomorrow.

Vanessa went on to laugh at what she considered Leonard's inability to take a joke, but the particular joke she had in mind would have been hard for anyone to take: She told Roger that Lytton gave Leonard some of Clive's letters, and "in one of them Clive said that as soon as Virginia was well enough he meant to start an affair with her! According to him it was obviously not meant seriously." Leonard's scolding of Clive promised, Vanessa thought, to "make the difficulties between them more marked than ever, but I must say I think Woolf rather an idiot to get into a state." Teasingly, Vanessa added a whimsical nude sketch (Figure 29), presumably of herself, to her letter (VB/FR, 7 April 1914).

Although Virginia's health was still uncertain, the Woolfs rented Asheham to Maynard Keynes and with the rental income traveled to St. Ives. Filled with anticipation and probably trepidation lest Stephen ghosts still be there or Leonard unresponsive, Virginia took him to Talland House. The Woolfs found the somewhat ramshackle holiday home "wonderfully done up and spick and span" (VWL, 2:48). Thus its "intimate sights & smells," which Vanessa thought "no one but the Stephens will ever really appreciate," had largely been erased (VB/VW, 16 April 1914). Nevertheless, Leonard empathized with the "nostalgic romance" the family felt for this place of "immaculate happiness," and returning to the site of so many joyous childhood memories briefly soothed Virginia's anxieties (LWA, 2:118–19).

Roger Fry had been more closely involved than Virginia probably knew with placating the authorities after her suicide attempt and advising Leonard about her health. Furthermore, he seems to have been waiting for her full recovery to extend this brotherly role into renewed artistic stewardship as well. Meanwhile, Vanessa's former dependence on Roger's stewardship now seemed self-defeating to her.

During a May 1914 bicycling trip around France, Vanessa assured Clive that Roger slept in a "neighbouring chamber" and that although bicycling was "a little hard on the cunt & on the muscles of my soft legs," little else was. Her letters to Clive dissociate her affections from Roger: "I cant give the account you wished for of the Great Man's conversation but you can imagine our topics. Art. Gothic Art. Omeganic Art. My Art. Roger's Art. Duncan's Art. Art of the Theatre, etc." Her sketch of herself and Roger bicycling would "convey more eloquently than words the spirit of our tour" (VBL, 164–65). Within a week, Roger was cycling alone, and Duncan and Vanessa were together, although as she wrote to Clive three months later, they were "as chaste as angels."[10] Despite Vanessa's weariness with not only bicycling but also talking about "Art," her sense of kindred spiritedness was developing into love for the singularly charming and unashamedly irresponsible Duncan, a painter, not a talker, and a homosexual at that.

The Woolfs spent the summer of 1914 at Asheham. Although Leonard asked Vanessa to visit, no accounts of her visit from 6 to 8 June seem to have survived (LW/JC, 21 May 1914; LWD). Under Leonard's care, the garden was beginning to flourish, and Virginia took considerable pleasure in it. She and Leonard made increasingly long walks over the downs, virtually treeless (thanks to the limestone soil) humpback hills rising like huge whales out of the undulating valleys. Often Virginia would take the train along the valley into a nearby village, and Leonard would walk or bicycle and join her. Leonard "bought V a bicycle," and they began occasionally to bicycle together to Seaford to swim (LWD). But despite such wholesome activities, Virginia still had not recovered completely. The "fixity of ideas about food" was still a problem, and Dr. Craig told Leonard that "she would go to pieces very rapidly if she were left to have her own ways." She could not be left alone but did not want anyone new to stay with her (VB/RF, 7 April 1914; LW/JC, 5 and 25 June and 21 May 1914).

Again Virginia tried her hand at needlepoint, but her distracted state of mind so interfered that Vanessa wondered whether it were "best to take no notice" of the "pieces of work" that Virginia gave her for her birthday (VB/LW, 14 June 1914). Apparently Vanessa's commitment to aesthetic honesty left no space for sympathy with Virginia's need—only nine months beyond a major breakdown and suicide attempt—for a word of thanks, if not a compliment.

When Leonard visited Bella in June, only the servants seem to have stayed with Virginia. Leonard wrote a contract for "Mandril Sarcophagus Felicissima Varrassima," which grew out of the new "system" of self-monitoring. With a flourish, Virginia signed the agreement, holding her to such commitments as a 10:25 P.M. bedtime (LWP, 16 June 1914). With his Mongoose in service to her great Brute, Virginia no longer feared that love would make her helpless and dependent. And with Virginia's health seeming to improve, Leonard had more time to devote to journalism and the Cooperative movement, on which he had become a self-taught authority.[11]

On 20 July, Leonard traveled to Keswick to open the discussion of the Cooperative movement at a Fabian conference. The night before, Leonard had recorded, "G d b n [good day bad night] Gave veronal 2 a m," but he did not make another contract for Virginia before leaving. From Keswick, he traveled to London with the then little-known Walter Lippmann. The two immediately established an empathetic relationship, talking about "Freud, psychoanalysis, and insanity" (LWA, 2:120).

There must have been a startling irony between this objective discussion and the (heretofore unnoted) near-tragic events at Asheham in his absence. After he left, Virginia took about twenty-five grains of veronal. With servants puttering about below and wood pigeons crooning in the trees, she lay down in one of the rooms overlooking the Ouse valley and went to sleep and slept and slept (see also Y, 469). When Leonard returned on 22 July, he found "V not v well owing prob to overdose veronal" (LWD). He must have summoned first Ka and then Vanessa, who arrived a week later and wrote the following accounts, first to Clive:

> Virginia seems very well & cheerful but the most extraordinary thing happened last week—I cant tell all the details in a letter but she took veronal by mistake for phenacetin—25 grains about. Leonard was just going away & went without knowing & the servants being complete fools, though they thought it rather odd that she should sleep for about 36 hours, took no steps. Eventually she slept it off & is none the worse but it seems almost incredible.

To Roger, Vanessa wrote of the event as a "most extraordinary adventure" in which Virginia took "another overdose of veronal by mistake for phenacetin!" She admitted, "Its too odd a story but there's no doubt it was quite an accident. She seems on the whole to be very well" (VB/CB, 29 June 1914; VB/RF, 29 June 1914). It is hard to imagine Virginia's taking twenty-five grains of any drug by accident.[12] Indeed, this overdose may bear some relation to her 1904 leap from one of Violet's windows: It was not serious enough to kill her but was serious enough to sound an alarm.

No wonder Leonard remembered that "in many ways 1914 and 1915 were years which we simply lost out of our lives, for we lived them in the atmosphere of catastrophe or impending catastrophe" (LWA, 2:119). Vanessa seems to have permitted herself little acknowledgment of potential emotional catastrophes. Her repressed feelings, however, erupted in a dream that transposed action for inaction. On the night after writing about Virginia's "extraordinary adventure" with veronal, Vanessa dreamed that Virginia "was ill & I looking after her." Reading old letters must have made her think of Virginia with admiration and empathy, but she did not tell Virginia about her dream (VB/CB, 30 July 1914). In fact, rather than "looking after" Virginia, Vanessa—arriving at Asheham on a Thursday—wanted to avoid a protracted visit and "get away before Saturday." The Woolfs expected her to stay on, however, and she could not "pretend there is any real reason why I shouldn't" (VB/RF, 30 July 1914). Vanessa's restless irritation at being expected to stay longer could not have been lost on Virginia.

Postimpressionism had seemed to symbolize a new era of enlightenment in the arts and human relations as well; Leonard had thought that "human beings might really be on the brink of becoming civilised." But that summer, the outbreak of World War I dashed such hopes (LWA, 2:20–21). Roger saw the war as confirmation of his worst suspicions: "I've known since Helen [that is, since his wife's mental illness] that the world was made for the worst conceivable horrors" (RFL, 2:301). Nevertheless, he attended the Neutrality League meeting in August. Leonard felt that the enemies of liberty and equality had caused the war and thereby had "postponed" the advent of true civilization "for at least a hundred years" (LWA, 2:21). He was enlisted by Sidney and Beatrice Webb for a "huge job" drafting a plan for preventing war (VWL, 2:53).

In August 1914, the Woolfs took a long holiday in the north, and Vanessa moved back into Asheham, where, she bragged to Lytton, she and Duncan could "think of other things [than the war], such as art and fucking." To Clive, however, Vanessa told a rather different story: He was welcome to join them and would "interrupt no love-making. Only Duncan is here & he as you know is impervious to my charms" (VB/LS, 28 July 1914; VB/CB, ? September 1914). Virginia's one letter to Ka Cox exclaiming about the beauty of the North Country and the sophistication of people who talked about "Thomson's poetry, and postimpressionism" sounds exuberantly healthy, but she seems to have been inconsistently so (VWL, 2:51). Leonard was still recording her health nightly and worrying that Virginia wanted to move back into Bloomsbury; his distrust of Bloomsbury made Vanessa wish that "Woolf didn't irritate me so." When the Woolfs returned to London on 15 September, both Vanessa and Roger visited Virginia. Vanessa reported that when she was there, Virginia "was very difficult & kept saying she would not do a rest cure. I suppose she will go on fighting all the time as she did before. I only hope Leonard will stand it as its the most awful strain." A year after attempting suicide, Virginia's health was still in peril.[13]

Seeking to ease Leonard's worries, Virginia began hunting for a house, not in Bloomsbury, but in Richmond—some thirty minutes from the heart of London—where they found an apartment. They argued with Vanessa over the lease to Asheham, which they wanted her to reassume, but she was angry about the "terrible shape" in which they had left the house. Indeed, Sydney Waterlow rented Asheham in December 1914 for several months and found it necessary to repair the roof, stove, and window joints; to have a sweep clean the chimney; and to spend weeks scrubbing cupboards. He deducted "a small fraction" of his costs from the rent and wrote to Leonard he was sorry to hear that Virginia was "aggrieved" at his doing so but that his wife was "aggrieved at her raising the question."[14]

With their domestic negligence now public knowledge, as self-defense, therapy, or an act of despair, Virginia tried her hand at being domestic, going to cooking classes (VWL, 2:55) and making omelets in the evening (LW/Janet Case, 29 December 1914). She read books but wrote few letters, no reviews, and no new fiction. In fact, Virginia may have feared that her brain was damaged and her life as a writer was over. If so, the linchpin of her identity was lost. Rather than sacrifice herself to domesticity, however, Virginia approved the publication of *The Voyage Out* and dedicated it to Leonard, but she was still (in December 1914) profoundly ambivalent about this public display. Although she could acknowledge objectively that *The Voyage Out* was not, after all, "pure gibberish," she felt no joy over its impending publication: "Yes, that damnable book is coming out" (VWL, 2:56).

Despite financial difficulties (the legacy of her illness) and Leonard's sensing the futility of his work (D, 1:22, 23), the year 1915 began auspiciously. If publishing fiction before an unknown audience could threaten self-assurance, Virginia knew of other writing that did not. She surmounted her anxieties as she had with the *Hyde Park Gate News* and her other early writing: At the beginning of 1915, she began a diary and found that "my writing now delights me solely because I love writing & dont, honestly, care a hang what anyone says. What seas of horror one dives through in order to pick up these pearls—however they are worth it" (D, 1:20).

The recognition that she loved writing for itself and not for others' reactions was liberating, and Leonard no longer needed to record the state of her health. The Woolfs' house hunting discovered a handsome eighteenth-century house (then divided in halves, called Hogarth and Suffield) in Richmond, which offered space, elegance, and easy access to London. While her novel was in press, Virginia acknowledged her debt to her literary forebears by writing to thank Thomas Hardy "for his poem about Father, & [for] his works!" (D, 1:22).

Meanwhile, Roger Fry's Omega Workshop printed a book of poetry by the art critic Arthur Clutton-Brock. Roger's daring to go beyond the visual arts into publishing may have reminded Virginia of her early work in printing bookplates and making bookbindings (Willis, *Leonard and Virginia Woolf as Publishers*, 11–12). At any rate, after several "merry and pleasing" birthday treats, she and Leonard vowed to "take Hogarth" (one of the halves of the Richmond house) and buy a printing press and a bulldog (D, 1:28). Being publishers would allow them to take control of their own publications. After some delays, the Woolfs were able to lease Hogarth House, but they did not buy a bulldog or, just then, a printing press.

In early 1915, the Woolfs read what would be the "Cardinal Manning" chapter of Lytton Strachey's *Eminent Victorians*. Virginia found it "quite superb" (VWL, 2:58). Leonard found it "very nearly perfect" and also saw how revisionary both Strachey's form and content were: "It's a great invention, this kind of biography & the material just fits your method" (LWL, 210). Perhaps Lytton's accomplishment heightened Virginia's fear that her first novel, despite all her efforts, was not a "great invention." Her diary, however, shows her more worried about Leonard's depression than her own, for he now doubted the efficacy of his work for a better world.

On the last day of January 1915, with *The Voyage Out* in press, Virginia decided to "start reading" *The Wise Virgins*, which had been published the preceding October. She "read it straight on till bedtime when I finished it." She told her diary, "I was made very happy by reading this: I like the poetic side of L. & it gets a little smothered in Blue-books, & organizations" (D, 1:32).

Leonard apparently intended to expose other characters, including himself, and to leave Camilla untouched, but Virginia may have felt that instead he exposed her untouchableness. Except when Harry says good-bye to her, Camilla's only substantive entrance into the last half of the book is in a letter Harry receives after he has committed himself to marry Gwen. Camilla's letter deals with marriage in terms too easily connected with Virginia: "It's the voyage out that seems to me to matter, the new and wonderful things. . . . I want them all. I want love, too, and I want freedom. I want children even. But I cant give myself; passion leaves me cold" (*Wise Virgins*, 231).

Just as coldness makes Camilla unable to love, Jewishness makes Harry unable to enjoy the artistic and intellectual world that the Lawrence sisters (the "wise virgins") inhabit. Apparently, Leonard's intention was to show that had he and Virginia been unable to overcome their differences, they, like Harry and Camilla, might have ended up living empty, separate lives. Thus in showing how their relationship might have ended but did not, Leonard offered an oblique celebration of their marriage. Virginia read the novel first in those terms when she declared that the poetic side of Leonard made her happy.

Later reflections, however, seem to have undermined that happiness. Reading about Camilla's remoteness in the novel that her husband began on their honeymoon must have renewed Virginia's fear of emotional and aesthetic inadequacy and her sense that another beloved had abused her, this time publicly.

Public exposure of her inadequacies (in her book and his) resurrected the hobgoblins that had bedeviled Virginia since her mother's death. On 17 February, eighteen days after reading *The Wise Virgins*, a month before the publication of *The Voyage Out*, and the day the Woolfs looked into buying a printing press (LWD), another "sea of horror" began to envelope her. Virginia's diary and their plans for a printing press were abandoned, and Leonard summoned Vanessa, who stayed with Virginia on the afternoon of 18 February while Leonard went into London to consult Dr. Craig. When he came back, he found Virginia ill with a bad headache. She had a restless night, and he gave her veronal to help her sleep.

By 22 February, Virginia was cheerful enough to write to Margaret Llewelyn Davies about their excitement over "our Printing Press" and to joke about saving Leonard from the pressure always to do good works (VWL, 2:59). But on 23 February, a manic bout of incoherent babbling took possession of her (VWL, 2:60, note). This may be the time that Leonard described:

> One morning she was having breakfast in bed and I was talking to her when without warning she became violently excited and distressed. She thought her mother was in the room and began to talk to her. It was the beginning of the terrifying second stage of her mental breakdown. (LWA, 123)

Two days later, however, Virginia was well enough to write to Margaret about her recovery and to thank her for saving Leonard by occupying him with good works while she was ill in 1913 and 1914. Virginia also told Margaret that although she hardly knew her, "I felt you had a grasp on me, and I could not utterly sink" (VWL, 2:60–61; MLD/LW, n.d.). In addition, she wrote a manic letter (which Nigel Nicolson calls "malicious" [VWL, 2:xvi]) to Lytton Strachey, proposing to buy a parrot for Clive "trained of course to talk nothing but filth, and to indulge in obscene caresses" (VWL, 2:61). Then Virginia retired to her bed, from which she dictated two letters to Margaret, but this time, rest could not ward off the upcoming trial.

The previous month's harmony disintegrated. Virginia often became incoherent, hysterical, and even violent. By 4 March 1915, her condition was bad enough to require nurses to move in (LWD). On 6 March, Vanessa visited her sister, and on 25 March, the day before the publication of *The Voyage Out*, Virginia was hospitalized. This was the only time after her 1913 suicide attempt that Leonard ever had her admitted, and he did so only during the week while he moved them to Hogarth House. As soon as he had settled in, Virginia was released, with four nurses to care for her. Leonard's diary entries again record how much Virginia slept, whether or not she needed medicine for sleep, when she was excited, and who visited. Ka Cox and Janet Case again stayed with Virginia when Leonard had to be away. Margaret Llewelyn Davies visited her frequently, but Vanessa did not come.

In early April, Vanessa went to stay at Eleanor House in West Wittering, with Duncan Grant, her children, and David Garnett (Bunny), with whom Duncan was having an affair. Her one letter to Virginia from there offered a blend of sympathy and reproach: "I was afraid you werent very well when I did see you but I didn't realize that you were going through any horrors. You poor little beast. Why didn't you tell me?" (VB/VW, n.d. [April or May 1915]). When Vanessa visited, Virginia had been in the care of nurses and obviously did not need to "tell" her sister she was not well.

Meanwhile, Roger Fry found it incredible that Vanessa so loved Duncan while Duncan was so attached to Bunny that she would include Bunny in her household in order to establish a semistable relationship with Duncan. Roger thought that Vanessa was humiliating herself, and he was furious that she imagined his now hopeless love for her akin to her hopeless love for Duncan (RFL, 2:383). He tried to escape these imbroglios by joining the Quaker Relief Fund in France and by painting the countryside there. But the war seemed pointless, and peace impossible.[15]

The war killed Bloomsbury friends as well as ideals. Rupert Brooke died before seeing battle, and Roger's friend Henri Doucet was killed in action.[16] Reading *The Voyage Out* in the midst of the war's devastation seems to have renewed Roger's faith that art could survive and preserve humanity. He wrote to Vanessa that *The Voyage Out* proved Virginia's genius. Vanessa thought the novel to be "extraordinarily brilliant" and reflected that novel writing was "art of quite a different sort from making a picture." Then she compared Virginia unfavorably with Jane Austen, but Roger seems to have reminded her that Austen did not write the only model for a novel. Feeling threatened, Vanessa supposed that "it's true that she has genius. I'm not sure I should call it so. Can Stephens produce a genius? I doubt it."[17]

Vanessa regularly received progress reports from Leonard of Virginia's manic condition and passed on the news to Roger and Clive. She told Clive that Virginia had attacked a nurse and was angry with Leonard. He, in turn, was afraid of "neighbours complaining of the noise" (VB/CB, 21 April 1915). Apparently Virginia, who had resented growing up in an environment in which one could never say what one thought, found release by screaming imprecations at anyone who disciplined, avoided, or (she thought) despised her.[18] Virginia's paranoiac ravings in 1904 and 1915 seem to have been exaggerations of the mood that the young Ginia had described as "furious and tantrumical" (PA, 69).[19] Her accusations partly explain why Vanessa kept her distance, but Vanessa's distance partly explains why Virginia was angry. Vanessa did not write to Virginia and, after 6 March, did not visit for nearly two months.

Margaret Llewelyn Davies accompanied Leonard to consult Dr. Craig on 28 April about Virginia's health, and Leonard dutifully sent Vanessa the news (LWD). She then wrote to Clive that she had had "another most dismal letter from poor Woolf. Things seem to be worse than ever as she has been sleeping very badly. It looks to me about as bad as it could be, & I think Woolf rather wants to see me." But then she changed the subject to her plans to stay on at Eleanor. She planned to go to London in "a few days to see Woolf," not Virginia. She told Roger that even though Virginia was "worse again," Leonard was determined not to institutionalize her (VB/CB, 29 April 1915; VB/RF, 30 April 1915). Keeping Virginia at home was an enormous emotional and financial drain on Leonard, but he nonetheless insisted on doing so.[20]

A new doctor named MacKenzie visited Virginia on 1 May 1915. She had had no sleep the night before, despite drugs, and was "too bad to realize him [the doctor] at all" (LWL, 212, 213). On 2 May, Vanessa told Roger, "The news of Virginia is a little better, but she still sounds pretty bad" (VBL, 177). On 4 May, Vanessa traveled to London and then to Richmond to visit Virginia for the first time since 6 March. Her effect on her sister was dramatic: Virginia had been "quieter & clearer" during the day, but after Vanessa came to tea, she entered a manic phase. Then on 5 May, she was "talking most of day." Leonard met Vanessa in London to see Drs. MacKenzie and Craig. On 6 May, Virginia lay out in the garden and took no tranquilizers, and on 7 and 8 May, she was quiet but incoherent from time to time. Finally, on 8 May, her mind was clear for a time, but then she became excited and angry. She also refused to eat and could not sleep (LWD). Vanessa did not return to Hogarth House.

On 10 May, Leonard wrote to Violet Dickinson that Virginia had "stopped the incoherent talking to a great extent. She is also sleeping splendidly without drugs." But later that day, Leonard noted that Virginia was "A w me" (angry with me). His laconic style allowed no elaboration, nor did Vanessa, when referring to Virginia's anger, suggest a reason. Circumstantial evidence suggests that Virginia might have been angry because Leonard refused to give her sleeping pills and sedatives. In this period, Leonard last recorded administering a sleeping pill on 30 April, the day before Dr. MacKenzie visited.[21] Also on 10 May, Leonard decided against further consultation with Dr. MacKenzie (LWL, 213). He gave Virginia a light tranquilizer with food on 11 May. Then, after more than two weeks without major drugs, on 16 May he capitulated and gave Virginia a sleeping pill, but to no avail. By 19 May, she was "exc to vio" (excited to violent) and remained so off and on for four days. On 20 and 21 May, Leonard did not see her at all (LWD).

While Virginia's very self seemed to be in a state of dissolution,

Vanessa remained at Eleanor to paint with Duncan and to make a home for him, her children, and Bunny.[22] In her two months there, she seems to have written only one letter to Virginia and one to Leonard (or he kept only these). In the letter to Leonard, Vanessa hoped that "she's still going on getting sleep" and then expressed her regret for the troubles Leonard was having with the nurses. In the rest of her long, self-absorbed letter, she spoke of how sorry she was that country living was not a permanent solution to life and that her painting kept being interrupted in London. Vanessa never even mentioned Virginia's name (VB/LW, 16 May 1915).

Vanessa wrote to Roger that she had heard from Leonard "very bad news of Virginia":

> She has been very violent again & has attacked one of the nurses of whom they now have 3. The consequence is they are getting upset & say they cant manage her & that she ought to go to an asylum. The doctor thinks so too & Woolf is afraid of being driven to it. He himself can do nothing as she is very angry with him now. It looks to me as if it might be inevitable & I know he will be very much upset. (VB/RF, 21 May 1915)

However upset he may have been, Leonard still refused to send Virginia to an asylum, however inevitable that outcome seemed to others.

Meanwhile, Roger had remained in France to assuage his pain over being exiled "outside the circle" of Vanessa and Duncan (RFL, 2:385). But after being arrested as a spy, he retreated from the war's absurdities. In June, he was visiting and painting with Simon and Dorothy Bussy in the south of France. On 24 May, Vanessa wrote to Roger that she was going to London to see Leonard, and on 26 May, Leonard went to Gordon Square to confer with her.[23]

Vanessa wrote to Roger that Leonard was "dismal" because "Virginia seems to go up and down at times being pretty reasonable & at times very violent & difficult." Leonard hoped to "hang on as long as possible . . . in the hope that she may get well enough to be able to go to some nursing home & not have to go to an asylum which he thinks might have a disastrous effect on her" (VB/RF, 27 May 1915). After passing on Leonard's news to Roger, Vanessa again retreated to Eleanor (VB/RF, 30 May 1915).

By 9 June, Vanessa reported to Roger that she had heard that Virginia was "quieter and more rational. She's still very angry with Leonard which I think depresses him very much." In another letter, which does not mention Virginia, Vanessa told Roger that she herself was ill and got "quite limp & exhausted very easily."[24] In mid-June, as Leonard later wrote to Violet Dickinson, Virginia "suddenly turned very violently against" him. Leonard went to a Woman's Cooperative Guild meeting in Liverpool on 14 June, but whether Virginia was angry with him for leaving or he left because she was too angry to see him is unclear.[25] When he returned, Virginia had "n g d & n g n" (not a good day and not a good night) (LWD).

When Ka Cox, Adrian Stephen, and his wife, Karin, visited on 22 June, Virginia was "rather exc w Adr."[26] (LWD). Ka reported, as Vanessa wrote to Roger, that she thought Virginia was "really getting better slowly." But, Vanessa continued,

> It sounds to me most depressing, as she seems to have changed into a most unpleasant character. She wont see Leonard at all & has taken against all men. She says the most malicious & cutting things she can think of to everyone & they are so clever that they always hurt. But what was almost the worst thing to me was a small book of new poems by Frances Cornford which has just come out which Virginia has annotated with what are meant to be stinging sarcasms & illustrations. They are simply like rather nasty schoolboy wit, not even amusing.

Concluding that Virginia had "worn out her brains," Vanessa assumed that Virginia, like J. K. Stephen and Helen Fry, had experienced a permanent dissolution of personality and so, like Helen, would be sent to an asylum (VB/RF, 25 June 1915).

Demons did still possess Virginia, howling messages of hate and fear into her brain. Although during rational periods, the demons were swept out of her mind, paranoid illusions concerning Leonard, Vanessa, the nurses, and food soon regained control. At age thirty-three, Virginia seemed to have completely lost touch with her sane healthy self. In 1913, she had been willing to kill herself rather than go to a rest home, but in 1915, the dissolution of her personality seemed to leave only two options for Virginia Woolf—a rest home or an asylum.

At last, however, Leonard's unwavering care, visits from Ka and others who clearly loved her, probably the nurses' vigilance, and Virginia's own sane, brilliant self chastened the demons. Margaret Llewelyn Davies seems to have been especially instrumental. She had sustained Leonard with sympathetic personal letters as well as innumerable letters on political organization, thanking him, for example, for all he had done to help women resist oppression. She and Ka had visited regularly, but lest "the mere sight of me" excite Virginia, Margaret did not stay alone with her.[27]

The many afternoons that Margaret spent sitting with Virginia and a nurse, chatting endlessly about women and politics, however, soothed Virginia until she ceased to be excited by or jealous of her, and they helped her reenvision a world outside herself. At last, not wanting "to talk any more about herself," Virginia asked whether she could give a course on English literature to Margaret's Women's Cooperative Guild. Imagining that her literary knowledge would be of use to poor women, as it had been in 1905, was restorative. Margaret told Leonard that Virginia called the prospect of teaching literature "a nice thing to think of."[28] That prospect marked the beginning of Virginia's reclamation of her self.

After mercilessly shredding her brain for more than three months, the manic period subsided.[29] By late July, Virginia had so reintegrated herself that Leonard no longer made records of her health. Vanessa could visit without having imprecations hurled at her; she came to tea once in July and twice in August. When the worst phase of the attack had passed, she wrote Virginia the first letter in three months, probably only the fourth in all of 1915.[30] She suggested that when she went to London, "perhaps I shall see you." Meanwhile, she promised to write.

Vanessa did not apologize for her withdrawal; she instead cast her rapprochement in aesthetic terms. She acknowledged, despite Virginia's inability to "criticize art in detail," that she was "sensitive to art in general" and hence "ought to be" pleased by pictures (VB/VW, ? July 1915). This attempt at consolation indicates, however, that Vanessa knew that her attacks on Virginia's supposedly limited visual sense had been a source of stress. In turn, Virginia courted her sister by trying to take up her art, copying paintings by Vanessa and Duncan, which she must have destroyed. As Virginia recovered, Vanessa now became the jealous one: "Besides seeing all the world of beauty & intellect now you also get letters from all the wits of our time—but still you havent enough" (VBL, 189, 188). Virginia did not reply.

Virginia's severe 1915 breakdown seems not to have been an isolated event but a culmination of troubles. Seas of horror had ebbed and flowed over the course of five years, as had Virginia's aesthetic and emotional confidence. Then they engulfed her in 1913 to 1915 for various reasons that were first physiological, the root cause being the chemical imbalance associated with manic-depression. More immediate causes were also physiological: Virginia suffered from eye problems and repeated toothaches, so her bad nights were very bad indeed. (For example, she once went sixty hours without sleep [LWL, 212]). Tranquilizers and sleeping pills seemed to help, but of course, they had side effects, of which the medical profession then was unaware, and they created a vicious cycle of dependence.

These physiological causes were linked in another vicious cycle as well. Several psychological factors may have provoked Virginia's hostility to Leonard. That is, his sexual expectations and the decision not to have a baby certainly exacerbated her sense of personal inadequacy. She feared that Vanessa abused her in private, and Leonard's second novel may have seemed to offer evidence that after building her trust, her husband had exposed her in public. Perhaps she worried that her love had once again been betrayed. Furthermore, if she resented Janet and Ka for being "put above" her and hated discipline and authority, Leonard with all his regulations guarding her health must have seemed simply to be a jailer.

Clearly, another major factor behind this breakdown, as Vanessa told Roger, was Virginia's fear regarding the reception of *The Voyage Out*. She had put herself under enormous pressure to reform the novel, but despite her exhaustive (and exhausting) revisions, the book would only, she feared, be ridiculed. Whereas Lytton was succeeding at reforming biography, she believed that she was failing at a comparable reformation. Furthermore, she had projected onto Rachel Vinrace's fatal illness and final hysteria many of her own fears and limitations. Hence, even her own book threatened to expose both her aesthetic and her emotional inadequacies.

Virginia's worst fear, as she told Ka, was that she had offended Vanessa by her flirtation with Clive and that Vanessa now despised both her and Leonard. Bloomsbury's being "down on" the Woolfs, thanks partly to Clive, might have provoked that suspicion, as would Vanessa's distance from her sister. Virginia wrote to Vanessa on 3 August 1914, telling her, "I do adore thee," but her next letter was only a brief note dating from perhaps October 1915. It was not until 5 April 1916 that she again wrote to her sister.

The evidence that Vanessa's distance from her sister was a major provocation for Virginia's disturbed condition has long gone unnoticed. From the beginning, Leonard discouraged criticism of Vanessa, apparently because he felt that it would only continue the Stephens' habit of always criticizing others and would make Virginia's troubles worse.[31] He kept

Vanessa posted and asked her to come when he thought it was necessary, but he did not complain about her neglect, and he discouraged Ka from doing so.

When Violet Dickinson got the "impression that Vanessa [had] washed her hands" of Virginia, however, Leonard actively intervened against such talk. To Violet, he maintained that Vanessa "cares enormously" and defended her absence by saying she was willing to come "whenever any big change had to be considered" (LWL, 212–13). Leonard's restrained comment suggests that although he would not openly criticize her, he too felt that Vanessa had washed her hands of her sister, in much the same way she had of Roger. Roger's liaisons with other women afforded him no "hope of my getting anything of what you gave me" (RF/VB, 9 July 1915). Vanessa wanted to hear no more of such declarations. Her silence made Roger conclude that she was probably incapable of "intentional cruelty," but after her behavior in 1915, he, like Virginia, had only too good a precedent for considering her "capable of any amount of indifference" (RF/VB, 11 May 1916).

Although the precariousness of Vanessa's own health could partially explain her distance from Virginia, her aloofness also indicated to Virginia that her sister now hated her. If her "vile imaginations" had implicated both Vanessa and Leonard, soon Leonard's constant, unwavering care proved his devotion. While she was recovering, he took Virginia (in a wheelchair) almost daily on walks through Kew Gardens and around Richmond. Even though he did not publicly acknowledge its offense, Leonard may have privately acknowledged his indiscretion in *The Wise Virgins*. In any case, he later refused to let it be reprinted.[32]

As she recovered, Virginia and Leonard must have agreed on a private pact, promising never again to betray each other to outsiders. Leonard never wrote another personal exposure of Virginia in fiction or prose. And even though snide remarks were always tempting, Virginia never again sneered at Leonard, as she had in the "penniless Jew" letter to Violet Dickinson or the honeymoon letter to Ka Cox. However emancipatory such criticism had been to the little Stephen girls who first made fun of their parents, after 1915, guilt, suffering, and wider empathy tarnished the pleasure, at least for Virginia.

Taking back the lease of Asheham, the Woolfs spent most of September and all of October 1915 there. Although a nurse still limited her activities, Virginia did "a great deal of reading and wandering about on the downs" (VWL, 2:69–70). Leonard accelerated his activities, traveling to

London to lecture to the Women's Cooperative Guild in October, for example. The Woolfs returned to Hogarth House in November. Before the year's end, Virginia visited Vanessa in London without Leonard, and Vanessa visited them twice at Asheham. She wrote to Roger: "I told her you would like to see her & she would like to see you. so I think you had better suggest something." While Virginia had feared jeers from the public and further teasing from Vanessa, now she talked to her sister "a great deal about her novel & was very anxious to know what you [Roger Fry] thought of it so I thought it safest to say simply that you thought it a work of genius which satisfied her" (VB/RF, ? ? 1915).

Roger did consider *The Voyage Out* a work of genius, but Virginia was afraid that it had not been true to her vision. To reconstitute her whole self, she needed to evict Vanessa from the center of her psyche and to continue the reformation she had begun with her first novel. For both of these enterprises, Roger Fry provided a model. That is, his slow recovery from Vanessa's dismissal demonstrated that one could continue loving Vanessa without being destroyed by her callousness. And his theories insisted that novelists could, like Cubists, make artistic wholes out of fragments of perception.

The Woolfs spent a quiet restorative Christmastime at Asheham, visited by several friends, among them Roger Fry. By then, Virginia no longer resented aesthetic discussions or her friends equating Postimpressionism with up-to-date sophistication. Rather, she was eager to resume the debate that she and Roger had begun in 1911. She wrote to him on Boxing Day, delighted with their talk, confident of her critique of his translations of French poems, and eager to "discuss everything under the sun before we die" (VWL, 2:73). Virginia seems to have suspected that if she could shift the rivalry between their arts into noncompetitive territory and if the brush could continue to teach the pen, such discussions might offer her a way out of the impasses she had faced since 1904.

By the close of 1915, Virginia had recaptured the normal life she then maintained until her death. Despite her bouts with mental illness, she survived to become one of the great modernists, enriched in part by the very seas of horror that had ebbed back and forth in her consciousness for nearly five years.

7

Aesthetic Emotions and Others of a Literary Nature

1915–1919

Beginning in late 1915 until her death, Virginia Woolf became her own best psychotherapist. Dreading the demons that threatened her writing self, she learned to defend herself against them. To do this, she learned to examine her moods and gauge their normality, and she set about making herself emotionally and perceptually whole, a daunting task without the aid of modern drugs like lithium. Virginia also knew that her illness had been a terrible burden on Leonard, but she refused to feel guilty.[1] When she later wrote about the creative "rapture" that "put the severed parts together" (MB, 72), Virginia could have been speaking of her psychological as much as her creative process.[2] For manic-depressives who have experienced the frightening split of their own egos, the desire to reclaim their wholeness is paramount. And those who become great artists or leaders typically exhibit tremendous strength of character. Certainly, Virginia Woolf did.

For Virginia, becoming whole meant readjusting the positions of Leonard and Vanessa in her emotional landscape, especially the terms of her relationship with Vanessa.[3] Instead of begging for affection, Virginia kept her distance, and instead of alienating Vanessa with accusations, Virginia tested new ways of charming her with playful ripostes. Finally, instead of perpetuating the battle between their arts, Virginia initiated an artistic rapprochement. She even tried to paint, although she told Nelly Cecil (but not Vanessa) that both painting and embroidery seemed "vapid" after writing (VWL, 2:64).

Like Roger Fry, Virginia believed that the aesthetic theorists to be trusted most were practicing artists.[4] Unlike Clive, who was exclusively a theorist, or Vanessa and Duncan, who were exclusively practitioners, Roger could be trusted because his views integrated theory and practice. Therefore, when she began reintegrating her self as a writer, Virginia sought his company.

After Christmas 1915, at Asheham, Virginia invited Roger to stay at Hogarth House while he was restoring what she called the "Hampton Court frescoes" at Henry VIII's favorite palace (VWL, 2:74). The Mantegnas he was restoring were not frescoes but nine linen hangings as monumental as their subject, "The Triumphs of Caesar." Painted about the time of Columbus's discovery of the New World, by the twentieth century they were as damaged as the ruins of ancient Rome. After receiving a royal commission to attempt to restore them, Fry first stripped off from the first canvas a disastrous overpainting done in William III's time and found Mantegna's original paint dissolved into buttery softness. Then before the war, with a young Omega painter, he had sporadically worked at repainting the canvas.[5] In early 1916, he moved in with the Woolfs and traveled daily the short distance to Hampton Court to try to complete (or perhaps repair) his "restoration."

Back at the handsome but rather staid Hogarth House, Roger renewed his conversations with Virginia about the relationship between literature and painting. He had developed a theory that content is "merely directive of form" and that "aesthetic quality" is a function of "pure form"; thus in intense poetry, content "has no separate value at all" (RFL, 1:362). Just as we do not evaluate a poem according to whether we approve of its "sense," we should not evaluate a painting according to whether we approve of its subject. This argument was more easily applied to poetry than to fiction, but it offered Virginia a further clue to how she might "enclose the whole" in her writing by subordinating content to form.

Still in rather precarious health, Virginia nonetheless was happy to be well enough "to write for a little every day" and to be able to notice shifting artistic values (LWD). When the *Times*'s art critic faulted Vanessa's exhibition at the Omega for being too beautiful, Virginia saw a victory for Postimpressionist values; then she gleefully imagined a "complete rout of post impressionism, chiefly because Roger" had supposedly defected to literature.[6] He lent the Woolfs two of his works on aesthetics (VWL, 2:78, 77–78), and he also "lent" them two servants, Nelly Boxall and Lottie Hope, who remained with the Woolfs for years, offering both support and frustration.[7]

The works that Roger lent to Virginia must have included his argument that art expresses "the imaginative life" rather than copies appearances (VD, 15). Just after Roger's visit, Virginia claimed in *The Voyage Out* that she had subsumed individual events into "a vast tumult" or pattern.[8] But she worried that such a diminution of subject matter might make the novel "too scattered to be intelligible" (VWL, 2:82), and so she began to write journal articles again. One of her few publications in 1916 shows how much she was torn between imitation and expression in art: She praised Ibsen for both representing "the prosaic look of things as they are" and expressing his poetic imagination (VWE, 2:69).

Whereas Virginia had been able to readjust Vanessa's position in her emotions, Roger still could not do so. Lowes Dickinson speculated that Roger got "nothing but pain from Vanessa. I suppose however you still *want* the pain" (GLD/RF, 31 March 1916). Between 1911 and 1915, Leonard, Ka, and Janet had similarly tried to advise Virginia not to become emotionally dependent on Vanessa. Unlike Roger, Virginia transformed self-awareness into self-therapy after her 1915 illness; unlike Roger, she did not write to Vanessa and no longer reproached her; and unlike Roger, she did not develop a "crust" against Vanessa's indifference (RF/VB, 15 June 1916). That is, Virginia waited for Vanessa to make the first move to reestablish their closeness.

In April 1916, when Virginia contracted influenza, Vanessa emerged from her self-absorption to send a peace offering, a "parcel of daffodils," from the country to Hogarth House. In rhapsodic hyperbole, Virginia interpreted the flowers as "an embrace." But if theirs was to redevelop into a balanced relationship, she instructed her "Dearest," "you have to write as well," a complaint Vanessa was hearing from Roger, too.[9] Virginia rewarded Vanessa's "embrace" with three more letters in rapid succession, the first correspondence since August 1914, except for a brief note. She offered gossip, a promise to critique one of Duncan's paintings, and the news that she was to review a book on Ruskin. In an attempt to build artistic bridges between them, Virginia asked her sister to "tell me something about Ruskin and Turner," neither of whom Roger admired.[10]

By this time, Virginia had learned how to avoid replicating the destructive terms of the rivalry between their arts, by complimenting Vanessa on her "very great gift for writing" (VWL, 2:88). At Eastertime, from Asheham—which was now exclusively the Woolfs'—she sent Vanessa another description calculated to secure affection: this time, a riotous account of the gossip she had heard during an excursion back "into the nineties" with Gerald Duckworth and Violet Dickinson. Virginia invited Vanessa to send Bloomsbury gossip in return. She also requested a design for a summer cloak, noting that her new dress made her a walking advertisement for the Omega, and demanded a letter (VWL, 2:91, 92). Virginia's readjusted relationship with Vanessa and with the visual arts took great courage, as they transcended the *paragone* on which so much of her private mythology had been founded.[11] Now the sisters might be able to live near each other without reanimating old destructive patterns, and so Virginia began hunting for a house for Vanessa near Asheham.

Even though her pulse was stronger, Virginia still needed to rest, drink lots of milk, limit her writing and reading, and get plenty of sleep (VWL, 2:89). This regime seemed to be helpful.[12] Leonard's loyalty during Virginia's protracted period of horrors proved that she need not fear losing him. When he had to be away, she wrote to him,

> I lie and think of my precious beast, who does make me more happy every day and instant of my life than I thought it possible to be. There's no doubt I'm terribly in love with you. I keep thinking what you're doing, and have to stop—it makes me want to kiss you so. . . . It will be joyful to see your dear funny face tomorrow. The Marmots kiss you. Yr MANDRILL. (VWL, 2:90)

The tenderness and physicality of that letter indicate that Virginia had come to accept and reciprocate Leonard's love, perhaps sexually.[13] Indeed, their monogamous closeness now served as a base for the Woolfs' considerable social and professional independence of each other.

As conscientious objectors, Duncan Grant and Bunny Garnett had taken up fruit farming in hopes of avoiding conscription, and Vanessa, in Virginia's words, was "keeping house" for Duncan in Suffolk (VWL, 2:83). All of Vanessa's energies were funneled into painting, taking care of her children and Duncan, along with the obligatory Bunny, and seeing that Duncan was not called up for service. She poured her resentment of Bunny into a portrait that sees him as a child with, as their friends gossiped, "a crimson nose" (Noel Olivier/James Strachey, 17 February 1916).

Virginia's house hunting for Vanessa in Sussex turned up Charleston, a typical country farmhouse that had been built in the eighteenth century and had served as a boarding house in the early twentieth century. (The upstairs bedrooms still were numbered.) It was large and private enough for Vanessa's ménage but had no electricity or telephone, and fireplaces were its only source of heat. In a big open valley with a walled garden and a pond, the house offered seclusion, charm, and the opportunity for Duncan and Bunny to work on a nearby farm (VBL, 201).

In 1910, when she had felt herself helpless against Vanessa's abuse, Virginia had complained and threatened a counterattack. In 1913, she had confronted Vanessa with various accusations about egoism and indifference. In 1915, she had worried about being despised. But by May 1916, Virginia had developed a better defense, another example of her strength and resilience. By mockingly assuring her sister that Charleston was so far from Asheham that "you wouldn't be badgered by us," Virginia was able to neutralize Vanessa's abusive manner.[14] Virginia was afraid, however, that Adrian and his wife, Karin, instead of Vanessa, would take Charleston (VWL, 2:110).

As the war dragged on, reformist impulses proliferated. Virginia claimed that Margaret Llewelyn Davies proposed a "new society or scheme on the telephone about every other day" (VWL, 2:78). Leonard published *International Government*, a "landmark in the study of international affairs" with an introduction by George Bernard Shaw (D. Wilson, *Leonard Woolf*, 67). Despite continuing to work for peace and economic justice, Leonard turned down a position as political secretary for "domestic" reasons, causing Adrian to inquire whether "the Goat" was ill again (AS/LW, 16 September 1916.) By then, Leonard's diaries record no more bad nights for Virginia, although the possibilities that Leonard might be called up for service and that zeppelin bombs might any night wound or kill them still made this an unsettling time. Virginia became "steadily more feminist, owing to the Times, which I read at breakfast and wonder how this preposterous masculine fiction [the war] keeps going a day longer." But she remained skeptical about the efficacy of female suffrage and of the peace proposals offered by Leonard, Margaret, and Bertrand Russell (VWL, 2:76, 78).

Clive Bell and Adrian Stephen had been classified as "conscientious objectors" and so were doing farmwork. Leonard had been exempted from military service, because both Virginia's health and his own trembling hands were "shaky" (VWL, 2:95). Despite Leonard's peace work, Vanessa censured "the Woolves" for not being sympathetic enough to conscientious objectors (meaning Clive but especially Duncan) (VB/RF, 1 January 1916). Consequently, rather than be accused of insensitivity to Duncan's vulnerability, Virginia adopted her sister's obsession.

Against rampant talk of jailing conscientious objectors, Virginia spent much of the summer of 1916 writing letters to help Duncan avoid conscription (VWL, 2:102). Roger Fry wrote to Lord Curzon asking for exemptions for all writers and artists, and he also wrote, for Vanessa's sake, a separate letter on Duncan's behalf.[15] When the news arrived at Durbins that both Duncan and Bunny had been granted noncombatant status—thanks largely, Vanessa wrote, to Roger's efforts—Roger was entertaining Leonard and Virginia. Virginia was "very gay and charming," knowing that she and Roger had passed Vanessa's litmus test. And Vanessa was so pleased that she invited Lytton Strachey for a weekend in Suffolk, with the news that "I am asking Virginia! but without very much hope of getting her" (VB/RF, 5 May 1916; RF/VB, ? July 1916; VB/LS, 9 June 1916). But she did "get" Virginia and Leonard too, on the weekend of 21 July. Vanessa reported that

> Virginia took up so much of my time when she was here that I couldn't get a moment to write. She was very well & very amiable—so was Leonard. I saw that if he was going to stay here one must try not to get him on one's nerves at once, so we were very polite to each other & got on quite easily. They were both made useful weeding. (VB/RF, 26 July 1916)

Missing (or covering over) Vanessa's feelings about Bunny and sunning herself in her sister's gratitude, Virginia pronounced the environment there "wonderfully harmonious" (VWL, 2:107).

With Duncan saved from conscription, Vanessa was indeed on better terms with those who had helped him. She wrote to Roger in July that "the Woolves are on us" but continued with the assurance that Virginia was "very well & very amiable. So was Leonard." The tone of that letter toward Virginia, Leonard, and him was so improved that Roger rejoiced: "It's ages since a letter of yours has given me pleasure and no pain as this did," but he still could not "endure the pain" of visiting Vanessa with Duncan there.[16]

Roger thus escaped to Paris, where he met Picasso and Matisse and rediscovered his own painting style—long suppressed, he said, to please Vanessa (RFL, 2:398).

Virginia now joked about her envy of, for example, Vanessa's "brats," and Vanessa opened some space in her circle for Virginia (VWL, 2:108). Amicable relations between the sisters encouraged Vanessa, rather than Adrian, to lease Charleston for a fall move and inspired Virginia to make the courtship novel she may have envisioned on her honeymoon into "another novel" about Vanessa (VWL, 2:109). Involved in the continuing complicated emotional entanglements with Duncan, Bunny, Roger, her sons, and her husband, Vanessa might have worried about a novel about her life had she not sincerely believed that Virginia did not understand its sexual dimension.

When Roger visited Virginia and Leonard in September at Asheham, he found "V perfectly delightful" and assured Vanessa that they had talked "much of you & always in tremendous terms." Then he played to her belief in Virginia's naïveté: "I hardly think she guesses anything of our relations." He did a sketch for a portrait of Virginia (apparently now lost), which he thought "rather good as likeness." Roger regretted to Vanessa that he got "likeness without character & you get character without likeness," as in her featureless portraits of Virginia. He seems to have worked on the one Mantegna canvas in the fall of 1916, again staying with the Woolfs at Hogarth House (RF/VB, 5 and 8 September 1916, 28 January 1917, and fall 1916).

Virginia continued to respond ironically to Vanessa's abuse: "Not being an artist, my feelings are not to be considered ha! ha!" (VWL, 2:118). Although this method of stealing Vanessa's fire promised harmony, a September event threatened to ruin their new good relations. Bunny Garnett, with Dora Carrington and Barbara Hiles (both students at the Slade School of Art), set out on foot from Lewes to look at Duncan's prospective home. Starting too late to get to Charleston, the trio broke into Asheham for the night while the Woolfs were in London. Virginia was offended, and Leonard outraged. Then Virginia invited Carrington, now working for Roger on the Mantegnas, to dinner. Virginia took some pleasure in finding herself, as cross-examiner, a formidable, even frightening, presence. Carrington called Leonard "Crusty the Wolf," but she also labeled him and Virginia as charming. Nevertheless, their break-in was not received well, especially as Virginia suspected Bunny of stealing her *Oxford Book of Poetry* from her bedroom (VWL, 2:123–24).

Vanessa, her childrem, Duncan, and Bunny moved into Charleston in October 1916. Vanessa believed that

> the Woolves have a morbid terror of us all—I cant think why. They seem to think we should contaminate the atmosphere and bring wicked societies into Virginia's life. . . . Surely the downs are wide enough for us all & they neednt fear a constant flow in & out of Asheham as long as Woolf is in it. Of course it might provide useful spare rooms when they are away.

Although there was no "constant flow" into Asheham, Quentin Bell remembers breaking in with Julian, their teacher, and her daughter and running around shouting "I am the ghost." (Asheham was supposed to be haunted.) Their "fault lay in breaking into the house" and leaving "appalling traces" (VB/LS, 24 October 1916; Quentin Bell to me, 15 March 1995; interview with Quentin Bell, 26 July 1993). The invasion earned the children "a stern rebuke from the Woolfs" (Bell, *Elders and Betters*, 116). Although Leonard was irritated on this occasion and disapproved in general of Vanessa's child-rearing practices, these pranks were not the sort of contamination that he most feared.

Virginia found a way literally to purchase Vanessa's affection: She began to buy, from the Omega, clothes, decorative objects, and paintings by Vanessa and Duncan. Thus Virginia was not only courting but supporting Vanessa; later she itemized her artistic purchases as expenditures "with a philanthropic element" (D, 5:192). After lunching with Roger at the Omega on 8 December 1916 (LWD), Virginia confessed to Duncan that even though she liked his paintings, especially a "green and blue one . . . which seemed to me divinely romantic and imaginative," the display otherwise disappointed her, and she feared saying "all the wrong things" to Roger (VWL, 2:130).

In late 1916, Virginia and Leonard decided to learn how to print, only to discover that lessons were available just to trade union apprentices. As Leonard ironically remarked, "The social engine and machinery made it impossible to teach the art of printing to two middle-aged middle-class persons" (LWA, 2:169). Also at this time, Virginia met her fellow writer Katherine Mansfield and became involved in Saxon Sydney-Turner's improbable courtship of the much younger Barbara Hiles.

The tragedy of war touched home when Leonard's brothers Philip and Cecil were wounded by the same shell and the brilliant Cecil died.

Virginia did not mention that loss in her letters at the time, but Leonard's sense of the "pitiless, useless slaughter in France" must have affected her (LWA, 2:142). She was depressed also because Dr. Craig wanted her to rest so that she would not lose weight (VWL, 2:131). When Ka Cox visited the Woolfs at Asheham for Christmas, one premise of the visit—set out in the letters from Ka and Janet Case to Virginia (LWP)—was that no one would refer to those apparently crazed episodes from the past, but Virginia remembered that Ka had "seen me mad" (D, 5:143) and feared that her demons were still lurking somewhere within her.[17]

In early 1917, Virginia presided over the Richmond Women's Cooperative Guild's meeting at Hogarth House at which Leonard spoke (LWD). At about the same time, Roger started an evening club at the Omega, intended to bring the arts together and to integrate members of a younger generation, like Carrington and Hiles, into "Old Bloomsbury." Such night events in London were considered a drain on Virginia's health, but she was intensely conscious of the differences between such artsy events and the more sober guild and adult suffrage meetings that she and Leonard attended. Roger was organizing an exhibition in which Omega artists "translated" masterpieces into Postimpressionist terms. Vanessa charged Virginia with making "detailed criticisms of all the works" (VB/VW, 10 May 1917). If she had studied Roger's rather Cubist "translation" of Raphael's *Bridgewater Madonna* (Figure 46), Virginia might have concluded that simplifications of form could, after all, convey emotions.

Virginia's relationships with her rivals always were prickly, and her friendship with Katherine Mansfield quickly foundered, although they had a "slight rapprochement." Virginia claimed that Katherine was "an unpleasant but forcible and utterly unscrupulous character" and therefore a fit "companion" for Vanessa. Then, feeling guilty about such remarks, Virginia confessed to Duncan that "one of the concealed worms of my life has been a sisters jealousy—*of* a sister I mean; and to feed this I have invented such a myth about her that I scarcely know one from tother" (VWL, 2:144, 146). Confessing fears, jealousies, and self-protective strategies was therapeutic, but Bloomsbury gossip only fomented further jealousies, and both Virginia and Katherine jeopardized opportunities for sharing their dedication to revisionary writing.[18]

All her life, Virginia took care with the appearance of her writing notebooks, binding, labeling, and covering them (however sloppily) with a variety of papers. For her, a notebook was not just a utilitarian container but an object with its own visual interest. Her lifelong concern with pens and their nibs, with papers and colored inks, testifies to a distinct consciousness of the "materiality of language" (Kristeva, *Language*, 18–40). Back in 1905, Vanessa had bought a "silver point press" on which Virginia had made at least two bookplates, although she had not mastered the art.[19] In March 1917, the Woolfs fulfilled their long-delayed plan of buying a printing press, so Virginia set about learning the art of printing and discovering how writing could be not invisible but opaque, a signifier in its own right.

The press, type, accessories, and an all-important pamphlet on how to print arrived at Hogarth House in late April. Virginia found the printing process "exciting, soothing, ennobling and satisfying," and fulfilling for them both.[20] Leonard had conceived of the press as physical therapy to take Virginia's "mind off her work," but printing offered Virginia aesthetic therapy as well (LWA, 2:169). Setting each unit of type with her own hands and arranging the "furniture" to demarcate empty spaces between words heightened Virginia's sense of a text's visual properties. Not as professional or as self-consciously ornate as the Kelmscott and Bodley Head Presses,[21] Hogarth Press publications captured much of the spontaneity of Postimpressionist painting.[22] In her post-Broussa revision of *Melymbrosia*, Virginia had conceptualized language as "blocks" (VO, 81). Now she experienced it literally as blocks of type, an experience that extended her "visual literacy."[23]

The press enabled Virginia to write "what one likes" because it promised to free her from the Gerald Duckworths of publishing. Near the end of her life, she asked, "Didn't we start the Hogarth Press 25 years ago so as to be quit of editors and publishers? Its my nightmare, being in their clutches: but a nightmare, not a sane survey" (VWL, 6:459). Without a publisher who could own her by abusing, rejecting, or converting her vision into his own, Virginia could explore the limits and the possibilities of prose, topics she talked over amiably in another rapprochement with Katherine Mansfield in the summer of 1917. Soon the press itself blossomed into a major vehicle for revisionist literature, art, politics, and ideas. However modernist, as a "recreation requiring hard work which produced a useful product," the press was also a decidedly Victorian enterprise (Willis, *Leonard and Virginia Woolf as Publishers*, 4).

In her review of a collection of essays by Arnold Bennett, Virginia focused almost exclusively on his "Neo-Impressionism and Literature."[24] Now Virginia repeated the question that Bennett had borrowed from Fry: "Is it not possible that some writer will come along and do in words what these men [the Postimpressionists] have done in paint?" That query heralded Virginia Woolf's first radical departure from what she, like both Bennett and Fry, now called "infantile" realism (*Times Literary Supplement*, 5 July 1917, 319; VWE, 2:130).

After hours spent almost daily on printing, by 13 July 1917, the Woolfs finished printing *Two Stories*—Virginia's "The Mark on the Wall" and Leonard's "Three Jews"—with woodcuts by Carrington.[25] Whereas Leonard's story is highly patterned, his subject—the alienation that Jews feel in England—is more important than the design. Various Bloomsbury letters show that his sense of alienation was justified. Just after "Three Jews" was published, Maynard Keynes wrote to Vanessa: "For tea I took Clive to Richmond to see Virginia (he seeing Hogarth House for the first time). When we arrived Ka was there but no Jew; nor did he appear at all, which was a great pleasure." In addition, the sisters' cousin Fredegond Maitland Shove wrote to Vanessa: "Three Jews—is not that rather too much of a good thing?"[26]

"The Mark on the Wall" is short and does not have, in any conventional sense, a subject at all. Announcing that "novelists in future" will leave "the description of reality more and more out of their stories," Virginia led the way in a piece that contrasts holistic meditation with purposeful activity (CSF, 85–86). "The Mark on the Wall" is about those very "rhythmic sequences of change" that, according to Fry, are determined by forces within the artwork (VD, 6). Furthermore, by dissociating space from action and from a focused character, "The Mark on the Wall" reflects an orientation rare in Western literature.[27]

The whole story is a narrator's contemplation of a mark on her wall, undertaken because she wants to "sink deeper and deeper, away from the surface" (CSF, 85). This point of view exemplifies, to quote Fry, "the complete detachment of the artistic vision" (VD, 7). For "The Mark on the Wall," Virginia earned Clive's assessment of "perfect" (CB/VW, 19 July 1917) and that most important tribute, Vanessa's praise: "No doubt you are a very good writer & I shouldn't think you had much to fear from the rivalry of Katherine Mansfield & Co" (VBL, 207). Virginia's entrepreneurial talents and Katherine's sense that it was "curious & thrilling" that she and Virginia were "after so very nearly the same thing" (Mansfield, *Collected Letters*, 1:327) prompted Virginia to ask Katherine for a story to print (VWL, 2:150). Katherine gave her the lengthy story "Prelude."

In the late spring of 1917, the Woolfs made various purchases for the press and received a gift of old bookbinding equipment from Virginia's cousin Emma Vaughan (Willis, *Leonard and Virginia Woolf as Publishers*, 22). Virginia's greater visual literacy and the success of "The Mark on the Wall" renewed her determination to search further for "some new shapes" or "a completely new form" for her prose (VWL, 2:167). She returned to an outlet begun before 1897 and renewed and then abandoned in 1915: keeping a diary. She began laconically but soon developed a new sort of diary, written freely or "indiscreetly" (D, 1:55) according to her "mood" (D, 1:79). She also drafted her second Postimpressionist story, "Kew Gardens," perhaps inspired by a letter from Katherine.[28]

Even though the war still threatened to wipe out Western civilization, the arts were arising remade from the ashes. In the heady atmosphere that produced Picasso and Matisse, Apollinaire and Proust, Diaghilev and Satie, Virginia wanted her aesthetic judgments clarified. She asked Clive why he thought "The Mark on the Wall" was good. She also tried to discover what Joyce was trying to do in *Portrait of the Artist as a Young Man*, but she was too bored to finish it (VWL, 2:167). Had she finished the last section, Virginia might have associated Stephen Dedalus's aesthetic with Roger Fry's. Joyce's Dedalus distinguishes between static and kinetic art in much the same way as Fry had replied to Shaw with the distinction between creation as "artistic" (like Roger Fry's and Virginia Woolf's) or "ethical" (like George Bernard Shaw's and Leonard Woolf's).[29] In "The Mark on the Wall," Virginia created a similar opposition between the female's detached artistic perspective and the male's active ethical stance.[30]

While Leonard was working with the Fabians to prevent another war, Roger in a lecture entitled "Art and Life" was telling those same Fabians how little social impact art has.[31] Although she was an active member of the Women's Cooperative Guild, holding meetings at Hogarth House and providing speakers, Virginia remained ambivalent about the choice between activism and aestheticism. She did not criticize Leonard's work, but she did question (as he did in low periods) its efficacy. She regretted that Janet Case now expected "all literature to go into the pulpit" (D, 1:213), claimed to have argued with Margaret Llewelyn Davies against a moral approach to art (VWL, 2:119), and suspected that writing like Bertrand Russell's did "no good" (VWL, 2:133).[32] Virginia's natural tendency, like Roger's, was to see the "morality" of art chiefly in its cultivation of the imaginative life. However, she feared that the "morality" of disinterested aestheticism could not match the "morality" of activist politics.

Such a feeling of inadequacy was dangerous to Virginia's psychological well-being. Later she facetiously voiced her feeling that the works by

Maynard, Desmond MacCarthy, Clive, and Leonard so eclipsed her own that she seemed a "mere scribbler; what's worse a mere dabbler in dreams" (MB, 204). Given the public outrage over the Postimpressionist exhibitions, Virginia had reason to believe that dabbling in fiction that was as devoid of character and plot as modern painting would invite ridicule, and ridicule might precipitate those terrible feelings of loss and helplessness. Hedging her bets, therefore, Virginia chose both to write a representational novel utilizing "literary" emotions and to compose short, experimental, nonrepresentational fictions appealing to "aesthetic" emotions.

In the summer of 1917, Roger Fry articulated the risk of experimentation: As art becomes "purer," it "appeals only to the aesthetic sensibility" (VD, 10). Under Virginia's influence, Roger began to think aesthetically of his articles "as writing,"[33] and Virginia summarized what he said to Vanessa:

> You see the worst of art is that it appeals solely to the aesthetic faculties, and one's extremely shy and snobbish about giving away ones deficiency or eccentricity about them. Of course, I committed myself all wrong to poor old Roger, about his cold hard little compositions and called them literary, which he said they aren't. (VWL, 2:195)

Virginia's sense of being at once "shy and snobbish" (what Duncan Grant later called shy and fierce) confounds inferiority and superiority, shy withdrawal and snobbish attack.[34] Diane Gillespie explains that Virginia "shrouded her raids on painting in all kinds of apologies and explanations" (*Sisters' Arts*, 104). Virginia also went on the offensive, attacking, for example, "poor old Roger" and his paintings to compensate for admitting herself to be "all wrong." Such responses replicated the personal terms of her *paragone* with Vanessa: If she's good, I'm not; or if painters are important, I'm not. When one of Vanessa's paintings was universally lauded, Virginia joked, "Thank God, I say, that she doesn't write" (VWL, 2:216).

Katherine Mansfield, however, did write. Clive reported to Vanessa that "Virginia has been up to her old wickednesses again. You remember that Sunday Maynard and I went to tea there and abused Katherine Mansfield; well, K. M. came to dinner the same evening and Virginia repeated all we said with interest." Later, with Virginia threatening to show to Mary Hutchinson Clive's letters to her, Clive observed, "What an extraordinary mischief maker she is" (CB/VB, August or September 1917; CB/VB,

1 September 1918). Of course, both Maynard and Clive encouraged Virginia's "wickednesses"—as long as they were turned on someone else.

Despite Virginia's vitality and sanity, Leonard still felt some need to guard her against Vanessa and Duncan. When Lytton Strachey was to visit at Asheham, Vanessa thought of meeting him there, "that is if we dare to face the Wolf but he's simply furious when Duncan or I appear—apparently he's quite mild to others" (VB/LS, 8 August 1917). Leonard knew how much Vanessa could hurt Virginia, and Duncan and Vanessa in turn were "conscious of a moral force in Leonard which they repudiated as narrow, philistine and puritanical" (Garnett, *Deceived with Kindness*, 109). Leonard sensed Vanessa's prejudice against him. Indeed, had Virginia read her sister's and friends' (especially Clive's and Maynard's) letters, they would have shown her that it was Leonard, not she, who was "despised."

Once Virginia dreamed that Vanessa and Duncan were "going off to live on a ranch in California because it was the only way to paint—and you were very bitter—didn't want ever to see me again" (VWL, 2:197). Her waking hours, however, were generally free of such fears. External worries were more pertinent: Necessities like meat and sugar, coffee and cocoa, milk and butter, coal and telephone service were in short supply. London and Sussex were vulnerable to bombing. In late September 1917, Leonard and Virginia walked to the top of the downs, where they filled a large handkerchief with wild mushrooms while waiting to meet Roger and Vanessa, and then they all walked back to Asheham together. At tea, Roger and Vanessa talked about the gunfire over London they had heard the night before (D, 1:54). As the war dragged on, Leonard again was called up for service and again was excused because of both his nervous tremor and the danger that Virginia might have another breakdown (LWL, 215, note). Leonard thought that the war was destroying "the bases of European civilisation" (LWA, 2:191), and Virginia concluded that words like *liberty* were now meaningless (D, 1:138).

Printing Katherine Mansfield's "Prelude" offered little respite from such gloom and threatened to rekindle their rivalry by making Virginia the handmaiden to Katherine's art. In Hogarth House—which Barbara Hiles described as "a beautiful house, built in a simple Georgian style, with wide windows and a large white porch"—Barbara, now the Woolfs' assistant printer, and Virginia set type together.[35] Virginia paid for Barbara's help by becoming her confidante in her determination to marry either Saxon Sydney-Turner or Nicholas Bagenal while keeping the threesome together as close friends.

Julia Prinsep Jackson, probably at Little Holland House, in a photograph perhaps taken by her uncle by marriage Lord Somers, ca. 1856. (Leslie Stephen's Photograph Album. Courtesy of the Mortimer Rare Book Room, Smith College) [1]

Julia Prinsep Jackson with her mother, Maria Jackson, ca. 1867. (Unknown photographer. Courtesy of the National Portrait Gallery, London) [2]

Julia Prinsep Jackson Duckworth Stephen, ca. 1878. (Stella Duckworth's Photograph Album. Courtesy of the Henry W. and Albert A. Berg Collection, The New York Public Library, Astor, Lenox and Tilden Foundations) [3]

Talland House, St. Ives, Cornwall. (Leslie Stephen's Photograph Album. Courtesy of the Mortimer Rare Book Room, Smith College) [4]

Virginia and Adrian Stephen playing cricket at Talland House, ca. 1886. (Leslie Stephen's Photograph Album. Courtesy of the Mortimer Rare Book Room, Smith College) [5]

Julia Stephen supervising her Stephen children's lessons, ca. 1890. (Leslie Stephen's Photograph Album. Courtesy of the Mortimer Rare Book Room, Smith College) [6]

Leslie and Julia Stephen in the Swiss Alps, 1889. (Leslie Stephen's Photograph Album. Courtesy of the Mortimer Rare Book Room, Smith College) [7]

Julia Stephen with her Stephen children in the drawing room at Talland House, ca. 1891. (Leslie Stephen's Photograph Album. Courtesy of the Mortimer Rare Book Room, Smith College) [8]

The Stephen children, Vanessa with an easel and Virginia with a book, ca. 1892. (Stella Duckworth's Photograph Album. Courtesy of the Henry W. and Albert A. Berg Collection, The New York Public Library, Astor, Lenox and Tilden Foundations) [9]

The Stephen/Duckworth family, 1894. Standing: Gerald Duckworth, Virginia Stephen, Thoby Stephen, Vanessa Stephen, and George Duckworth; sitting: Adrian Stephen, Julia Stephen, and Leslie Stephen. (Leslie Stephen's Photograph Album. Courtesy of the Mortimer Rare Book Room, Smith College) [10]

The family after Julia Stephen's death, 1895. (Stella Duckworth's Photograph Album. Courtesy of the Henry W. and Albert A. Berg Collection, The New York Public Library, Astor, Lenox and Tilden Foundations) [11]

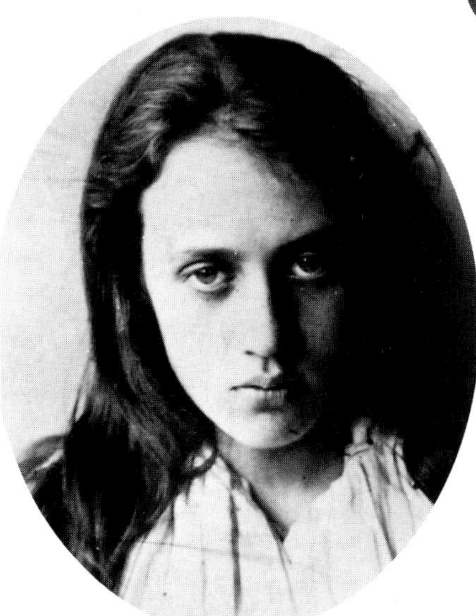

Vanessa Stephen, ca. 1896. (Stella Duckworth's Photograph Album. Courtesy of the Henry W. and Albert A. Berg Collection, The New York Public Library, Astor, Lenox and Tilden Foundations) [12]

Virginia Stephen, ca. 1896. (Stella Duckworth's Photograph Album. Courtesy of the Henry W. and Albert A. Berg Collection, The New York Public Library, Astor, Lenox and Tilden Foundations) [13]

Virginia Stephen's description and drawing of a sunset, 1899. (Warboys Journal. Courtesy of the Henry W. and Albert A. Berg Collection, The New York Public Library, Astor, Lenox and Tilden Foundations) [14]

Virginia Stephen and Clive Bell at Studland, 1910. (Courtesy of the Tate Gallery Archive, Vanessa Bell Photographic Collection, U24) [15]

Vanessa Bell, *The Post-Impressionist Ball* [with Virginia and others], ca. 1911. (Courtesy of Anthony d'Offay Gallery, London) [16]

Duncan Grant, *Virginia Stephen*, 1911. (Courtesy of The Metropolitan Museum of Art) [17]

to be back in town on Thursday
in fact I must for Grafton
business (really must this time)
so I shall make a great effort
to get down Thursday night even if
I don't get beyond Lewes.

That's the shape of your breast
when you're lying down ~~the only~~
~~thing you~~ I send it because its
one of the things you can only
enjoy ~~by~~ through me

*R*oger Fry, Vanessa Bell's torso, ca. 1912. (Courtesy of the Tate Gallery Archive, 8010.5.607) [18]

*V*irginia Stephen at Studland, September 1911. (Courtesy of the Tate Gallery Archive, Vanessa Bell Photographic Collection, B23) [19]

Roger Fry, *London Zoo*, 1911. The painting was designed to fit around the corner of a mantel. (Courtesy of the Tate Gallery) [20]

Duncan Grant, *Bathing*, 1911. (Courtesy of the Tate Gallery) [21]

Vanessa Bell, *Studland Bay*, 1911. (Courtesy of the Tate Gallery) [22]

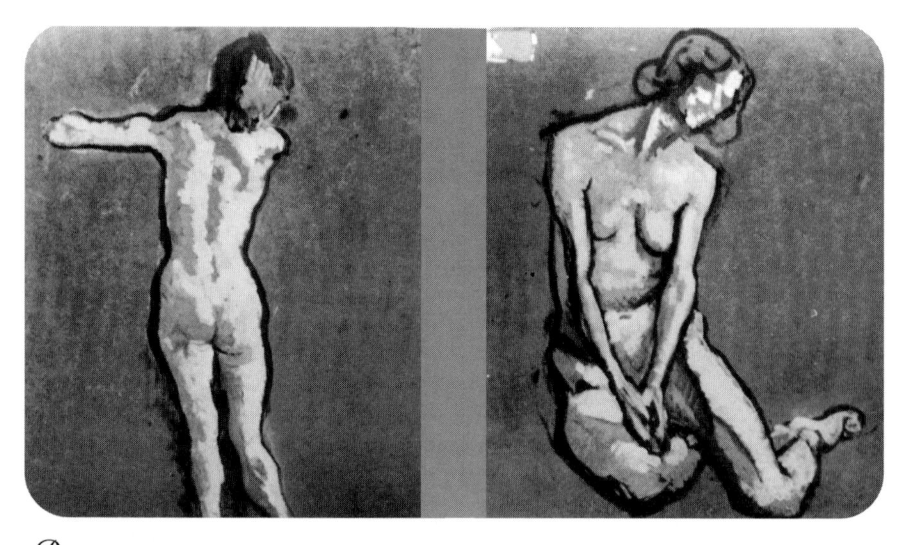

Roger Fry, *Vanessa Bell* [on reverse sides of the same canvas], ca. 1912. (Courtesy of Sandra Lummis Fine Art) [23]

Vanessa Bell, Roger Fry sketching, with Virginia Stephen in the background, 1912. (Courtesy of the Tate Gallery Archive, 8020.2.20) [24]

Vanessa Bell, *Virginia Stephen*, 1912. (Courtesy of the National Portrait Gallery, London) [25]

Roger Fry and Virginia Stephen at Asheham, 1912. (Courtesy of the Harvard Theatre Collection, Harvard College Library) [26]

Roger Fry, *Virginia Stephen*, ca. 1911. (Private collection, Great Britain, on loan to the Fitzwilliam Museum, Cambridge) [27]

Virginia Stephen and Leonard Woolf at George and Margaret Duckworth's house, Dalingridge, 1912. (Private collection) [28]

VBRF 96(4)

him it was obviously not meant seriously
which I should think must be true. But
Woolf took it quite seriously & you c
imagine the result! He spoke to m
about it yesterday & today came & sp
to Clive. Himself I think it will ma
the difficulties between them more ma
than ever. But I must say I think h
rather an idiot to get into a state ab
it considering that Clive never sees him
now & totally won't for a long time t
But think of Lytton's folly in showi
him the letters! How it's one of our
embroilments.
Now I must stop as I have a lot to do this
afternoon - shopping & tea with Ot. I'll wri
& tell you about that.
Your VB

*V*anessa Bell, sketch in a letter to Roger Fry, 7 April 1914. (Courtesy of the Tate Gallery Archive, 8010.8.130) [29]

*R*oger Fry, *Vanessa Bell*, fall 1918. (Courtesy of Sandra Lummis Fine Art) [31]

*D*uncan Grant, *Vanessa Bell*, n.d. (Courtesy of Tony Bradshaw and the Bloomsbury Workshop) [30]

*R*oger Fry, *The Stocking*, from the Omega publication *Original Woodcuts by Various Artists*, 1918. Copy six from a limited edition of seventy-five. (Collection of the author) [32]

*V*anessa Bell, *Nude (The Tub)*, from the Omega publication *Original Woodcuts by Various Artists*, 1918. Copy six from a limited edition of seventy-five. (Collection of the author) [33]

*D*uncan Grant, *The Tub*, from the Omega publication *Original Woodcuts by Various Artists*, 1918. Copy six from a limited edition of seventy-five. (Collection of the author) [34]

Maynard Keynes believed that "Virginia and the Jew, by the way, allege much approval of the match" between Barbara and Nicholas. Maynard also wrote of his plans to give a "demiparty" with Virginia, Duncan, and Elizabeth Asquith, who was daughter of former prime minister H. H. Asquith and later became Princess Bibesco. His plan shows how much Virginia's wicked tongue was now a given, perhaps one to be manipulated, in Bloomsbury society. Maynard said the object of his party was to make Elizabeth Asquith feel like "she's aspiring to a society she's not fit for. So I am going to egg Virginia on to rag her to the top of her bent" (JMK/VB, 31 January 1918). By this time, Virginia's ragging method—as when she passed on her friends' criticisms to Katherine Mansfield—led Katherine to conclude: "To Hell with the Blooms Berries" (Mansfield, *Collected Letters* 1:326).

In late 1917, Clive Bell "suggested a great historical portrait group of Bloomsbury," and Roger proposed that he paint "Lytton, Maynard, Clive, Duncan, me, you, Virginia, Mary [Hutchinson, Clive's mistress], Molly, Desmond. Is there any one else that ought to be in? P'raps Walter Sickert coming in at the door and looking at us all with a kind of benevolent cynicism" (RFL, 2:423). (The artist Sickert's more "literary" mode of painting made him cynical about Bloomsbury experimentation.) Given Maynard's references to Leonard as simply "the Jew," Roger's omission of Leonard might have been a demonstration of Bloomsbury's anti-Semitism, but it seems more likely to have reflected Roger's prevailing interest in the visual arts.[36] (He also omitted E. M. Forster.) Roger felt that the sine qua non of Bloomsbury was devotion to the new art. Because Leonard worked for social and political, rather than aesthetic, reformation, to Roger (at least in 1917) he was "ethical," not "artistic," and hence not true Bloomsbury.

While working on her long, traditional novel in October 1917, Virginia attended an exhibition that Roger had organized, with paintings and sculptures by Brancusi, Gaudier-Brzeska,[37] Doucet, Marchand, Vlaminck, and others, including Vanessa, Duncan, and Roger himself (D, 1:61). She liked to think of painters as "very little self-conscious" (D, 1:69), even though Roger's approach to painting was so conscious and intellectual that Virginia was uneasy talking with him about art (D, 1:75). Then one evening's conversation about "literature & aesthetics" freed her: "Roger asked me if I founded my writing upon texture or upon structure; I connected structure with plot, & therefore said 'texture.' Then we discussed the meaning of structure & texture in painting & in writing." Despite feeling unsure about whether "texture" or "structure" was the "right" answer,

Virginia discovered that she could voice ideas without lapsing into shyness or snobbishness or worrying that she had said "the wrong thing" (D, 1:80).

Virginia had claimed that living in Richmond made London more romantic, but it was also more inconvenient (VWL, 2:72). One London excursion involved looking at pictures and buying a coat at the Omega and then having tea and a party at Gordon Square with "a great many mop headed young women." Virginia left at ten, returned to Hogarth House, and found Leonard "testy, dispiriting, & tepid." She awoke the next day bitter with her old "sense of failure & hard treatment" and buffeted by wave after wave of depression. Late in the afternoon, she and Leonard walked together in a cold wind along the Thames, rehashing their differences. She seems to have convinced him that his worries about her taxing herself were now unnecessary. Virginia and Leonard agreed on the need for positive "illusions" regarding their careers in order to guard against fear of failure.[38] She happily recalled that on this evening the illusions returned and "were going merrily till bedtime, when some antics ended the day" (D, 1:73). In that mood of restored happiness, the "antics" certainly seem sexual.

Virginia bragged to Duncan of "a terrifically exciting and somehow sexual dream." She and Leonard slept in the same room, and he tried applying "the Freud system to my mind," although Virginia does not mention the results (VWL, 2:144, 141). In this contented period, Leonard seems to have so mellowed that visits between the Woolfs and the "Charlestonians" became easy and frequent. Virginia wrote in her diary about the way that marriage now fulfilled her: "as if marriage were a completing of the instrument, & the sound of one alone penetrates as if it were a violin robbed of its orchestra or piano" (D, 1:70). In late 1917, Virginia finally convinced Leonard that she was stable enough to attend one of Lady Ottoline Morrell's elaborate house parties at Garsington Manor. Virginia found her sympathy to be more with the Morrells than with those who accepted their hospitality and gossiped behind their backs.[39] From Katherine Mansfield's perspective, however, Virginia was the instigator of gossip. Katherine wrote to another friend, "I am sorry the Wolves are at Garsington. There will be a rare bone dragged into the light before they are gone" (Mansfield, *Collected Letters*, 1:334).

Vanessa now could confess to Duncan that her jealousy of Virginia was founded on admiration (Spalding, *Vanessa Bell*, 172–73), just as Virginia had acknowledged to him how harmful her jealousy of Vanessa had been (VWL, 2:146). By this time, Clive's attraction to Vanessa did not make Virginia "jealous as once it did, when the swing of that pendulum carried so much of my fortune with it: at any rate of my comfort" (D, 1:86). That reassessment might have been prompted by Clive's privately printed *Ad Familiares*, which circulated to friends at Christmas his 1909 poem to Virginia with its references to "gay delights" and "ravished heights." In return, Clive's book received Virginia's self-examination, Leonard's displeasure (D, 1:95), and Roger's amusement (articulated in a poem of his own about Clive's "wanton ways") (RFL, 2:422–23), and Virginia ordered a copy, improbably enough, for Margaret Llewelyn Davies. At Christmas, it took the Woolfs five hours to travel from Richmond to Asheham, "owing to Fog & frost," as Virginia recorded in her "Asheham Diary."

The Great War, with its introduction of automatic weapons, tanks, tear gas, and air bombing of civilian targets, introduced a new level of human depravity, adding urgency to Leonard's efforts to end war. On the Woolfs' return to Hogarth House at the beginning of 1918, one evening when Bob Trevelyan came to dinner, an air raid forced their dinner party into the cellar (D, 1:93–94). Virginia maintained that air raids were boring, simply "because one must talk bold & jocular small talk for 4 hours with the servants to ward off hysteria."[40]

As with his marriage, Roger's great affair of the heart had briefly united domesticity, passion, and devotion to art, a combination he despaired of ever finding again. Virginia still facetiously characterized herself and Roger as rival lovers, and Vanessa tried to counteract that characterization (VWL, 2:182). For example, she invited Virginia to join her when Roger visited, assuring her that he "has a foolish admiration for you" (VB/VW, 27 September 1917). But Virginia declined lest her visit be "rather hard on Roger" (VWL, 2:183).

Roger was gradually learning not to declare his love for Vanessa, and Virginia bartered for letters of affection from her, asking, for example, after several pages of gossipy news: "Do you really love me? How often a day do you think of me?" (VWL, 2:157). She congratulated herself for being "one rung higher on the ladder [of Vanessa's affection] than Roger," and he regretted being lower on the ladder than Duncan (VWL, 2:233). When Duncan was at Charleston, Roger often visited the Woolfs at Asheham, hoping that Vanessa would visit him there.

Although Virginia continued pretending to pose Roger and Clive as her rivals for Vanessa's affection, oddly enough she seems to have had no such feelings about Duncan. She helped him avoid conscription. She liked his pictures and joked about their indecency (VWL, 2:84). She could make sexual and scatological jokes with him (VWL, 2:144–46). To her, he was "like the first born of the human race, not yet entirely conscious" (VWL, 2:228). He had had affairs with Lytton Strachey, Maynard Keynes, and Adrian Stephen and was continuing the one with David Garnett.

Thus when Vanessa became pregnant, the child's paternity was a matter of such curiosity it seems that no one (except Roger and Clive) knew that Vanessa and Duncan were sometimes lovers until she told Virginia (perhaps in May 1918 [VWL, 2:245, 263, note]). Duncan's attachment to Vanessa seemed less threatening to Virginia than Clive's or Roger's had been because she did not know how much Vanessa needed him. Rather, she gossiped to Violet Dickinson about her sister's bohemianism: "Nessa seems to have slipped civilization off her back, and splashes about entirely nude, without shame, and enormous spirit" (VWL, 2:147). Duncan's quick sketch (Figure 30) captures not only Vanessa's nonchalance about nudity but also an astonished sadness.

Virginia and Leonard most often went into London together from Hogarth House (their "city" home), did their separate errands, and met later at the 1917 Club in Soho, the "centre of life" (VWL, 2:210). It had been founded largely by Leonard to mark the Russian Revolution's promise of a classless society and an end to imperialism. From there, they would go to concerts, exhibitions, lectures, or events such as the victory rally for the 1918 passage of the suffrage bill—the speeches for which Virginia thought platitudinous and boring (D, 1:124–25).

More to her taste was an encounter with Roger that delayed her arrival at the 1917 Club and made her "bristle all over with ideas, questions, possibilities which couldn't develope [*sic*] in the Charing Cross Rd." (D, 1:134). Apparently these possibilities imagined an aesthetic that resolved the *paragone* in terms of a "novel of still life . . . about substance and texture, with a design of men and women indicated like a fresco upon a wall" (*Times Literary Supplement*, 25 April 1918, 195; VWE, 2:239). In the *Times Literary Supplement*, Virginia called for a fictional "arrangement which is satisfying and harmonious in itself" and praised a form dissociated "from any feeling of pity or pleasure aroused by the fortunes of the characters" (*Times Literary Supplement*, 16 May 1918, 231, and 12 December 1918, 620; VWE, 2:246, 336). But when the *Times* rejected two of her articles, she consoled herself by writing *Night and Day* (D, 1:127, 136).

Although pity and pleasure in theory can be separated from the aesthetic response, Virginia's own displeasure was another matter. On 14 April 1918, Harriet Weaver, according to Leonard "a very mild blueyed advanced spinster," brought to the Woolfs portions of James Joyce's *Ulysses* that "raised a blush even upon such a cheek as mine," as Virginia joked. She claimed to worry about having this indecent book in the next room when the Women's Cooperative Guild was about to meet at Hogarth House (VWL, 2:231).

When Virginia joined Roger for a night at Durbins, Roger read to her from a book by Proust whose title she forgot and from his translation of the *Lysistrata*.[41] The next day, the two traveled back to London together and stopped at Gordon Square, where Vanessa brought forth "a small parcel about the size of a large slab of chocolate. On one side are painted 6 apples by Cézanne." Maynard Keynes had just returned from Paris with his newest purchase—Cézanne's *Pommes*.[42] Virginia found the artists' excitement comic:

> Roger very nearly lost his senses. I've never seen such a sight of intoxication. He was like a bee on a sunflower. . . . The artists amused me very much, discussing whether he'd used veridian or emerald green, and Roger knowing the day, practically the hour, they were done by some brush mark in the background. (VWL, 2:230)

In her diary, Virginia ruminated: "What can 6 apples *not* be? I began to wonder" (D, 1:140). Her imagination did not retain a visual impression of the picture itself with its *seven* apples, nor did she realize that Cézanne's asymmetrical arrangement would be hard to achieve with six apples. Rather, her perspective was more theoretical: If six apples do not mean six apples, can they mean anything and everything? If art does not represent something, what does it do?

Equally disorienting was what Virginia first termed the "filth" in *Ulysses* (D, 1:140). To Roger, she acknowledged Joyce's "highly developed" method, which "leaves out the narrative," but she also denigrated his subject: "After all the p——ing of a dog isn't very different from the p——ing of a man" (VWL, 2:234). She craved "the chance of being thoroughly enkindled" by Roger "rather oftener" (VWL, 2:234–35). When Proust and Joyce were leaving out narrative and Cézanne was painting the most mundane of subjects, Roger's interest in "representation, reality, & so on" no longer seemed arcane and irrelevant.

In May 1918, when Roger picnicked with the Woolfs near Asheham, he and Virginia exchanged complaints about Vanessa's indifference.

Virginia spent a lovely night at Charleston, "with my window open listening to a nightingale," and concluded that "May in England is all they say—so teeming, amorous, & creative." After long talks with Vanessa, mostly about servants, Virginia had the pleasure of conceding that Roger's complaints were now "more genuine" than her own. Roger's petulance spilled over onto Clive, who, Roger said, did not even "know about pictures." Virginia and Roger agreed that Clive lacked insight into literature: "His book is stout morality [as opposed to aesthetics] & not very good criticism" (D, 1:152, 151). Therefore, even though Clive had named as "our three best living novelists—Hardy, Conrad, and Virginia Woolf," she could not take comfort (*Potboilers*, 11).

Even while she was writing a highly representative novel, Virginia Woolf's 1918 reviews (nearly one a week for the *Times Literary Supplement*) show her dissociating herself from conventional representational values and assuming almost a collaborative campaign with Fry for those very experiments that would appeal not to the readers' referential involvement but to their aesthetic sensitivity. She insisted that the "material of life" must be "limited and abstracted" before words can deal with it and questioned "whether likeness to life is the prime merit in a novel." Moreover, Virginia suggested that life is "a much more ubiquitous presence than one is led by the [materialist] novelist to suppose" (*Times Literary Supplement*, 10 January 1918, 18; VWE, 2:208, 209). Roger complained about Clive's cribbing his ideas and Virginia agreed with him (D, 2:10–11). Roger, however, made no such complaint about Virginia's borrowings, for he was pleased that his aesthetic might buttress the reformation of modern fiction.[43]

On 1 July 1918, Virginia sent Vanessa a draft of the experimental "Kew Gardens" and promised her sister "an account of my emotions towards one of your pictures, which gives me infinite pleasure, and has changed my views upon aesthetics." But her developing aesthetic raised uncertainties: "Its a question of half developed aesthetic emotions, constantly checked by others of a literary nature." Virginia's aesthetic emotions continued to conflict with traditional "literary" emotions and associations (VWL, 2:257). Vanessa named "Kew Gardens" "a great success" and pointed out that her three women in conversation would "almost but not quite do as an illustration" for it. Given Vanessa's amusement with the prospect of reading Virginia's "theories on aesthetics," Virginia promised them only "for your derision."[44]

On 15 July 1918, Virginia visited the National Gallery alone. As her diary entry indicates, she seems to have applied Roger's aesthetics to her own reactions to pictures: "I insist (for the sake of my aesthetic soul) that I don't want to read stories or emotions or anything of the kind into them; only pictures that appeal to my plastic sense of words make me want to have them for still life in my novel" (D, 1:168). Clearly, on this expedition, Virginia accepted the irrelevance of a "literary" response to painting. Equally clearly, she was ready to express in words some of the properties of paint.

With great bravura, Virginia tried a new ploy for securing Vanessa's affections, sending a lengthy commentary on her gallery visit, although she still worried that her aesthetic feelings were "undeveloped." This worry had its comic, domestic side, as she vindicated her taste by replacing an irritating yellow Omega chair cover with eighteenth-century embroidery. Nevertheless, she assured Vanessa that pictures did give her "new ideas on art" (VWL, 2:259–60, 269). When Duncan's pictures prompted Carrington to tell Virginia that one day at Charleston was more valuable than "any attractions you literary people could offer me!" Virginia laughed, replying that Carrington was simply in a "tipsy condition."[45]

As Vanessa's pregnancy advanced, Virginia became more adoring, and Roger more bitter. He felt that her pregnancy marked the "completeness of yr. belonging to D," whereas he was "nowhere" in her emotional universe, and so he found it impossible to "unselfishly rejoice with you." His ambivalence is graphically illustrated in two sketches of Vanessa, one maternal and gentle and the other hard, almost grotesque (Figure 31). Vanessa tried to console him by comparing his relationship with her with her frustrating relationship with Duncan. But Roger was only offended: "He gives you far more than you even conceive it possible to give me." He felt it terrible that Vanessa crushed his love "so unconsciously so constantly" (RF/VB, 15 June and 7 August 1918). Despite this jealousy and resentment, however, he remained her friend.

Once again, Roger and Virginia competed for Vanessa's affection, both trying to find her new servants and offering to care for her children. Virginia was delighted less, it seems, with the prospect of a baby than with the "divine pleasure" of having Vanessa need her (VWL, 2:255). Vanessa seems to have interpreted Virginia's offers to include lending her their servants and then to have blamed Leonard for retracting that supposed offer (VBL, 215). Of Virginia's innumerable letters and time-consuming efforts to help her find servants, Vanessa observed, "Virginia of course makes wild offers of help but I know is really too wild" (VB/RF, 28 August 1918). Her dismissiveness did not bode well.

Nor did the signs bode well for Virginia's reformation of the novel when she was writing the traditional *Night and Day* while Joyce and Proust and Katherine Mansfield were startling the public with new experiments in fiction. To make matters worse, "The Prelude" took from the fall of 1917 to July 1918 for the Woolfs to print. They interrupted the process to print a small posthumous volume of Cecil Woolf's poems, which Virginia thought lacked "the vigour of my particular Woolf" (D, 1:124). Knowing that Katherine was "extremely ill," Virginia wanted to believe her to be "the very best of women writers—always of course passing over one fine but very modest example" (VWL, 2:241). She began visiting the ailing Katherine weekly (D, 1:222) but feared herself eclipsed by Katherine's writing, which "had the living power, the detached existence of a work of art" (D, 1:167).

Given the enormous success and notoriety of Strachey's *Eminent Victorians*, which Katherine claimed was "quite measureless to man" (Mansfield, *Collected Letters*, 2:258), Virginia later wondered, "Am I jealous? Do I compare the 6 editions of Eminent Victorians with the one of The Voyage Out?" (D, 1:238). Obviously she did. Clive Bell exclaimed, "Virginia's unconcealed jealousy of Lytton's success rather shocks me" (CB/VB, ? June 1918). Now Katherine was hailed as the best of women writers; Lytton had invented a new form for biography; Leonard was asked to run for Parliament (D, 1:173), had created a new genre of "documentary journalism" (D. Wilson, *Leonard Woolf*, 124), and was working to rid the world of war; Roger's essays were changing the contemporary understanding of art; Vanessa's and Duncan's paintings were selling well; and Vanessa was pregnant. By comparison, Virginia once again saw herself as a failure. She had published only one—almost forgotten—novel, one experimental story, and many—but usually anonymous—journal articles and reviews.

When André Gide visited Roger in August 1918, they talked mostly about literature, and Roger recommended "The Mark on the Wall" to the French writer, especially for its "power of artistic detachment from life" (RF/VB, ? August 1918; RF/VW, 19 September 1918). He planned to translate it into French in collaboration with the poet and dramatist Charles Vildrac. When Roger complimented Virginia, she appropriated his terminology from a series of French lectures on plastic design (entitled

"Quelques peintres français modernes"), suspecting herself of "a perverted plastic sense" working "itself out in words for me" (VWL, 2:285).

Virginia's "plastic sense of words" (D, 1:168) caused Roger to write, "You're the only one now Henry James is gone who uses language as a medium of art, who makes the very texture of the word have a meaning and a quality really almost apart from what you are talking about" (RF/VW, 18 October 1918). Two weeks later, Virginia described life as "an immense opaque block of material to be conveyed by me into its equivalent of language" (D, 1:214). As with the Postimpressionists, equivalency meant not referentiality but analogy. But such theories were too heady for Virginia's work in progress, the highly referential *Night and Day*.

Despite her interest in mathematics, Katharine Hilbery, the heroine of *Night and Day*, is a Vanessa figure (perhaps an inside allusion to Leonard's Katharine in *The Wise Virgins*), but this Katharine feels a Virginia-like guilt for all that her privileged background has secured for her.[46] When Katharine visits Ralph Denham's family, she first lets the ugliness and chaos of the household adversely affect her judgment of him. But then when she finds herself arguing against Ralph and for one of his little brothers, she suddenly recognizes the intellectual vitality and openness of his comparatively impoverished background. Katharine implores Ralph not to romanticize her as some superior being (ND, 373–83). When Ralph begins to appreciate her for who she really is, Katharine decides to marry him. Her courage in ignoring family and class expectations in her marriage matches Virginia's in hers.

Conceived on the Woolfs' honeymoon—as was Leonard's second novel—*Night and Day* is both more hopeful and more forgiving than *The Wise Virgins*. That is, whereas Leonard's Harry, trapped into marriage with a boring and silly woman, loses the remote and beautiful Camilla, Virginia's Katharine loses her remoteness in her love for the intelligent and good Ralph. Not only the principals but also many other characters in *Night and Day* have real-life analogues: Mrs. Hilbery is patterned after Virginia's "Aunt" Anny, Thackeray's daughter and Leslie Stephen's sister-in-law from his first marriage. Mr. Hilbery is most like Stephen himself. Nevertheless, Virginia maintained that in writing, "one gets more and more away from the reality" (VWL, 2:406), but later in her diary she wondered "whether I too, deal thus openly in autobiography & call it fiction?" (D, 2:7).

Night and Day's two major romantic triangles, with Katharine as a common side to both, and its superb rendition of dialogue and city and country settings continue the tradition of the nineteenth-century English novel. It is "literary" in both personal and political associations even as it contrasts an aesthetic response with a sexual or "literary" one. When Mary Datchet goes to the British Museum to look at the Elgin Marbles, her emotions are "not purely aesthetic, because, after she had gazed at the Ulysses for a minute or two, she began to think about Ralph Denham. . . . The presence of this immense and enduring beauty made her almost alarmingly conscious of her desire" (ND, 82). But Katharine, who feels no such desire, will marry Ralph, and Mary (like Margaret Llewelyn Davies) will probably remain unmarried. Of all Virginia Woolf's fiction, only this novel invites its readers to read it first for its "literary" associations with early-twentieth-century class and political conflicts, especially in regard to women's suffrage; and only this novel follows a traditional marriage plot.

Paramount among Virginia's writing strategies was avoiding the psychological disasters that had accompanied the composition and publication of *The Voyage Out*. She later explained:

> After being ill and suffering every form and variety of nightmare and extravagant intensity of perception . . . when I came to, I was so tremblingly afraid of my own insanity that I wrote Night and Day mainly to prove to my own satisfaction that I could keep entirely off that dangerous ground. . . . Bad as the book is, it composed my mind. (VWL, 4:231)

Her strategies for keeping off dangerous ground involved reversing much that she had said about and done with *The Voyage Out*.

When Virginia began her first novel, she claimed to be reforming it, but when she began the second, she claimed nothing at all. The idea for the novel first came to her on her honeymoon (Heine, "Virginia Woolf's Revisions," 435) and was resurrected in the summer of 1916 (VWL, 2:109 and note), and she began writing in earnest in early 1917. But Virginia hid the existence of this novel even from her diary. Her first reference came only in March 1918 when she had written more than 100,000 words. She also did not explicitly mention the novel in letters until April 1918, when she mentioned that she was writing about Vanessa (VWL, 2:232).

Virginia spent seven years writing her first novel but less than two on the longer second one. Her writing was interrupted by innumerable reviewing tasks, attempts to help Vanessa, headaches, and a bout with influenza that "divorced" her from her pen (D, 1:119). Although she wrote rapidly, she claimed that she "had to write for 10 minutes at a time, with only half my brain working."[47] Virginia accentuated her difficulties with *Night and Day* to minimize risk, for "its the curse of a writers life to want praise so much, & be so cast down by blame, or indifference" (D, 1:214).

To avoid being so cast down, Virginia made a concerted effort to underplay, undercut, and even undermine her own accomplishments. As compensation for her sense of inadequacy, she recorded Leonard's becoming editor of the *International Review*: "At any rate today I am the wife of an Editor" (D, 1:190). When she told her diary that she had finished *Night and Day* on 21 November 1918, she also remarked that her announcement fell "a little flat" after the news that Leonard might be secretary to the English representative to the League of Nations conference (D, 1:221). Walking by the Thames in Richmond, the Woolfs watched the gulls and discussed Virginia's "fit of melancholy." She was "divinely reassured by L. so that [that evening in Hogarth] here I sit comfortable & secure; once more established in that degree of belief [in her writing] which makes life possible" (D, 1:223).

Just as she was completing *Night and Day*, Virginia began her "artistic education again." She went to a Rodin exhibition and distinguished between her aesthetic and literary judgments: "I didn't think at all highly of [the statues], except from the literary point of view." Her earlier hedging had evolved into confrontation with a central issue in modernism: If "literary" associations were sacrificed to formal values in literature, wouldn't the result be dehumanized? (VWL, 2:284). To Vanessa, Virginia described the "desire to describe" the Rodins as a "torment." To Roger, she wrote that the desire to describe artifacts from the museums "became like the desire for the lusts of the flesh" (VWL, 2:284, 285).

For once, perhaps because the armistice had at last arrived, Roger's concerns were as much political as aesthetic (D, 1:220). Nevertheless, he had no intention of politicizing art. Virginia said that she envied him his independence of judgment, but her envy was partly self-protection, thereby acknowledging that for her, envy and self-protection, like shyness and snobbism, were related (VWL, 2:295, 296). Meanwhile, at Roger's suggestion, Leonard had written to the poet T. S. Eliot about publishing his poems (LWL, 279). When Eliot came to call, Virginia was disconcerted to find his judgment (of Joyce especially) so different from hers, but she and Leonard saw in Eliot's poems an experiment comparable to the one Virginia was embarking on, and they agreed to publish them.

Albeit overtly discouraging Roger's adoration, Vanessa showed herself secretly gratified by it when she asked Virginia to pass on his declarations of love (VBL, 220). Virginia admitted that she and Roger "loved and sang and despaired" together over Vanessa. Sympathizing with Roger's frustrated love, Virginia felt more kindly toward his paintings and so turned the charge of artistic insensitivity against Vanessa: "Why are you artists so repetitious; does the eye for months together see nothing but roofs?" (VWL, 2:300). Laughing together about those values he labeled "philistine" and she termed "masculine," Virginia and Roger got on "very well now; more genuine & free than we were, under the shadow of Gordon Square" (D, 1:225), when he was thought to represent the "dreaded upper world" of respectability and authority (MRF).

Although Virginia and Roger now agreed "on many points," separating aesthetic from "literary" emotions was no easy matter. Naturally responding to human interest rather than to plastic design, Virginia observed "that to like the wrong thing, or fail in sufficiently liking the right jars on [Roger], like false notes, or sentimentality in writing" (D, 1:228). Terms such as *right* and *wrong* seem ironic here and convey little of Virginia's earlier aesthetic inferiority complex.

When Roger published the Omega's *Original Woodcuts by Various Artists*, Virginia judged it "very magnificent but fearfully expensive" (VWL, 2:296–97). This collection showed Vanessa, Duncan, and Roger all improvising on a common motif: a woman in a tub. Roger's *The Stocking* (Figure 32) is the most realistic and sensual. Vanessa's *Nude* (Figure 33) isolates the woman from the tub and, indeed, any pretext of an environment.[48] Duncan's *The Tub* (Figure 34) features a woman, seen from the rear, in a tub, but his principal interest is in the pattern of recurring curved shapes. For Virginia, these radically different renderings of the same topic would have reemphasized the irrelevance of subject matter, the significance of treatment.

Roger accused Vanessa of being "such an old armadillo there's no forcing you out of yr protective armour of indifference." Recognizing the appropriateness of this metaphor, Vanessa wrote to him that while Duncan was in London, Bunny so pestered her with questions that "at last I had to become armadillo with a vengeance & pretend to be asleep" (RF/VB, 19 November 1918; VB/RF, 21 November 1918). Vanessa's armadillo-like defenses usually prevented her from admitting either need or gratitude. But the packages of cakes, soups, shortbreads, anchovies, sausages, and other edible delicacies that Virginia sent to Charleston made Vanessa call her a "wicked, wicked, extravagant & monstrous ape," and Virginia's letters, bubbling with gossip, made Vanessa "wish peace came oftener to distract you from your duties and make you write to your poor neglected sister" (VBL, 217). Peace was both public and private, as arrangements concerning illustrations for "Kew Gardens" for a time made harmonious and "almost daily letters necessary" between the sisters (VBL, 219).

Arrangements for the birth of Vanessa's baby were complicated by the isolation of Charleston, the lack of servants, and the presence of Bunny and the two boys (now aged ten and eight). Vanessa asked Virginia to keep the boys, even though she feared being in Leonard's bad graces if the children ruined Virginia's health (VB/VW, 19 December 1918). On Christmas Day, 1918, Vanessa gave birth to her third child, a daughter. Vanessa and Clive agreed that she would be raised as his daughter, with Duncan's fatherhood kept secret even from the baby. Virginia wrote to Ka delightedly of her niece: "As you politely say beauty and talents seem inevitable, considering her Aunt" (VWL, 2:309).

The Charleston situation became indeed "really too wild," with curious little boys, an incompetent doctor, and further melodrama. Bunny wrote to Lytton that the baby's "beauty is the most remarkable thing about it. I think of marrying it; when she is twenty I shall be 46—will it be scandalous? Its brothers were very restless last night and made secret plans to find out if it was being born and how." The doctor nearly killed the baby "by giving her carbolic" (David Garnett/LS, 25 December 1918; VB/Margery Snowden, 26 February 1919).

Virginia had been pleased that two of the Olivier sisters, former Neo-Pagans in Rupert Brooke's group of friends, were planning to become doctors, as she could get in "speaking distance of you, which is quite impossible with the ordinary male doctor" (CS, 88–89). Now a physician, Noel Olivier not only could be spoken to but could be enlisted. She sent her friend Dr. Marie Moralt to Charleston; she rescued the baby but then found herself the target of Bunny's sexual desires. (Apparently Duncan had left them alone together, knowing that the basically heterosexual Bunny planned to seduce her.) Noel Olivier reported the story:

> Poor Moralt was seized with terror at the thought of his gigantic size & strength & determination, but luckily she had no inclination to submit to anyone she found so repulsive. So after a melée in which (apparently) her hands only were covered with slobber—she fled from the room, leaving him amazed. . . . Next

> morning Vanessa was delighted with Moralt's account . . . and declared that it was the first time a virgin had escaped him in that house. (NO/James Strachey, 2 January 1919)

After the rowdiness of the brothers, the near death of the baby, and the near rape of the doctor, Vanessa was desperate for help. Leonard fetched the boys from Charleston to Asheham on 28 December (LWD). Planning to return from Asheham to Hogarth House, Virginia envisioned an elaborate series of city activities with the boys, including the zoo, the Ballets Russes, and tea with Roger at the Omega. She planned to teach Julian, who she thought possessed "the Stephen gift of language," the Greek alphabet. Both boys were to write "essays and stories for me to judge, so we shall be very literary, and I hope to persuade Quentin to be a writer and not a painter" (VWL, 2:312, 219, 312).

Virginia's plans, however, were interrupted by a toothache that kept her awake at night. She had the tooth extracted, but bleeding and headaches followed. Even though they were physiological in origin, headaches could signal the onset of an affective episode. So, to guard against such a recurrence, Virginia wisely stayed in bed for eleven days, under her doctor's orders to rest.

Both Woolfs assured Vanessa that the headaches had nothing to do with the children, but then Julian had a nightmare one night and "dashed in at midnight" to Virginia, destroying her prescribed rest. Vanessa despaired: "Apparently Virginia is ill & Leonard wants me to take the children away from them." When the Woolfs deposited the boys, they also lent Vanessa their servant Nelly, although Leonard, Vanessa later confided to Roger, "was furious at her coming & insisted on her going back the minute I could let her" (VB/RF, 7 January and 18 February 1919).

Roger and his daughter, Pamela, began to call Charleston "Wuthering Heights." With Vanessa busy with her baby, the servants became rude, slovenly, and incompetent. One was caught stealing. The "absolute horror and chaos" at Charleston culminated in Vanessa's discovery that the housekeeper-governess was writing a novel in which she "simply takes down our conversations verbatim." Vanessa's commitment to openness did not extend to having her private life published.[49]

Roger wrote to Virginia, "Life has been very vorticist of late & I feel a little giddy." Vanessa was insisting that "the sooner I'm broken in to the baby the better," and he had to admit that "of course she'll have her way—she always does." Roger, their father, and various current inhabitants of

Gordon Square tried to entertain the boys. Duncan came up from the country and took them to see the Tower of London. He tried to follow Virginia's cultural and literary scheme by taking them to see the British Museum and Shakespeare's *Twelfth Night*. Finally, however, as Quentin Bell remembers, "We proved too much of a nuisance and were sent back to Charleston" (*Elders and Betters*, 7). To Virginia, Vanessa emphasized Roger's beneficence: He "has been making all kinds of arrangements for [the boys'] & yours & my benefit" (RF/VW, 11 January 1919; VB/VW, 11 January 1919). Such remarks could make Virginia feel that she had failed as both a sister and an aunt.

Soon the baby, now named Angelica, was gaining weight, and Virginia's headaches ceased. She described her illness visually: "If I were a painter I should only need a brush dipped in dun colour to give the tone of those eleven days. . . . But painters lack sub[t]lety; there were points of light, shades beneath the surface, now, I suppose, undiscoverable" (D, 1:239). Virginia needed the net of language to capture those points of light.

In England, peace seemed to have brought only strikes and further deprivations (VWL, 2:325). In Russia, the promise of the Russian Revolution faded when the czar and his entire family were assassinated. And at Charleston, Bunny Garnett found Marxist rhetoric a convenient excuse for freeloading (D, 1:183). Virginia knew that he drained Vanessa's psychological and financial resources, but she did not know that he had made a habit of seducing houseguests or that he was Duncan's preferred sexual partner. Virginia was irritated—Roger infuriated—by his presence in Vanessa's household (RF/VB, 7 February 1919). Neither, however, seems to have known of his vow to marry the baby or of his attempt to rape the doctor. Vanessa was irritated chiefly with Virginia, whose considerable efforts to find servants for Vanessa were not appreciated:

> I heard such stories from Nelly—most interesting accounts of the Woolf menage—also accounts of the many times Virginia had lost me the chance of excellent servants, hearing which made me feel less scrupulous about borrowing Nelly for a short time & also determined to do no more servant hunting through Virginia. (VB/RF, 18 February 1919)

Vanessa did not explain how Virginia ran off servants while trying to find them.

Vanessa had accused her sister of being as affected as Lady Ottoline Morrell, but "on a minor & shabbier scale" (VBL, 191). She again associated the two, telling Lytton that "Ottoline must be dropped. Virginia is all very well but not Ott." Vanessa's disaffection with Virginia had revived to a point that Roger told her he thought an attack on Virginia "wld. have soothed yr. moments of utmost irritation" (VB/LS, 9 March 1919; RF/VB, 19 May 1919). Given Virginia's willingness to endanger her own health by hunting for servants and lending out Nelly, Vanessa's peevishness seems a churlish revival of the Stephen family's habit of criticizing others. But her aloofness only goaded Virginia to continue "bombarding" Charleston with food and presents (VB/VW, 13 May 1919). Over the years, her gifts to Vanessa included occasional £100 notes, a bed, a fur coat, a refrigerator, and a lifetime of dazzling letters. Virginia assumed that Vanessa at least appreciated her letters.

Eight years earlier, Virginia had despaired: "To be 29 and unmarried—to be a failure—childless—insane too, no writer" (VWL, 1:466). In 1919, though childless, she was married, blessedly sane, and content. She described marriage as a "renewal" for both Woolfs. But she still had not progressed very far on "my own particular search—not after morality, or beauty or reality—no; but after literature itself" (D, 1:214).

Approaching her thirty-seventh birthday and remembering that the ever energetic Roger was fifty-two, Virginia reassured herself that vitality and work could continue after fifty. She evaluated her friends and their work: Despite Lytton's remarkable contribution to modern biography, she thought he lacked creativity. Indeed, she faulted the Stracheys for not embarking on such adventures as Roger's Omega Workshop, the Stephens' living arrangement at Brunswick Square, and the Woolfs' Hogarth Press (D, 1:234–36). Although she imagined the pleasure that Virginia Woolf at fifty would take from rereading her diary, she did not include it or her fiction in her tally of adventurous accomplishments.

Even though the Woolfs were hoping to be able to publish Virginia's future novels themselves (VWL, 2:353 and note), her "poor old sluggard" (D, 1:250) *Night and Day* was sent to Gerald Duckworth in March 1919. At thirty-seven years old, Virginia had entombed her demons away from her sane flourishing self. Only her book reviews and one iconoclastic fictional piece suggested the direction in which her imaginative powers were growing. In her letters, Virginia was self-effacing: "Nobody cares a hang what one writes, and novels are such clumsy and half extinct monsters at the best" (VWL, 2:391). But to her diary, Virginia acknowledged that she had enjoyed writing *Night and Day* and that it was a more mature book than *The Voyage Out*. She compared well with her contemporaries in terms of originality and sincerity, but she could not yet compare with Proust and Joyce. Her problem, with both form and philosophy, was the process of "discarding the old" when she was unsure about what could replace it. Nonetheless, Virginia's secrecy and modesty had so placated her demons that she even wondered whether someday she would escape the quandary of needing to write while at the same time fearing public reaction, the "curse of a writers life" (D, 1:257, 259, 214).

In her review of Dorothy Richardson's *The Tunnel*, Virginia objected that although her fellow experimenter had discarded the novel's "old accepted forms," she had not replaced them with a new shapeliness or structural design (*Times Literary Supplement*, 13 February 1919, 81; VWE, 3:10–12). Virginia appropriated this painterly concept for fiction, but she did not apply it uniformly to painting. She also told Vanessa that the relatively realistic Sickert was her "ideal painter" (VWL, 2:331). Such a proclamation could be taken as evidence of the shakiness of her aesthetic conversion, as a declaration of her aesthetic independence, or as a perverse affront to her Postimpressionist friends.

At the end of February 1919, the Woolfs went to Asheham for a long weekend and received the disturbing news that their lease would not be renewed. Despite Leonard's advice about not becoming too attached to a house, Virginia would have despaired over losing the lovely Asheham were it not "for the devil of starting something new in me." As always, writing rescued her from depression. On 4 March 1919, Leonard returned to Hogarth House, and Virginia spent the night at Charleston, "by no means a gentleman's house," where "the atmosphere seems full of catastrophes which upset no one" and her tiny niece examined the fire "meditatively, with a resigned expression" (D, 1:248–50). After she had sampled the chaos of Vanessa's domestic circle, Roger found himself "oddly jealous" (RF/VB, 11 March 1919).

A dinner party in March 1919 at the Isola Bella restaurant reminded Virginia that Roger was still grieving over losing Vanessa to Duncan (D, 1:260), whom Virginia now could call "in a sort of way, my brother in law" (VWL, 2:341). At the "brilliant affair in the Bohemian style, with a great deal of wine, & talk of books & pictures, & a general air of freedom & content," Virginia and Leonard, Roger, Vanessa and Duncan, Clive and Mary Hutchinson talked "a great amount of Athenaeum gossip, all secretly delighted with our own importance" (D, 1:259–60).

Katherine Mansfield's husband, Middleton Murry, had become the editor of the *Athenaeum*. His "most brilliant list of contributors on record" included Virginia and Roger, who seem to have launched a concerted campaign for revisionism in the verbal and visual arts (D, 1:259–60). Although both of them had several articles published in the *Athenaeum* throughout 1919 and during part of 1920, neither Virginia nor Roger remained so sanguine about Murry and his "male atmosphere." Virginia came to the conclusion that the "orthodox masculine thing," which Murry embodied, resembled "stupidity" (D, 1:265).

Virginia jokingly proposed Duncan as art critic for the *Athenaeum*, aided by Vanessa's judgment and Virginia's writing, but Virginia and Roger were more nearly collaborators (VWL, 2:341). Roger wrote that Jean Marchand "becomes a creator, and not a mere adapter of form" ("Jean Marchand," 178–79; VD, 195–98). Virginia criticized a study of fiction for its "didactic style," simplistic notions of realism or romanticism, and failure to rethink the assumptions of the "incipient realist" ("The Anatomy of Fiction," *Athenaeum*, 16 May 1919; VWE, 3:44). Roger reiterated that the most valuable aesthetic emotions "emanate from purely formal relations," and he continued to distinguish between "specifically aesthetic emotion" and "literary" responses ("Explorations at Trafalgar Square," 211, and "Art and Science," 434–35; VD, 55–59).

This distinction could also be applied to the two Woolfs' attitude toward H. G. Wells. Leonard had launched his *International Review* in 1919 with a stellar group of contributors, even inducing Wells to serialize his latest novel. Leonard praised Wells not only for the beauty of his writing but also for its "large truth" (LWL, 280 and note). But even while Leonard was serializing Wells, Virginia's essay "Modern Novels" (later revised as "Modern Fiction") dismissed Wells, Bennett, and Galsworthy as materialists who wrote about "unimportant things." She lamented that current novelists failed to capture or contain the "essential thing" but followed "a design which more and more ceases to resemble the vision in our minds." She argued that art should resemble not the merely material world but a more subjective "life or spirit, truth or reality" (VWE, 3:30–37), as Roger had said the Postimpressionist painters expressed, "by pictorial and plastic form [,] certain spiritual experiences" (VD, 166). Virginia Woolf thus articulated in aesthetic terms the image of primal oneness that had haunted her from infancy. The mind receives "myriad impressions" that figure "as the semi-transparent envelope, or luminous halo, surrounding us from the beginning of consciousness to the end." Seeing events "commonly thought big" as unimportant, she agreed that all creative artists must "get behind the forms of actual objects, to create them afresh as it were from within." The "incipient realist" who does not get behind the forms is little more than a quack.[50]

Such revisionist theories extended even to Virginia's diary, for "there looms ahead of me the shadow of some kind of form which a diary might attain to." She envisioned an elastic all-embracing shape, another image of primal unity, that was nevertheless "not slovenly" (D, 1:266). On 5 May, she marked the twenty-fourth anniversary of her mother's death. Against that recollection, she reflected on her "most melodious time" with Leonard at Asheham after Easter and on their happiness, constituted for them both especially by their increasingly promising work (D, 1:269).

The Woolfs themselves (having lost Barbara to marriage, predictably with Nick Bagenal, in early 1918) hand set and covered in paper Murry's *The Critic in Judgment*, Eliot's *Poems*, and Virginia's "Kew Gardens," with woodcuts in the last by Vanessa; the three were published as a set on 12 May 1919. Virginia herself typeset Eliot's highly allusive and fragmented poems.

An anonymous review "cobbled" together by both Woolfs—perhaps written mostly by Leonard—articulated the revisionism that Eliot had achieved and Virginia was testing in her short fictional pieces (VWL, 2:373, 437). The Woolfs praised Eliot's poems in Roger-like formalist terms for their "careful juxtaposition of words" and their Leonard-like "even more careful juxtaposition of ideas," and they associated innovations such as Eliot's with Postimpressionism (*Athenaeum*, 20 June 1919, 491).

"Kew Gardens" was the last of Virginia's fiction that the Woolfs hand printed (Willis, *Leonard and Virginia Woolf as Publishers*, 28–33). As verbal art stripped of "literary" associations, it is not as "thingified" as a story about cut glass and furniture would be, but it does reduce four couples to a pattern like that "of falling words." The scene is as visually ambiguous as a Postimpressionist painting; even the narrative voice has to ask, "What were those shapes?" (CSF, 93, 94). In "Kew Gardens," Virginia seems to test Roger's current theory that "art is a blasphemy. We were given our eyes to see things, not to look at them." Thus we display "a very considerable ignorance of visual appearances" (VD, 33). A major achievement of "Kew Gardens" is making us look at the sensuous language; another is structuring the narrative so that pattern, rather than "reality," is paramount; and still another is removing from the center the human beings and foregrounding space. Thus Virginia created verbal art that was as blasphemously "nonliterary" as the visual arts.

She felt shaky about this departure from representation, but she "had Roger's praise of Kew Gardens by the way, so I suppose I'm still safe" (D, 1:273). She also had Katherine Mansfield's private praise, but Katherine's review in the *Athenaeum*, though praiseworthy, condescended with the idea that Virginia's world was "on tiptoe" with such terms as *exquisite*, *lovely*, and *poise*. It is no wonder that Katherine asked Virginia to "forgive" her for such comments (*Athenaeum*, 13 June 1919, 459; Mansfield, *Collected Letters*, 2:324). Then a laudatory review in the *Times Literary Supplement*—which brought in a flood of orders for "Kew Gardens" through the mail slot on the front door of Hogarth House—diminished the sting of Katherine's review. At last, writing had secured for the joyous Virginia the "pleasure of success" (D, 1:280).

The success of "Kew Gardens," however, was nipped by another dismissal of Virginia's visual literacy. She had taken great pleasure in using Vanessa's illustrations for the front and end matter of "Kew Gardens," even telling her sister, "So I suppose, in spite of everything, God made our brains upon the same lines, only leaving out 2 or 3 pieces in mine" (VWL, 2:289). Vanessa's many circular images do recall Virginia's image of consciousness as a "luminous halo," and the collaborative effort promised to demonstrate the "same lines" of their perception.[51]

But when the volume was printed, Vanessa objected to the Woolfs' reproduction of her images and threatened never to illustrate any more of Virginia's stories: "Nessa & I quarrelled as nearly as we ever do quarrel now over the get up of Kew Gardens." But rather than let old demons come out of their cupboards, Virginia withstood the "critical blast of Charleston" by offering Vanessa the chance to supervise the second edition (D, 1:279, 280). She also let Vanessa know that Roger's printer praised the Woolfs' layout and that another artist thought it "perfect." Finally, Virginia vindicated herself by asking Roger, rather than Vanessa, to arrange the new edition, shortly followed by a reprint of one thousand copies of "The Mark on the Wall" (VWL, 2:364, 380, 371 note, 367 note). The extent to which this quarrel threatened Virginia's sense of self is suggested by her means of retaliation: She bought an odd round house atop a hill in the center of Lewes, in order to give herself "value in my eyes" after Vanessa's attack (VWL, 2:369).

Despite being directors of the Omega, Vanessa and Duncan had begun taking commissions independently, and their defection was hastening the Omega's slippage into financial ruin. Roger was bitter, although finally "magnanimous & forgiving" (D, 1:228). Both Roger and Virginia had achieved some emotional equilibrium concerning Vanessa. Neither now demanded more affection than she was prepared to offer. And Roger considered it a "great compliment" to Vanessa that he had "thought it worthwhile to go on through all those years" to achieve a harmonious friendship (RF/VB, 28 August 1919).

Meanwhile, the emotional triangle at Charleston "had grown unbearably tense." Even though he saw "unconscious malice" in Vanessa and found her "ungenerous to those for whom she felt no affection," Bunny propositioned her in Duncan's absence. Vanessa simply refused him, lest they upset Duncan. He already was jealous of Bunny's heterosexual affairs, and the two men often fought. In addition, Duncan took no responsibility for his daughter but thought of all three children simply as Vanessa's.[52] Clive bore financial responsibility for his sons and acted as titular father to Angelica, even taking her to visit his parents as if she were their granddaughter; otherwise Vanessa, with a changing cast of servants, held her household together alone. Of all her guests, only Roger and Virginia seem to have understood the appropriateness of bringing her food. Vanessa had found Duncan's amorality a welcome relief from Roger's residual Quakerism, but now she suffered for her choice, cloaking her distress in stoic silence.

While his bitterness toward Vanessa subsided, Roger's irritation with "the cultured" boiled. The privileged classes were raising money to buy a Gauguin for the National Gallery when he was still being vilified "for having a show of him" nine years before (RFL, 2:448). But there were compensations: Gide, Picasso, Matisse, Vildrac, and Derain all visited with Roger during this time. Picasso took considerable interest in Omega crafts, but even he could not rescue the floundering enterprise. While in London, Derain designed ballet sets, which Roger applauded for "entire subversion of all standards of verisimilitude and probability" ("Scenery of 'La Boutique fantastique,'" 466).

On 23 May 1919, at Gordon Square, the Bells, with the help of Roger and Duncan, gave an after-ballet reception for Picasso and Derain in what Clive told Vanessa he hoped would be "a great success in the half bohemian, half mandarin style" (CB/VB, 3 May 1919). Since the party did not begin until 11:00 P.M., Virginia did not attend, but she did see the Ballets Russes in June 1919 (LWD) and experienced firsthand another testimony to Postimpressionism's transformative influence on the arts.

In this environment, Virginia invited herself to tea and dinner with Roger, even though she feared he would be "entertaining Picasso and Lady

Cunard." To others' fulsome praise for Roger, she added her own: "O if we could all be like Roger!" (VWL, 2:356). Virginia must have voiced such praise publicly, for Katherine gossiped that Virginia "wrote to me the other day telling me that Roger was good enough to be pleased with me (!!)." The other half of this mutual admiration was equally irritating: Privately, Katherine wrote that Roger "thinks that Virginia is going to reap the world. . . . After a very long time I nearly pinned a paper on my chest, 'I too, write a little.' But refrained" (Mansfield, *Collected Letters*, 2:334, 333). Although the mutual admiration between Roger and Virginia bothered Katherine, it was an outgrowth of their friendship, their shared devotion to Vanessa, and their shared aesthetic.

Through their articles in the *Athenaeum* and elsewhere, Virginia and Roger established a joint essentialist aesthetic that united the efforts of pen and brush.[53] Roger criticized the "literary" English aesthetic taste because it required a knowledge of literature and history in order to understand painting ("Art and Science," 434; VD, 55–59). Two months later, Virginia reviewed an exhibition of traditional, historical, and sentimental paintings at the Royal Academy. Like Roger, who condemned the snob value of art's "symbolic currency" (VD, 50), Virginia dismissed the paintings' "power to make the beholder more heroic and more romantic."[54]

Virginia was, however, more conflicted than her review implied. In a letter to Vanessa, she displayed philistine inclinations, confessing that she actually liked such "literary" paintings as *Cocaine* and *The Wonders of the Deep* (VWL, 2:378). By rejecting Postimpressionist values, she seems to have been reacting to Vanessa's attack on the visual layout of "Kew Gardens." In print, however, she proved her aesthetic credentials by reviewing art and dismissing the very "literary" approach that she had confessed to Vanessa was partly hers. She ended her review facetiously by lamenting, "I must leave it to Mr. Roger Fry to decide whether the emotions here recorded are the proper result of one thousand six hundred and seventy-four works of art" (*Athenaeum*, 22 August 1919; VWE, 3:93). Of course, he would say that such emotions were not the "proper" result of works of art.

Roger did not even accept the emotions captured in Leonardo da Vinci's masterpieces, insisting that in the *Paragone*, Leonardo's emphasis on painting's ability to reproduce nature had resulted in the subordination of formal elements to psychological illustration. He argued that Leonardo's interest in "the minuter and even more psychologically significant movements of facial expression demanded a treatment which hardly worked for aesthetic unity." Roger claimed that "European art has hardly yet recovered from the shock." Thus whereas Virginia attacked the pernicious influence of materialism in fiction, Roger attacked Leonardo and "literalism and illustration" in painting (VD, 128, 129).

After Virginia's impulsive house purchase in the center of Lewes, the Woolfs looked for a more suitable country home. They had often walked across the River Ouse from Asheham to the post office in the village of Rodmell and then down a narrow twisting lane lined with chalk-bound flint and brick walls over which steep slate gables and even an ancient low thatched roof appeared. They had passed by the church and looked at an unprepossessing vine-covered cottage called Monk's House (Figure 35). There was barely room for a few flowers to nestle between the flint wall and the frame house, but behind it, the view was almost limitless, and the orchard "was always beautiful" (L. Woolf, in Noble, *Recollections of Virginia Woolf*, 249). Monk's House looked back on meadows and pastures sloping down to the River Ouse, which slowly wound south into the English Channel. Across the Ouse to the east were Asheham and Charleston and the rolling downs the Woolfs so loved. And now Monk's House was for sale. The Woolfs were able to dispose of the round house (at a profit of £20), and on 1 July 1919, they bought Monk's House.

After a final dinner party at Asheham on 24 August, with Pernel Strachey, E. M. Forster, Maynard Keynes, Mary Hutchinson, Clive, Vanessa, Duncan, and Roger as guests, the Woolfs moved on 1 September. Virginia's story "The Haunted House" is an elegy to the graceful, romantic Asheham House and a tribute to the loves it nurtured.[55] Protected by trees and walls from the fierce winds roaring over the downs, Monk's House itself was a plain, inward-turning cottage. The house was "lower than the garden outside, so that one stepped into it rather as one steps into a boat" (Garnett, *Deceived with Kindness*, 109). Although considerably less elegant than Asheham or Hogarth House, Monk's House grew on Virginia "after the fashion of a mongrel who wins your heart" (D, 1:302). Like Asheham, Monk's House had a history, "a quiet continuity of people living" whose tranquillity cheered Virginia (LWA, 2:196).

Peace might have renewed business at the Omega, but Roger was too weary of its burdens, so he cut his losses and closed it down. Then he cheered himself up by traveling through France in the fall of 1919. Roger claimed to find in Aix-en-Provence more interest in art than in "the whole of London" (RF/VB, 3 November 1919; RF/VW, 26 October 1919).

During this time, both Virginia and Roger began to reexamine their formalism. Reviewing "Modern French Art at the Mansard Gallery," Roger virtually retracted his dismissal of the "literary." Echoing the Woolfs on Eliot, he saw in Picasso "the possibility of a new kind of literary painting. Ideas, symbolized by forms, could be juxtaposed, contrasted and combined almost as they can be by words on a page." Roger went on to "translate" a Survage painting into prose.[56] His "translation" follows an "ekphrastic" tradition of description: It begins, "Houses, always houses, yellow fronts and pink fronts jostle one another, push one another this way and that way, crowd into every corner and climb into the sky." Writing in what he called "the Virginian manner," Roger's words reflect particular images and offer, despite his theory, a referential ekphrastic translation. He concludes "that Survage is almost precisely the same thing in paint that Mrs. Virginia Woolf is in prose," for modern literature approximated the "same kind of relationship of ideas as Survage's pictures give us!" ("Modern French Art at the Monsard Gallery," 724).

Virginia wrote to Roger that she "enjoyed immensely" his sense of the arts' shared enterprise, and she invited him to write more "translations" from paint to words for their press.[57] Then Virginia displayed a perverse pleasure in praising three paintings by "an early Victorian blacksmith" she had bought with the house.[58]

Dedicated to Vanessa, *Night and Day* was published on 20 October 1919. Virginia invested that event with enormous psychological and professional significance. At twenty-two, Virginia had warned that she might "give up writing," for "I only get criticism and abuse." Then she had threatened to "give up literature and take to art" (VWL, 1:170). At thirty-seven, Virginia threatened that if *Night and Day* were "pronounced a failure, I dont see why I should continue writing novels." She knew that such fears were "the usual writers melancholies," yet they nevertheless were real (D, 1:297). Unlike earlier worries, Virginia's 1919 apprehensions offered an opportunity for self-consolation. She examined her depression and reflected that "at one's lowest ebb one is nearest a true vision. I think perhaps 9 people out of ten never get a day in the year of such happiness as I have almost constantly; now I'm having a turn of their lot" (D, 1:298). Soon she found herself "more excited & pleased than nervous" (D, 1:307).

Vanessa anticipated the novel in solipsistic terms. As she told Roger, "I am the principal character in it & I expect I'm a very priggish & severe young woman, but perhaps you'll see what I was like at 18" (VBL, 205). When the novel finally came out, she informed Roger, "Clive says its a

Victorian novel—only subtler" (VB/RF, 24 October 1919). Although Vanessa surprised Virginia by stating that the novel's setting gave her "the horrors," Virginia still complimented her sister on her knowledge of writing. And Virginia acknowledged her debt: "I think I'd rather please you than anyone, if only because I feel that its all your doing if I have any wits at all" (VWL, 2:393).

Night and Day brought unexpected results. Sibyl Colefax (later Lady Colefax) said that she would like to entertain Virginia at Argyll House, a center of political and cultural society. Virginia had described herself in comparison with the aristocracy as a "wretched, illdressed, inkstained Suburban" (VWL, 2:238). When she considered Mrs. Colefax's invitation, Virginia's immediate reaction was, as she told Roger, "I have no clothes" (VWL, 2:396 and note).

Book reviews were similarly unexpected, but most were gratifying. Katherine Mansfield's review of *Night and Day*, however, seemed calculated to wound Virginia. She described the book as "a novel in the tradition of the English novel" that showed Woolf "unaware of what has been happening" in current artistic experiments (*Athenaeum*, 21 November 1919). Virginia suspected "spite" as the cause of such a dismissal.[59] *Night and Day* was not written like "Jane Austen over again," even though it was traditional (D, 1:316).

Roger had facetiously wondered, "Is this mighty stream of invention all turned aside [from painting] into the modern novel?" ("Allied Artists at the Grafton Gallery," 626–27). Now he was traveling through France, visiting André Gide, hoping (unsuccessfully) to meet Renoir before he died, and denouncing the bourgeoisie, which he found "the same everywhere & everywhere equally repugnant to me" (RF/VW, 26 October 1919).

Fearing that Roger would echo Katherine's indictment of *Night and Day*, Virginia sent his copy to his London address, "not daring to send a large, old fashioned, high minded English novel to Roger in the south of France" (VWL, 2:395). But Roger did not find the novel "old-fashioned." Virginia told Janet Case that he sent her four pages of commentary that made her "tremendously relieved" (VWL, 2:416).

For the next eighteen years, no one would be able to accuse Virginia Woolf of writing old-fashioned fiction. But another charge leveled in 1919 haunted her for the rest of her life: Vanessa implied that Virginia did not "know anything about human nature" (VWL, 2:263). Margaret Llewelyn Davies seems also to have said something comparable, for Virginia concluded, "You'll never like my books, but then shall I ever understand your Guild? Probably not" (VWL, 2:399). Katherine privately decided that Virginia possessed an exquisite sensibility that saw all "that a bird must see—but *not humanly*" (Mansfield, *Collected Letters*, 2:333–34). Morgan Forster made a similar complaint publicly. Because Virginia's main characters are so rational, Forster thought them flat and passionless. Virginia tried to dismiss Forster's complaint as a product of his limitations, but the critics offered so many opinions that "the wretched author who tries to keep control of them is torn asunder" (D, 1:310). The accusation of "some narrowness—some lack of emotion" was infuriating to Virginia, especially when she had offered such a generous answer to *The Wise Virgins* (D, 1:313).

The year 1919 ended with Virginia taking self-reflexive pleasure in rereading her diary and finding that it had "grown a person, with almost a face of its own" (D, 1:317). Even though both the Woolfs had been ill, Virginia wrote, "I daresay we're the happiest couple in England" (D, 1:318). On New Year's Day, they huddled before a fire at Monk's House during a storm, and Virginia wrote to Ka about her reaction to the criticism of *Night and Day*: "Old creatures have crept from what I supposed to be their graves to hiss at me." These "creatures" were her demons, raised by the fear that people despised her characters, if not her, but she successfully reburied the creatures. New Year's found Roger at Charleston cooking enough *boeuf en daube* to feed residents and guests for days. When the storm passed, the Woolfs hiked over the downs together; Leonard pruned and tied fruit trees (he said he "pruned trees and my finger" [LWD]), and Virginia read his *Empire and Commerce in Africa*, "to my genuine satisfaction" (VWL, 2:410).

Leonard's editorship of the *International Review* ended with the journal's demise. Then in 1920, he became editor of the *Contemporary Review*'s international supplement. Virginia's reputation, too, was growing, as American publishers vied to publish new editions of both her novels. Seeing Vanessa, however, again undermined her sense of achievement, as she wrote to Ka Cox (now Ka Arnold-Forster): "I crept home a good deal abashed." But she was not cast down and managed "under the shelter of Leonard, to fan a subdued glow" and restore a positive self-image (VWL, 2:410).

Later, Virginia assessed her own accomplishment in painterly terms: *Night and Day* "taught me a great deal, or so I hoped, like a minute Academy drawing: what to leave out: by putting it all in" (VWL, 6:216). She made herself "copy from plaster casts, partly to tranquillise, partly to learn anatomy" (VWL, 4:231). Like the academy painter who learns anatomy and representational rendering before beginning to paint expressionistically, Virginia Woolf mastered her craft even while she prepared to transcend it.

As Leonard Woolf explained, "Before she broke the mould, she thought she ought to prove that she could write a novel classically in the traditional mould" (LWA, 2:400). These explanations echo Roger's defense of the Postimpressionists, who proved themselves masters of representation before they began their nonrepresentational experiments ("Grafton Gallery—I," 332; VD, 167–68). Proving herself in command of the traditional form would protect Virginia Woolf against accusations that she could not "draw" character and plot.

Virginia had successfully subdued her demons, reestablished her always volatile relationship with Vanessa, and endured public exposure through two very personal novels. She had even survived Morgan Forster's criticism of her characters and Katherine Mansfield's dismissive review of *Night and Day*.

But even though Virginia was now a proven novelist, she was not yet a reformer of the novel. First she needed to test how content in fiction could be directive of form. She needed to complete her "new experiment" with aesthetic emotions (D, 1:297). By practicing those emotions in her short fiction pieces and other emotions "of a literary nature" in *Night and Day*, she could graft them to create, as an "intelligent" correspondent predicted, "the forerunner of a new species of book" (VWL, 2:400).

8
Real Discoveries and the Right Path

1920–1923

Virginia Woolf began her diary for 1920 by describing at Monk's House how the "winter down & meadow" took her "breath away at every turn. Heres the sun out for example & all the upper twigs of the trees as if dipped in fire; the trunks emerald green; even bark bright tinted, & variable as the skin of a lizard." In this world, "human beings have figured less than the red berries, the suns & the moon risings" (D, 2:3, 5). Her later diaries are saturated with descriptions like these, which are far more sensuous and harmonious than those in her early diaries, which had been shaped by some felt obligation to analyze. While still at Monk's House, Virginia also read Leonard's *Empire and Commerce*, with "genuine satisfaction, with an impartial delight in the closeness, passion, & logic of it; indeed its a good thing now & then to read one's husbands work attentively" (D, 2:5).

In the five years since the seas of horror had separated her brain's ways of apprehending, Virginia had opened her rational self more and more to visual experiences: viewing pictures; developing her aesthetic talents; wrapping her words, as it were, around the visible universe; and conceptualizing language in spatial terms.[1] Furthermore, she now could think of drawings in non-"literary" terms. For instance, in a January 1920 review of a book of caricatures entitled *Personalities*, she faulted the drawings for being "literary" and "not aesthetic"; she even argued that the drawings' effect would have been improved if viewers had had no knowledge of their subjects and hence no "human or literary susceptibilities to placate." In the same article, Virginia resurrected the *paragone* as facetious envy, despairing of words and longing "Oh to be a painter!" (*Athenaeum*, 9 January 1920; VWE, 3:163–66).

In this mood, the more visual *Voyage Out* seemed "a direct look ahead of me," a "more gallant & inspiriting spectacle" than *Night and Day* (D, 2:17). As she revised *The Voyage Out* for the American edition, Virginia made it less "literary" and more aligned with her current experiments in fiction (Heine, "Virginia Woolf's Revisions," 400).

Rather than be "swept into the vortex" at Gordon Square around which Vanessa, Adrian, and their families and friends lived when in London, the Woolfs, after five years of leasing, purchased Hogarth House (with the adjoining Suffield House) in Virginia's name (D, 2:8, note). They also owned Monk's House, which Virginia now found "tiny compared with [Vanessa's Charleston], like a nut shell" (VWL, 2:444), perhaps because a low-beamed ceiling cramped the ground floor. In addition, Virginia still owned a share in 22 Hyde Park Gate. Although they were now substantial property owners, the Woolfs continued to live very frugally, as Leonard was determined not to repeat his father's improvidence. With Clive's help, Vanessa leased Charleston and owned only a share in 22 Hyde Park Gate.

At first, Charleston had provided only the barest necessities "but not an ornament"; Virginia had felt her life "padded" in comparison (D, 1:270). By 1920, however, Charleston could no longer be said to lack ornament. Vanessa and Duncan had begun painting spirited decorations on almost all available surfaces. Now their dwelling, like their living styles, was exuberant and daring. Charleston's past existence as a boarding house seems to have determined its destiny. Besides Vanessa, Duncan, and three children, their maids, nurses, and governesses (sometimes with their own children) all lived under the same roof. The largest bedroom was reserved for Clive Bell, who spent much of his time there writing, seeing his sons, and serving as a titular father to Angelica. Maynard Keynes frequently retreated to Charleston to write. In fact, to young Quentin Bell, Vanessa, Duncan, and Maynard formed a virtual "triumvirate" (*Elders and Betters*,

96). As Vanessa had insisted, Roger had been quickly "broken in" to little Angelica's existence, and he now was more willing to visit, sometimes even when Duncan was there. Most weekends brought a host of other guests, often including Clive's mistress, Mary Hutchinson.

Determined that her children not suffer the repressions that had characterized her childhood, Vanessa encouraged all three of them—in the name of "beauty" or "health"—to play outdoors at Charleston naked (Bell and Garnett, *Vanessa Bell's Family Album*). Angelica did not question her mother's judgment, especially when, for Vanessa and Duncan, nakedness "vaguely recalled something like *l'Age d'or*, by Ingres, and, I suppose, the ancient Greeks" (to me, 11 August 1995). Vanessa presided over her kaleidoscopic household with amazing calm and good sense. By contrast, Virginia's two homes and ordered life seemed restrained, safe, almost dull.

Meanwhile, Roger—his venture with the Omega Workshop having nearly bankrupted him—had been forced to sell the home he had designed to his own love and taste. He then set up housekeeping with his sister Margery in the London suburb of Camden Town, where Virginia joined him for an overnight stay on 18 January 1920. She found Roger "shrunken, aged?" but supposed that she projected from her depressed sense that "he didn't much care for N.&.D." She speculated that disappointment with the English people's disregard for his painting (as well as the failure of the Omega and the need to sell Durbins) made him "testy" about Clive's cribbing his ideas, but on that point, she thought his resentment justified.[2]

Having completed the traditional *Night and Day*, Virginia Woolf was now free to test the limits of aesthetic emotion. The non-"literary" writing she had begun with "The Mark on the Wall" culminated in even more radical experiments, mostly written in 1920. Her one-page "Monday or Tuesday" is a meditation, probably on a square of marble. "String Quartet" releases the narrator's mind to flights of imagination; in a remarkable synesthesia of musical and visual images and of words considered as words, the story testifies to the power of disinterested perception. The most extreme of these experiments consists of two paragraphs, one headed "Blue" and the other headed "Green," offering diverse associations produced by the colors, but not by an identifiable scene.[3]

Virginia's "Unwritten Novel" illustrates the powers but also the dangers of freeing the imagination from referential constrictions. It ends with a paean to actuality, the "adorable world" (CSF, 121). The implication is that aesthetic emotions can exist in harmony with, rather than in opposition to, "literary" emotions, a combination that suddenly made Virginia's first truly experimental novel a possibility. Her diary for 26 January 1920 records her excitement:

> The day after my birthday; in fact I'm 38. Well, I've no doubt I'm a great deal happier than I was at 28; & happier today than I was yesterday having this afternoon arrived at some idea of a new form for a new novel. Suppose one thing should open out of another—as in An Unwritten Novel—only not for 10 pages but 200 or so—doesn't that give the looseness & lightness I want: doesnt that get closer & yet keep form & speed, & enclose everything, everything? My doubt is how far it will enclose the human heart—Am I sufficiently mistress of my dialogue to net it there? For I figure that the approach will be entirely different this time: no scaffolding; scarcely a brick to be seen; all crepuscular, but the heart, the passion, humour, everything as bright as fire in the mist. Then I'll find room for so much—a gaiety—an inconsequence—a light spirited stepping at my sweet will. Whether I'm sufficiently mistress of things—thats the doubt.

Determined not to indulge "the damned egotistical self; which ruins Joyce & Richardson," she charted "a path for me" very different from the paths that such activists as Leonard Woolf, Bernard Shaw, and Maynard Keynes were following (D, 2:13–14).

During the war, Maynard Keynes's work for the Treasury had alienated him from many of his pacifist Bloomsbury friends, but his *Economic Consequences of the Peace*, published in 1919, with its attack on the political aggrandizement of the Allies and their shortsighted determination to make Germany pay for its aggression, recertified his Bloomsbury credentials. After a private showing of Duncan's paintings, at which Virginia found herself adoring the pictures and marveling at the old group's eminence, walking to tea with Roger at the *Burlington Magazine* offices, Virginia heard Maynard say that his sales were in the fifteen thousands. Because Adrian also had felt himself overshadowed by Bloomsbury (Virginia suspected that he meant her), she fought not to let Maynard's success likewise overshadow hers (D, 2:18).

Virginia's belief in the value of art over activism was further challenged by the Fabian and social historian Beatrice Webb, who seems to have accused Virginia of keeping Leonard from running for Parliament when "we want men of subtle intellect." Virginia wondered, "What is 'right' & who are 'we'"? (D, 2:20). Rather begrudgingly, Virginia characterized Maynard's writing as "a work of morality, I suppose" (D, 2:33).

Leonard, with H. G. Wells, was now working for the League of Nations, work that Virginia more easily conceded to be moral. When Wells invited the Woolfs for a weekend visit, knowing that Leonard was "married to Virginia Woolf," he confessed, "I used to write novels in the dear old past. Will she come?" However materialist and ethical Wells's prose may have seemed to her, Virginia seemed pleased by the invitation.[4]

In March 1920, Madge Vaughan briefly reentered Virginia's world. Back in 1904, when she had been exiled with her cousin Will Vaughan and his lovely wife, Madge, in Yorkshire, Virginia had suspected that Madge was becoming a conventional matron. The events of 1920 only confirmed her suspicions. Madge wanted to rent Charleston for a family holiday while Vanessa was abroad, but in what Vanessa called an "almost incredibly impertinent" letter, Madge said that she could not bring Will and their children to Charleston until she spoke with Vanessa about Angelica's parentage. Vanessa found herself "half amused and half furious." In a wonderfully spirited letter, she asked, "Why on earth should my moral character have anything to do with the question of your taking Charleston or not? I suppose you don't always enquire into your landlords' characters" (VBL, 235). Like most of their relatives, Madge was thereupon summarily "dropped" by the sisters.

The month of March was momentous for old Bloomsbury. Virginia had found a form for her diary that suited her so well that she could "fancy [that] old Virginia, putting on her spectacles to read of March 1920 will decidedly wish me to continue. Greetings! my dear ghost; & take heed that I dont think 50 a very great age. Several good books can be written still; & here's the bricks [her diary entries] for a fine one" (D, 2:24). That March also saw the Woolfs delighting in their walks in Kew Gardens where the magnolias were beginning to flower (LWD).

At the time, Virginia was writing "Freudian Fiction" (even though she still claimed to know nothing about Freud) and distinguishing between the brain's scientific and artistic sides (*Times Literary Supplement*, 25 March 1920, 199; *Contemporary Writers*, 152–54). In the same month, she heard "Roger's speech at the [1917] Club & my first effort—5 minutes consecutive speaking—all very brilliant, & opening the vista of that form of excitement not before glimpsed at" (D, 2:21). The aesthetic vista that she and Roger opened to the others at the 1917 Club must have included the possibility of prose, like painting, that was not representative of particulars but embodied a vision. Also in March, Molly MacCarthy started the Memoir Club. Its ostensible purpose was for members to read aloud chapters from their autobiographies, but the ulterior motive was to prod Desmond MacCarthy into writing the brilliant novel expected of the man supposedly destined to be Henry James's successor (Bell, 1:103; D, 2:23, note).

At the first Memoir Club meeting, Roger showed his composition skills, even though he was too objective and controlled for Virginia's taste. Vanessa, however, demonstrated why she usually kept her emotions tamped down. She began "matter of fact: then overcome by the emotional depths to be traversed; & unable to read aloud what she had written" (D, 2:23). This pattern was almost repeated at the second Memoir Club meeting: "Leonard was objective & triumphant; I subjective & most unpleasantly discomfited." Virginia was afraid that her memoir (lost now) was "egotistical sentimental trash" that created "a kind of uncomfortable boredom on the part of the males" (D, 2:26), who, following the "masculine point of view which governs our lives" (CSF, 86), valued the English public school habit of keeping a stiff upper lip, of never—no matter what—baring one's soul.

In April, Virginia went with Roger to a showing of African carvings. She admitted to finding them "dismal and impressive" but nevertheless recognized that "something in their style might be written" (VWL, 2:429). The next day, Roger Fry's essay "Negro Sculpture" appeared in the *Athenaeum*, elaborating the "discourse" that Virginia had heard at the gallery: "Without ever attaining anything like representational accuracy [African sculptors] have complete freedom." He argued that the English lack the autonomous "power to create expressive plastic form" that Africans possess innately.[5] Immediately after considering the expressive freedom of the African carvings, Virginia had her first vision of *Jacob's Room*, her initial attempt to expand the method of her short experimental fictions into a novel form.

Even while she was planning this new novel, an important critic for the *Times* scoffed at her review of Henry James's letters (*Times Literary Supplement*, 8 April 1920; VWE, 3:198–207) because, although anonymous, it supposedly illustrated how "the most immaculate of women . . . *will* sentimentalise their men friends" (D, 2:29, note). Despite debilitating tremors of failure, Virginia seems not to have worried much previously about being patronized as a woman. Now, however, she suddenly realized that whatever she wrote might be considered "pretentious" and that she would be dismissed as a mere "woman writing well." Accordingly, fearing that her "unwritten novel" "will certainly be abused," Virginia's progress on *Jacob's Room* was momentarily stalled (D, 2:29–30).

The Woolfs spent many long weekends and holidays walking, gardening, and writing at Monk's House. Returning to Hogarth after these breaks, an ever increasing book list kept them printing most afternoons. Sometimes Virginia was content living in Richmond, as she liked returning to Leonard there after visiting Vanessa at 50 Gordon Square (which was now Adrian's home: Vanessa now leased the upper part, and Maynard had taken over 46 Gordon Square). Virginia also liked "continuing our private life, unseen by anyone." Although Vanessa seems to have virtually ceased writing to her, she must have agreed to telephone.[6] When she failed to do so, Virginia became "fidgety" (D, 2:36, 38). She missed seeing Vanessa's children and wanted "them to want to see me. Thats how it begins, and we all know how it ends" (VWL, 2:429). Such a comment on her emotional needs served partially to liberate her from them.

In April and May 1920, Vanessa and Duncan traveled through Italy, visiting the art critic and connoisseur Bernard Berenson at his villa, I Tatti.[7] In her letters, Virginia needled Vanessa about both her aloofness and her sense of superior taste, telling her that she planned to "paint the house bright yellow. As you'll never come here, I can give my taste full rein" (VWL, 2:432).

In 1920, when Desmond MacCarthy became literary editor of the *New Statesman*, Roger and Virginia began to shift much of their reviewing from the *Athenaeum* to that magazine. Perhaps that was why the *Athenaeum* "adopted Katherine as their writer of genius." Although she claimed not to be jealous (VWL, 2:430), Virginia confessed to Roger that she was.[8] Roger told her that her confession was healthy and assured her that she need not feel jealous because only she had "any idea of what the essential texture of prose should be" (RFL, 2:486). Typically, Roger explained in aesthetic terms of prose "texture," and Virginia reacted in emotional terms to Katherine's dismissive review of *Night and Day*. When Virginia next met Katherine, she adopted a self-dismissive pose, only to hear Katherine praise "Kew Gardens" and even *Night and Day*. Nonetheless, after all Virginia's work as the compositor and Leonard's work as the printer of "Prelude," Katherine moved on to larger presses (Willis, *Leonard and Virginia Woolf as Publishers*, 26). Nevertheless, despite their rivalries, defections, and mixed signals, these two dedicated female writers developed, at least for the moment, a "certain understanding" (D, 2:44–45).

On 28 June 1920, Virginia dined with Vanessa in London and found her own "attempt[s] at sensation" bested by Vanessa's story of "nothing less than the death of a young man at Mrs Russell's dance." Some of the guests had gone out to sit on the roof, and one young man had moved away, "perhaps to light a cigarette, stepped over the edge, & fell 30 feet onto flagstones." Adrian had tried to be of help, but the man died in an ambulance. This "strange event—to come to a dance among strangers & die—to come dressed in evening clothes, & then for it all to be over, instantly, so senselessly" intrigued Virginia (D, 2:51). That juxtaposition of a party and a death would come to fruition toward the close of her fourth novel, *Mrs. Dalloway*.

On 21 July, when Leonard was unwell, Virginia attended a party that Vanessa gave at 50 Gordon Square. Virginia was developing a beguiling and teasing party persona, but she wondered how artificial her talk really was (D, 2:54). The next day, Virginia and Leonard traveled to Rodmell for a two-month stay in Monk's House, but Virginia went up to London alone twice to see Katherine Mansfield. On her visit of 2 August, Katherine asked Virginia to review her book *Bliss*, but Virginia declined. "Despicable as I am, I find myself liking to hear her underrated" by Clive and Vanessa. She also had the odd experience of spending a night at 46 Gordon Square and sleeping in what were now Maynard's quarters. Virginia was unable to remember who "had his room when we lived there—how many years ago?" The "ease & rapidity of life in London" impressed her, especially the proximity of friends—"Roger, Duncan, Nessa, Clive, & so on" (D, 2:55).

On 23 August, Virginia returned to London from Rodmell to say good-bye to Katherine, who was about to travel to Mentone, on the French Riviera, for her health. Regretting that she had liked hearing Katherine underrated, Virginia reflected that a "woman caring as I care for writing is rare enough I suppose to give me the queerest sense of echo coming back to me from her mind the second after I've spoken" (D, 2:61). Back at Monk's House, Virginia continued to write *Jacob's Room* every morning, "feeling each days work like a fence which I have to ride at, my heart in my mouth till its over" (D, 2:56). Leonard gardened, continued his political writing, and gathered a collection of stories based on his Ceylon experiences, to be called *Stories from the East*.

Virginia still reinforced Vanessa's prejudice: "My command over artistic language is still too poor to let me launch out," and "art criticism, as you observe, is not my strong point" (VWL, 2:428, 472). Apparently, Vanessa did not read the sophisticated art criticism Virginia was writing. Nor did she know how much pride Virginia took in talking with painters or in making a metaphor "about paint with a yellow glaze on it" to describe the field and downs between Rodmell and Charleston (D, 2:58). Vanessa gloated that when Arthur Clutton-Brock, the art critic of the *Times*, visited, Virginia "couldnt get in a word ag[ainst] Clutton Brock."[9] Virginia reluctantly admitted that she resented the Clutton-Brocks being at Charleston (D, 2:57).

Although in his early art criticism, Roger had emphasized the dramatic idea behind a painting, by 1920 he wanted to "disentangle our reaction to pure form from our reaction to its implied associated ideas" (VD, 92, note). How one might disentangle one's reaction to a novel's "pure form" from its drama and human associations and whether one should do so were matters over which Virginia puzzled as well. Her everyday musings kept returning to such aesthetic issues. She worried that formalist experimentation might be only a cover for failure to understand people, especially those of different classes, and she wondered, for example, how an old blind beggar woman, who held "a brown mongrel in her arms & sang aloud," came "to be there, what scenes she can go through, I can't imagine. O damn it all, I say, why cant I know all that too?"[10]

Roger responded to Virginia's suggestion that he write more prose poems for the Hogarth Press by sending her a short collection. Her influence—what he called "my attempt in the Virginian manner" (RFL, 2:500)—surfaces in his images: "The freezing crests of waves in an ocean the colour of creme de menthe"; "The nursery firelight on a child's bare legs"; "A clot of foam turning endlessly in a mountain stream." Many of these so-called poems, however, are theoretical aphorisms that are not at all Virginian: "Only those capable of asceticism understand luxuries—the wise man hates both"; "To apprehend perfectly the sensation of a single moment would be life eternal" (RF/VW, ? June 1920). Fascinated with Mallarmé's poems because they are minimally referential, Roger was translating them for a Hogarth edition, and Virginia thought his translations captured the strange feeling of the originals (VWL, 2:439).

The Woolfs' press now was growing so rapidly, "like a beanstalk" (VWL, 2:434), that they had either to hire a new assistant or to abandon their publishing venture (VWL, 2:439–40; Willis, *Leonard and Virginia Woolf as Publishers*, 25). So they hired Ralph Partridge, who later married Dora Carrington, although (in this odd ménage) she really was in love with Lytton Strachey, and Lytton really was in love with Ralph.

Leonard began writing a weekly foreign affairs column for the *Nation*.

He also was working for the Labour Party's Irish and Anti-Slavery Committees, the League of Nations, the Fight Famine Council, the Cooperative movement, and the Peace with Ireland Council (LWD). For her part, Virginia at last could dictate her conditions to the *Times Literary Supplement*, to write only lead articles on topics of her choosing. She congratulated herself: "To have broken free at the age of 38 seems a great piece of good fortune—coming in the nick of time, & due of course to L. without whose journalism [or the income from it] I couldn't quit mine." But she was appalled to find that Lytton, Clive, and Mary Hutchinson thought that her immortality would rest on her letters rather than on her fiction, including "poor Jacob" (D, 2:63, 66, 63).

The possibility that Virginia would successfully open realms of experience new to language was jeopardized again when in the patronizingly titled *Our Women*, published in September 1920, Arnold Bennett argued that "women's indisputable desire to be dominated is . . . a proof of intellectual inferiority" (*Our Women*, 106). Now it became increasingly apparent that even as Virginia was forging her way toward a new form for a new novel, even as she was taking up Bennett's plea for a writer to do in words what Postimpressionist painters had done in paint, when such dazzling experiments were in the hands of a woman, Bennett and his ilk considered them to be merely pretentious. To make matters even worse, Desmond MacCarthy, writing in the *New Statesman*, agreed with Bennett. Then T. S. Eliot told Virginia that among living writers, he most admired "Wyndham Lewis & Pound.—Joyce too," completely neglecting Virginia's "claims to be a writer." At first, she prided herself for not letting Eliot sap her self-confidence, but she did not write in her diary for six days (D, 2:67).

During this time, Virginia spent a night at Charleston, where she found Maynard there looking repulsive, like a "gorged seal." Quentin Bell describes Maynard as an "acquisitive eater" and remembers that Clive objected not only to Maynard's table manners but also to his "claim to omniscience," bolstered by his recent financial successes (*Elders and Betters*, 97). Virginia suspected that she was biased against Maynard because he had read neither of her novels. She joked about the emotional strain Charleston had often imposed, noting that despite the tensions, this time she so enjoyed her visit that Leonard found her the next day "neither suicidal nor homicidal" (D, 2:69).

Eliot's judgments had stung her more than she first realized, for when she returned to her diary, she noted defensively that "what I'm doing is probably being better done by Mr Joyce." She even wondered "what it is that I am doing" and judged herself "distanced by L. in every respect." Virginia fought this crisis of confidence by writing a "counterblast" to Arnold Bennett (D, 2:69) and Desmond MacCarthy, contending that external restraints rather than incapacity had held women back over the centuries. Like Leonard, she argued against oppression: "The degradation of being a slave is only equalled by the degradation of being a master." At the end of this second exchange of public letters, Desmond capitulated, promising that "if the freedom and education of women is impeded by the expression of my views, I shall argue no more."[11]

Despite her assertive and principled response, at bottom Virginia had made the appalling discovery that many men, as she later wrote, secretly despise women (D, 2:149). Since childhood, Virginia had been afraid that her best friend or sister despised her. Now the notion that half the human population despised the other half thoroughly disheartened her. She wrote to secure affection and describe those aspects of experience usually ignored by male novelists. By the early 1920s, however, she recognized that the masculine point of view threatened both enterprises.

In *Jacob's Room*, Virginia was inspired not only by the African carvings but also by the typesetting of the book itself, which forced her to think about the placement of words on a page. Indeed, her typescript shows that she conceived the book in terms of separate set pieces focusing on physical places. "St Pauls," for example, even has its own numbering (MHP/B12). Accordingly, Virginia decided that all but the last one-page chapter in *Jacob's Room* should be broken up into discrete blocks of print. These blocks of varying size are sometimes logically and sequentially connected, but most often, each is a separate non sequitur. The white space surrounding these blocks further emphasizes the discontinuity of human experience.[12]

By September 1920, Virginia's "experimental mood" had taken her through about half of *Jacob's Room* (D, 2:67), although Virginia became fearful that experimentation could lose more than it gained. She declared the aestheticism of the French novel that Roger gave her as trivial compared with Hardy's literary density (VWL, 2:440). Roger's insistence on the plasticity of French, Spanish, Italian, and African, but not English, art began to seem suspicious. Now his talk of "significant harmony" emerging from a "sequence of relations"[13] alone began to wear on Virginia's literary sensibilities.

In the story "Solid Objects," Virginia commented on formalism, dehumanization, and Roger's "The Artist's Vision":[14] A man begins collecting objects, which he places on a mantelpiece to admire aesthetically, as

Vanessa and Duncan must have done when they painted pictures of solid objects on their mantelpieces.[15] The man's obsession seems to be a direct application of Roger's notion that "every solid object is subject to the play of light and shade, and becomes a mosaic of visual patches" (VD, 36). The man becomes so fascinated with the "play of light and shade" on his objects that he gives up his friends and his active life (he had intended to run for Parliament). The story ends with him in complete isolation, contemplating his collection of solid objects. The intention of this story is to ridicule all who privilege the merely aesthetic over the ethical, the artistic over the practical.

But no matter how much Virginia may have ridiculed disinterested contemplation, she was no more comfortable with activism. Leonard's bid for Parliament alarmed her lest "L. might devote himself to a cause" (VWL, 2:435). She defied Beatrice Webb's admonition by doing "my utmost to ruin his career" (D, 2:71) yet then deemed her diary trivial because she was "always talking about people, never about politics" (D, 2:92).

Although Virginia's generally buoyant letters present the person that she knew her friends and family adored, her diary in the fall of 1920 tells a different story, one of depression threatening. On 25 October, she thought about why life was "so tragic; so like a little strip of pavement over an abyss. I look down; I feel giddy; I wonder how I am ever to walk to the end." Her panic was spurred by the old demons: "having no children, living away from friends, failing to write well, spending too much on food, growing old." And she felt isolated: "Like a lantern stood in the middle of a field my light goes up in darkness." Although the comparative isolation of Richmond enabled the huge amount of work the Woolfs undertook, it also meant losing touch with Virginia's family: The "labour of going to London is too great. Nessa's children grow up, & I cant have them in to tea, or go to the Zoo." These were feelings that could catapult her into the abyss. In 1915, imagining teaching literature had rescued her from mania. In 1920, although she still regretted living so far away from Vanessa and her family, writing carried her safely along that strip of pavement over the abyss of depression: "To write Jacob's Room again will revive my fibres" (D, 2:72, 73). After another quarrel with Vanessa and Duncan, she declared her independence: If they "tell me lies—very well, I don't go near them till I'm asked." Admitting that they might not even notice her absence, "in all that shindy of children &c," Virginia immediately retracted and reassured herself of the "immense value" of her visits to Vanessa (D, 2:75).

Not only to make up for her first "sentimental" Memoir Club reading but also to reestablish her value at least as an entertainer, Virginia wrote and delivered on 17 November a "fearfully brilliant" paper that played to her friends' antiestablishment sentiments. (The memoirs were "written for friends who took a humorous rather than a reverential view of eminent Victorians" [RF, 22].) This paper was "22 Hyde Park Gate," her exposé of George Duckworth's pomposity, conventionality, and frustrated sexuality. It was the memoir that ended with the proclamation that George Duckworth was not only the Stephen girls' caretaker but "was their lover also" (MB, 177). Virginia reflected that Leonard's presentation was "much more impressive with so much less pain" (D, 2:77 and note), whereas she had neglected her diary and her novel for this carefully honed, dazzling piece of humor, written to vindicate herself before the Memoir Club and to please Vanessa as well.

By December 1920, Roger's *Vision and Design* had been published: "Everyone's book is out—Katherine's, Murry's, Eliot's," whereas Virginia, nearing her thirty-ninth birthday, had published only two novels. She confessed: "In my heart I must think [Katherine Mansfield] good, since I'm glad to hear her abused" (D, 2:78, 79). Virginia reminded herself "that both Mr & Mrs Woolf slowly increase in fame" (D, 2:80) and again claimed not to be jealous of Katherine's success (VWL, 2:454). She first adopted the same dismissive resentful attitude toward the "sumptuous" *Vision and Design*, judging it "rudimentary compared with Coleridge" (D, 2:81), but then she wrote to Roger, "I'm in the middle of your book and fascinated" (VWL, 2:450). In her review, Virginia called *Vision and Design* "probably the most important art criticism of our time," and she regretted that "there is to-day no literary critic who does for literature what Mr. Fry does for painting" (Review of *Vision and Design*, by Roger Fry, 28–29). Later Vanessa told Roger that "Virginia talked most enthusiastically about your book. I think you can quite believe all she says. She was full of its praises" (VB/RF, 30 August 1921).

Near the end of 1920, because Vanessa had "put off" an invitation to the Woolfs, Virginia weighed the advantages of a fire at Hogarth House against Bloomsbury society. She looked forward to seeing the garden and taking "a soft grey walk" from Rodmell. At Charleston, Roger asked Vanessa to bring with her from London supplies for a house party so that he could cook another "boeuf en daube which'll last for days off & on." He hoped his desire to see her would not seem "excessive" and assured her that his feelings were safely "this side [of] idolatry" (RF/VB, 30 or 31 December 1920). Virginia and Leonard spent Christmas at Monk's House, Virginia still sharing Roger's idolatrous need for Vanessa. If she got to Charleston, she probably also shared his *boeuf en daube*. Virginia began 1921 with a letter to Vanessa that again satirically inverted the terms of their relationship: "How you must be wanting to hear from me!" Even if she became Vanessa's neighbor in Gordon Square, she offered the ironic assurance that they need not see each other (VWL, 2:453).

Seriously worried that she might not see Katherine again, Virginia wrote to her in February that despite their differences, "we have some of the same difficulties." She admired the "transparent quality" of Katherine's prose and explained why she "lashed out" at Arnold Bennett—because it was "important that women should learn to write" and Bennett's assumptions would inhibit their trying. Virginia ended by pleading, "Please Katherine, let us try to write to each other" (CS, 127–30).

Gathering her energies for a last "sprint" to the end of *Jacob's Room* (D, 2:89), Virginia took a brief respite from writing, traveling in March to Manchester, where Leonard lectured at the university "on Africa & the League" (LWD). Virginia visited a Manchester art gallery stocked with popular nineteenth-century "literary" paintings. They proved the absurdity of what Roger called the "representation of particulars" ("a lamb with 8 separate whiskers," for example), and Virginia could gleefully notify Vanessa that "your art is far more of a joke than mine" (VWL, 2:458). The next week, when the Woolfs set out for Cornwall, Virginia wondered why she was "so incredibly & incurably romantic about Cornwall" (D, 2:103). The answer was the beauty and also the memories, "Well, Leonard, & almost 40 years of life, all built on that, permeated by that" (D, 2:103). *Permeated*, the operative word, suggests the ideal fusion of self and world.

In 1921, Virginia's formalist and intensely visual short fictional pieces were published together as *Monday or Tuesday* for the Woolfs by a Richmond printer (D, 2:96, note). These pieces appeal to aesthetic emotions through such formal elements as rhythm and design. A month before their publication, Vanessa, who had designed the cover and furnished four woodcuts, "mercifully" approved these stories (D, 2:98). Her designs were heavily impressed on cheap paper, but Vanessa seems not to have complained, even though she and Maynard Keynes often shared gibes at the poor quality of the Hogarth Press's printing.[16]

As the publication date neared, pressure mounted on Virginia. Again she worried that the reviewers would be patronizing and dismissive (D, 2:98), and she viewed herself "a failure as a writer. I'm out of fashion" (D,

2:106). Rather than being dismissive, however, Kenneth Burke offered "condescending praise" (D, 2:118). Then, after the *Times* failed even to notice the work's radicalness, Virginia considered "never writing any more—save reviews" (D, 2:106). Although this was not so drastic a plan as earlier threats to give up writing for painting, it does suggest the extent of her anxiety.

In *Art*, Clive Bell oversimplified Roger Fry's aesthetics and announced that the "representative element . . . is irrelevant" (27). Roger had been so dogged by critics who simply equated him with Clive that in 1919 he protested to the editor of the *Burlington Magazine*: "Whatever Mr Clive Bell may have said, I personally have never denied the existence of some amount of representation in all pictorial art. . . . What I have suggested is that the purer the artist, the more his representation will be of universals and the less of particulars." (In these terms, *Monday or Tuesday*'s concern with universals like the nature of perception, rather than with particulars like individual personalities, renders these experiments in verbal abstraction very pure.) Roger admitted that he had used the word *representation* rather carelessly, but he defended himself in almost Emersonian terms against a foolish consistency because modernist aesthetics had to grapple with such new and difficult problems.[17]

In 1920, a committee of the Royal Academy visited Hampton Court and drew "up a violent report against" Roger Fry's rather haphazard attempts at restoration. Thus in 1921 he "came to stay again" with the Woolfs for about two weeks (LWD). He made one last desperate attempt at restoring what he now called "my Mantegna" and was glad to be relieved of this albatross of a responsibility so he could "begin to paint seriously again" (RFL, 2:507–8).

Virginia now found Roger had "the nicest nature among us," for he was "entirely without meanness." She could discuss her ambitions and fears with him and try to counter her feeling that she had lost her public just as she "was becoming more myself. One does not want an established reputation, such as I think I was getting, as one of our leading female novelists" (D, 2:107). Even though Roger assured her that she was "on the track of real discoveries" (D, 2:109), critics who dubbed her the "ablest of living women novelists" (D, 2:115 and note) seemed determined to belittle her achievement.

In 1919, Leonard (with Bloomsburian objectivity) had judged Lytton Strachey's current work as "weak & mechanical, as if his method had already become a trick and his style a formula" (LWL, 280). Then, in her

1920 "experimental mood," Virginia judged Lytton's *Queen Victoria* as too conventional (D, 2:65). Consequently, given her jealousy of and guilt over Katherine Mansfield and her resentment of the lavish praise heaped on Lytton's *Queen Victoria* (dedicated to Virginia, under her stipulation that Lytton use her "name in full" [D, 2:87]), Virginia faced the "question of praise & fame" (D, 2:106).

In any case, Virginia did not want to be jealous of Leonard's success with *Stories from the East.* She thus carefully analyzed her emotions so she might "medicine" herself (D, 2:108–10). Unlike painting or its implied equivalent, "making patchwork quilts or mud pies," the intense pressure of experimental creation in *Jacob's Room* called "upon every nerve to hold itself taut" (D, 2:129); therefore, following a pattern established in that 1908 dream of her father's disapproval of her novel, Virginia planned a less exacting "Reading book," a collection that would become *The Common Reader* (D, 2:120).

Word of Katherine Mansfield's literary triumphs again made Virginia question Katherine's genius (D, 2:87). But news that Middleton Murry had had an affair with Princess Elizabeth Bibesco (née Asquith) and was publishing a story by her in his *Athenaeum* while Katherine might be dying (D, 2:91) aroused Virginia's sympathy and guilt. Their rivalry was doubled because the *Nation* was going to merge with the *Athenaeum*, a prospect that distressed Leonard, as to him Murry represented "the literary underworld" (LWA, 2:147). Leonard's doubts mounted when Murry attacked the Hogarth translation of Gorky's *The Notebooks of Tchekhov.*[18]

In May, while Vanessa and Duncan were in Paris to see the famous exhibition of Dutch paintings at the Orangerie, Roger wrote to Vanessa about a scheme that suggests how loyal and forgiving members of old Bloomsbury could be. Despite her public quarrel with Desmond about women's intelligence, Virginia, Leonard, and Roger took up Molly MacCarthy's campaign to get her husband to write a book; they even attempted to "trap Desmond's talk by a shorthandist." Roger wrote to Vanessa from the train to Richmond "loaded with bottles of Chablis which it's hoped will let him loose. It's very like Henry James's story of the Coxon Fund." They "got Desmond well on after two bottles of Chablis and he told some of his stories in great style and Miss Green (whom Virginia calls the chest of drawers—it's a terribly exact description) managed to take it all down" (RFL, 2:509, 510). The triumph, however, produced only "the dullest thing you can imagine to read." Virginia entertained Vanessa with the story but cautioned her not to tell Desmond "in one of your strange moods of expansive hilarity" (VWL, 2:471).

Absolved of his Mantegna responsibility, Roger was caught up in a whirl of other activities. He gave a lecture to the Royal Institute of British Architects entitled "Architectural Heresies of a Painter," and although "all the younger architects were on my side," he once again outraged the establishment. Roger also organized the "Nameless Exhibition," at which paintings were displayed (and sold) anonymously. The press treated the show "as a game played against the critics" (RFL, 2:511, 510); Virginia, however, took it seriously. She put herself "in training" for it, and despite misspelling Van Gogh (offering a phonetic spelling of the Dutch pronunciation "Van Goch" here and in *Jacob's Room* [40]), she told Vanessa she could "rather plume" herself on identifying artists' styles (VWL, 2:469–70). She purchased the *Farm* by the London painter Frederick Porter (RFL, 2:511 and note) and sympathized with Roger when Clutton-Brock of the *Times* "pounced upon him."[19]

Meanwhile, when Roger entertained her children in Camden Town, he wrote to Vanessa about two-year-old Angelica's filling her mouth with "rather squashy cake" and talking "incessantly & most eloquently but almost unintelligibly" (RF/VB, 23 and 25 May 1921). Vanessa said she knew no one other than Roger to whom she could entrust her children, "certainly none of my relations! not Virginia nor the Stephens" (VBL, 249–50).

Because Virginia was becoming "rather well known," she conceded that "now L may have considerable success" (D, 2:98). Clearly, for her psychological balance, Leonard's "right" to success was conditioned by her own success. Leonard's story "Pearls and Swine" was named one of "the great stories of the world," and an American agent said that he could guarantee Leonard £3,000 a year if he would continue writing fiction about the East. Leonard refused (LWA, 2:250–52). Virginia first reacted in the pattern begun with Vanessa, but then she reminded herself how "idiotic" she was to let Leonard's success make her "immediately think myself a failure" (D, 2:116 and note).

Virginia was joking when she said that her "feelings as an artist were ravished" (VWL, 3:470), but they were indeed nearly ravished when on a visit to Gordon Square, she had to defend *Night and Day* against Maynard Keynes. Having at least begun to read her second novel, he accused her of too many dull details. She explained that "you must put it all in before you can leave out." In any case, Maynard still thought her memoir of George was "the best that you ever did." Virginia's confidence was dashed, "for if George is my climax I'm a mere scribbler" (D, 2:121).

Roger was known to believe that experimental writing did not offend the public as much as experimental painting did.[20] *Monday or Tuesday* proved him correct, for rather than being outraged, as at the Postimpressionist exhibitions, the public simply ignored it.[21] Few people other than Roger, T. S. Eliot (D, 2:125), Leonard, and possibly Vanessa realized the significance of her experiment. But Virginia learned from it the methods that would make her novels among the great achievements of modernist prose. Years later she recalled:

> I shall never forget the day I wrote The Mark on the Wall—all in a flash, as if flying, after being kept stone breaking for months. The Unwritten Novel was the great discovery, however. That—again in one second—showed me how I could embody all my deposit of experience in a shape that fitted it—not that I have ever reached that end; but anyhow I saw, branching out of the tunnel I made, when I discovered that method of approach, Jacobs Room, Mrs Dalloway etc. (VWL, 4:231)

At the time, however, Virginia suspected that readers such as Middleton Murry considered her book to be merely "trivial" (D, 2:125). Murry's wife, Katherine Mansfield, was indeed dying, and again the devils of loss and guilt danced around Virginia, even after so much health and happiness and accomplishment.

Virginia spent the summer of 1921 battling headaches and sleeplessness (again with sleeping medicine). Leonard kept temperature charts showing that at the times of her 1913 and 1915 breakdowns, her temperature had vacillated wildly. In June 1921, it was almost as low as it had been in May 1915 (LWD). Virginia abandoned her diary, took to her bed, and wrote no letters. Progress on *Jacob's Room* stalled. Roger visited at Hogarth House in June when "V not well" (LWD). On 5 July, Leonard recorded an ominous sign—"V rather worried" (LWD)—as moderate depression and mania seesawed in Virginia's brain (Caramagno, *Flight of the Mind*, 309). The Woolfs spent the last half of June at Monk's House and the first half of July at Hogarth House. Roger visited them at least twice in July, bringing along his woodcuts to carve (LWD). Then on 18 July, Leonard hired a car to take Virginia back to the country. Only then was she able to sleep without medication.

On 20 August, Leonard went to Charleston to tell Vanessa about Virginia's illness. Not well herself, Vanessa wrote to Roger the next day:

> I suppose I ought to console myself by thinking of Virginia except that it never is any consolation to think of people worse off than oneself. She sounds to me still pretty bad as whenever she begins to get better her monthlies come on & she gets bad again. She can't do anything at all.

Then she wrote to Virginia: "Leonard told me I was to write to you. Its unfortunate for you, you havent got a literary sister now. Its with the greatest difficulty as you know that I take to the pen."[22] Vanessa's excuse for not writing was not comforting, especially given Virginia's ongoing concern for Vanessa's health, the multitudes of letters she wrote to her "Dearest" sister, and Vanessa's extensive letters to Roger, Clive, and, when they were apart, Duncan.

The frequency of Roger's visits suggests that like Leonard, he was very worried about Virginia. Roger was the one person, other than Leonard, who could be counted on to endorse her writing vision without jealousy and hence to help her battle the demons that again beset her: "All the horrors of the dark cupboard of illness once more displayed [themselves] for my diversion." For two months, Virginia did not write (D, 2:125 and note). She felt "chained to my rock: forced to do nothing; doomed to let every worry, spite, irritation & obsession scratch & claw & come again" (D, 2:132).

Finally, without a chemical antidote but with amazing strength of character, Virginia returned the worries and horrors to their cupboard. She made a life-enabling vow "that this shall never, never, happen again." She also pointed out some of the consolations: "to be tired & authorised to lie in bed is pleasant" after "scribbling 365 days of the year as I do." She was grateful for "the fundamental security" of her life with Leonard, compared with "the old fearfully random condition." Furthermore, the "dark underworld has its fascinations as well as its terrors." Virginia claimed not to envy Vanessa her "summer of dissipation" and to crave nothing "but quiet & an active brain" (D, 2:125, 126). Quietness might, however, be upstaged by "an infinite capacity of enjoyment horded [*sic*] in me, could I use it" (D, 2:133). By September 1921, Virginia experienced the salvatory "recovery of the pen; & thus . . . I felt reborn" (D, 2:134).

Roger left for France in the fall of 1921. As Vanessa readied herself and her entourage to join him in Saint-Tropez, Virginia asked her, "Have you forgotten me?" She cited indifference but relied on familiar leverage: "You'll miss my letters though in France." She urged Vanessa to write and also to send a sketch of "La Tropez" (VWL, 2:483, 494). Yet, however much she feared that her sister had abandoned her, Virginia remained Vanessa's staunch ally. When their cousin Dorothea Stephen (daughter of Leslie's brother, James Fitzjames Stephen) objected to Vanessa's living with a man who was not her husband, Virginia refused to see Dorothea until she understood that Virginia approved of Vanessa's living arrangements (VWL, 2:492). Apparently, Virginia did not know that Duncan was now having affairs with a succession of male lovers (Bunny Garnett having found other lovers, including a wife). Even though the Charleston situation was neither harmonious nor comforting, Vanessa hid her distress from Virginia and everyone else.

Virginia told Roger that she did not write to Vanessa "on principle" (VWL, 2:485) because Vanessa did not write to her. She resumed a comic sense of competition with him. Replying to an elaborate letter (RFL, 2:513–15), Virginia complained that she could not endure his writing so well: "Would you like it if I dashed off a little sketch of the eclipse of the moon last night, which entirely surpassed your great oil painting of the Rape of Euridyce—or whatever it is?" While both knew that such "great" subjects, though commonly thought big, did not necessarily make "great" art, Virginia was defensive about the absence of grandeur in *Jacob's Room* (VWL, 2:484, 485).

In "a state of abject admiration at the Virginian style," Roger examined the *paragone* they played out in their letters: "I know quite well that this queer business of writing is quite as odd and peculiar as ours of putting paint on. It only looks like ordinary writing because we both use words but we use them quite differently and certainly I'm not taken in by the similarity" (RFL, 2:515). Defending her intoxication with words, Virginia resurrected the old notion that any dissimilarity between the arts would, from Roger's point of view, redound to the credit of the visual arts (VWL, 2:489–90).

Despite her guilt over not appreciating Katherine's fiction, Virginia privately admitted that reading it made her want to "rinse my mind," perhaps in Dryden (D, 2:138). Her feelings were further complicated by her deepening distaste for Middleton Murry. In late 1920, Virginia had been pleased to report that Roger had "broken" with the *Athenaeum* (D, 2:74). Murry's article "Mr. Fry Among the Architects" (*Nation and Athenaeum*, 27

August 1921) was an "orthodox masculine" retort to Roger's defection that placed Leonard in a very awkward position. Despite his now being editorially aligned with Murry, Leonard was trying to expose him publicly (VWL, 2:477–78). Roger felt that neither Murry's "under-world" nor the "old cultured" could tolerate Bloomsbury's disinterested freedom of thought (RF/VW, 5 September 1921). In fact, Murry was one of the "horrors" of London that made Roger want to "live in Provence and carry on a copious and constant correspondence with Virginia Woolf" (RFL, 2:514). The Woolfs shared his assessment of Murry, and Virginia advised herself that they "must go on doing what we like in the desert Roger says, & let Murry climb the heights" (D, 2:139).

Although Virginia's diary records considerable anxiety about writing "in the desert," her letters hardly mention her dedicated work on her remarkable new novel. Then on 4 November 1921, Virginia wrote "the last word of Jacob." She still needed to "furbish up" this novel, but her mind was ready for a respite, although not a rest (D, 2:141, 142). Her demons were safely locked up in their dark cupboard; her writing genius was liberated. Planning her collection of essays, Virginia knew that as soon as she began, another novel would tempt her away, "so that the only question appears to be—will my fingers stand so much scribbling?" (D, 2:142).

The Hogarth Press continued to grow with new, more professional equipment (VWL, 2:487), and Leonard's political obligations became increasingly absorbing. In 1921, he campaigned for a seat in Parliament representing the universities:

> Whenever the train stops he gets out [Virginia claimed] and makes a speech on International politics, or the Cooperative Movement. He will drop into my arms at 6 tomorrow morning in a state of coma I imagine, having travelled all night. Then we are in the frenzy of our Xmas publishing season—the first edition of Roger's woodcuts sold out in 2 days, and another to be printed, folded, stitched and bound instantly.[23]

Leonard now wrote

> a monthly article in the Contemporary; and he also writes for the Webbs, for the Nation, for the Herald, for the Statesman—in short he is becoming, not rapidly, but what alarms me more, solidly, one of our public men. God knows—he may be Member [of Parliament] for 7 universities next month. (VWL, 2:500)

In addition to his journalism, Leonard began writing his massive political inquiry *After the Deluge* and advising the Labour Party.

When Roger spent a "delightful evening" with "the Wolves" (RF/VB, 17 December 1921), Vanessa grew suspicious of Virginia's habit of characterizing her: "Wasn't I given all the dull domestic virtues with never a spark to lighten them? You see how suspicious I am—but not of you."[24] Vanessa's suspicions were unwarranted, for Virginia did not belittle her sister's domesticity but instead shared Roger's adoration. They also shared "an atmosphere of illusion" in which "Roger always sees masterpieces ahead of him & I see great novels" (D, 2:150). Roger now said he preferred being alone, except for the company of his sister Margery and Leonard and Virginia (RF/VB, 30 December 1921). He sketched the Woolfs and planned a portrait of them (RFL, 2:529–30).

Probably during this period, Virginia read "Old Bloomsbury" to the Memoir Club. She laughed about the revolutionary new openness about sex that characterized the group's discussions sometime after Thoby's death. When Lytton Strachey pointed to a stain on Vanessa's dress and asked, "Semen?" Virginia remembered that "with one word all barriers of retience and reserve went down. A flood of the sacred fluid seemed to overwhelm us." She reflected that

> the loves of buggers are not—at least if one is of the other persuasion—of enthralling interest or paramount importance. But the fact that they can be mentioned openly leads to the fact that no one minds if they are practised privately. Thus many customs and beliefs were revised.

These revisions shifted from sexual to aesthetic interests after the advent of Roger Fry, who seemed to have "more knowledge and experience than the rest of us put together" (MB, 196, 197).

The Woolfs spent Christmas 1921 and New Year's 1922 at Monk's House, where "the wind blew from every quarter at the top of its voice, & great spurts of rain came with it, & hail spat in our fire, & the lawn was strewn with little branches, & there were fiery sunsets over the downs, & one evening of the curled feathers that are so intense that one's eyes see nothing for 10 seconds afterwards" (D, 2:155). Vanessa came back from France "for a fortnight, & then off again." With an eye on her fortieth birthday, Virginia reflected on the foolishness of tying her emotions to Vanessa. Leonard now was central, for "one person one must have, like air to breathe." During her second attack of influenza during the new year, "Nessa came again. How painful these meetings are!" Vanessa's complaint was that "no one had mentioned painting to her in the course of three weeks." Virginia reminded herself, "donkey that I am," how absurd it was to let Vanessa undermine her sense of self. Nevertheless, she began to suspect that her life, in comparison with her sister's, was "settled & unadventurous" (D, 2:156, 157, 159).

In 1922, Virginia suffered a number of debilitating illnesses. In the first part of the year, she was largely incapacitated by influenza. She also had a heartbeat so irregular that it threatened early death (VWL, 2:498, note). Consequently, she was caught in a vicious cycle in which illness kept her from writing and not writing depressed her: "& I at the prime of life, with little creatures in my head which won't exist if I dont let them out." While Katherine Mansfield's *The Garden Party* was expected to burst "upon the world in glory next week; I have to hold over Jacob's Room till October; & I somehow fear that by that time it will appear to me sterile acrobatics" (D, 2:161). By comparison, Adrian Stephen, not needing illusions of fame, seemed "fortunate" to Virginia (D, 2:162).

Vanessa was in Paris scheming a move there. Despite illness and depression, Virginia joked to her about Roger's unwillingness to visit her during the day because "every hour of daylight is sacred to art. So he sacrifices my society, which mayn't be long for this world, to some livid teapot" (VWL, 2:505). The difference between the spirited tone of that letter and the condition in which Roger found her—"only a little talk seemed to tire her & Leonard & I had to relapse into chess"—suggests just how valiant was Virginia's acceptance of her "mission in life": amusing Vanessa. Vanessa told Roger that she had not heard about Virginia's condition but would write to Leonard (RF/VB, 28 February 1922; VB/RF, 2 March 1922). A postscript to Virginia's spirited letter reveals her sense of Vanessa's glamorous persona: "'Look at me now—only sixpence a year—lovers—Paris—life—love—art—excitement—God! I must be off.'" Virginia admitted, "This leaves me in tears" (VWL, 2:505, 506).

Vanessa's self-absorption continued to constrict her world. She created what her daughter later called a "circle of safety" around herself (Garnett, *Deceived with Kindness*, 23). Maynard Keynes was in the process of courting the Russian ballerina Lydia Lopokova, who had come to England with Diaghilev during the war. Vanessa candidly advised him that no one could "introduce a new wife or husband" into the "existing circle." She said she was not criticizing Lydia, but "we feel that no one can come into the sort of intimate society we have without altering it." The problem was coexisting at Charleston for the summer: "We would much rather you were at Charleston of course, but I'm afraid you may be forced to choose between us and Lydia" (VBL, 266–67). All to often, Virginia felt herself the loser in such either/or choices.

Virginia used an odd metaphor—"my dose of phenacetin"—to describe a "mildly unfavourable review of Monday or Tuesday." She developed a "little philosophy" to drug herself against the ravages of depression. This philosophy was stoic: If the "pain over my heart suddenly wrung me out like a dish cloth & left me dead," she would be indifferent "except for L." It was self-aware: "I write what I like writing & there's an end on it." It was realistic: "I'm not going to be popular." It was confident: If years passed before *Monday or Tuesday* was appreciated, she still would be proud of its "queer individuality" (D, 2:166–69). She could challenge Vanessa's indifference: "Is the shade of death between us, and is this my valediction to one who played a part in youthful jaunts and jollitries? [*sic*]" (VWL, 2:520). She expressed her resentment by having Jacob call painting a "stupid art" (JR, 127). With increasing resilience, Virginia effectively medicated herself against the recurrence of the manic-depressive syndrome: She turned her "queer individuality" on a new story, "Mrs. Dalloway in Bond Street."

Illness and toothaches forced Virginia to cancel an Easter visit with H. G. Wells (VWL, 2:520 and note) and further interrupted her writing. That spring was the worst "on record," a full "27 days of bitter wind, blinding rain, gusts, snowstorms, storms every day" nearly ruined the Woolfs' spring retreat to Monk's House (D, 2:175). Virginia enjoyed rereading much of English literature for her reading book, but perusing her rivals in the reformation of the modern novel—Mansfield, Joyce, and Proust—again sapped her self-assuredness. When asked to contribute to an "album of admiration" for Proust, Virginia asked Roger to collaborate with her.[25] She told him that she was at once stimulated, inspired, and disheartened by Proust:

> But Proust so titillates my own desire for expression that I can hardly set out the sentence. Oh if I could write like that! I cry. And at the moment such is the astonishing vibration and saturation and intensification that he procures—theres something sexual in it—that I feel I can write like that, and seize my pen and then I can't write like that. Scarcely anyone so stimulates the nerves of language in me: it becomes an obsession. (VWL, 2:525)

Given Proust's achievement, Virginia's pride in her "queer individuality" and experimentation in *Jacob's Room* ebbed.

Ill himself, Roger suggested that he and Virginia "have a real heart to liver talk. For there are you fainting in coils like any Victorian Miss on your couch at Richmond and here am I groaning with indigestion in bed and in Camden Town . . . it would be much better if we could wheel our beds alongside and pass the day in gossip and in ribaldry" (RFL, 2:525). In May 1922, Virginia had influenza again and, having to miss hearing Roger lecture, "prayed all night that it might be a complete failure." Instead, it was, she heard, brilliant, "a triumphant success" (VWL, 2:527). Even Vanessa, usually so impatient with theory, thought that Roger's Raphael lecture was "frightfully good" (VB/CB, ? June 1922). The lectures provided "a rendezvous" for Bloomsbury socializing and theorizing (D, 2:178). Virginia requested copies of the lectures and again urged Roger to write on modern painting for the Hogarth Press (VWL, 2:539, 546).

Virginia continued to confess to Roger the mixed admiration and self-doubt that Proust inspired: "Well—what remains to be written after that?" Again she described the pleasure of reading Proust as physical (VWL, 2:565–66). Proust stretched the nonrepresentational powers of language as she did, but Joyce seemed to her to be merely naturalistic. If Vanessa had forwarded the Paris gossip, Virginia's aesthetic judgment might have been reinforced by social judgment. At a party at the Majestic, Clive

> had the opportunity of shaking hands with—Marcel Proust, Stravinsky, Picasso & James Joyce. James Joyce was so much the great man that he arrived in the middle of supper—at 2 in the morning—not dressed; but as no one in the room, except our host & hostess, [illegible] & myself, had ever heard of him, his entrée fell rather flat. And his entrée was all there was to it, for, whether he was drunk or shy I know not, but he said not one word throughout the evening. (CB/VB, 22 May 1922)

If Joyce's social demeanor and writings were unlike hers, Proust's seemed so similar that Virginia had to reexamine her achievement. The examples of Picasso and Rembrandt likewise challenged Roger's aesthetic assumptions. One of his gifts as a theorist was that he kept puzzling over and testing his own theories and their implications. In 1912, he had acknowledged that "the logical extreme" of Postimpressionism would be giving up "all resemblance to natural form" for a "purely abstract language of form—a visual music," as exemplified by Picasso's current paintings (VD, 167). By 1921, Roger hoped that Picasso would return from abstraction to representation "with the varied and enriched pictorial vocabulary which his abstract painting has given him," and he found such a balance in T. S. Eliot's poetry (*New Statesman*, 29 January 1921, 503–4, and 14 May 1921, 158–59).

Overlooking his earlier attack on Leonardo for psychologizing painting, Roger claimed that although he had emphasized formal values, he had never discounted subject matter. For example, the greatness of Rembrandt—whom he called the "Shakespeare" of painters—could not be explained in formalist terms alone. Being himself deeply moved by the dramatic situations in such paintings as *The Marriage of Tobias* and *Bathsheba*, Roger knew that their effect was heightened by the stories they told. With Rembrandt, "literary" elements did not distract one (as they did in, say, history painting) from apprehension of the aesthetic achievement; instead, they supplemented that achievement. Although one could appreciate Rembrandt's formal achievements alone, such a response diminished the power of paintings in which formal and "literary" elements actually did exist in harmony.

On 21 June 1922, the Woolfs heard Roger's Rembrandt lecture,[26] which must have reinforced Virginia's sense (so comically illustrated in "Solid Objects") of the limitations of formalist aesthetics, even when her reviews had become distinctly formalist. She praised D. H. Lawrence for his "extraordinary sense" of "the colour and texture and shape of things." She found absurd the "literary" assumption of Coventry Patmore (a friend of her grandmother Maria Jackson) that "there is no true poem or novel without a moral." She praised the Russian novelists in a phrase that, as Andrew McNeillie notes, "might also have derived from an essay on Post-Impressionist portraiture" for their ability to create "characters without any features at all" (*Times Literary Supplement*, 2 December 1920; VWE, 3:271; *Times Literary Supplement*, 26 May 1921; VWE, 3:310, xiv; *Nation and Athenaeum*, 1 December 1923; VWE, 3:386).

As her ambivalence about formalism and about the picture *Cocaine* revealed, Virginia could prefer stories to proportion and design. In "Solid Objects," she rejected the formalism that characterized her Postimpressionist stories, especially the abstract "Blue & Green." In July 1922,

Virginia went to an exhibition of Jean Marchand's works at the Carfax Gallery and found herself "reading [Roger] aloud" (VWL, 2:539). Marchand's gift was germinating new shapes out of the ordinary material of life (VD, 196, 197), as Virginia was doing in *Jacob's Room*. Nevertheless, "premonitory shivers" about the reception of that experimental novel returned, and Virginia worried, "Now what will they say about Jacob? Mad, I suppose: a disconnected rhapsody" (D, 2:177, 179).

In the *Times Literary Supplement*'s lead essay for 20 July 1922, "On Re-Reading Novels" (a review of Percy Lubbock's *The Craft of Fiction* beside several classic English novelists), Virginia Woolf wrestled, as she had earlier done most explicitly in "Modern Novels," with the problem of craft and design. She acknowledged that novels offer more distractions from and "temptations" toward "literary" responses than do other genres because they invite us to identify with the characters and to compare fictional events with actual occurrences. Her summary of Lubbock's method of registering "literary" impressions and then perceiving "the form itself" resembled the way that Roger had come to combine a "literary" and formalist response to art. Nevertheless, no matter how often she had used the term *form*, Virginia here objected to it: "This word 'form,' of course, comes from the visual arts, and for our part we wish that [Lubbock] could have seen his way to do without it." She argued that "the 'book itself' is not form which you see, but emotion which you feel." This point sounds similar to Roger's early defense of the Postimpressionist painters, who "do not seek to imitate form, but to create form; not to imitate life, but to find an equivalent for life." Again preparing a way for herself, Virginia expected "the novel to change and develop as it is explored by the most vigorous minds of a very complex age."[27]

Ralph Partridge was proving to be an unsatisfactory assistant at the Hogarth Press. Virginia had always disliked his masculine sense of power (D, 2:75), and both Woolfs were furious one day when he left the tedious cleaning of the type to Leonard (D, 2:160). Ralph complained to Virginia about Carrington's infidelity to him, whereupon Virginia said that she would have left him, had she been Carrington, after his infidelity to her (D, 2:177). The Woolfs sought, through Roger and others, to find a replacement, but the question of Ralph and the press continued to rankle for the rest of the year.

While she worked to finish *Jacob's Room* in the summer of 1922, Virginia took time off for two major events, one political and one social. In June, she and Leonard attended a Cooperative Conference at Brighton, over which Margaret Llewelyn Davies presided. And in July, Virginia went alone to one of Lady Ottoline Morrell's extravagant house parties at Garsington Manor. The talk there centered on Aldous Huxley's thinly disguised exposé of Ottoline and her guests in *Crome Yellow* and his rather lame apology. Virginia found this experience of high society sometimes trivial, sometimes flattering, and always stimulating (D, 2:179–81), her response whenever invitations from the aristocracy arrived.

After Virginia finished refurbishing *Jacob's Room*, Leonard called it "a work of genius," but he objected to Virginia's voicing "no philosophy of life." He also suggested that her characters "are ghosts . . . puppets" and that she should use her "'method,' on one or two characters next time. & he found it very interesting, & beautiful, & without lapse (save perhaps the party) & quite intelligible." Virginia was pleased: "There's no doubt in my mind that I have found out how to begin (at 40) to say something in my own voice; & that interests me so that I feel I can go ahead without praise." With that great assertion of aesthetic confidence, Virginia planned her next novel, "Mrs D[alloway] (who ushers in a host of others, I begin to perceive)." Her self-protective little philosophy this time involved writing criticism and studying Greek, so "when Jacob is rejected in America & ignored in England, I shall be philosophically driving my plough fields away" (D, 2:186, 189). Continuing the agricultural metaphor, she named this pattern "rotating my crops" (D, 2:198).

The Woolfs spent August and September at Monk's House, worried that Virginia might have pneumonia. Finally, on 24 August, for the first time they took a bus from Rodmell that dropped them on the road by Charleston; as Virginia told her diary, "One must respect civilisation." Vanessa, who "concentrates upon one subject, & one only, with a kind of passive ferocity" that Virginia found alarming, wanted only to discuss with Leonard her relationship with Mary Hutchinson (D, 2:195).

The arrival of proofs for *Jacob's Room* raised old specters, as Virginia feared that her radical experiment did not have "much bearing upon real life." To compensate, she vowed to "write something good: something rich, & deep, & fluent & hard as nails, while bright as diamonds" (D, 2:199).

The Woolfs' press affairs remained complicated and time-consuming. They hired a female assistant, who before long was seduced by Ralph Partridge. Despite such problems, since the idea of turning to Gerald Duckworth or to Leonard's publisher filled Virginia "with horror and misery" (LWA, 2:235), the Woolfs convinced Gerald to abandon his option on *Jacob's Room* and then contracted with a commercial printer to handle this large job. In this way, they could insist that Virginia's spatially fragmented novel be printed just as she had written it. They asked Vanessa to design the dust jacket. Virginia tactfully requested changes in the design and, like Duncan and Roger "whose hearts are always wrung for you," worried about Vanessa's health (VWL, 2:543).

Virginia was still trying to goad Roger into readying his Mallarmé translations for the press (VWL, 2:565). As an initial trial, he finished his translation of "Herodiade" for T. S. Eliot's *Criterion*. In addition to translating "quite literally," he explained his aim as "a rhythmic effect" that would at least recall the original poem (*Criterion* 1 [January 1923]: 119). In poetry, as in painting and in music, Roger cared most about rhythmic repetition and pattern, even when the very act of translation affirmed the importance of content.

Jacob's Room was published on 27 October 1922. The first reactions were discouraging, for the book and its dust jacket were labeled "reproachfully post-impressionist"; they were "almost universally condemned by the booksellers, and several of the buyers laughed at" Virginia's third novel (LWA, 2:241). She sent Roger a copy, asserting that "it has *some* merit, but its too much of an experiment. I am buoyed up, as usual, by the thought that I'm now, at last, going to bring it off—next time." Virginia suggested, "Why don't you come back and explain it?—you are the only person who ever does" (VWL, 2:573). Roger's reply echoed Flaubert and, although ostensibly about her letters, had relevance for her fiction: "What I like is your way of telling about nothing at all—that's the real test of style."[28]

Jacob's Room is not about "nothing at all," but it might seem so to a reader turning to it from, say, Dickens or Thackeray. It progresses, as a bildungsroman would, in chronological order from Jacob's childhood to his years in college, to life in London, through romantic and sexual entanglements and a trip to Greece, to his early death in World War I; yet it is structured by formal relations rather than plot. Although the book is filled with named characters, none of them (including Jacob) is described in detail. Woolf deletes conflict, plot, even continuity, from this novel. And yet she does manage to "enclose everything" by submerging Jacob's story in descriptions of people and places and ideas that define Jacob and his world. Quentin Bell noted that they do so just as a plaster cast replicates the figure within it (Introduction to JR, xiv). Virginia Woolf's own metaphor of a "spiritual shape, hard yet ephemeral" (JR, 45) offers a sense of the vaporous but entirely real world that encloses and shapes Jacob. Among friends who saw the novel mimetically, as a life of Thoby, only Carrington grasped its achievement: "You would make an amazing painter. Your visions are so clear & well designed."[29]

This novel offers an intense awareness of the shaping power of human perception. Names of places when traveling on the Underground, for example, are just names, but one name means "shops where you buy things, and houses, in one of which, down to the right, where the pollard trees grow out of the paving stones, there is a square curtained window, and a bedroom" (JR, 67). From a distance, "nothing could appear more certain from the steps of St. Paul's than that each person is miraculously provided with coat, skirt, and boots; an income; an object" (JR, 66), but a closer perspective erases such certainty. Although "nature and society between them have arranged a system of classification" to prevent chaos (JR, 69), Virginia refuses to classify. She presents the "reality" of the unknown, a reality that recalls her feelings near the beginning of the century when Thoby's activities (as when he stayed in separate quarters, probably to enjoy the nightlife in Paris in 1904) were beyond her ken.

Virginia Woolf invokes the chaos, the blankness, the flux, and the insubstantiality of experience. She describes the world Jacob moves through, even the blind beggar woman with her dog (JR, 67), but here she does not seek to understand the woman; instead, Woolf celebrates her *thereness*. She emphasizes simultaneity, indeterminacy, and the improbability of associations, the "chasms in the continuity of our ways" (JR, 96). Explaining that "spaces of complete immobility separated each of these movements" (JR, 99), Virginia conveys those spaces in both words and the fragmented visual arrangement of the text on the page.

In *Jacob's Room*, Virginia Woolf created a new, open-ended form, which Phyllis Rose defines as decidedly feminine.[30] It is about as non-"literary" as a substantial narrative could be. In the novel, Woolf explains the aesthetic assumption (rather like Roger's explication of Marchand) behind it: "To see the flash and thrust of limbs engaged in the conduct of daily life is better than the old pageant of armies drawn out in battle array upon the plain" (JR, 163). Jacob's war experiences are omitted from the book, and his death is treated only in a one-page final chapter in which his mother and friend sort through his things. The book ends on a minimalist note: "'What am I to do with these, Mr. Bonamy?' She held out a pair of Jacob's old shoes" (JR, 176). This minimalism emphasizes the patterned surface of language and the repetition of symmetrically placed motifs (such as recurring references to Mrs. Flanders's chickens and eighteenth-century houses). Minimalism also makes *Jacob's Room* an evocative antiwar novel.

Alex Zwerdling calls it a "satiric elegy," for "there is nothing grand about Jacob; the sacrifice of his life seems perfectly pointless, not even a cautionary tale" (*Virginia Woolf and the Real World*, 73). I agree but further emphasize Woolf's remarkable aesthetic and personal achievement. She writes that although it is "what we live by—this unseizable force" of life, "they say that the novelists never catch it" (JR, 156). Miraculously, Virginia Woolf's "spiritual shape" manages to catch something of that force.

The "they" in the previous citations hints at what seems to be the one serious problem with *Jacob's Room*: its uncertain voice. The unidentified "they," the narrator sometimes speaking in the first person and sometimes acknowledging the limits of her perception, seem to intrude into the global shape of this novel. Woolf saw her problem elsewhere. She defined *Jacob's Room* in Fry-derived terms: "The effort of breaking with complete representation sends one flying into the air. Next time I shall stick like a leech to my hero, or heroine" (VWL, 6:501). In a second letter, she defined her aim in *Jacob's Room* as "the effort of breaking with strict representation" (VWL, 2:588), yet she wondered, "How far can one convey character without realism? That is my problem—one of them at least" (VWL, 2:571). The young writer Gerald Brenan wrote to Virginia that "the word for writers today is Renunciation" (GB/VW, 30 April 1922). She replied, "I don't see how to write a book without people in it," and insisted, "It is not possible now, and never will be, to say I renounce. Nor would it be a good thing for literature were it possible" (VWL, 2:598).

Here Virginia showed herself struggling with a problem similar to the one Roger addressed in his Rembrandt lecture. Both believed that art should break with the representation of particulars because "reality" was not to be found in mere surface appearance and because merely representational art invited appreciation of the subject matter, not of the art itself. But both Virginia and Roger had reservations about abstraction, whether in Picasso or in Virginia Woolf. And both liked stories, whether in Rembrandt or in Hardy. Both believed that art should not renounce what Virginia called the "adorable world." The problem for both of them was how to abandon representational fidelity to surface appearances without renouncing "reality."

Responsibility to "reality" was an artistic issue for Roger and Virginia; for Leonard, it was an ethical one. This division reemerged in another debate concerning Bernard Shaw. Leonard praised Shaw for improving society. Virginia objected that Shaw "only influenced the outer fringe of morality" and maintained that the "human heart is touched only by the poets" (VWL, 2:529). This division was most apparent in the difference between Leonard's politically and Virginia's artistically reformist writings, in Leonard's running for Parliament (and losing respectably [D, 2:103; VWL, 2:586]), and in Virginia's claim not to take part in political movements or parties (VWL, 2:514). She repeatedly tried to relinquish the job of getting speakers for the Women's Cooperative Guild (VWL, 3:54). Regarding Leonard's activities, she concluded that "politics seems worse to me than mistresses" (VWL, 2:585).

After Roger returned from France and read *Jacob's Room*, the Woolfs dined with him in Camden Town, and "I was praised whole-heartedly by him, for the first time; only he wishes that a bronze body might somehow solidify beneath the gleams & lights—with which I agree" (D, 2:214). This was hardly the first time that Roger had praised Virginia's work, but perhaps this was the first time she could believe that his praise was wholehearted. Roger wrote to Vanessa that both his career as a painter and Virginia's as a novelist were turning. He imagined their competition as a tie: "Perhaps we shall *arrive* together a neck & neck race" (RF/VB, 1 November 1922).

Nevertheless, Roger's notion that her book needed a "bronze body," like Leonard's reservations about her characters, indicated something insubstantial in her achievement. Virginia maintained that *Jacob's Room* would be a pointless experiment unless she could "push on further next time" (VWL, 2:581). She pushed on by planning to expand "Mrs. Dalloway in Bond Street" into a novel, probably inspired by Vanessa's tale of death at a party. Virginia's original plan was to have her protagonist "kill herself, or perhaps merely to die at the end of the party."[31] Instead, she suddenly developed the brilliant foiling design of this novel in which Septimus Smith kills himself and Clarissa Dalloway lives. In her diary, Virginia planned the bold arrangement as "a study of insanity & suicide: the world seen by the sane & the insane side by side—something like that" (D, 2:207).

Virginia assigned herself a challenge: both to renounce representation and to substantiate characterization. Like Roger, she puzzled over the problem. She admitted to the painter Jacques Raverat how much of her time and energy went into "thinking about literature." Then she returned to the old tease: "Do you maintain that one can think about painting?" (VWL, 2:554). Later after assuring Jacques that Roger did praise his paintings, Virginia added: "We are so lonely and separated in our adventures as writers and painters. I never dare praise pictures, though I have my own opinions" (VWL, 2:592). Exchanges with Jacques, as with Roger, continued Virginia's attempt to overcome the separation between the arts.

Toward the close of 1922, Virginia met "the lovely gifted aristocratic" Vita Sackville-West at Clive's. She thought that Vita was "florid, moustached, parakeet coloured, with all the supple ease of the aristocracy, but not the wit of the artist." Vita made her feel "virgin, shy, & schoolgirlish" (D, 2:216, 217). When Virginia and Vita dined alone five nights later, Virginia's feelings began to change. She may even have begun to think of Vita as a substitute for Vanessa, for about this time Vanessa informed her sister of plans to spend half of each year in France. Virginia responded with "black despair. Must you really become wandering again? Six months in France—six at Charleston—I shall never see you." In vain, she requested "a letter of love" (VWL, 2:594, 596). Virginia wrote a "Punch and Judy" sketch to be performed by Vanessa's children in an attempt to be as essential to their lives as Vanessa and now they were to hers.

Neither Virginia nor Leonard kept any diary records of their 1922 Christmas alone at Monk's House, but Virginia's letters suggest no more "black despair." On Christmas Day, writing to Gerald Brenan, she anticipated visiting him in Spain: "My eyes are entirely grey with England—nothing but England for 10 years; and you can't imagine how much of a physical desire it becomes to feed them on colour and crags—something violent and broken and dry—not perpetually sloping and sloppy like the country here." She went on to write at some length about art and the crusading work her generation had been given. She was convinced that "there are no teachers, saints, prophets, good people, but the artists," who risk humiliation and despair to create some new completeness. Thinking of the horrors recurring "every 10 years," Virginia at forty seems to have been bracing herself against another onslaught (VWL, 2:597–99).

At the beginning of 1923, such an onslaught almost came. A visit to Vanessa left Virginia in "one of my moods, as the nurses used to call it." She felt her life bare without children, pallid besides Vanessa's colorful existence. But she reflected wisely on middle age and the inhibitions that marriage imposed (D, 2:221–22). When Katherine Mansfield died on 9 January 1923, Virginia delayed writing in her diary for a week. Then she ruminated,

> One feels—what? A shock of relief?—a rival the less? Then confusion at feeling so little—then gradually, blankness & disappointment; then a depression which I could not rouse myself

> from all that day. When I began to write, it seemed to me there was no point in writing. Katherine wont read it. Katherine's my rival no longer. . . . And I was jealous of her writing—the only writing I have ever been jealous of. This made it harder to write to her; & I saw in it, perhaps from jealousy, all the qualities I disliked in her. (D, 2:226, 227)

Saddened by death, besieged with guilt for letting jealousy interfere with friendship, Virginia became sick with a fever and remained largely bedridden for the rest of the month, although she did attend Roger's lecture in mid-January.

The Woolfs shared their anger over Ralph Partridge's trying to "pickpocket" their press and start a rival press with Lytton. They also shared their depression. Ralph and Lytton had been disloyal; Leonard's position with the *Nation* was threatened; and Virginia had failed to understand the toll that illness had taken on Katherine (D, 2:224–32). But then the Woolfs' spirits jointly lifted when Ralph Partridge left the press, Leonard was appointed literary editor of the *Nation* (a position Virginia had tried to secure for T. S. Eliot), and Virginia abandoned journalism for fiction.[32]

After an elegant dinner party at Maynard's, 46 Gordon Square, in which Virginia enjoyed sitting by the painter Walter Sickert, she felt certain that her generation knew how to enjoy themselves as their parents' had not. Later, however, when the Woolfs stayed at Clive's to avoid commuting, Virginia "innocently" found the sounds of Mary and Clive's lovemaking alarming (D, 2:223–25).

That February, the Woolfs enjoyed a surprise visit at Hogarth House from Vita Sackville-West and her husband, Harold Nicolson. Virginia was intrigued by the possibility that Vita, "a pronounced Sapphist," might "have an eye on me. . . . Snob as I am, I trace her passions 500 years back, & they become romantic to me, like old yellow wine" (D, 2:235–36). Even though she had observed of her marriage "one person one must have, like air to breathe" (D, 2:157), Vita's interest suggested other possibilities.

Virginia had suffered from sleeplessness, and the beginning of 1923 had been mostly a "very unpleasant quarter" for both Woolfs, but by mid-March, Virginia's melancholy was lifting.[33] She developed a newly enabling factor in her "little philosophy" after Lytton Strachey, looking at a photograph of Julia, remarked, "'I don't like your mother's character. Her mouth seems complaining.'" The moment was revelatory: "A shaft of white light fell across my dusky rich red past" (D, 2:239). It highlighted what Virginia had never before allowed herself to admit: her longed-for mother's coldness and bitterness.

Virginia had entered adulthood suspicious and largely ignorant of the visual arts. Conversely, Vanessa seems to have remained mostly ignorant of the verbal arts. For example, in February, Vanessa was taken to the theater to see *Oedipus Rex*. She found the drama "appalling" and wrote to Roger that she had never heard the Oedipus tale before. Her son Julian saw her ignorance as disgracing the memory of Sir Leslie Stephen. Perhaps in hopes of filling in the gaps in Vanessa's literary education, Clive and Virginia proposed starting a reading society to read old plays aloud, as they had in 1908 (VB/RF, ? February 1923; VB/RF, 18 February 1923).

Leonard's new position and the flourishing press offered the Woolfs such financial security that for the first time since their honeymoon they could travel abroad (Willis, *Leonard and Virginia Woolf as Publishers*, 34). In the spring of 1923, they spent nearly a month in Spain, visiting with Gerald Brenan, who appointed himself their guide. Virginia was delighted with the rugged, dry landscape, the earthy colors, and the persistent talk. Although himself something of a "callow and arrogant youth," Brenan was impressed with both Woolfs' friendliness and their lack of condescension. He found that in conversation, Virginia's "irony took on a feminine and one might almost say flirtatious, form." Typically, she would attack "the painters: how fantastic they were, how only the smallest details interested them, how they argued about everything." Then, with Bloomsburian frankness, Virginia admitted to Brenan "how incomplete she felt in comparison to her sister Vanessa, who brought up a family, managed her house, and yet found plenty of time left to paint."[34] On her way home, Virginia stayed on in Paris for three days without Leonard. She went to the Louvre, where, she told Vanessa, she liked only the Poussins. She also met Roger's friend the painter Maria Blanchard, "who said that Vanessa Bell was one of the few women to take art seriously" (VWL, 3:32). She revealed her lack of artistic self-confidence, however, by suggesting that Maria simply reflected what Roger had said rather than having a true appreciation of Vanessa's art.

Roger had intended to take a trip to Spain coinciding with the Woolfs' visit, but illness delayed him. He found himself "marooned in Nancy [France], tied by my intestines which have rebelled again." Vanessa and her "brats" stayed at Monk's House in the Woolfs' absence, a tenure for which she thanked them with a painting, being "not eloquent enough with a pen to express thanks." Vanessa told Virginia that to keep the "brats" under control, she had made "Leonard out to be a terrible ogre & you, though capable of weak kindness, a she-dragon when roused" (VBL, 269). Vanessa wrote that being at Monk's House had been "the greatest blessing," but Virginia lapsed into their old rivalries: "I'm sorry you don't like Rodmell nearly as much as Charleston, but it aint my fault" (VWL, 3:33 and note).

Later, when Vanessa and Duncan followed in Virginia's footsteps in Spain, Roger joined them. Despite Roger's poor health, Vanessa wrote to Virginia that he was "so energetic that no one can keep up with him." She admitted to Roger that "perhaps really its too difficult for the three of us to travel together," for (then she shifted into third person) "old associations one thought one had forgotten come to the surface again in spite of oneself." Typical of his dissociation of aesthetic judgment from such "literary" matters as those memories and jealousies, Roger wholeheartedly praised Duncan's painting in his introduction to the Hogarth Press's *Duncan Grant*. He followed the line of argument he had used in his lecture on Rembrandt, that the "very idea of invention in painting implies a literary or representational element" (Rosenbaum, *Bloomsbury Group*, 228). His jealousy of Duncan and Vanessa finally eased when later that year, Roger began an affair with a French woman who, when she was in a disturbed condition, reminded him of Vanessa at Broussa, for she "has an instinct that I'm the one person who can save her" (VB/VW, 4 June 1923; VB/RF, 22 June 1923; RF/VB, 1 October 1923).

Virginia's and Roger's Spanish excursions resulted in publications that each pretended to reverse the terms of the *paragone*. Virginia published an essay entitled "To Spain" in which, Leonardo-like, she proclaimed, "Blessed are painters with their brushes, paints, and canvases. But words are flimsy things. They turn tail at the first approach of visual beauty" (*Nation and Athenaeum*, 5 May 1923; VWE, 3:363). Praising "To Spain," Roger pointed out the absurdity of Virginia's desire "to paint when you can do a landscape like that over the mountains" in words. He suspected that "words as you use them . . . give me more of what the Germans call *stimmung* (and I wish to God you'd invent an English word for it) than painting can ever do."[35]

Roger's own Spanish essays, published in the *Nation*, evolved into the Hogarth Press's *A Sampler of Castille*. The book is a literate travelogue, complete with remarkable pencil-and-ink drawings and written in a style that Roger's son later decided was "as near to having a conversation with him as can be achieved [in 1976] forty-two years after his death"

(Rosenbaum, "Conversation," 135). While Virginia regretted not being a painter who could capture the beauty of the landscape, Roger wondered, "How he could hand over to you, through language, the faintest image of a single moment's physical sensation?" But he found an answer, at least in the abstracted "rhythmic contour" of the Sierra that challenged language more than paint, for its beauty was "of so abstract, so remote a kind, so suggestive of infinity . . . that its very beauty becomes, as it were, metaphysical. It, too, invites to renunciation" (*Sampler of Castille*, vi, 41). For both Virginia and Roger, "renunciation" meant abandoning the futile attempt to make art replicate the particulars of nature.

Back in London, although she suspected herself of liking the wrong things, Virginia felt that Roger's painting had entered a new stage (VWL, 3:40). She was pleased when "dear old Nessa returned shabby, loose, easy; & 44, so she said" (D, 2:246). She teased Vanessa: "I do think its time Roger ceased to be in love with you, and [ceased] to resent my presence. After all I'm your only surviving sister, and knew you long before he did" (VWL, 3:67). Still, Virginia called the *Sampler* a "perfect triumph" and again asked Roger to write about literature. In this period of affinity, Virginia concluded that Roger's mind and being were infinitely more subtle than Clive's and that age (Roger was almost fifty-seven) even seemed to be enriching him (VWL, 3:80).

In June 1923, Leonard and Virginia were again invited by the Morrells to Garsington, where Leonard stayed for one night, and Virginia for two. Virginia pondered how to describe the weekend, with "thirty seven people to tea; a bunch of young men no bigger than asparagus; walking to & fro, round & round; compliments, attentions." Photographs from that occasion with Lytton Strachey (Figure 36), Goldsworthy Dickinson, Lord David Cecil, Edward Sackville-West, and others all show Virginia as the center of attention. She now smoked rather dramatically while making spirited conversation.[36] Figure 36 is unusual also because it caught Virginia showing her teeth. (She rarely did so in photographs lest her upper plate of false teeth show.[37]) Her lace shawl and large hat show her sense of drama in dress, which led her, she feared, sometimes to excesses. Later, she pronounced much of the talk at Garsington "wearisome" and recalled the "slipperiness of the soul," specifically her own when she wrote snidely in her diary about Lady Ottoline after praising her to her face (D, 2:243–45).

With Adrian and Karin Stephen undergoing psychoanalysis, Virginia had a onlooker's opportunity for reflection. If Adrian was "a tragedy," Virginia felt responsible: "I should have paired with him, instead of hanging on to the elders" (D, 2:242); he had been "suppressed as a child," and apparently it was her fault (VWL, 3:43). Of course, she too had been neglected by her elder siblings, but even as early as the *Hyde Park Gate News*, Virginia's art had enabled her to reconstruct herself and compensate for her loss. By 1923, both she and Vanessa were relatively well known (D, 2:246) and hence had no need to resent the other's art and talent. By now, their rivalry could be treated mostly as a joke, as when Virginia said she liked to hear Vanessa praised, "chiefly because it seems to prove that I must be a good writer" (VWL, 3:34). Given such harmony, the prospect of living near each other in London (as they had not done since Virginia's marriage) was tempting, although Leonard still feared that London would unsettle the good health that Virginia had achieved. For Virginia, who now felt herself imprisoned in Richmond (D, 2:250), the possibility of moving back to Bloomsbury was liberating.

Having established her own confident and self-assertive voice in her reviews, Virginia was upset to find how vulnerable her fiction was to counterattack. In an article pointedly titled "Is the Novel Decaying?" Arnold Bennett (reacting to her attack on his "materialism") criticized her characterization in *Jacob's Room*. She summarized Bennett's argument: "I haven't that 'reality' gift." Then, rather than be humiliated, Virginia proclaimed herself impervious to criticism because, finally, she wrote for self-fulfillment: "Now I'm writing fiction again I feel my force flow straight from me at its fullest. . . . This is justification; for free use of the faculties means happiness" (D, 2:248–49).

After publishing "Mrs. Dalloway in Bond Street," Virginia admitted that her next novel had become "the devil of a struggle." The design was "so queer & so masterful. I'm always having to wrench my substance to fit it" (D, 2:249). Given the repeated criticism of her characterization in *Jacob's Room* (mild from Leonard and Roger, scathing from Arnold Bennett), her next challenge was to create believable characters within the framework of her highly original design.

Virginia considered Roger "incomparably more generous than one could suspect Christ to be, should Christ return, and take to painting in the style of Cézanne at the age of 56" (VWL, 3:51). Even though she had set T. S. Eliot's *The Wasteland* "with my own hands" (VWL, 3:56), Virginia felt herself free of Eliot's influence and concluded, "This I must prize, for unless I am myself, I am nobody" (D, 2:259).

In 1920, Roger stressed the need to "disentangle our reaction to pure form" from our reaction to content (VD, 92, note), whereas in the spring of 1923, Virginia similarly emphasized that we must "disentangle" literature from life. She did so with Jane Austen's *Emma*, asking the reader to "dismiss" likeness to life and human values ("How It Strikes a Contemporary," *Times Literary Supplement*, 5 April 1923; VWE, 3:356). She hypothesized that without "literary" associations, the reader could dwell "upon [Austen's] more abstract art," the design rather than the story.[38] Virginia carefully consulted Roger's analysis of Marchand's ability to build out of the "ordinary things of life" expressive, unified forms.[39] But despite all the evidence of Roger's influence, in 1923 Virginia's notion that "unless I am myself, I am nobody" (D, 2:259) prevented her from acknowledging her debt.

As usual, the Woolfs spent August and September at Monk's House. When they called at Charleston, although "thinking quite well of ourselves, we were not well received by the painters." That is, Vanessa, Duncan, and Roger were so absorbed in painting a still life that they hardly noticed their guests, re-creating the situation that surely had inspired Virginia's "Solid Objects." Virginia found herself feeling close to the four-year-old Angelica, who "is sensitive—minds being laughed at (as I do)" (D, 2:260).

In early September, Maynard Keynes, having accepted Vanessa's ultimatum and chosen Lydia Lopokova over Charleston, invited the Woolfs to join Lydia and him at Studland.[40] Virginia noted that she made "a splash of my own" and that she and Leonard were so old they were greatly respected by the young George (Dadie) Rylands and Raymond Mortimer (D, 2:267). Looking young themselves, she and Leonard seem particularly happy together in a snapshot taken on that holiday (Figure 37). In fact, she seems more physically at ease than he does. Her skirt, blouse, sweater, and flat pumps show how much fashions had been simplified in the eleven years since the Woolfs' marriage (Figure 28).

Their happiness added poignancy to an experience of potential loss in October. Virginia bicycled to Lewes to meet Leonard on his return from a day in London. But when Leonard did not arrive on even the "last likely train," Virginia left Lewes in despair, thinking, "Now the old devil has once more got his spine through the waves." She became rigid, believing that "reality" had been unveiled. Feeling at once outcast and ennobled by her desperation, she frantically decided to travel to London to find Leonard. Then, just before she was to board a train, he appeared. She felt a "physical feeling, of lightness & relief & safety." There had been "something terrible behind" her panic that left an ongoing pain, yet she did not show her feelings or tell Leonard about her fear, perhaps because he might think her instable (D, 2:270–71).

By the close of 1923, after extensive pleading, Virginia convinced Leonard that she was mistress enough of her own health to maintain it in the volatile ambience of Bloomsbury. Still, the news that Adrian and Karin were getting a divorce distressed her, not for love of Karin, but over the blame that both Adrian and his analyst laid on Adrian's childhood. Virginia's conversation with Adrian affords some insight into some of her own difficulties. Partly she identified with Adrian, being amused to hear that Vanessa's sympathy quickly evaporated over the prospect of his frequent visits. Like Adrian, Virginia felt the "old servile" feeling and "the desire for praise, which he never gets." Like him, she felt that Sir Leslie's work on the *Dictionary of National Biography* had damaged their childhood. But like Karin, she felt "all I used to feel [from Adrian]: the snub; the check; the rebuke; the fastidiousness; the lethargy." Virginia also regressed into feeling as rejected by him as he did by her: "The old futile comparisons between his respect for Nessa & his disrespect for me came over me, that made me so wretched at Fitzroy Sqre" (D, 2:277).

Unable to influence Adrian's and Karin's lives and trying not to reanimate the dynamic of rejection and counterrejection with Vanessa, Virginia vowed not to boast during her Christmas visit to Charleston. Preparing for the Woolfs' visit, Vanessa told Roger that she dreaded having to endure their discussions about the *Nation*, for it "really bores me stiff, all this talk about writing no one can possibly remember in a year's time or less" (VB/RF, 29 December 1923). Rather than bore, Virginia entertained her sister. She and Quentin (then thirteen) collaborated on a comic life of Vanessa that left Vanessa "almost overcome." She insisted that "not a single word of it's true of course but I admit it made me laugh and the pictures [by Quentin] are really lovely" (VBL, 275). Virginia did not jeopardize that goodwill by bragging of triumphs like being "accepted in America" (D, 2:282, 278).

Virginia continued her "mission" of entertaining Vanessa by writing a first version of *Freshwater*, a comic play about their great aunt the photographer Julia Margaret Cameron. The play had been "spirited fun" to scribble (D, 2:251) in the summer of 1923, but by late fall, revising it for a Christmas production was such an effort and distraction that Virginia, to the disappointment of friends and family, abandoned the project (VWL, 2:72–73, 75).

Meanwhile, Virginia entertained the painter Jacques Raverat with a remarkable series of letters. Jacques had married Gwen Darwin, one of the Neo-Pagans of Ka Cox's age. Now he was said to be dying, and even though she barely knew him, Virginia responded out of her own psychological strength. Her letters convey not only empathy and a real interest in comparing their arts but also a determination to avoid the guilt she had felt after Katherine Mansfield died. A letter from another Neo-Pagan, Frances Cornford, indicates that Virginia was overtly trying to compensate for past sins. In the depths of her 1915 illness, she had annotated and illustrated Cornford's poems in a manner that Vanessa had found unamusing, even nasty. In 1923, Virginia wrote a letter to which Cornford replied, "I've never had things said about my poems that I appreciated so much.[41] If Virginia needed to deepen her characterization in her new novel, this period of emotional munificence was her best entrée into a richly human universe.

Throughout the last of 1923, Virginia continued to "rotate her crops" between the *Common Reader* and *Mrs. Dalloway*. In her "most tantalising & refractory of books," she now could "dig out beautiful caves behind my characters" combining "humanity, humour, depth." Unity would emerge from disunity, for "the caves shall connect" (D, 2:262, 263). Her "so queer & so masterful" design would not be abstract or depersonalized but humanized through a "tunnelling process" into the essence of her characters (D, 2:272), so that this novel would indeed "enclose the human heart" (D, 2:13). Disheartened to hear "theres Picasso in painting & no one to match him in writing" (D, 2:264), although perhaps just as glad not to hear Joyce's name mentioned, Virginia was matching Picasso's Cubist compositions with the "remarkable" new form of *Mrs. Dalloway*, which juxtaposes the psychological volumes of negation and affirmation in Septimus and Clarissa, who never meet.[42]

In 1921, Virginia had proved that she had the amazing strength of character to "medicine" herself against the demon depression. In *Monday or Tuesday*, she had made real discoveries and charted a right path. In *Jacob's Room*, she had followed that path and created a holistic spiritual shape surrounding her characters. Despite facing horrors in 1920 and again in 1921, illness in 1922, and near despair briefly in 1923, she had risen above these troubles with renewed physical strength and spiritual stability and integrity. At the end of 1923, with the innovative design of *Mrs. Dalloway* drawn, she was ready to humanize it.

Back in 1904, a virtual renaissance in life and art had flowered in the Bloomsbury section of London. After nearly nine sheltered years in Richmond, Virginia was excited about the prospect of replanting herself in the soil that still nourished intellectual, political, and artistic debates. Bloomsbury was also, of course, a hotbed of the sort of social intrigue that Virginia loved to observe and also to cultivate. The Bloomsbury friends' commitment to tolerance and openness nurtured all sorts of social and sexual associations; thus, with Vanessa so often transplanted to France, new relationships might flourish there. Virginia was eager to begin house hunting in Bloomsbury. And she also was ready to dig into her past for the memories that would help explain her self and to bask in the artistic and personal glories that seemed at last, as she neared her forty-second birthday, to be ready to come to fruition.

9
A New Form for a New Novel

1924–1928

When the new year began, Virginia discovered 52 Tavistock Square, virtually "next door" (D, 3:46) to Gordon Square, which Virginia now saw as a "rabbit warren" (D, 2:23) of Stracheys, Stephens, and Bells. (Karin, Adrian's estranged wife, kept 50 Gordon Square, with Clive upstairs; Vanessa had rooms above Maynard Keynes at 46 Gordon Square; and Adrian was boarding with James and Alix Strachey at 41 Gordon Square.) The Woolfs quickly leased 52 Tavistock Square, a tall Georgian row house ample enough to house them, a solicitor's firm, and the Hogarth Press, now employing two new assistants.

A ground-floor extension at the back of the house served as both a storage room for the press and a studio for Virginia. Surrounded by crates of books, beneath a skylight, hunched in a wicker armchair beside her typewriter and a small gas fire, she would compose essays and novels while packers intermittently appeared to cart off books. Richard Kennedy's sketch (Figure 38), although it dates from a few years later, captures the inconvenience of the studio from its inception (*Boy at the Hogarth Press*, 22).

Tavistock Square lacked Gordon Square's intricacy of paths and flower beds, but its plane and copper beech trees and smooth lawn offered a peaceful retreat from the traffic beyond it on Woburn Place, an extension of Southampton Row.

Impressed with the lively decorating that Vanessa and Duncan did for others, Virginia was able to "force L. into the outrageous extravagance of spending £25 on painted panels, by Bell & Grant." The panels depicted curtains pulled aside to reveal still lives of a giant vase, mandolins, and fruit. Bold hatchmarks both flattened and suggested perspective.[1] The panels proclaimed the Woolfs as avant-garde as Maynard Keynes and others who commissioned work from Bell and Grant. The Woolfs also took advantage of "the resources of civilisation," chiefly in modern utilities (D, 2:293).

Hogarth House held nine years of memories, particularly of Virginia's 1915 bouts with mania, of the birth of "that strange offspring" the Hogarth Press, and of being "exquisitely happy" (D, 2:283). Nevertheless, the Woolfs left Hogarth in mid-March 1924. Vanessa immediately gossiped to Lytton:

> Virginia is actually living in Tavistock Sq. Life in Bloomsbury in consequence has already become charged with rumours of every kind. I dont know how long she will stand it but unless she collapses as I expect in about 6 months the rest of us will have to found a colony in another quarter. Its very lively while it lasts however. (VB/LS, 17 March 1924)

But instead of leading to collapses and recolonizations, the Woolfs, Vanessa soon wrote, "seem very flourishing" (VBL, 277–78). Many harmonious evenings, reminiscent of "old Bloomsbury," ensued.

On 31 March 1924, Roger Fry arranged a dinner for the London Group of Painters and invited Osbert Sitwell and Virginia Woolf to speak (D, 2:300, note). Sitwell thought Virginia was so nervous that he assumed she had agreed to speak only because Fry was "one of her oldest friends. If so, what happened was the more astonishing." Sitwell spoke first, and then Virginia announced the marriage of the arts:[2]

> The next quarter of an hour was a superb display of art and, more remarkable, of feeling, reaching heights of fantasy and beauty in the description of the Marriage of Music to Poetry in the time of the Lutanist, and how, in the coming age, Painting

> must be similarly united to the other arts. It was a speech beautifully prepared, yet seemingly spontaneous, excellently delivered, and as natural in its flow of poetic eloquence as is a peacock spreading its tail and drumming. Somehow I had not foreseen this bravura. (Sitwell, *Laughter in the Next Room*, 24, in Stape, *Virginia Woolf*, 50–52)

Despite such bravura, Virginia still questioned her vision. Worrying that novels should, after all, begin with character and conflict, she feared that her own efforts were "too verbose too little dramatic." To counteract such fears, "a delicious idea comes to me that I will write anything I want to write."[3] Clearly, she was weighing artistic conventions against Roger's beliefs that artists, like African sculptors, had "complete freedom" and that Virginia herself was "essential" to "liberty & freedom in literature" (RF/VW, 9 August 1922).

When five-year-old Angelica Bell was hit by a car on a Bloomsbury street, Virginia once again found herself an onlooker to suffering. She felt separated from Vanessa's anguish as through a pane of glass, knowing that "people who don't talk" keep their feelings in stoic reserve. She was afraid that "death & tragedy had once more put down his paw, after letting us run a few paces" (D, 2:299), but tragic "reality" let them loose this time. Angelica was scarcely hurt, and Virginia's overflow of sympathy for her niece and sister bore fruit. When Angelica next was sick, Virginia got "the longest letter I've had from you [Vanessa] this ten years, stuffed with tit bits such as I most relish" (VWL, 3:102).

Tragedy's paw pounced elsewhere and challenged Virginia's powers of sympathy. She and Roger dined together in early June 1924 and were to have gone to the theater, but his story of the recent suicide of Josette Coatmellec, the French woman who had loved him, eclipsed the idea of going. Insanity, rejection, and now death had taken beloved women from him. Roger wrote down the "story of Josette" and concluded, "God how one has to pay for those little glimpses of happiness" (RF/VB, 23 April 1924). Virginia remarked in her diary,"How long can Roger love a woman without driving her mad?" (D, 2:303), but in his presence, her irony vanished. Her "great sympathy and understanding" did help him cope with his "immense" loss (RFL, 2:554).

That summer, Dadie Rylands came to work at the flourishing Hogarth Press. The Woolfs spent another weekend with the Morrells, and "treading close on Garsington [was] the enamelled Lady Colefax, actually in this room, like a cheap bunch of artificial cherries." Virginia cautioned herself: "Aristocrats, worldlings, for all their surface polish are empty, slippery, coat the mind with sugar & butter, & make it slippery too" (D, 2:305). Nonetheless, she was becoming increasingly infatuated with the aristocracy, especially Vita Sackville-West. Vita took her to visit Knole, her ancestral home, a "conglomeration of buildings half as big as Cambridge I daresay." There "you perambulate miles of galleries; skip endless treasures—chairs that Shakespeare might have sat on—tapestries, pictures, floors made of the halves of oaks; & penetrate at length to a round shiny table with a cover laid for one" (Vita's father, Lord Sackville). Realizing that she had trouble keeping her "human values & my aesthetic values distinct," Virginia was deeply impressed at the same time that she knew Knole was big enough to house "all the desperate poor" of a London slum (D, 2:306–7).

Meanwhile, if Vita had a sexual "eye on" her, Virginia had a literary one on Vita and asked her to write a story for the press (D, 2:235). So, while on a walking tour in the Dolomites with Harold Nicolson, Vita wrote the long story *Seducers in Ecuador*. On 13 September, she spent the night with the Woolfs at Monk's House, which she characterized as "spartan" (Glendinning, *Vita*, 140). Although Virginia had initially thought of Vita as "incurably stupid," she was impressed by the "beauty and fantasticallity of the details" in Vita's story (D, 2:239). Such artistic respect offered a foundation for emotional entanglement: Virginia was "indeed touched, with my childlike dazzled affection for you, that you should dedicate [*Seducers in Ecuador*] to me" (VWL, 3:131). Vita's claim that Virginia liked people "through the brain better than through the heart" soon manipulated Virginia into trying to prove Vita wrong (Sackville-West, *Letters*, 51; VWL, 3:138).

The familiar opposition that the Woolfs saw between the aesthetic and the ethical recurred and evolved into yet another debate that summer about George Bernard Shaw, whose activist writing Leonard endorsed and Roger denigrated while Virginia hedged (VWL, 3:123). Leonard's own activism involved weekly meetings with Labour members of Parliament, most of whom came from working-class backgrounds and hence knew "occupational and industrial problems" but not "finance, economics, education, [or the] international or imperial affairs" about which Leonard advised them. To his regret, however, the first Labour government (1924) betrayed its promise of bringing social and economic equity to Britain. After regular advisory sessions with Prime Minister Ramsay MacDonald,

Leonard developed a contempt for the leadership, which, he felt, "was a disaster not only for the party [whose control lasted for less than a year] but also for Britain." He blamed treacherous and stupid Labour leaders for "the barren wilderness of the 1930s and the howling wilderness of the war" (LWA, 2:164–65, 247–48).

Given the failure of intellectuals and politicians to make a better world, Virginia again saw art as the guarantor of civilization. She wrote to Jacques Raverat that writers should try "to catch and consolidate and consummate (whatever the word is for making literature)" splashes of association, as painters do. She argued that rational, activist, materialist, masculine, "railway line" ways of thinking missed the subjective, fluid, feminine, painterly nature of actual experience.[4]

The Sussex countryside was an established retreat for much of old Bloomsbury. The Woolfs spent another August and September at Monk's House. Vanessa and her entourage were at Charleston. And Maynard Keynes was leasing Tilton, the farmhouse nearest to Charleston. When Virginia, Leonard, Maynard, and Lydia visited at Charleston, Virginia accidentally called Lydia "Rezia," revealing that Lydia's Russianisms were her source for Rezia Smith's Italianisms in *Mrs. Dalloway* (D, 2:310).

Alone at Monk's House, Virginia remembered her panic of the year before when Leonard was late coming back from a day trip to London, and she felt "the old wound twingeing" (D, 2:313). But she took great pleasure in the Rodmell postman's request that Leonard talk to the local Labour Party about the League of Nations. Thinking about her own diary entries, Virginia decided that the range of her thoughts "confirms me in thinking that we're splinters & mosaics; not, as they used to hold, immaculate, monolithic, consistent wholes" (D, 2:314).

Still smarting from Arnold Bennett's criticism of *Jacob's Room*'s characters and campaigning for prose that reflected the mosaic nature of experience, Virginia read a paper, "Character in Fiction," at Cambridge in May. In the version published in T. S. Eliot's *Criterion* in July 1924, Virginia said, "Such, I think, was the predicament in which the young Georgians found themselves about the year 1910. . . . And so the smashing and the crashing began" (*Criterion*, July 1924, 426).

When she revised "Character in Fiction" as *Mr. Bennett and Mrs. Brown*, the first of the Woolfs' series of pamphlets called Hogarth Essays, Virginia altered this passage into a much more explicit tribute to Roger Fry: "In or about December, 1910 [the opening of the first Postimpressionist exhibition], human character changed." She saw this change as a shift in patriarchal values that affected "religion, conduct, politics, and literature. Let us agree to place one of these changes about the year 1910." As Virginia indirectly praised Fry, she directly and spiritedly attacked Bennett. Exploring "the question of reality which Mr Bennett raises," she distinguished between "lifelike" and "real" characters by redefining "reality" as "vision." Those writers whom she had earlier called "materialists" look around an actual Mrs. Brown: Wells to a utopian society, Galsworthy to reformist goals, and Bennett, with "all his powers of observation," to surroundings. Each, then, deals in "literary" associations and fails to capture the essential Mrs. Brown. Furthermore, however revolutionary Joyce, Strachey, and Eliot may be, none of them captures the vision of their characters (CE, 1:330, 321, 330, 319, 325, 335).

In addition to theoretical importance, *Mr. Bennett and Mrs. Brown* carried tremendous psychological importance. The Virginia Woolf who had gone into a near-suicidal depression over the thought of public criticism had replied to it by taking the offensive. Her perceptive argument and sharpened prose neatly eliminated her Edwardian and wounded her Georgian competitors. And even though she did not mention herself, Virginia had spunkily prepared a way for her own attempts to capture the essence of "Mrs. Brown."

The second Hogarth Essay was Roger Fry's *The Artist and Psycho-Analysis*, originally a lecture for the British Psychological Society. In it, Roger distinguished between Freud's notion of art as wish fulfillment and his own sense of it as detached "contemplation of formal relations." Classic novels do not "represent wish-fulfillment" but offer "a peculiar detachment from the instinctive life." Pleasure in reading them emerges from the "recognition of *inevitable sequences*" (Fry, *Artist and Psycho-Analysis*, 4, 12).

Before the essay was published, Clive Bell stole Roger's ideas from his lecture and published them that fall in the *Nation and Athenaeum*, for which Leonard was the literary editor. Roger told Vanessa that "everything down to the quotation was bagged from my lecture. I think it very foolish of him because no one will think I've got it from him." Roger protested the "theft" to Leonard.[5]

From Monk's House, Virginia consoled Roger that *The Artist and Psycho-Analysis* "fills me with admiration and stirs up in me, as you alone do, all sorts of bats and tadpoles—ideas, I mean." Trying to smooth his ruffled feathers, she pointed out that Clive offered only "a mere snap shot of your argument. . . . It was a pity Leonard didn't send it him back; but it has become a joke almost—Clive's cribbing." Half admitting that it was

Leonard's responsibility to reject Clive's article for obvious plagiarizing, Virginia begged Roger to "forgive the poor Woolves your devoted admirers" (VWL, 3:132–33).

Trying to defuse this contretemps, Virginia rescued the term *form*, which she had claimed to jettison two years earlier in her review of Percy Lubbock's *The Craft of Fiction*. While revising it, she told Roger that she was "puzzling . . . over some of your problems: about 'form' in literature." Roger maintained that "the esthetic emotion is an emotion about form" (*Artist and Psycho-Analysis*, 7). Now Virginia, too, theorized that form was "emotion put into the right relations," but then she announced that her use of form (a veritable paraphrase of Roger's) has "nothing to do with form as used of painting" (VWL, 3:133). Virginia's heavily revised version of "On Re-Reading Novels" argues that by form "we mean that certain emotions have been placed in the right relation to each other."[6]

Roger still believed, as he had in 1913, that the artist "can create forms entirely corresponding to his feeling";[7] thus "our aesthetic sense is continually aroused and satisfied by the succession of inevitable relationships" (VD, 58). Perhaps concerned that she, too, had usurped Roger's ideas, Virginia did not publish the revised review of Lubbock's book during her lifetime. However, she compensated for her borrowing by indirectly testifying to Roger's influence. She reminded her readers that important avant-garde paintings had been ignored or ridiculed until a decade and a half earlier, when Roger had taken on the reeducation of English taste (*Nation and Athenaeum*, 18 October 1924; VWE, 3:448).

About this time, Middleton Murry deplored the absence of plot among the "most original minds" of their generation, including Virginia Woolf, and declared that the modern novel had reached an "*impasse*." But rather than launch a counterattack, Virginia persevered in her own writing. By mid-October, she had finished *Mrs. Dalloway* and knew that through this deeply human novel, she had "exorcised the spell wh. Murry & others said I had laid myself under after Jacob's Room" (D, 2:317).

Late in November 1924, the Play-Reading Society (Virginia, Leonard, Vanessa, Clive, Duncan, and Roger) gathered at Vanessa's London studio to read Dryden's *Marriage à la Mode* (LWD), the sort of Restoration drama whose risqué underpinnings appealed especially to Vanessa. Virginia's worries that Vanessa might move away and Leonard might take an extended trip to Ceylon proved as unfounded as her writerly confidence was well grounded. When retyping *Mrs. Dalloway*, as she had done with *The Voyage Out*, Virginia used a "wet brush" metaphor for joining "parts separately composed & gone dry" that conveys both a respect for and an understanding of painting technique.

Virginia considered *Mrs. Dalloway* as "the most satisfactory" of her novels but was afraid that the reviewers would find it "disjointed because of the mad scenes not connecting with the Dalloway scenes." Plunging "deep in the richest strata of my mind," she thus connected those characters in a painterly manner. Both Septimus Smith and Clarissa Dalloway are described in birdlike terms. Each character recalls the same lines from *Cymbeline*. Septimus's madness, which results from the numbing experience of war and guilt over his inability to feel grief, recalls Virginia's initial reaction to her mother's death. Clarissa Dalloway feels intensely even such simple experiences as walking the streets of London. In this way, the novel contrasts unhealthy and healthy, insane and sane, experiences of the imagination. Whereas Clarissa's life is guarded and careful, Septimus flings his away, bringing death (or the news thereof) into her party. By making a shaped whole out of such diversity, Virginia exulted, "I can write & write & write now: the happiest feeling in the world" (D, 2:323).

In its mosaic quality and its exploration of the phenomenon of consciousness itself, *Mrs. Dalloway* is a painterly novel. Vanessa, Duncan, and Roger painted common motifs—objects on mantelpieces or women in tubs (Figures 32, 33, and 34)—each in different styles. The Cubists painted objects or motifs from several different angles at the same time. In *Mrs. Dalloway*, Peter Walsh has the ability to "see round things" (MD, 159), and he also makes Clarissa "see herself." But she thinks he is being critical and wonders, "Why not risk one's one little point of view?" (MD, 168). "One's one little point of view" is exactly what Virginia Woolf risks in her distinctly feminine, nonauthoritarian mode of writing, the characters express a variety of points of view about themselves and one another.[8] The doctors, with their devotion to "proportion" and "conversion," are the enemies of the individuality and freedom that the novel endorses.[9] The various characters agree only in their negative assessments of the doctors; they see everyone else so differently that the reader cannot make simplistic monologic judgments about any of them.

By abandoning the "old accepted forms," Virginia achieved, as she had required Dorothy Richardson to do, their "shapeliness" by means of an intricately formal design indebted to Fry's theories[10] regarding Postimpressionist form. She grafted her abstract design onto a complex of complicated characters treated with such extreme referentiality that their real-life analogues can be identified and Mrs. Dalloway's movements through

London on the day of her party can be precisely mapped.[11] Thus Virginia Woolf managed at the same time to give up simple representation and not to renounce the actual "adorable world" (CSF, 121). She thereupon achieved freedom, retained shapeliness, and avoided abstraction.

When both Roger and Vita dined with the Woolfs at Tavistock Square on 19 December 1924, Virginia's personal and aesthetic loyalties were set against each other in a volatile new combination. Roger (now getting "grumpy" [VWL, 3:150]) objected to the sloppiness of Vita's talk about art. The discussion was "all very thorny until that good fellow Clive came in," who managed to conciliate and divert Vita from talking about art (D, 2:325). Virginia was irritated by Roger's thorniness, and he by what he took to be snobbism in her. She had learned a lesson: not to mix carelessly the company of Vita and Roger again.

For Christmas in the country, Virginia and Quentin collaborated on the "Dunciad," a life of Duncan, with (presumably) apologies to Alexander Pope. Presents included a bottle of Spanish wine from Vita and a "superb" present of a painting from Vanessa. Dadie Rylands had left the press, and the Woolfs were entertaining their newest assistant, but with the River Ouse overflowing, the residents of Monk's House and Charleston did not meet on Christmas Day (VWL, 3:149; D, 2:327).

In January 1925, in a long entertaining letter, Virginia told Jacques Raverat about Roger's praise for Jacques's painting: "the best worth having." Trying to distract Jacques from pain, she passed on the latest gossip and described various people, including the "violently Sapphic" Vita, who had once eloped with another woman. With teasing bravura, Virginia told Jacques a secret: "I want to incite my lady to elope with me next" (VWL, 3:154–56). With the possibility of a new friend adding romance and intrigue to her life, Virginia could admit to Jacques that Vanessa "often shocks me by her complete indifference to all my floating loves and jealousies." She also could forgive: "With such a life, packed like a cabinet of drawers, Duncan, children, painting, Roger—how can she budge an inch or find a cranny of room for anyone?" (VWL, 3:164).

Despite Vanessa's warning that his marriage would end their friendship, Maynard Keynes married Lydia Lopokova in 1925. Purchasing a ninety-nine-year lease of Tilton, Maynard set himself up as a Sussex country gentleman and coerced Vanessa into accepting Lydia at least at the periphery of her circle. Vanessa's response to having the Keyneses bring the world of politics, ballet, and fashion so close to Charleston was to post a sign at the gate bearing only the word OUT (Bell, *Elders and Betters*, 96–97).

Roger at last was on easy terms with Duncan and he adored Angelica, so there was a "cranny of room" for him at Charleston. There Vanessa still encouraged Angelica to "go about naked." Indeed, Angelica enjoyed such freedom "until the moment when a photo was taken!" (to me, 11 August 1995). One such picture (Figure 39) shows Roger and Duncan fully clothed and Angelica looking decidedly embarrassed. In another photo (Figure 40) with Stephen Tomlin, who sculpted a bust of Virginia (Figure 66), and Lytton Strachey, Angelica is sitting comfortably on Clive's lap. Clive continued to play father to her, even (since she was born on Christmas Day) reassigning his birthday to her (D, 4:44, note). Even though Virginia felt almost as crowded out as the Keyneses did from Vanessa's emotional territory, she was thrilled that Vita seemed to be opening emotional space for her.

Since sharing the early stages of *The Voyage Out* with Clive, Virginia had shared her works in progress with no one; she usually finished her long works before showing them even to Leonard. In fact, she had hardly acknowledged writing *Night and Day* until it was published, and she had been only slightly more open about her progress on *Jacob's Room*. Despite being "a little morbid about people reading my books" (VWL, 3:154), Virginia violated her instinct for self-preservation and in January 1925 bravely sent Jacques proofs of *Mrs. Dalloway* to read on his deathbed. Her kindness produced serendipitous results: Jacques dictated "a letter about Mrs Dalloway which gave me one of the happiest days of my life. I wonder if this time I have achieved something?" She feared not, compared with Proust, who managed to be "as tough as catgut & as evanescent as a butterfly's bloom." When she heard of Jacques's death, "the siege of emotions began," but Virginia resisted those old terrors and experienced "no leavetakings, no submission" (D, 3:7). Generosity had delivered Virginia from "submission" to the demons of helplessness, loss, and guilt.

The Hogarth Press continued to flourish. As Frederic Spotts writes, "With a flight of inspiration that was to typify their publishing, [the Woolfs] brought out T. S. Eliot's *The Wasteland* and also the first English-language translations of works by Gorki, Dostoevsky, Bunin, Tolstoi and Freud." The press published the *Collected Papers of Sigmund Freud*, and five Hogarth authors won Nobel Prizes (LWL, 266, 269). The financial security that resulted from their brilliant publishing decisions enabled the Woolfs to travel to Cassis to visit Vanessa in the spring of 1925.

With her fame slowly increasing, Virginia could now "suppose I might become one of the interesting—I will not say great—but interesting novelists" (D, 3:12). *The Common Reader* and *Mrs. Dalloway* were published in rapid succession. At first neither sold well, although they were generally praised in the press. *The Common Reader* was a collection of essays that illuminated literature and theory to her readers through a light, talkative style. (Gerald Brenan later wrote that he could not "read a page of *The Common Reader* today without her voice and intonation coming back to me forcibly" [*South from Granada*, 139–40].)

Mrs. Dalloway was not so accessible, but a number of sophisticated friends understood its significance. Morgan (E. M.) Forster, whose earlier criticism of Virginia's characterization had so disturbed her, pleased her by praising *Mrs. Dalloway* (D, 3:24). Goldsworthy Lowes Dickinson thought that it had more "thickness" than *Jacob's Room*. Gerald Brenan noted that only Virginia and Joyce wrote modern literature, although Joyce was her superior at characterization (GLD/VW, 22 June 1925; GB/VW, 8 June 1925; MHP). Against that familiar charge, Virginia consoled herself that the sales of both books were picking up and would make enough money to install "*two* waterclosets" (VWL, 3:241) at Monk's House. Nigel Nicolson remembers Vita's story that when Leonard used the water closet, Virginia proudly, almost gleefully, told Vita, "Hear!" when he flushed it. Both Woolfs, Vita said, ran up the stairs from time to time for the sheer fun of flushing their new toilet (interview with Nigel Nicholson, 23 July 1993; Glendinning, *Vita*, 163).

On 25 April 1925, the *Nation and Athenaeum* published Virginia's "Pictures." This essay credited painting's gift to the modern novel and essentially (albeit temporarily) resolved the *paragone* by calling for a scholarly treatise on

> the flirtations between music, letters, sculpture, and architecture, and the effects that the arts have had upon each other throughout the ages. Pending [such a study], it would seem on the face of it that literature has always been the most sociable and the most impressionable of [the arts]; that sculpture influenced Greek literature, music Elizabethan, architecture the English of the eighteenth century, and now, undoubtedly, we are under the dominion of painting. (MOE, 173)

A decade and a half earlier, Virginia had believed that painting was the inferior art and hence had nothing to teach literature. By 1925, however, she had reversed her opinion. Now when she spoke of the "dominion" of painting over literature, she meant that modern painting had shown writers how to transcend appearances in order to present other, more essential realities. No painter was "more provocative to the literary sense" than Cézanne, because he stirred "words in us where we had not thought words to exist." But "painters lose their power directly they attempt to speak . . . [for] a story-telling picture is as pathetic and ludicrous as a trick played by a dog" (MOE, 176).

By their very silence, non-"literary" paintings stimulate writers to fling words, like nets, over new realities: "Were all modern paintings to be destroyed, a critic of the twenty-fifth century would be able to deduce from the works of Proust alone the existence of Matisse, Cézanne, Derain, and Picasso." Virginia explained that the writer does not learn "anything directly from painting," that painting shows writers, like Proust (and of course herself), how to make "hard, tangible, material shapes of bodiless thoughts" (MOE, 173, 174).

Having thus directly admitted the dominion of painting, Virginia could not help but end with a gibe at painters, for "the most extreme of penalties, the most exquisite of tortures—[is] to be made to look at pictures with a painter" (MOE, 178). Virginia seems to have written "Pictures" both to refute Brenan on character and to acknowledge (directly) painting's influence and (indirectly) Fry's (*Nation and Athenaeum*, 15 April 1925; MOE, 173–78). But in *Mrs. Dalloway*, Virginia herself had written a painterly novel that successfully captured—through the marriage of realistic referentiality and formalist purity—the language of thought and the images of feeling, the vision that she had sought for so long.

Cross-pollination between Virginia's and Roger's criticism developed as he attempted her light touch with weighty topics. For example, Roger's mock narrative on London statues takes a Virginian delight in suggesting that God had taken vengeance on Londoners by giving them so many bad statues ("London Statues," 293–95, 730–31). Vanessa does not seem to have reacted to Virginia's art criticism; she said essays bored her (VB/RF, 29 December 1923).

Virginia had kept her vow to "push on further next time" on the path that *Jacob's Room* (VWL, 2:581) had charted. After *Mrs. Dalloway*, she was already pushing on toward a more overtly painterly and at the same time "literary" novel set at Talland House in St. Ives (D, 3:18–19). This novel, entitled *To the Lighthouse*, would touch the dual core of both sisters' emotional beings.

Once again in aesthetic rapport with Roger, thanks to "Pictures,"

Virginia's personal rapport with him nearly dissolved. There were two main causes: First, Roger at last established a lasting relationship with a woman, Helen Anrep, about whom Virginia felt considerable ambivalence,[12] and second, Virginia was increasingly attracted to Vita Sackville-West, to whom Roger felt genuine hostility. He asked Leonard to pass on to Virginia that he now knew that "*all* aristocrats are virtuous but incredibly boring"; hence he refused "to suffer them any more" (RFL, 2:569). Virginia concluded that Roger had "grown so surly and incorruptible, biting aristocrats at sight, that I can only have him here with the greatest precautions" (VWL, 3:187–88).

If Virginia and Roger did not agree on aristocrats, they did agree about *Mrs. Dalloway*. Virginia had a long talk with him and then wrote to the critical Brenan that Roger "gave me an entirely different view of Mrs D from yours." Whereas Brenan thought that Septimus had no function, Roger termed Septimus "the most essential part of Mrs D: And this I certainly did mean—that Septimus and Mrs Dalloway should be entirely dependent upon each other" (VWL, 3:189). Roger saw in the "very beautiful" *Mrs. Dalloway* a "search towards another kind of novel—a novel poem" (RFL, 2:562). Whereas Roger's term suggested how much *Mrs. Dalloway* defied the novel form, Virginia pondered "an idea that I will invent a new name for my books to supplant 'novel'. A new —— by Virginia Woolf. But what? Elegy?" (D, 3:34).

In the summer of 1925, Virginia was especially contented with Leonard, who settled wrangles at the press and immunized her against slights and fears. She reflected that the "immense success of our life, is I think, that our treasure is hid away" (D, 3:29–30). One afternoon when the Woolfs were strolling in Hyde Park, however, they found themselves no longer "hid away" when an amateur photographer snapped them (D, 3:26 and note). His photo shows them handsomely dressed and Virginia looking rather shy at this intrusion (Figure 41). She was more assertively protective of her literary self, however. Lytton Strachey's criticism of *Mrs. Dalloway* stimulated her "working fighting mood" (D, 3:32). When Janet Case suggested that Virginia showed more interest in form than substance, Virginia defensively retorted that the artist cannot "possibly separate expression from thought in an imaginative work" (VWL, 3:201).

Heated debates among members of the British Psycho-Analytical Society came within Virginia's purview that summer. While the Hogarth Press was publishing James and Alix Strachey's translations of Freud, the Stracheys and Adrian and Karin Stephen were finding themselves increasingly attracted to the theories of early childhood propounded by Melanie Klein. As a rival to Freud's Oedipal complex, Klein offered a mother–infant plot focusing on the centrality of the mother's breast (rather than the father's phallus) in the infant's self-concept. In July, after Klein lectured in Karin Stephen's house to an enthusiastic society, Adrian visited Virginia. Since Klein's theories offered explanations for his various yearnings and unhappinesses, Adrian probably expounded on them to Virginia, although she recorded only "cancer" as their topic of conversation (D, 3:36). Possibly recognizing that the loss of her mother rather than not having children explained her desperate need to communicate, Virginia found the term *elegy* more and more apt for her sense of narrative longing; hence she began to shift the focus of her new novel from the father- to the mother-figure.[13]

Even as Virginia was congratulating herself that "I can go through a tussle of emotions peaceably that two years ago even, would have raked me raw," the specter of illness again intervened (D, 3:39). On 19 August 1925, Leonard and Virginia bicycled from Monk's House to Charleston (LWD) for a heady gathering celebrating Quentin's fifteenth birthday, but after much good food, wine, and talk, Virginia nearly fainted. Seeing her rise wobbily from the candlelit table, Vanessa and Leonard quickly jumped up and led her away (Bell, 2:114). Afterward, the Keyneses drove the Woolfs back to Rodmell. Although "enfeebled," Virginia was able, while recuperating at Monk's House, to do a little work on the great tripartite novel *To the Lighthouse*, which grafted aesthetic debates into its very subject matter (D, 3:38 note, 39).

Vanessa gave a very antitheoretical lecture at the Leighton School (which Quentin attended), in which she made the Romantic argument that technical skill for the painter is not so important as "the spontaneous and fresh embodiment of vision and emotion."[14] At a French literary conference at Pontigny, Roger found Virginia's work discussed with "mysterious awe" (RF/VW, 12 September 1925). Maintaining that he was "about the only person in the world I wished to see," Virginia also promised him to "forego [*sic*] aristocratic society in future" (VWL, 3:208). Just as she considered writing on "something like Painting and Writing," she continued reconciliation of the *paragone* by urging Roger to write about literature for the press (VWL, 3:203–4, 208–9).

Confessing to Roger that she called him "Crusty," Virginia threatened to put him "into a book one of these days"—a promise that she was fulfilling even then by making Lily Briscoe in *To the Lighthouse* a proponent of his theories. She also told Roger that the Woolfs were "lying crushed under an immense manuscript of Gertrude Stein's" (VWL, 3:208, 209). Even though Roger had helped Stein find her first English publisher and may have recommended the Woolfs to her, his opinion of her experiments differed radically from his opinion of Virginia's.[15] He admitted that "in my giddy middle age," he had tried to take seriously Stein's writings on "Matisse & Picasso but after that I gave up and came to think she was just trying to see what might turn up if so very fat a lady stood on her head" (RF/VW, 24 September 1925). Stein took the artist's "complete freedom" beyond Fry-endorsed formalism—even beyond Virginia's experiment in "Blue & Green"—into radical dissociation with what Virginia had called the "adorable world."

The Woolfs returned from Sussex to London in early October, but Virginia was too incapacitated, after her fainting spell, to galivant around London. That fall, she reflected that it was "strange indeed that illness has not taken its place with love and battle and jealousy among the prime themes of literature." In any case, Virginia theorized about the literary advantages of being sick: "In illness the make-believe ceases" and truth can be spoken (CE, 4:195, 193, 200, 196). She complained that Roger was too egocentric to sympathize with her illness, but Vita did.

Vita's letters, the second set of passionate courtship letters that Virginia received in her lifetime, did not frighten Virginia as Leonard's had thirteen years before. Since girlhood, Virginia's declarations of love had led Vanessa to accuse her of "Sapphic" tendencies, which Virginia expressed more explicitly in her flirtatious early letters to Violet Dickinson. Now Virginia was being invited to act on those tendencies. Over the years, Virginia's love for Leonard seems to have become more physically affectionate and occasionally erotic. But the promise of erotic love with a "violently Sapphic" woman (VWL, 3:155) was newly thrilling.

The sudden competition from Vita, along with worries about Virginia's health, loosened "the earth about the roots" and provided Vanessa an opportunity to be newly forthcoming with sisterly affection: "Nessa wants to have us—Indeed, I have seen more of her & Duncan than for many a day" (D, 3:47). Vanessa now lavished on her sister, as she had not since 1910, "the maternal protection which, for some reason, is what I have always most wished from everyone." Since her earliest days, Virginia had "reason" (according to Klein's theories) to need mothering. She received it from Leonard and now Vanessa and, in a "more clumsy external way," Vita (D, 3:52). News of Vita's planned departure in early 1926 to join her husband, Harold Nicolson, who was assuming a diplomatic post in Persia, added poignancy to these two women's romantic entanglement. Virginia was "filled with envy and despair. Think of seeing Persia—think of never seeing you again" (VWL, 3:217).

Since childhood, Virginia's emotional and artistic confidence had been riddled by worries that Vanessa did not love her and that Vanessa's art was superior. Even Leonard's sustaining love often carried with it (in Virginia's low moments) the suspicion that his work was more important than hers. But neither Vita's love nor her art posed such worries. Although Vita was a successful novelist, poet, and nonfiction writer, her art did not challenge Virginia's. And unlike the mother and the sister of "naughty little Ginia," Vita did not withhold love. In fact, she pursued Virginia ardently. Hence, Vita's love promised release from the convoluted and torturous emotional and artistic dynamic of the *paragone* that had begun long ago in the Stephen nursery.

When Madge Vaughan died, Virginia felt that she, like her husband, Will, had become so conventional that they buried only "a faggot of twigs at Highgate."[16] Even though she shed no tears, Madge's death reminded Virginia that if "Leonard, Nessa Duncan Lytton, Clive Morgan [were] all dead" her life would be too empty to savor. But even her own death seemed closer as she neared forty-three and wondered, "How many more books?" (D, 3:46, 48, 51).

Vita's tributes were becoming increasingly erotic, for "these Sapphists love women; friendship is never untinged with amorosity" (D, 3:51). Before Christmas, feeling both apprehensive and excited, Virginia accepted an amorous invitation to stay for two nights with Vita at her country home, called Long Barn. As an experienced lesbian lover, Vita could congratulate herself for the aggressiveness with which she pursued Virginia and thus laid "the train for the explosion which happened on the sofa in my room here when you behaved so disgracefully and acquired me forever" (Sackville-West, *Letters*, 238). Whatever Virginia and Vita did together, it opened a new dimension of eroticism and romance into Virginia's life, as her description of Vita's "voluptuousness" (D, 3:52) makes clear.[17]

Leonard joined Virginia for her third night at Long Barn. He encouraged her to write to Vita in appreciation, and so thanks to both of them, Virginia "wound up this wounded & stricken year in great style" (D, 3:51–52). Leonard seems to have been pleased to see Virginia eroticized by Vita; certainly he was not shocked by homosexuality, nor was he afraid of losing Virginia's love. Frederic Spotts's perspective is that Virginia's infatuation with Vita did not alter her affection for Leonard, who was genuinely fond of Vita.[18] With Julia Stephen, Leslie Stephen, and Vanessa Bell, love and loss had always been intertwined. But with Leonard, love was a secure platform from which Virginia now realized she might launch fresh emotional flights.[19]

At the end of 1925, while renovations were under way at Monk's House, the Woolfs stayed at Charleston with Vanessa, Clive, her children, and (until Christmas Eve) Roger. They spent one enthralling evening reading from Virginia's 1905 early Bloomsbury diary and another amused by Quentin's and Virginia's outrageous "The Messiah," supposedly a life of Clive. Emboldened by her new love and perhaps wanting to make mischief, Virginia invited Vita for lunch at Charleston on Boxing Day. She assured her, "I am sorry to say that my dear old friend Crusty Roger who has been talking of the Gulf Stream, Rembrandt, instinct, sex in chickens, since Dawn, will not be here" (VWL, 3:225). All of Virginia's current lovers and beloveds, along with Julian and Quentin Bell (Angelica dined separately), sat together on Omega chairs with artists' wild decorations everywhere and (probably) a staid ebonized cabinet from 22 Hyde Park Gate sitting in a corner. As disparate as the furniture, the guests created a volatile lunch. Afterward, sitting in the new studio that Roger had designed for Vanessa, Virginia "held forth in her usual style . . . very amusing but also most uneasy." Vanessa admitted that Virginia's talk was "brilliant" but "curmudgeonly" enough that she found it exhausting (VBL, 287–88).

Nigel Nicolson thinks that his mother was actually afraid of Vanessa; at least he was. Vita returned to Charleston probably only once during Virginia's lifetime, perhaps because, Nicolson speculates, this visit was so "charged," but of course he was not there. (Nigel was not quite nine at the time, and children did not eat with adults.) Quentin Bell remembers no difficulties but only being struck by Vita's beauty and being involved in an argument after lunch. After Vita left, Virginia and Clive exclaimed over Vita's beauty and lineage. Then Leonard turned to Julian Bell and observed, "What snobs they are." The argument lasted for the rest of the afternoon.[20] Leonard left Charleston on 27 December, and Virginia joined him the next day because, Vanessa assumed, she could not "bear to be parted from [Leonard] a moment" (VBL, 287). But Virginia certainly had left him in order to be with Vita, so perhaps her refusal to stay on at Charleston was an attempt to compensate.

Back at Tavistock Square and sick with German measles in early

1926, Virginia finished the essay "On Being Ill," which testifies to some of the creative advantages of forced retirement from society. During this time, she neglected her diary and wrote almost exclusively to Vita. Vanessa must have written to Vita too, amusing Virginia with a replication of their old rivalries (VWL, 2:230). After Vita left for Persia, her letters spoke of her "terrible and chronic homesickness" for Virginia (Sackville-West, *Letters*, 103).

Both gratified by that testimonial and freed from emotional upheavals, Virginia soon found herself "blown like an old flag by my novel." Her rapid progress on *To the Lighthouse* confirmed "that I was on the right path" (D, 2:59). Writing conquered moments of depression: "I am old: I am ugly. I am repeating things. Yet, as far as I know, as a writer I am only now writing out my mind." By April 1926, Virginia had finished the first part of her new novel and was contemplating the radically experimental second part, "the most difficult abstract piece of writing—I have to give an empty house, no people's characters, the passage of time, all eyeless & featureless with nothing to cling to" (D, 3:67, 76). Writing this section involved observing as carefully as a painter does a still life, a landscape, or an interior.

With Vita away in Persia, Vanessa seems to have resumed her inexpressive ways. Secure in Vita's, as well as Leonard's, love, but again alienated from Vanessa, Virginia was emboldened to voice what she had long felt but dared not say: "You are a scandal to sisterhood not to have written—Everything in the way of affection is always left to me." She threatened, "The time will come when exhausted nature sleeps." If Vanessa could excuse her silence by saying that she could not find a pen, Virginia could warn that her own supply of pens would eventually run out as well (VWL, 3:255). The old fear that Vanessa abused her was confirmed when Roger, whom she saw more of now that Vita was gone, "last night upset me by saying that Nessa finds fault with my temper behind my back" (D, 3:76). Actually, Roger was being tactful, for the tone of many of Vanessa's comments about Virginia was much more caustic than "finding fault," as Virginia discovered late in her life.

When the two current assistants at the press objected to Leonard's autocratic methods, Vanessa and Duncan—in their usual catty ways—spread the story of Leonard's harshness. Virginia maintained that the assistants had taken one of Leonard's jokes too seriously and that Vanessa's accusations were "wildly unjust" (VWL, 3:257 and note). Now Vanessa saw Duncan and herself as victims: "All the blame is put on us." Vanessa also objected to Virginia's "disgracefully" publicizing Vanessa's plans for a dining club. (Playing an old game, Vanessa threatened to start a club only for painters [VB/VW, 19 or 20 April 1926]).

In May, a general strike brought English commerce to a standstill. Virginia and Leonard organized petition campaigns urging the government to negotiate with the workers. Vanessa assumed that Virginia was at the "hub" of the strike (VB/VW, 15 May 1926). Certainly Virginia's diary concentrated on it, although she supposed "all pages devoted to the Strike will be skipped, when I read over this book." The bickering and backbiting between the government and labor convinced Virginia that "really I dont like human nature unless all candied over with art" (D, 3:85). The strike made Roger conclude, "It's a disgusting animal is man & I sometimes think specially white man." When the strike ended, Roger and Helen, Virginia and Leonard, and Clive and Mary dined together, with Virginia "in her grandest vein" (RF/VB, 12 May 1926).

Virginia had developed a hyperbolic conversational style, with Leonard sometimes providing a "backbone" for conversation that she ornamented with barbed sallies and hilarious exaggerations. After dinner, the party returned to Clive's and were joined by others (including Madge Vaughan's daughter Janet) who had gathered signatures for the petition. Virginia begged for details about Janet's experiences riding a bicycle in London traffic. Then the talk turned to art, provoking Janet's memory that "if you could get [Virginia] and Roger Fry arguing—the ball went backwards and forwards" and offered "superb" entertainment (Stape, *Interviews and Recollections of Virginia Woolf*, 10; Noble, *Recollections of Virginia Woolf*, 98). Probably Roger and Virginia were spiritedly debating the role of the aristocracy in culture, for shortly thereafter, Roger began an essay attacking snobbism's pernicious influence on art (RF/VB, 25 May 1926).

In just over a month, Virginia finished the first version of the midsection of *To the Lighthouse* (D, 3:88), the "eyeless & featureless" writing that applied the method of "Blue & Green" to an empty house. Flanked by characters in the first and third sections, the midsection is saved from abstraction and offers a radical inversion of single-voiced, authoritarian, masculine, or monologic narration.[21]

Partly because his friend the French aesthetician Charles Mauron was going blind, Roger promoted Mauron's career as a translator. He gave him a sketch of his, entitled "Mosquitoes," to translate and arranged for Mauron to translate E. M. Forster's novels into French. Then in the summer of 1926, Roger asked Virginia to allow Mauron to translate a portion of her work in progress. The experimental midsection, "Time Passes," of *To the*

Lighthouse, was brief and self-contained enough for Virginia to permit Roger to send it to Mauron. Taken alone, however, "Time Passes" seemed to Roger too abstract, especially when he was considering the necessity of "psychological volumes in the visual arts" (RFL, 2:594).

When Vita returned from Persia in May, she and Virginia were shy and rather disillusioned with each other, but Virginia expected their relationship to become more solid, "more lasting than the first rhapsody" (D, 3:88). With its recent improvements, Virginia now thought Monk's House a "perfect triumph" (D, 3:89) where she could proudly entertain houseguests. After Virginia recovered from a "nerve exhaustion headache" (VWL, 3:272), Vita spent two nights alone with her at Monk's House in June 1926. Less reckless than she had been when courting her, Vita now began to worry about the fragility of Virginia's body and mind (Raitt, *Vita and Virginia*, 159–60).

Soon Virginia's social and aesthetic life resumed in London when Edith Sitwell invited her to meet Gertrude Stein, and she saw a very successful exhibition at the Leicester Galleries that included works by Bell, Grant, Fry, and Porter.[22] Virginia complimented the "combination of pure artistic vision and brilliance of imagination" in Vanessa's paintings and criticized the "problems of design on a large scale." She selected a Charleston landscape as Vanessa's masterpiece and fathomed that "the problem of empty spaces, and how to model them has rather baffled you" in a large picture of Angelica. She lightly renewed old jokes about Vanessa's being domestic, having children, and hence not deserving artistic prominence and fame. Then Virginia declared, "I will not let you two bitches [Vanessa and Duncan] have the laugh of me any longer." But in case her remark might disrupt the prevailing harmony, Virginia retracted it: "I see why you laugh at me for writing about painting" (VWL, 3:270–71).

Vanessa had not reacted to Virginia's brilliant art criticism in "Pictures," but she certainly reacted (ignoring the criticism) to this personal praise: "I wont say you're an infallible critic, though Duncan can tell you that as we pace the length of the Giudecca [in Venice] I ruminate again & again on your words & wonder if I should recast all my intentions regarding my art in consequence of them." She even suggested that Virginia, instead of "unbalanced though amiable zanies like Mr Ede," write art criticism for the *Nation* (VBL, 295–96). But what should have been a victory over their personal *paragone* became instead an argument about Ede, the art reviewer for the *Nation* whom Vanessa had recommended, thus planting Leonard "with a complete imbecile" and then "complain[ing] of

[his] idiocy." Liberated to say what she felt and even risk loss, Virginia contrasted her admiration for her sister's gifts with Vanessa's boredom with everything connected with Virginia (VWL, 3:274–75). From Paris, Vanessa complained that Vita reacted to her like "an Arab steed looking from the corner of its eye on some long eared mule—But then you do your best to stir up jealousy between us" (VB/VW, 16 June 1926). However mockingly, Vanessa for once was jealous of Virginia's loved one.

As Vanessa and Duncan expressed their artistic vision on more and more of the woodboxes, mantels, furniture (Figures 42 and 43), walls, bathtubs, and even chamber pots at Charleston, Virginia, recovering from influenza at Monk's House, solicited Vanessa's decorating skills. She wondered whether she would be "allowed some rather garish but vibrating" colors for the mantelpiece (VWL, 3:273). Nigel Nicolson writes that Vita "never ceased to despair about the lack of visual taste in either of the Woolves." He explained that Monk's House was "a mess." There were saucers of cats' and dogs' food about, and the stairs were so cluttered with books and old newspapers that one could hardly mount them. The only tasteful furniture came from the Omega. Quentin Bell remembers that Virginia was unsure of her taste and that the Bells were "sniffy" about her lack thereof (Nicolson, in Lee, *Cézanne in the Hedge*, 86; interview with Nigel Nicolson, 23 July 1993; interview with Quentin Bell, 26 July 1993). Virginia preserved "her own unaccountable predilection for green paint," and "possessions came to [the Woolfs] awkwardly, haphazardly, from the naive portraits that were sold with the house to the modernist armchair that was the laughing stock of arty friends" (Julian Bell [son of Quentin and Olivier], "Monk's House and the Woolfs," in McQueeney, *Virginia Woolf's Rodmell*, 24, 23).

Such friends not only laughed at her decorating taste, but sometimes ridiculed Virginia's taste in clothes as well, thereby reinforcing her "clothes complex" (D, 3:81). On 29 June, in the presence of a rather elegant group at Clive's in Gordon Square, Clive "laughed at my new hat, Vita pitied me, & I sank to the depths of gloom." Clive had "said, or bawled rather, what an astonishing hat you're wearing!" Even Duncan "told me it was utterly impossible to do anything with a hat like that." Conversation about her attire was "humiliating"; "Leonard got silent"; and Virginia found herself "as unhappy as I have been these ten years" (D, 3:91). Fortunately, this unhappy cloud blew over for a time. Clive took her to an extravagant lunch, and Virginia bought herself a handsome new dress (VWL, 3:281), but her self-image had been cracked.

In July, the Woolfs traveled to Dorchester to have tea with the aged Thomas Hardy, who pleased Virginia by telling her how Leslie Stephen had defended his writing against a philistine public (D, 3:97). When autographing her copy of *Life's Little Ironies*, even though Hardy misspelled "Woolf" as "Wolff," he seemed to Virginia to be the consummate "Great Victorian" (D, 3:100).

While Leonard worked in London, Virginia spent the night of Monday, 26 July, with Vita at Long Barn. On Tuesday, Vita then drove Virginia with her gift, a spaniel puppy the Woolfs would call Pinker or Pinka, to Rodmell. Virginia must have felt intense emotional upheaval at Long Barn, for when Leonard arrived (driven back to the country by Clive and Julian Bell [LWD]), he found Vita and the puppy in fine shape but Virginia completely enervated. Later Virginia described her situation as "a whole nervous breakdown in miniature." She sank "into a chair, could scarcely rise; everything insipid; tasteless, colourless. Enormous desire for rest." Indeed, she was so physically exhausted that she lost her "power of phrase making," the core of her identity. When her sensibilities returned the next Saturday, Virginia analyzed her condition. She still lacked the "'making up' power" and felt "no desire to cast scenes in my book," but her "curiosity about literature" was returning, as was her curiosity about herself; for she not only conducted her own self-analysis, but also wanted to "make a looking glass with shell frame." By Sunday or Monday, the "power to make images" signaled the return of her personal confidence and creativity (D, 3:103, 104).

Cautioned by this incident, Vita wrote to Harold Nicolson that she would be more "sagacious" with Virginia because she was now less physically tempted and more worried about Virginia's mental health (Raitt, *Vita and Virginia*, 160; Glendinning, *Vita*, 165). Virginia theorized that "the Married Relation" depended on periods of automatic unexamined routine that prepare for "a bead of sensation (between husband & wife)." She compared these "moments of great intensity" with Hardy's "moments of vision" (D, 3:105). The sequence of the humiliating discussion of her attire, Vita's withdrawal, and Virginia's speculations about a meaningful marriage may have meant that again helplessness, loss, and guilt, as well as simple exhaustion, triggered her minor breakdown.

During August and September 1926, the Woolfs hibernated at Monk's House. Virginia took a break from her novel to work on a most remarkable Hogarth publication, *Victorian Photographs of Famous Men and Fair Women by Julia Margaret Cameron, with Introductions by Virginia Woolf and Roger Fry*. Her contribution is a wonderfully comic sketch of her great-aunt's eccentricities, vitality, talent, and capacity to give and to love. Virginia tells us that Cameron began her photographic career almost accidentally at the age of forty-eight when her daughter and son-in-law gave her a camera.[23] Over the next fourteen years until her death in 1879, she was an indefatigable photographer of the great figures of the Victorian intellectual world. Typical of Virginia's scene making is the following: "Dressed in robes of flowing red velvet, she walked with her friends, stirring a cup of tea as she walked, half-way to the railway station in hot summer weather. There was no eccentricity that she would not have dared on their behalf." She continues with an ironic sentence about friends who "suffered the extreme fury of [Cameron's] affection." Virginia also mentions her aunt Caroline Emelia Stephen in this sketch: "The variety and brilliance of the society [Cameron] collected caused a certain 'poor Miss Stephen' to lament: 'Is there *nobody* commonplace?'" In her sketch of her great-aunt, Virginia displayed powers of life writing that had lain fallow for almost twenty years, powers she amplified during the rest of her life. She even began a "life" of Julian Bell, speculating about the effect of merging the Bell and Stephen inheritances. But this topic, as she had discovered back in 1908 with her "character" of Clive, was so sensitive that she abandoned Julian's life after five typed pages (MHP/A18).

Roger Fry's treatment of Cameron's photographs praised her insight into the period in which the women "were fashioned by the art of the day" (as Julia Jackson had been) and the men were fashioned by a "walled garden" that cultivated genius. He termed that Victorian garden "an atmosphere wherein great men could be grown to perfection—or rather in which men of distinction could be forced into great men" (Cameron, *Victorian Photographs*, 10). Later, writing about Ruskin, Virginia said much the same thing, that the nineteenth century "liked their great men to be isolated from the rest of the world. Genius was nearly as antisocial . . . as insanity" (CE, 1:205). Perhaps in this period, Roger sketched Virginia with Mary Hutchinson behind her on the cover of his "autobiography" for a Memoir Club reading (Figure 44). Virginia looks graceful, well dressed, and at ease.[24] The combination of his life writing and Virginia's image proved to be prescient.

Even though Roger's article on Seurat had been "suppressed" by the strike, he expected Leonard to publish it in the summer of 1926. His conviction was that in art, as in Seurat's paintings, "nothing of the original theme, of the thing seen, remains untransformed, all has been assimilated and remade by the idea." Through the quality of "transmutation" or "transformation," Roger reconciled the formal and the "literary" and allowed for referentiality in the work of art.[25] Virginia seems to have read the article at Monk's House during the summer and wondered, "If art is based on thought, what is the transmuting process?" (D, 3:102).

The reconciliation afforded by the concept of transmutation is implied in the difference between the dualistic title *Vision and Design* and the monistic *Transformations*, Roger's second major collection of essays, which was published in 1926. In this, he complained "that as a nation our aptitudes for literature are developed out of all proportion to our aptitude for the other arts" (26). He explained that in reacting against his early emphasis on the "dramatic" content in a work of art, he had overemphasized "the pre-eminence of purely plastic aspects, and almost hinted that no others were to be taken into account." Furthermore, he promised to develop his aesthetic ideas to reconcile the dramatic and the plastic, or the literary and the formal (10), and to explicate the "perfect fusion of the two elements" (154). The notion of fusion solved a theoretical dilemma for Roger; for Virginia, it spoke to a longing "deep in the richest strata of my mind" (D, 2:323). That summer, she transformed her highly personal "literary" subject matter into the formal symmetry of *To the Lighthouse*. She also made the painter Lily Briscoe a Virginia/Vanessa figure who articulates Roger's theories and exposes Mr. Ramsay's aesthetic insensitivity.[26]

By 5 September, Virginia was facing the formal problem of bringing together Lily Briscoe and Mr. Ramsay for "a combination of interest at the end" of *To the Lighthouse*. She found this novel "subtler & more human" than *Jacob's Room* or *Mrs. Dalloway*, as she had concentrated her lyricism in the middle section, where it did not interfere with the textual design (D, 3:106–7). Virginia planned to rest her mind from this intense creativity by writing a book on the theory of fiction.

In mid-September, however, after eleven years, a horror "like a painful wave swelling about the heart" threatened to engulf Virginia. She was enough her logical self to analyze her own *State of Mind* (as she labeled it):

> I watch. Vanessa. Children. Failure. Yes; I detect that. Failure failure. (The wave rises). Oh they laughed at my taste in green paint! Wave crashes. I wish I were dead! I've only a few years to live I hope. I cant face this horror any more—(this is the wave spreading out over me). (D, 3:110)

Virginia faulted herself for "intense depression": "Why have I so little control? It is not creditable, nor lovable. It is the cause of much waste & pain in my life." Again she analyzed the catalyst for her depression: "Nessa humming & booming & flourishing over the hill; & one night we had a long long argument. Vita started it." Later Leonard "admitted that my [Virginia's] habits of describing him, & others, had this effect often.[27] I saw myself, my brilliancy, genius, charm, beauty (&c. &c—) . . . diminish & disappear. One is in truth rather an elderly dowdy fussy ugly incompetent woman vain, chattering & futile" (D, 3:111).

All the old demons had come back to haunt her: loss (of Vanessa's, Vita's, and Leonard's affections); guilt (over criticizing not only others but also Leonard and hence violating whatever agreement the two had reached after her 1915 illness); and helplessness (before Vanessa, who laughed at her taste in green paint, others who laughed at her hat, and a public who would think that her novel was a failure). But by this time, Virginia knew how to stand up to her demons. She drew up a three-part plan for working, entertaining, and traveling. She and Leonard considered how their new money should be spent. Leonard had grand plans for the garden, whereas Virginia did not want household purchases or travel jeopardized. One purchase they both agreed on was a gramophone; in fact, Maynard Keynes cattily claimed that Leonard saw it as "the greatest addition to his life since marriage" (JMK/VB, 28 March 1927).

Virginia recognized the advantages in her "plunge into deep waters," for "one goes down into the well & nothing protects one from the assault of truth." Facing her true self was interesting, "though so acutely unpleasant." In consequence, she made a happy vow—renewing the one she seems to have made after the 1915 horrors—to "be much more considerate of L.'s feelings; & so keep more steadily at our ordinary level of intimacy & ease: a level, I think, no other couple so long married, reaches, & keeps so constantly" (D, 3:112). This restoration was "mystical" as well as psychological: Her "plunge" had been

> frightening & exciting [for] in the midst of my profound gloom, depression, boredom, whatever it is: One sees a fin passing far out. . . . Life is, soberly & accurately, the oddest affair; has in it the essence of reality. I used to feel this as a child—couldn't step across a puddle once I remember, for thinking, how strange—what am I? &c.

This image recalls "the old devil" who "got his spine through the waves" that evening when Leonard seemed lost (D, 2:270). The fin seems to have meant the shape of ultimate truth that Virginia defined nihilistically, as shark rather than porpoise, and understood most clearly after plunging into deep waters. The dilemma was how to convey the shape of "reality" through her art.[28] Virginia could "hazard the guess" that this state of mind might be "the impulse behind another book." Some three years later, rereading this passage in her diary, she noted that this was perhaps the moment of inception for *The Waves* (D, 3:113).

Virginia did not give up her social life, but she did give up diary writing in November to finish her novel, which she now found "easily the best of my books" (D, 3:116 note, 117). Meanwhile, Leonard reviewed *Transformations*, praising Roger for his knowledge and imagination and iconoclasm. Theorizing about the connections between painting and literature, Roger had criticized Clive for dissociating them, but Leonard thought he saw an about-face in Roger's determination to connect the visual and verbal arts (*Nation and Athenaeum*, 27 November 1926, 304). That is, he seems to have been projecting onto Roger the trajectory Virginia's thinking had followed.

After a visit from Vita at Tavistock Square that seems to have renewed their erotic relationship, Virginia looked back at their "spirited, creditable affair, I think, innocent (spiritually) & all gain, I think; rather a bore for Leonard, but not enough to worry him." Virginia could now regard death as "active, positive," but the "one experience I shall never describe" (D, 3:117). It would not be the only one. In *Jacob's Room* when Jacob takes Florinda into his bedroom, the narrator, like the letter from Jacob's mother, remains outside the closed door. Both death and sex remained undescribed by Virginia's various narrative voices.

In this late 1926 period of artistic and emotional exuberance, "one has room for a good many relationships" (D, 3:117). After a Christmas visit to Ka and Will Arnold-Forster in Cornwall, the Woolfs returned to London. Virginia received a Christmas letter from Vita telling her about the nine-foot-wide four-poster bed at the vast ancestral Knole, which she invited Virginia to come and see for herself (Sackville-West, *Letters*, 157). With Duncan away in Cassis with his mother, Vanessa invited Virginia, Leonard, and Roger to join her at Charleston early in the new year.[29] Instead, however, Roger visited the Woolfs in London (LWD), and then they spent a few days at Monk's House.

Back in London later in January 1927, when Vanessa heard that Duncan was ill in Cassis, she and Virginia "kissed on the pavement in the snow" as Vanessa left to care for Duncan. Virginia was deeply gratified by Vanessa's returning willingness to show affection for her: "We are very intimate—a great solace to me" (D, 3:124), especially when Vita was about to leave for Persia again. Virginia had a chance to see Vita's four-poster bed when she and Vita spent two nights together at Knole in mid-January 1927, but both were strangely silent about this visit. Virginia's diary instead talks about her relief at Leonard's praise for *To the Lighthouse*, which he called a "psychological poem" (D, 3:123). Vita spent the morning with Virginia before departing for Persia on 28 January 1927.

In Cassis, Vanessa found Duncan recovering and wrote the sort of letter that Virginia had spent a lifetime craving. Vanessa thanked Virginia: "As usual in a crisis—& in ordinary life too though you pretend not to think so—I depended on you and Leonard." She also warned, "Don't expend all your energies in letter writing on her [Vita]. I consider I have first claim." Virginia responded with a series of delightful letters to which Vanessa responded, "You can imagine how Duncan & I suck your letters dry & what intense amusement & pleasure they give." She also recommended that Virginia see an exhibition of paintings by Fry and Porter (VB/VW, 26 January and 5 February 1927). Eleven years earlier, Vanessa had all but ceased writing letters to Roger and Virginia, causing each of them major distress. Now when Roger stopped writing, Vanessa reacted by asking Virginia to advise him that "I am still alive & tell him to read my last letter & be as cutting as you can & make him feel really uncomfortable for never writing to me. Am I to suppose that he prefers Helen?" (VB/VW, 13 February 1927).

As soon as she had completed her novel, Virginia's imagination leaped forward. She needed to finish her book on fiction and then indulge in an "escapade after these serious poetic experimental books whose form is always so closely considered." She thought of kicking up her heels and writing a "whole fantasy to be called 'The Jessamy Brides'" in which "sapphism is to be suggested" and her "own lyric vein is to be satirised. Everything mocked." Afterward, she would compose a "very serious, mystical poetical work which I want to come next" (D, 3:131).

The *New York Herald Tribune* invited Virginia on an expense-paid trip to New York, but Leonard's fare would have cost more than sending both of them to Italy or Greece, perhaps with Roger (D, 3:124). Despite Vanessa's warning to Helen Anrep that she was "doubtful as to how [Roger] and the Woolves will get on as travelling companions, having had experience myself of both him and Virginia in that capacity" (VBL, 310), Roger was undeterred, and the Woolfs may have decided to travel to Greece with him.[30]

Virginia was about to dedicate her fifth novel to Roger, but her plans for such a public declaration were undermined by a nemesis from the past: After checking the proofs of Mauron's translation of "Time Passes," Roger explained to Marie Mauron in December 1926:

> To tell the truth I do not think this piece is quite of her best vintage. I have noticed one peculiarity. She is so splendid as soon as a character is involved—for example the old concièrge is superb—but when she tries to give her impression of inanimate objects, she exaggerates, she underlines, she poeticizes just a little bit. Several times I felt it was better in the translation, because in translation everything is slightly reduced, less accentuated and in general better.[31]

Roger must have offered the same critique to Virginia, to which she reacted at first depressively and then defensively. In her diary on 12 February 1927 she wrote, "I may note that the first symptoms of Lighthouse are unfavourable. Roger, it is clear did not like Time Passes" (D, 3:127). Unlike the dedication of *The Common Reader* to Lytton, in which she reciprocated his dedication to her, a dedication to Roger would signal debt or deference, perhaps dependence. Furthermore, if he "did not like" the book honoring him, she would have found herself in an intolerable dreadful fix. Virginia therefore withdrew the dedication, and on 28 February, she concluded self-protectively, "I'm glad I didn't dedicate my book to Roger" (D, 3:129).

Virginia was looking forward to a "festive" time when Vita returned but was "nervous" about Harold's reaction (VWL, 3:325). As if to signal a new bisexual identity, she had her hair cut short. She felt "happier, wiser, serener, cleverer a thousand times shingled than haired" (VWL, 3:331). When she automatically started to twist her hair into a coil to be pinned up into an old-fashioned bun, Virginia discovered with delight that she need not bother. Her haircut also gave her new confidence in elegant company (D, 3:127).

The Woolfs decided to travel neither to America, because of the expense, nor to Greece, since Roger wanted to bring along Helen Anrep

(VWL, 3:341). Instead, they chose Vanessa and France. Vanessa wrote to Virginia that they were wise not to go to Greece, for Roger would "have driven you into an asylum & Leonard after you." Psychological and aesthetic harmony were, as always, aligned. Vanessa was "very much flattered & interested by your criticisms on my paintings." Anticipating Virginia's visit, she at last could write, "I daresay you won't be half as much pleased to see me as I to see you" (VB/VW, 13 and 17 March 1927).

After a period of creativity in which Virginia's brain was "ferociously active" and she wrote with "never a word wrong for a page at a time," the Woolfs joined Vanessa, Duncan, Clive, Julian, Quentin, and Angelica in Cassis, "a strange resurrection of us all abroad" (D, 3:132). Sometime during this period, Virginia set out to capitalize on Vanessa's now-acknowledged need for her. She began demanding kisses from both Vanessa and Angelica. Vanessa's "mute embarrassed dislike of the whole demonstration" had no effect on Virginia, who would "demand her rights" until Vanessa offered a kiss as if to buy her sister off so that she could withdraw into herself again (Garnett, *Deceived with Kindness*, 107).

Before the Woolfs left Cassis on 6 April to travel through Italy, Virginia negotiated a bargain healthier than demanding kisses: Vanessa would barter a painting in return for long letters about Italy from Virginia (VWL, 3:360, note). When Vanessa and Angelica held back kisses and Vanessa suggested that Virginia would not have liked having children, Virginia countered with an attack on the "maternal passion." This passion, of which she had experienced so little as a daughter and none as a mother, was, she claimed, "unscrupulous" in Vanessa, for she "would fry us all to cinders to give Angelica a days pleasure" (VWL, 3:365 note, 366). Vanessa tried to sound less unscrupulous, objecting to her sons' "male conversations at meals" and Angelica's ambition to become an actress. She assured her sister, "I paint my pictures for you to choose from" (VBL, 312–14).

Virginia kept her bargain with lengthy letters focused on gossip, customs, scenery, and Italian architecture and painting. She liked Raphael but noted that the artists' judgments "slightly inhibit my art criticism" (VWL, 3:367). Nevertheless, the sisters' bargain worked astonishingly well until the Woolfs returned to London. Then Virginia had the mixed pleasure of learning that her rapturous letters from Italy had so delighted Vanessa that she read them aloud to the group gathered around her. In this way, Vanessa aligned herself with Virginia in entertaining others, but in an odd replay of the dynamics of 1910, Clive then made the mistake of relating to friends the substance of Virginia's letters, thereby infuriating her (D, 3:134, note).

Virginia regretted, "I'm afraid I shall have to make an end of our agreement and lose my picture," for the whole contretemps took her "back to the old days!" (VWL, 3:368).

Vanessa maintained that she had read aloud nothing indiscreet and insisted, "I really dont feel to blame. It was very difficult not to read aloud bits of your letters" (VB/VW, 4 May 1927). Virginia wondered why Clive got so much pleasure out of telling stories about the Woolfs: "There must be some obscure jealousy at work I think. He grudges, not your affection for me, which doesn't exist, but mine for you—Or he wants to parade his knowledge of our affairs" (VWL, 3:370).

Virginia's threatened breach of their bargain inspired Vanessa to drop the pose of being too ineloquent to write. She penned a marvelous letter beginning, "It is a work of absolute heroism to write to you." She described a giant moth fluttering around her that she attempted to chloroform, as the Stephen children had done with moths and butterflies at St. Ives years before, until it finally died, "rather the worse for wear." The atmosphere was like old times, with Roger starting several pictures at once, but also newer times, with Vanessa asking Virginia to write more on the maternal instinct. Vanessa fumed against Helen Anrep's plans to settle in Sussex, dramatically noting, "The moths die around me." She closed with the assurance: "I'll decorate Rodmell or a gramophone or anything you like and be most generous with my works if only you'll continue to do your part" with letters (VBL, 314–16).

Considering the maternal instinct, Virginia found herself proud enough of Quentin to appropriate him. Even though both Bell sons supposedly preferred Shaw to Shakespeare, she maintained to Vanessa that "probably I am almost as spotted with the maternal taint as you are." Virginia asked Vanessa about "Raphael and Michael Angelo," feared that Vanessa would "laugh at the painting bits in the Lighthouse!" and complimented her sister's way with words: "What a terrific letter!" (VWL, 3:371–72). Tensions subsided as Vanessa further placated Virginia by insisting that Clive thought Virginia a brilliant letter writer and that she preferred Virginia's descriptions of their friends to the friends themselves (VB/VW, 19 May 1927). Then as if in apology, Clive conceded to Virginia that Leonard "seems to have grasped the mystery of your health" (CB/VW, 31 May 1927).

The major balm for these tensions, however, was the publication of *To the Lighthouse* on 5 May 1927, the thirty-second anniversary of Julia Stephen's death. Virginia wrote a wonderfully comic letter to Vanessa imagining the difficulties that Vanessa and Duncan would have reading it (VWL, 3:375–77). Instead, both "poor dumb creatures" read the book in three days, "a record I should think" (VB/VW, 19 May 1927). Virginia's previous novels had flattered, amused, and even impressed Vanessa, but *To the Lighthouse* at last touched Vanessa's heart. She declared that even if Virginia never forgave her for reading her letters aloud, even if Virginia snubbed her, she still would praise this novel:

> All my pride was humbled & I was eating dust at your feet in any case & all owing to the Lighthouse. I dont flatter myself that my literary opinion is really of any interest to you . . . [but] you have given a portrait of mother which is more like her to me than anything I could ever have conceived of as possible. It is almost painful to have her so raised from the dead. . . . [Reading was] like meeting her again with oneself grown up & on equal terms & it seems to me the most astonishing feat of creation. . . . So you see as far as portrait painting goes you seem to me to be a supreme artist.

Self-conscious about such a display of emotion, Vanessa continued, "You can put it down to the imbecile ravings of a painter on paper." She even assured her sister that she and Duncan "didn't laugh at the bits about painting" (VBL, 316–18). Virginia found Vanessa's reaction "sublime" but also an "almost upsetting spectacle" (D, 3:135).

Perhaps Vanessa's reaction was "almost upsetting" because it failed to note the book's judgment of Julia. If the analysts following Klein are correct that it is only "when the loss has been acknowledged . . . that re-creation can take place" (Segal, "Psycho-Analytic Approach," 199), Virginia had lost her cherished images of her parents to a good end. If Lytton's criticism of Julia's photograph and Adrian's discussion of Klein's theories had been revelatory, writing this novel had helped Virginia understand how much of her life had been a futile attempt, like Lily's, to be "one with the [maternal] object one adored" (TL, 51). Thus she "did for myself what psycho-analysts do for their patients" (MB, 81).

Virginia could jettison the sainted mother image and admit that despite all her charm, Julia had cared less about young women's talents than that "they all must marry" (TL, 49). Mrs. Ramsay's sense that James would never be so happy again replicates the melancholy that led Julia to suppose that cheerful little Thoby would "never be so happy again."[32] Mrs.

Ramsay's reluctance to express her love surely echoed Leslie Stephen's writing to Julia: "I know what you think though you dont like to put it into words" (LS/JPS, 3 January 1895). Virginia, too, had needed Julia's love put into words and actions. She had needed and at last found, as Lily does, freedom from coercive gender roles that limit men and belittle women. Now she could expose Leslie Stephen's "insatiable hunger for sympathy" as a mask for his determination to make women "'submit to me'" (TL, 151, 170). By mourning her sainted mother and esteemed father, Virginia could re-create them honestly and exorcise their haunting legacy in ways that Vanessa seems to have missed.

Vanessa's reaction was upsetting also because it was so entirely personal. She told Clive that it was "fascinating as a portrait of my parents" and "extraordinarily true" (VB/CB, 20 May 1927). Virginia then asked her to assess it as a work of art, and Vanessa promised, "I shall find plenty more to say on the aesthetic questions later—just to keep you attentive" (VB/VW, 27 May 1927). These aesthetic questions were the ones that Roger had been pressing since 1910. Even though his name is not publicly linked to this novel, there are private references: One of the Ramsay's children is named Roger; the *boeuf en daube* that transforms the dinner party into a festive occasion was Roger's culinary speciality; and his definition of pictorial unity as "a balancing of the attractions of the eye about the central line of the picture" (VD, 22) surely defines both Lily's final picture and Virginia's novel. In addition, *To the Lighthouse* explicitly illustrates, embodies, and discusses Roger's aesthetic theories.

Mr. Bankes's favorite picture is valued for its associations with the place where he spent his honeymoon (TL, 53), but Lily Briscoe shows him that such "literary" associations are irrelevant to true art and that the aesthetic question is "one of the relations of masses, of lights and shadows" (TL, 53). "Mr. Bankes was interested. Mother and child then—objects of universal veneration, and in this case the mother was famous for her beauty—might be reduced, he pondered, to a purple shadow without irreverence" (TL, 52). Lily explained to Mr. Bankes that she "did not intend to disparage a subject which, they agreed, Raphael had treated divinely" (TL, 176).

In 1917, a Madonna by Raphael, known as the *Bridgewater Madonna* (Figure 45) had been on view in London at Bridgewater House, where private collections of art had been made available to art lovers like Roger Fry since the early nineteenth century.[33] In the translation exhibition of 1917, which Virginia had seen, Roger had painted a rather Cubist version of the

Bridgewater Madonna (Figure 46). Mr. Bankes comes to recognize that Lily's reduction of the mother and child is not disparaging or irreverent, any more so than was the simplification of Raphael in which Roger captured the Christ child's movement and the Virgin's tenderness in a modernist focus on geometric forms and essential elements.

If Roger's Cézanne-inspired painting was the inspiration for Lily's more radical reduction of the Madonna and child icon to a triangular purple shape, that simplification also inspired Virginia's reduction of a decade of family history into two parallel masses with a corridor connecting them. Indeed, with the phrase "two blocks joined by a corridor," Virginia drew such a shape in her notes:[34]

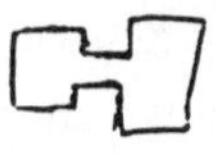

This diagram, following Roger's notion of balancing the attractions of the eye around a central line (VD, 22), shows Virginia creating a new spatial form for a novel.

Memories of several pictures inform *To the Lighthouse*. Many of the photographs in Leslie Stephen's and Stella Duckworth's albums show Julia posing on the steps leading down from one of the French doors at Talland House. In one (Figure 47), with Virginia hovering behind her, Julia is sitting by Adrian, before turning to pose in profile as photographers had long ago taught her. By taking herself out of the tableau of Mrs. Ramsay and James in the first section, "The Window," Virginia not only simplified the image, but made an emblem of Julia's devotion to Adrian. In the second section, "Time Passes," Virginia explored the "problem of empty spaces and how to model them," which she had said baffled Vanessa; this meditation on space could have been inspired by any number of Vanessa's interiors. And Roger's Cubist rendition of Raphael's *Bridgewater Madonna* served as a model for the painting of Mrs. Ramsay and James that Lily begins in "The Window" and ends in the final section, "The Lighthouse."

In addition, *To the Lighthouse* can be viewed as a triptych. The first and last sections, "The Window" and "The Lighthouse," are parallel, not just because many characters reappear and the actions planned in the first section are completed in the last, but also because motifs, images, words, and concerns from the first section recur in the last. For example, the "gust of wind" that fills the sails parallels the "gust of life" that invigorates the dinner scene. Lily's question "What does it mean?" (TL, 145) echoes Mrs. Ramsay's "What have I done?" (TL, 82). References to Mr. Ramsay's boots, "The Charge of the Light Brigade," Cowper's "The Castaway," the "fatal sterility of the male" (TL, 37), the beauty of a butterfly's wing "clamped together with bolts of iron,"[35] and the pattern on the tablecloth, along with many other repetitions, recur with variations in both sections. Mrs. Ramsay's efforts to "compose" a dinner party are transmuted into Lily's efforts to harmonize a painting. With life "now strong enough to bear her on again," Mrs. Ramsay gathers her energies "as a sailor not without weariness sees the wind fill his sail and yet hardly wants to be off again." Renewed, she makes of the party something that "partook . . . of eternity" (TL, 84, 105). As the wind literally fills the sails of the Ramsay boat, Lily similarly gathers the strength to exchange "the fluidity of life for the concentration of painting" (TL, 158). In "The Window," to help Mrs. Ramsay, Lily had to renounce her intention to resist the code decreeing that female conversation must sympathize with male egos. In "The Lighthouse," Lily resists Mr. Ramsay's enormous pressure to sympathize with him in his grief. In the impasse between his need and her determination, she happens to notice his boots. Then the "blessed island of good boots" (TL, 154; see Figure 7) reconciles them and sets the stage for the book's formal resolution, which pairs his reaching the island and her finishing the painting. By such paralleling and interweaving of motif, Virginia's triptych may be apprehended not only sequentially as one reads but also statically as one views a painting.[36]

Virginia imported Roger's aesthetics into the heart of her novel, and he elaborated on them in *Cézanne* by distinguishing true art from both verisimilitude and association and also abstraction. He was very specific about the process of transformation in aesthetic creation by which actual objects are "reduced to pure elements of space and volume" and then reintroduced into the world of familiar appearances, so that they "retain their abstract intelligibility, their amenity to the human mind, and regain that reality of actual things which is absent from all abstractions" (*Cézanne*, 58–59). Having similarly united the abstract and the actual, Virginia could now feel "almost an established figure—as a writer." Yet even her triumph hinted at her lingering irrational insecurity: "They dont laugh at me any longer" (D, 3:137); even Arnold Bennett thought that her "character drawing has improved."[37]

Roger had had reservations about "Time Passes," but he praised both the design and the humanity of *To the Lighthouse* as

> the best thing you've done, actually better than Mrs Dalloway. You're no longer bothered by the simultaneity of things and go backwards and forwards in time with an extraordinary enrichment of each moment of consciousness. . . . My own private suspicion (which is not you must note an appreciation) is that its rather a great book. It's certainly the most intensely human thing you've done.

He also said, "Well I wish intensely that you had dedicated it to me. I think I must exact a copy with a private dedication. But I should have been enormously proud if my name had got linked on to that book." As he had done with all Virginia's work, Roger promoted it to acquaintances both at home and abroad.[38]

Virginia had sacrificed her intimacy with Roger for intimacy with Vita, who had arrived back in England on the day that *To the Lighthouse* was published and accompanied Virginia to lecture at Oxford. Virginia described to Vanessa Vita's aristocratic manner, and she said that she regretted that Vanessa would "never succumb to the charms of any of your sex." Duncan at least was "hermaphrodite, androgynous, like all great artists" (VWL, 3:381), including, she jokingly implied, Virginia but not Vanessa.

In 1904, when her father died, Virginia had lost the opportunity to "tell him how one cared" (VWL, 1:133). Her guilt for not doing so in part caused her first serious breakdown. In 1927, she shied away from telling the world how much she owed Roger and then felt guilty for remaining silent. She wrote to him:

> I am immensely glad that you like the Lighthouse. Now I wish I had dedicated it to you. But when I read it over it seemed to me so bad that I couldn't face asking you. And then, as it happened, that very day, I met you somewhere,—was so overcome (did you guess it?) by your magnificence, splendour and purity (of intellect, not body) that I went home and was positive it was out of the question—dedicating such a book to such a man. Really therefore the not-dedication is a greater compliment than the dedication would have been—But you shall have a private copy, if you'll accept it. What I meant was (but would not have said in print) that besides all your surpassing private virtues, you have I think kept me on the right path, so far as writing goes, more than anyone—if the right path it is. (VWL, 3:385)

In that same graceful letter, Virginia echoed Roger's theory that "in proportion as an artist is pure he is opposed to all symbolism" (Fry, *Artist and Psycho-Analysis*, 16). She insisted on a formalist reading of her novel: "I meant *nothing* by The Lighthouse. One has to have a central line down the middle of the book to hold the design together." She promised to reread his criticism and described the "wave of Fry worship" she had encountered at Oxford, proclaiming, "Really Roger, if you go on like this they'll be making a Christ of you within a century." Then she made a commitment to "set about the Fry memoir which I have it in my mind (as you Quakers say) to do before I die" (VWL, 3:385–86).

After sixteen years of resisting and absorbing Roger Fry's influence, Virginia Woolf had offered two consolations: a private dedication acknowledging his "dominion" and a public tribute in the form of a projected "Fry memoir." She offered lavish private testimony to his influence but public silence about that influence; privately, she expressed guilt about her failure to make a public declaration. And immmediately, Virginia developed one of her debilitating headaches.

After a three-week bout with headaches, lying in the garden at Monk's House, Virginia "rhapsodised" about a "play–poem idea" that finally became *The Waves* (D, 3:139). She was "deeply touched" that Adrian and Karin had "patched" up their relationship (VWL, 3:377). And as Adrian's career as a psychoanalyst became more and more a probability, she noted with pride that "we Stephens mature late. And our late flowers are rare & splendid" (D, 3:141).

With Quentin Bell, the Nicolsons, and others, the Woolfs traveled to Yorkshire during the night of 28 June 1927. At 3:30 A.M., their train was met by buses that carried them out to a hillside for the best view of the total eclipse of the sun (Glendinning, *Vita*, 177–78). Virginia wore a coat with a fur collar, the one in which she was also photographed as part of the American publicity for *To the Lighthouse* (Figure 48).

When the sun rose normally, the party first thought itself "cheated," but then "suddenly the light went out." The experience was elemental: "We had fallen. It was extinct. There was no colour. The earth was dead." The experience made Virginia appreciate the "light & colour" normally taken for granted (D, 3:143–44). Even though it lasted for about twenty-four hours, the excursion took no toll on Virginia's restored health.

This late-blooming flower, however, was suspicions that Vita was being unfaithful to her. Virginia warned Vita to be "a careful dolphin in your gambolling, or you'll find Virginia's soft crevices lined with hooks" (VWL,

3:395). Attempting to stir up jealousy, Virginia even claimed that she had been "on the telephone imploring Clive to come back to me," as in 1910. When she and Leonard were about to speak for the first time on the BBC, Virginia wrote to Vita, "I shall yawn & say, My God to think of Vita going to bed with Mary! This will be broadcast & ruin the chastity of 12 million homes."[39] Like her sister, Virginia transposed emotions and aesthetics; she privately condemned Long Barn (where she had spent a night alone with Vita) as unoriginal, despite its elegance, whereas Monk's House showed life and intensity, "some cutting edge" (D, 3:144–46).

Back in London, Virginia was invited to Vanessa's studio to meet with the artist Walter Sickert, who was reputed to have "the greatest admiration for Virginia" (VB/CB, ? July 1927). Soon the Woolfs and Vanessa were talking of "nothing but cars" (D, 3:146). With the proceeds from *To the Lighthouse*, Virginia and Leonard bought a "nice light little shut up car in which we can travel thousands of miles." The car, a Singer, gave them "an additional life, free & mobile & airy to live alongside our usual stationary industry" (D, 3:147, 151).

Vanessa also seemed at last financially secure after many lean years (D, 3:152). She bought an automobile about two weeks before the Woolfs bought theirs and soon referred to their "rival cars."[40] Both the Woolfs took driving lessons, and their new car greatly increased their activities. Virginia wrote about driving in London, where Vita gave her lessons (Glendinning, *Vita*, 179). Virginia claimed that she was "competent to drive alone in the country," that her gear shifting was "very good," and that she would "rather have a gift for motoring than anything" (VWL, 3:401, 400). But after she drove through a hedge, Virginia gave up driving.[41] Leonard began making almost daily records of the distances he drove in the country. Traveling around Sussex with him, Virginia observed the "vast enterprise" of human activity with something other than "the usual writers care for the aesthetic quality." Not only did she seem to lack driving skills, but she feared she also lacked a "comprehensive magnificent statesmanlike mind" that could turn such a panorama into art (D, 3:155–56).

Undeterred, Virginia wrote an essay to herald an experiment of which she alone was capable. She hypothesized about the unsuitability of various literary genres for the modern world. Then she speculated about a new nameless sort of novel that "will give, as poetry does, the outline rather than the detail. It will make little use of the marvelous fact-recording power. . . . For under the dominion of the novel we have scrutinized one part of the mind closely and left another unexplored."[42]

Meanwhile, as if to compensate for their altercation nine years earlier over the layout of "Kew Gardens," Virginia invited Vanessa to design the illustrations for the third edition. In another twist of the *paragone*, Vanessa created designs so lavish that they dominate Virginia's prose.[43]

Campaigning for a new prose form (and planning to create such a form herself), Virginia also needed to establish how outdated certain traditional ways of judging literature were. In her query "Is Fiction an Art?" she faulted E. M. Forster for valuing a simplistic notion of "life" more than a formalist notion of art: "the humane as opposed to the aesthetic view of fiction." In the affirmatively retitled "The Art of Fiction," Woolf found Forster's *Aspects of the Novel* outdated. Thus the English novel "might become a work of art" if English novelists cut themselves adrift from "preposterous formulas which are supposed to represent the whole of our human adventure." They would be well served by the example of Henry James, who created "patterns which, though beautiful in themselves, are hostile to humanity." Roger had said that only Henry James and Virginia Woolf were aware of language as a medium in itself (RF/VW, 18 October 1918). Now Virginia accused Forster of neglecting "the medium in which a novelist works. Almost nothing is said about words." Considering the design of a novel, she wondered, "How are we to take a stick and point to that tone, that relation, in the vanishing pages, as Mr. Roger Fry points with his wand at a line or a colour in the picture displayed before him?" ("Is Fiction an Art?" *New York Herald Tribune*, 16 September 1927; "The Art of Fiction," CE, 2:53–55). Although a critique of the aesthetic naïveté of most English writers and of Forster's theories, "The Art of Fiction" is also a defense of the formalist designs of Virginia's major fiction and a harbinger of the extreme Postimpressionist work she was envisaging.

Just as Forster offered a simplistic sense of "life," so did the traditional biographer: "I have never forgotten . . . my vision of a fin rising on a wide blank sea. No biographer could possibly guess this important fact about my life in the late summer of 1926 [1927]: yet biographers pretend they know people" (D, 3:153). Ten years after Roger Fry had planned a "great historical portrait group of Bloomsbury" (RFL, 2:423), on 20 September 1927, Virginia Woolf decided to "sketch here, like a grand historical picture, the outlines of all my friends." She mentioned Gerald Brenan, and then "Vita should be Orlando, a young nobleman. There should be Lytton. & it should be truthful; but fantastic. Roger. Duncan. Clive. Adrian. Their lives should be related."[44]

Thinking of such a "historical picture," Virginia wrote "The New Biography" as an indication of another direction in her writing. Ostensibly a review of Harold Nicolson's *Some People*, "The New Biography" raises questions comparable to those in "The Art of Fiction." This essay is at once a query about the art of biography and a manifesto of its potentials. Virginia observed that "biography is the most restricted of all the arts" but that invention could transform those restrictions "if we think of truth as something of granite-like solidity and of personality as something of rainbow-like intangibility" and combine the two; however, biographers "have for the most part" failed to weld the two "into one seamless whole" (CE, 4:221).

Virginia devoted a whole section of her "review" of Nicolson's book to Lytton Strachey's contribution to the art of biography and praised his imaginative *Elizabeth* over his more accurate *Victoria*, because in "the *Elizabeth* he treats biography as an art; he flouted its limitations." Virginia ended with praise of neither Strachey nor Nicolson but with a call for a biographer who could capture a "queer amalgamation of dream and reality, that perpetual marriage of granite and rainbow." In the hands of such a "subtle and bold enough" writer, biography need not be the lesser art (*New York Herald Tribune*, 30 October 1927; CE, 4:221–35).

In 1908, Virginia had planned "to write a very subtle work on the proper writing of lives" (VWL, 1:325). Later she talked about writing a "character" of Roger (VWL, 3:209) and a "Fry memoir" (VWL, 3:386). But for nearly two decades, she had abandoned biographical theory and practice, except in her reviews, until her sketch of Julia Margaret Cameron. By the fall of 1927, Virginia was wondering whether the formalism that had shaped *Jacob's Room*, *Mrs. Dalloway*, and *To the Lighthouse* might also reshape the art of biography.

By 5 October, Virginia's "grand historical picture" had taken a bizarre shape: "a biography beginning in the year 1500 & continuing to the present day, called Orlando: Vita; only with a change about from one sex to another" (D, 3:161). For a "treat," she began *Orlando* and did "nothing, nothing, nothing else for a fortnight; & am launched somewhat furtively but with all the more passion upon Orlando: A Biography. It is to be a small book, & written by Christmas." While writing it, she was "in the thick of the greatest rapture known to me" (D, 3:161).

Creating *Orlando* expressed at both a psychological and a literal level Virginia's desire to reclaim Vita. The composition was tinged with loss, as Virginia increasingly suspected Vita of "infidelity" with Mary Campbell, who was staying with her husband, the poet Roy Campbell, in a cottage at

Long Barn. She warned Vita that *Orlando* was "all about you and the lusts of your flesh and the lure of your mind (heart you have none, who go gallivanting down the lanes with Campbell)" (VWL, 3:427–29). Vita was both "thrilled and terrified" at the prospect of *Orlando* (Sackville-West, *Letters*, 238), which Virginia was writing "half in a mock style very clear & plain," balancing "truth & fantasy" (D, 3:162).

In late October 1927, Vita and Virginia went together to Knole to choose illustrations for *Orlando*. During their visit, Virginia entertained Vita by narrating its fantastic plot, but she did not let Vita read her manuscript (Nicolson and Sackville-West, in Stape, *Virginia Woolf*, 77, 78). After hedging for months, on 10 November Vita made a tearful confession to Virginia that her suspicions about Mary Campbell were correct. Virginia waited for ten days to note in her diary that "I made Vita cry the other night." Then after another ten days, with "Nessa Vita Clive" all away from her, she planned various London social events to "steal a march on that depression" (D, 3:165 and note).

Clive was writing on the august topic of civilization itself. Virginia praised his manuscript but asked him to "leave out your compliment to me" (VWL, 3:438). That compliment, a dedication, implied an intimacy that no longer existed between them. Virginia's great compliment to Vita more elaborately attempted to reinstate a fading intimacy, for writing *Orlando* gave her an excuse to command Vita's presence and to find out private details about her life. Suzanne Raitt explains that "beneath the desire to compliment and to flatter, so evident in *Orlando*, lay a more sinister impulse to punish and to hurt." Thus humor became a disguise for aggressiveness. And writing this fantastic "biography" of her friend without letting Vita read any of it until it was published was a ploy that allowed Virginia to take possession of Vita's life (Raitt, *Vita and Virginia*, 18, 31–35).

Virginia equated regaining Vita with writing *Orlando*: "Should you say, if I rang you up to ask, that you were fond of me? If I saw you would you kiss me? If I were in bed would you–I'm rather excited about Orlando tonight: have been lying by the fire and making up the last chapter" (VWL, 3:443). Although such letters may have fueled the creative and emotional fires, Virginia's diary, always self-protective, quenched them. *Orlando* was intended as "a joke; & now rather too long for my liking . . . & too frivolous for a serious book" (D, 3:177). With *Orlando* thus expanding, at the end of 1927 Virginia was still working on its third chapter.

At this point, Virginia found herself feeling very sentimental about young Angelica (aged nine on Christmas Day). Nonetheless, she realized that "yet oddly enough I scarcely want children of my own now. This insatiable desire to write something before I die" would not allow for the interruptions by or the physical presence of children (D, 3:167). Despite her vicarious maternal feelings for Angelica, Virginia preferred "the old childish feeling that [she and Vanessa] were in league together against the world." Her love and gratitude remained almost filial: "How proud I am of her triumphant winning of all our battles" (D, 3:168).

In this mood of harmony and generosity, with her new affluence and a desire to compensate for the loss of Vita, Virginia "overwhelmed" Vanessa with Christmas presents, as Vanessa wrote to "My dearest Woolves" (VBL, 325), who spent Christmas alone at Charleston (most of the Charlestonians were with Clive and his mother). There "snow all day" (LWD) covered the fields. Alone for three nights at Charleston, the Woolfs had ample opportunity to examine this bizarre farmhouse, where more walls, doors, mantels, bedsteads, woodboxes, and bathtubs were art objects covered with abstract designs and humorous, usually nude, figures. On her return, Vanessa invited Roger and Virginia to visit together, and Virginia and Leonard had lunch at Charleston on New Year's Day, 1928 (VB/VW, 27 December 1927; LWD). In harmony with her sister, Virginia assessed her own faults and determined again to be less self-centered (D, 3:168–69).

Once again emotionally close, the sisters were about to become physically distanced. Vanessa had discovered La Bergère, a ruined stone cottage in the vineyards near Cassis. She had it remodeled and obtained the lease of what would become "another Charleston in France" (Garnett, *Deceived with Kindness*, 67; VBL, 290). Vanessa told Roger, not Virginia, that only her need to be near Virginia and him kept her from living in France all year round (VB/RF, 25 January 1928). One sure way to purchase Vanessa's affection was continuing the pattern of comic, dismissive remarks about others, so Virginia joked about trying to disrupt the affair between Roger and Helen Anrep (VWL, 3:452) and maintained that Roger said Helen was even more in love with Vanessa than he was: "What a complicated and indecent association the three of you make together! Do you have Helen one way and Roger t'other?"[45] With a touch of petulance, Virginia later admitted that Roger and Helen seemed to be an "invincible and triumphant" couple (VWL, 3:467). But her amusing and insistent letters had achieved their goal. Vanessa decided that the sisters were in "telepathic touch."[46]

Virginia's jokes supposedly masked her disapproval of Helen and her "rather underworld" milieu (VWL, 3:384), but neither Helen nor Roger was fooled. Virginia's manner conveyed the sense that Helen was too silly to be listened to, and her snobbism hurt Helen and offended Roger (interview with Igor Anrep, 4 June 1994). Roger turned his thoughts into a Memoir Club paper, "Culture and Snobbism," that objected to the aristocracy's influence on artists. Virginia retorted that Roger himself was "a good deal run after" by wealthy art collectors and patrons (VWL, 3:467). He did attend "highbrow" parties, write for the *New York Times*, lecture on the BBC, accept commissioned portraits, preside over the twenty-fifth anniversary dinner of the *Burlington Magazine*, and lecture in Monte Carlo. Graduate students in art history came from America to interview him. Vanessa saw his eminence as a reclamation of the stature he had held when they first met: "Is it true that you have returned to the year 1910 & are going into smart society?" (VB/RF, 17 March 1928).

By March 1928, despite headaches, Virginia finished *Orlando*, a testimony to her love and loss and her commitment to the art of biography. Ironic references to the normal biographer's craft suggest the book's thematic background. At the beginning of chapter 2, for example, Virginia notes that thanks to documents, she has been able "to fulfill the first duty of a biographer, which is to plod, without looking to right or left, in the indelible footprints of truth" (O, 65). Later, her ambivalence surfaced:

> To give a truthful account of London society at that or indeed at any other time, is beyond the powers of the biographer or the historian. Only those who have little need of the truth, and no respect for it—the poets and the novelists—can be trusted to do it, for this is one of the cases where truth does not exist. (O, 192)

The entire book harks back to those questions of "life" or "truth" that had plagued Virginia since almost her first precocious writings. As Orlando over the centuries tries desperately to write in the fashion of the ages (and Virginia herself satirizes the styles of different periods), *Orlando* exposes the artificiality of literary constraints. Orlando's sex change at the end of the eighteenth century gives Virginia the opportunity to explore the artificialities of gender constraints. These two themes finally become one: Just as the doctors in *Mrs. Dalloway* try to enforce their wills through doctrines of proportion and conversion, so Virginia shows in *Orlando* that genre and gender conventions over time constrict the free expression of unique personalities.

Virginia was confident that at forty-six she was "as experimental & on the verge of getting at the truth as ever" (D, 3:180). After she had completed *Orlando*, she and Leonard set out to join Vanessa and Duncan with Julian, Angelica, and Clive in Cassis, driving for the first time abroad in their Singer car, beside which Vanessa photographed them (Figure 49). Vanessa, who described the clutch on her car in potent sexual terms (VBL, 327), named their Singer a "hermaphrodite" (VB/VW, 24 February 1928) or an "Old Umbrella" (VBL, 338) of a car, a description justified, it seems, by the Singer's many inadequacies (LWL, 232).

Vanessa had asked Roger in two separate letters to come to Cassis after "the Woolves" had left. Roger regretted missing "a fascinating spectacle because you [Vanessa] set Virginia off on her grandest flights" (VB/RF, 17 March and ? ? 1928; RF/VB, 21 March 1928). Back from his lecture at Monte Carlo, he visited Dorothy (Lytton's sister) and Simon Bussy. Then despite Vanessa's instructions to stay away until the Woolfs left, Roger joined the Bell entourage while the Woolfs still were there. And he, rather than Vanessa, set Virginia off on a grand flight. She described to Quentin Bell, then in Munich, Roger's decline into old age:

> Roger is fast crumbling, like a lump of sugar in hot milk; that is to say, he remembers nothing and invents everything. Sometimes his pocket handkerchief has eleven knots in it by nightfall; and he and Helen sit on the edge of the bed till the small hours untying them and trying to remember which is which. But his decline is beautiful as the sunset, and it is certain that he will be alive when you and Julian are old, old men, prodding each other in the back with umbrellas. (VWL, 3:480)

The naughty little Virginia Stephen had grown up to be the genius Virginia Woolf, who retained her wicked tongue. She used it, as Vanessa had said she did as a child, as both weapon and entertainment, in this case to ridicule Roger and amuse Quentin. Leonard had warned her that such outlandish descriptions could irritate him and offend others (D, 3:111). But humiliations and exposés were, Virginia knew, the way to Vanessa's heart. Thus after being worn out by her mother-in-law's conversation, which "never follows a line; is always about people; starts anywhere; at any moment," Virginia replicated that conversation in an "epistolary *tour de force*" (D, 3:194 and note) for Vanessa's entertainment (VWL, 3:523–26). To her diary, but not to Vanessa, Virginia acknowledged that intimacy with the old lady was distressing because it called up "the terrible threat to one's liberty that I used to feel with father, Aunt Mary [Fisher] or George [Duckworth]" (D, 3:194).

Leonard's mother never saw or heard Virginia's elaborate transcription of her conversation, but most of Virginia's letters were not so safe. Clive had been the perpetrator of such trouble the year before; this time he was the victim. He accused Virginia of "making merry at his expense" in her letters to Vanessa. In a part of her letter labeled "*Private*," Virginia naïvely asked Vanessa whether she repeated anything from her letters (VWL, 3:486). She then instructed Vanessa to begin locking up her letters and again claimed to be "jealous of Roger being with you" (VWL, 3:489, 490). At last, suspecting that Clive "rummaged" through her drawers after Virginia's letters, Vanessa had to admit, "I think you must be right about Clive's reading your letters." To make peace, Vanessa wrote that "the Woolves are considered a very devoted couple [in Cassis] you'll be glad to hear" (VB/VW, 22 and 26 April and 3 May 1928). Virginia let her know that Angelica was becoming "essential to me." That is, if she became deaf or ill (still Virginia's test of love), she imagined "only Angelica will be kind to me" (VWL, 3:490).

When Virginia returned to England, she found that she had won for *To the Lighthouse* the Prix Femina, which she described to Vita as "the most insignificant and ridiculous of prizes" and to Julian as a "dog show prize" (VWL, 3:479, 491). Clive sensed her feelings, writing that she was "full of her prize and slightly ashamed of it at the same time" (CB/VB, 20 May 1928). At the ceremony, Virginia felt herself "ugly in cheap black clothes." Elizabeth Robins, the American actress who had popularized Ibsen in England before the turn of the century, crept up to Virginia to talk about her memories of Julia Stephen: Julia "never confided. She would suddenly say something so unexpected from that Madonna face, one thought it *vicious*" (D, 3:183). Although by "vicious" Robins seems to have meant only worldly wit, when Virginia reported to Vanessa on "the extreme beauty combined with viciousness of mother" (VWL, 3:498), she was making a judgment of Julia, a healthy counter to the insecurities that Julia's sainted image usually dredged up.

Even though Leonard thought that Clive's argument was simply "wrong," Hogarth Press published his *Civilization*, dedicated, against her instructions, to "Dearest Virginia" (D, 3:184 and note). At last Clive admitted that he had read her letters, and he apologized. Virginia elided two perspectives: Clive's—"He cant help these outbursts, which date back to old horrors in the past"—and then her own: "As I am also scarred and riddled with complexes about you [Vanessa] and him, and being derided and insulted and sacrificed and betrayed, I don't see how we can hope for a plain straightforward relationship" (VWL, 3:501).

Virginia had a weapon against being derided and insulted. Quentin Bell remembers that the "running banter between Charleston and Monk's House" was a playful version of (to use my term) the *paragone*. Julian's interests were both literary and political, which, in his mother's terms, made him "violently anti-artist at the moment" (VBL, 312). He sent Virginia his poems, which she thought he took "a little too much to heart" (VWL, 3:490), for criticism. Quentin's interests, however, were chiefly artistic. Virginia insisted that painting was a "low art" and teased him that it would be "terrible to throw in your lot with those painters."[47]

Quentin, who was studying painting in Paris under Jean Marchand, went with Roger and Helen Anrep to Munich to see the galleries. Roger wrote to Vanessa: "Quentin [is] pretty constantly at our side 'picking up words of wisdom' as he puts it" (RF/VB, 20 May 1928). Quentin's artistic bent further aroused Virginia's competitiveness: "How in Gods name can you be content to remain a painter? Surely you must see the infinite superiority of the language to the paint?" She urged him "throw up your career, for God's sake" and become a writer (VWL, 3:493). She even wondered whether Roger might be "giving up painting altogether?" (VWL, 3:496).

Putting aside her "weapon" when Vanessa "mercifully" returned from Cassis, Virginia felt as if she had reclaimed her mother:

> My earth is watered again. I go back to words of one syllable: feel come over me the feathery change: rather true that: as if my physical body put on some soft comfortable, skin. She is a necessity to me—as I am not to her. I run to her as the wallaby runs to the old kangaroo. (D, 3:186)

Despite looking forward to "adventuring on the streams of other peoples lives" (D, 3:187)—thinking of Julian Bell's rather than Vita's life—Virginia acknowledged, "Yes & I have no children of my own; & Nessa has; & yet I dont want them any more, since my ideas so possess me; & I detest more & more interruption" (D, 3:189). Clive wrote that *Orlando* echoed Proust in its "capricious use of time" (CB/VW, 15 October 1928). Leonard reacted "more seriously than I had expected," but Virginia seems to have decided that it was more a flawed novel than an experimental biography. Referring to the genre's "fact-recording power," she hoped "never to be accused of [writing a novel] again" (D, 3:185).

Virginia had found out enough about the facts of Vita's life and had come through the emotional ups and downs of the last years an impressive woman, usually proud of her haircut, clothes, homes, and style of life. Her self-therapy had enabled her at last to see her writing as an expression of longing, not so much for children, but for lost maternal love. She was emotionally stronger for surviving loss, for giving generously, and for having loved Vita and reclaimed Vanessa.

Now future writing projects lined themselves up in polar positions: translating a picture or writing "biographies of living people," "something abstract poetic next time—I dont know," or "closely reasoned criticism" (D, 3:185). Having invented a new form for the novel through the patterning of *Mrs. Dalloway* and *To the Lighthouse*, Virginia wanted both to push the novel form beyond factuality and to write biography and criticism. Thus the artistic and ethical division that had been largely a rhetorical choice for so many years became a writerly opposition between a "mystical poetical" work and a foray into fact and even politics.

10
Wild Ideas and Assiduous Truths

1928–1934

Toward the end of the 1920s, the division between Virginia's ways of framing the world pulled her farther in first one direction and then another: between the artistic and the ethical, fiction and biography, autonomy and relevance, formalism and feminism, and also between Vita and Leonard, Vita and Vanessa, Roger and Leonard, and even Roger and Vita. Virginia's binary ways of thinking, being, and loving all connected with the issues posed by the *paragone*. Whether thinking of love or art or public responsibility, Virginia always debated which was truer, more valuable. She had accepted Roger's premise that literature should be as mute, or non-"literary," as Postimpressionist painting. Now she often adopted nearly opposing Leonard-like (and Leonardo-like [see Appendix C]) terms to talk of literature's obligations to the factual world.[1]

The summer of 1928 had been "very animated," with many guests at Monk's House, but the country also offered "a sanctuary; a nunnery," a "religious retreat" enabling a "consciousness of what I call 'reality': a thing I see before me; something abstract; but residing in the downs or sky."

Virginia felt that an acute sensitivity to that "reality" was uniquely hers (D, 3:196). While she explored the countryside on foot, collecting images for her word hoard, Leonard traversed an ever widening territory by automobile. He was running something of a taxi service for Vanessa and the Bell children. He frequently made day trips from Sussex to London and back, experimenting with different, carefully recorded routes (LWD). Virginia stayed at Monk's House, relentlessly turning out journal articles, "nest eggs at the bank," and a talk on women and fiction (D, 3:198).

Virginia's theories about women and fiction crystalized after Desmond MacCarthy made another of his condescending public remarks. In August, he wrote that "female novelists should only aspire to excellence by courageously acknowledging the limitations of their sex (Jane Austen and, in our own time, Mrs Virginia Woolf, have demonstrated how gracefully this gesture can be accomplished)." He went on to advise female novelists not to attempt "masculine standards" (D, 3:195–96, note).

Infuriated that this man, whose friends had tried so hard to cultivate his ambitions as a novelist—although to no effect—could speak so condescendingly about her actual accomplishments as a novelist, Virginia attacked him. Desmond then tried to compensate by praising her writing's "butterfly lightness," but that compliment only angered and depressed her even more by suggesting that her great achievements with modern fiction would forever be considered, from the odious masculine point of view, lightweight (D, 3:195, 197). Later, when Desmond dropped by and destroyed the Woolfs' Saturday walk with his self-indulgent chitchat, Virginia sourly reflected on the "egotism of men" (D, 3:204).

Suddenly, for Virginia, the balance between aesthetic and ethical, formal and social values now shifted in favor of the ethical and social. After the publisher of Radclyffe Hall's *The Well of Loneliness* withdrew the book because of government prosecution for its lesbian theme, Leonard and Morgan Forster organized protests. Virginia was irritated that Morgan found "Sapphism disgusting: partly from convention, partly because he disliked that women should be independent of men" (D, 3:193). But the censorship issue was too important to be caught in the net of her irritation, so she cosigned with Morgan a courageous letter that appeared in the *Nation and Athenaeum* (VWL, 3:530).

Vanessa was alarmed that Virginia might testify at the hearing, as she considered her "perfectly unreliable and can be made in cross-examination to say every sort of incriminating thing about the circles she moves in" (VBL, 340). Virginia, however, was glad not to testify, as she was not an expert "in obscenity, only in art" (D, 3:207). Leonard, who was protesting also the confiscation of D. H. Lawrence's poems during this period, noted that he "gave evidence," presumably on the censorship issue (LWD). Actually, Virginia shared Vanessa's and Leonard's reservations about Hall's art. Vanessa thought it "dull as ditch water" and "very sentimental" (VBL, 340), and Leonard faulted in print *The Well of Loneliness* for its clichés and lack of design, but at the same time he vehemently objected to its being censored (*Nation and Athenaeum*, 4 August 1928, 593).

Just before the publication of *Orlando*, planning a trip to France alone with Vita, Virginia wrote about her "alarming holiday in Burgundy." She was "afraid—she may find me out, I her out" (D, 3:197). Although Virginia and Leonard quarreled briefly about her leaving him, he stoically drove Virginia and Vita to the ferry. Vita had the new experience of riding on second-class trains in France, as Virginia made all the arrangements.[2] In France, the women observed an old customer in a bookstore in "a rhapsody about Proust," which they decided no Englishman would have permitted himself. When they strolled through a town fair, the French people delighted them by showering them with confetti. Vita and Virginia bought green corduroy coats with buttons that looked like hares, pheasants, and partridges. They were gamekeepers' coats, but Vita later remembered that Virginia preferred to think of them as poachers' coats (Noble, *Recollections of Virginia Woolf*, 165–66).

While she was in France, Virginia's anger at Desmond erupted into a "heated argument about men & women" with Vita, who found Virginia "curiously feminist" because she "dislikes the possessiveness and love of dominance in men. In fact she dislikes the quality of masculinity. Says that women stimulate her imagination, by their grace & their art of life." After that explosion, however, Virginia seems to have thought about how unpossessive and undominant (at least to her) Leonard was. The "poacher's" coat was a gift for him, and every day during the holiday, she (as Mandrill) wrote him (as Mongoose) affectionate, seemingly sexual letters. At first, she was "very much upset because she heard nothing from Leonard." After several trips to check the mail, Virginia telegraphed him, apparently very much afraid that she had offended, perhaps even lost, him.

Given Virginia's fatigue, worry, and guilt, Vita assumed Leonard's role and "made her go to bed at 1/4 to 10." During a thunderstorm, Vita went to Virginia's room to comfort her and found her frightened enough to worry about death and extinction and to talk about science, religion, immortality, and the "ultimate principle." On their last night in France, Virginia read Vita her good-natured recollection of "old Bloomsbury," a topic that stimulated a lengthy conversation about Thoby, perhaps the first in the nearly twenty-two years since his death. Vita's diary records no erotic moments with Virginia, although, of course, she recorded selectively. The Mongoose, who seems to have spent most of his time gardening (LWD), hoped that the Mandrill would not "make a habit of deserting me" (LWL, 234). Virginia did not. She and Vita did not air their grievances or expose their cooled feelings and thus "did not find each other out."[3] Their holiday having "flashed by," Virginia was "glad to see Leonard again" (D, 3:199).

After the holiday, it seemed that all London was "beginning to twitter" about *Orlando* (CB/VB, 5 October 1928). Vita's mother's efforts to suppress and discredit *Orlando* piqued public curiosity and made a succès de scandale of Virginia's mock biography (Glendinning, *Vita*, 206–7). When *Orlando* was published (on 11 October 1928, which is also when the narrative itself ends), with pictures of Sackville ancestors and recent photographs of Vita herself, Virginia discovered that she had written not only a mock biography but a successful love letter as well. Again (as with *To the Lighthouse* and Vanessa), her language had purchased maternal love. Vita reported being "dazzled, bewitched, enchanted, under a spell. It seems to me the loveliest, wisest, *richest* book that I have ever read, —excelling even your own Lighthouse" (VWL, 3:573–74). A gratified Virginia replied by telegram on 12 October 1928: "Your biographer is infinitely relieved and happy" (CS, 241). Harold Nicolson wrote to Virginia that Orlando "really *is* Vita—her puzzled concentration, her absent-minded tenderness. . . . She strides magnificent and clumsy through 350 years" (VWL, 3:548, note). As both a love and a punishment letter,[4] *Orlando* succeeded, but as a fulfillment of her plan to "revolutionise biography in a night" (VWL, 3:429), Virginia felt that it failed.[5]

Imagining her way into the life of a person who lives for several centuries as both male and female had focused Virginia's attention on the debilitating effects of female gendering. That awareness and her horror at Desmond's notions of the "limitations of their sex" culminated in October 1928. Then Virginia traveled to Cambridge, first with Leonard, Vanessa, and Angelica (almost ten years old) and then with Vita to give two talks at Newnham and Girton (the women's colleges) on women and fiction. Contrasting the sumptuous luncheon served to the Woolf party by Dadie Rylands in his delightful Carrington-decorated rooms at King's College with the meager fare at the women's colleges, Woolf made a spirited plea for women's financial independence: "I blandly told them to drink wine &

have a room of their own" (Rylands, in Noble, *Recollections of Virginia Woolf*, 174; D, 3:200).

The witty, informal talks delighted her audiences, who found Virginia "formidable" (Stape, *Virginia Woolf*, 14). Revised and published as *A Room of One's Own* the following year, these talks investigate the reasons for women's poverty and absence from history. Virginia called for female emotional and financial independence, perhaps echoing Melanie Klein by associating maternal poverty with the failure to nurture (Abel, *Virginia Woolf*, 96–98). She traced the workings of masculine and feminine perspectives and hypothesized an androgynous "state of mind" beyond binary divisions.[6] After the publication of *Orlando* and then *A Room of One's Own*, Virginia could no longer remain a little known, but highly respected, novelist. She was now a public celebrity.

As a middle-aged man, Leslie Stephen had exhibited many androgynous qualities, playing with his children and "mothering" the brood, especially when Julia was absent. But in his old age, Leslie's various weaknesses and sadnesses had turned into petulant self-centeredness that Virginia now identified as masculine. On his birthday, she decided that if Leslie had lived longer, "his life would have entirely ended mine. What would have happened? No writing, no books;—inconceivable. I used to think of him & mother daily; but writing The Lighthouse, laid them [to rest] in my mind." She finally had faced her parents' weaknesses and thereby freed herself from their debilitating prescriptions for female behavior. Nevertheless, she could now understand Leslie "more as a contemporary" and therefore, perhaps, forgive him (D, 3:208).

Roger wrote to Virginia that fall from France, insisting that there is "a good deal of thought in painting a picture but it's of the kind that has nothing to do with words" or abstract ideas, such as Surrealists tried to paint (RF/VW, 9 October 1928). Virginia replied to Roger's peace offering by adopting a very different persona from the woman enamored of Vita and her aristocratic connections. She faulted herself for the highbrow company she kept and he distrusted. She sent him *Orlando*, ostensibly because she had "taken the liberty of mentioning you in the preface" (VWL, 6:523–24). The dedication of *Orlando* to Vita only acknowledged Vita as its hero/heroine and beloved; it did not imply aesthetic debt, as the withdrawn dedication of *To the Lighthouse* might have. Virginia compensated in an ironic way in *Orlando*: "To the unrivalled sympathy and imagination of Mr Roger Fry I owe whatever understanding of the art of painting I may possess" (O, vii).

Virginia assessed *Orlando* as both a quick and a brilliant book, but it was not one in which she had explored new forms, as she felt impelled to do. Her new endeavor, which became *The Waves*, "was to be an abstract mystical eyeless book: a playpoem." But she worried whether there might "be affectation in being too mystical, too abstract; saying Nessa & Roger & Duncan & Ethel Sands admire that: it is the uncompromising side of me; therefore I had better win their approval—" (D, 3:203). Although Virginia was afraid that it would be an "affectation" to court their approval, she did so anyway, apparently because each person on that list was a painter.

Roger at sixty-two feared "never [to] be able to do the work I still feel I have in me." He was seeing a doctor who promised him "50% more energy within a year. Helen says I shd. go off like a rocket" (RF/VB, 17 November 1928). Thanking Virginia for the gift of *Orlando*, he described her genius, so different (as he had said in his introduction to *Julia Margaret Cameron*) from the nineteenth century's notion of the need to cultivate genius in men: Her genius instead followed a "certain gushing impetus like an instinctive act" (RFL, 2:631). Roger saw *Orlando* as a fulfillment of Virginia's earliest Postimpressionist stories, which freed her spirit, "simply revolving & scintillating in space & eternity."[7] He had reservations only about *Orlando*'s ending. Virginia agreed with him on that point and with his suggestion that Mauron translate *Orlando* into French (VWL, 3:562). She also admitted what the dedication of *To the Lighthouse* would have established and what the preface to *Orlando* skirted: "Lord you dont know what a lot I owe you!" (VWL, 3:562).

The reestablished friendship between Virginia and Roger reestablished aesthetic affinities as well. Having promised herself to write "closely reasoned criticism" rather than feminist tracts, she echoed Roger in the study she was calling "Phases of Fiction": "By cutting off the responses which are called out in the actual life, the novelist frees us to take delight, as we do when ill or traveling, in things in themselves" (CE, 2:82). Absorbed in writing that long essay and haunted by rhapsodic ideas for "The Moths," even Virginia was objectively "a little disquieted" that her mind "never stops reading & writing" (D, 3:210). Subjectively, of course, she knew that writing was her means of becoming herself.

In late 1928, Virginia joined Vanessa in holding Tuesday evening parties in London, reminiscent of the old Thursday evenings, except that "too many people press to come" (D, 3:211). Before Christmas, Vita left England to join Harold, now posted at the embassy in Berlin. With Vita gone, the Woolfs dined with Roger and Helen Anrep on Christmas Day in their flat, filled with Omega furniture and pottery and Postimpressionist paintings. Roger wrote to Vanessa that they had declared this Christmas "the best they'd ever had" (RF/VB, 28 December 1928). Virginia announced that they were "impressed almost to tears by his charm." With manic hyperbole, she insisted that "Roger is the only civilised man I have ever met, and I continue to think him the plume in our cap; the vindication, asseveration—and all the rest of it—If Bloomsbury had produced only Roger, it would be on a par with Athens at its prime" (VWL, 3:566).

After a post-Christmas week at Monk's House, Virginia reflected, as she often did throughout the 1930s: "I am impressed by the transitoriness of human life to such an extent that I am often saying a farewell—after dining with Roger for instance; or reckoning how many more times I shall see Nessa" (D, 3:218).

While Virginia was working on "Phases of Fiction," *Orlando* was selling very well.[8] This success, she calculated, came from writing "exteriorly." In another compensatory artistic move, she planned now to dig into the subjective realm, even though she did not like having to lose the naturalness, ease, and accessibility of external writing. Her plan was to "saturate every atom," to "give the moment whole," which might be a "combination of thought; sensation; the voice of the sea" (D, 3:209). When a third printing of *Orlando* was ordered, she could tell herself, "Anyhow my room is secure. For the first time since I married 1912–1928—16 years—I have been spending money. The spending muscle does not work naturally yet. I feel guilty" (D, 3:212). Since 1922, Leonard had been comparing their actual yearly expenses with their estimated expenses. They now spent more money on the house and car, but their clothing expenses (£25 for him, £40 for her) remained constant. Even though she dressed now with considerable elegance, Virginia's "spending muscle" was still rather cramped.[9]

Old rivalries supplanted new affinities at the beginning of 1929: "Orlando is recognised for the masterpiece that it is. The Times does not mention Nessa's pictures [on exhibition at the New Burlington Galleries]. Yet, she said last night, I have spent a long time over one of them. Then I think to myself, So I have something, instead of children, & fall comparing our lives." Secure in her talent and finances, Virginia overcame her tendency to tally up accomplishments and instead turned to "what I call, inaccurately, ideas: this vision" (D, 3:217). In early January 1929, Virginia had "written myself out of breath—30,000 words ["Phases of Fiction"] in 8 weeks." The issues she had been thinking about since 1908—when she vowed to reform the novel—crop up in that essay. In a letter to Vita,

Virginia explained that after Tolstoy had exhausted the possibilities of "sex and realism," the modern novelist had "to break away" (VWL, 4:4), presumably toward detachment and experimentation with form.

In early 1929, the Woolfs joined Vita Sackville-West and Harold Nicolson in Berlin. Virginia planned that she and Vita could, with tact, "spend a good deal of time alone together." She distinguished between Virginia and "Potto," her beast name (a potto is a lemur) for herself in her relationship with Vita: "And you'll be loving and kind to Potto wont you? And kind to Virginia?" (VWL, 4:5). Later joined by Vanessa, Duncan, and Quentin, Virginia and Vita seem to have had little time to be "loving and kind." Instead, Vita withdrew from Potto, as she had fallen in love with another woman in Berlin. While dining together in a romantic tower restaurant, Vita told Virginia about her most recent infidelity (VWL, 4:8 and note; interview with Nigel Nicolson, 23 July 1993). Having found out Vita again, Virginia was shaken, but she told no one.

Quentin Bell quotes Vanessa's elaborate account to Roger of the tensions of that visit, with Leonard refusing to attend a luncheon that Harold had arranged and with almost everyone arguing during a "thundery" evening about a Russian film's anti-British slant. Vanessa saw the tensions in social terms: "The Nicolsons seem to me such an unnecessary importation into our society that I can only leave Virginia to deal with them" (Bell, 2:141–42; VBL, 341–42). Roger enjoyed Vanessa's amused tale of the "Woolves" and Nicholsons, but he offered rather severe criticism of Leonard's "pedantic" democracy on the grounds that if Leonard disapproved of wealthy aristocrats, he should not hobnob with them. He also theorized that Virginia's sensations "are so instantly transmuted by her imagination into her special key that she knows nothing about what has caused the sensations. P'raps that's one of the great differences between the poetic and the plastic imagination" (RF/VB, 23 January 1929, and 25 May 1926). The distinction between the poetic and the plastic is quintessential Fry, but the claim about Virginia's living in an imaginative world of her own seems to have been a fairly standard assessment by those who knew her and failed to realize at what intensity Virginia conducted her inner life or that she saw the "walls, the pictures and the Venus against the pear tree" rather than the fly on the floor (VWL, 4:199).

Virginia returned from Berlin overmedicated against seasickness, exhausted, saddened, too ill to work. Vanessa was horrified "to hear of the results of my drugging you, though indeed I suppose most of the results must be put down to the rackety life you were leading in Berlin" (VB/VW, 4

February 1929). Leonard concluded that the cause of Virginia's relapse was not the seasickness drug but "overdoing it" in a week of late nights, when even two consecutive late nights could upset the balance in Virginia's health. He had recognized the danger but had decided, as he later wrote to Vita, before they left that it had to be risked, as he did not want to spoil the holiday for Virginia with his "continual nagging" (LWL, 236). Leonard apparently did not know about (or he chose not to mention) another reason for this relapse: Vita's devastating revelation to Virginia that once again she loved another.

Virginia may have felt guilty about betraying Leonard's devotion for the fickle affections of a woman whom Leonard (on this occasion), Vanessa, and Roger all saw as a representative of the evils of class and privilege. All the factors that could precipitate disasters were again aligned as they had not been since 1915. The problem, therefore, was to prevent a relapse like that of 1915. Virginia took to her bed and led an invalid life. She did not confide in her diary, but Leonard recorded in his diary the first time she was able to walk in the square, when she was "rather better," when "not so well" (LWD). Virginia resurrected the childhood equation between being sick and being loved, as she regretted that she had not become ill in Berlin (VWL, 4:8), where Vita would have been obligated to take care of her. But in planning the "angular shape" of her radically subjective work, Virginia reflected that even though "old age is withering us," in herself "forever bubbles this impetuous torrent," which, after six weeks, rescued her from depression and overflowed into even bolder writing. Furthermore, money could, like love, free her from the old feeling of helplessness: "no more poverty" (D, 3:219).

Clive Bell still did his best to stir up rivalries, remarking about Duncan's sensational current exhibition: "Does it vex Virginia? What about Roger?" (CB/VB, 20 February 1929). But Virginia now considered her rivalry with Vanessa to be a thing of the past. She could even joke to Vita about Vanessa's ingenuousness: "I told Nessa the story of our passion in a chemists shop the other day. But do you really like going to bed with women she said—taking her change. 'And how d'you do it?' and so she bought her pills to take abroad, talking as loud as a parrot" (VWL, 4:36). Despite her amusement and despite Vanessa's departure (for four months in Cassis), Virginia now believed that "life draws us together" (D, 3:219). Vanessa generously supplied the Woolfs with various art objects for their home. She explained from Cassis: "One gets all the benefits of the Lighthouse or Orlando for a few shillings—or even free—so why shouldn't you get more benefits such as they are out of me?" Then she rather snidely added, "Leonard may perhaps see the force of this argument" (VB/VW, 17 April 1929).

From Paris, Vanessa wrote to Virginia that Roger, although ill with sciatica or perhaps arteriosclerosis, was so preoccupied with his latest enthusiasm that "he forgot all about his inside—in fact he led a life which would have killed most young men & throve on it. We left him proposing to paint at the Louvre & make expeditions to Versailles, etc." (VB/VW, 9 March 1929; VB/VW, 17 April 1929). The sisters' harmonious relationship was threatened as Vanessa rather bitterly supposed that if Virginia were "going into society again that you'll be more of a celebrity than ever" (VB/VW, 29 April 1929).

When Charles Mauron arrived in London, "entirely unprovided for" by Roger, Virginia was exasperated (VWL, 4:47), but the Woolfs came to the rescue of "dear old rapscalliony Roger" (VWL, 4:49) by giving a party for Mauron at Tavistock Square. To Quentin, Virginia described Julian Bell's jousting with Fry and Mauron and "vociferating at the top of his voice—Whats pure poetry? Whats association? Whats abstract? Whats concrete?" She ironically summarized the party's intellectual effect: "Mauron and Roger last night confuted Leonard and Oliver [Strachey] and proved beyond a doubt the non-existence of everything but an Idea" (VWL, 4:56, 57). Roger saw the evening as "a good old apostolic talk about existence and the absolute or relative meaning of good," in which talk he and Charles convicted Oliver and Leonard "of being mystics to their horror" (RF/VB, 16 May 1929). The verdict was that activism was not practical and down-to-earth but platonic and idealistic. Virginia found the argument dry and Roger "a little old—to my mind he needs Nessa to fertilise & sweeten him" (D, 3:224).

If Roger set the terms for that "apostolic" debate, Leonard's interests set terms for a debate with another visitor from abroad. In London, negotiating about the American publication of *A Room of One's Own* and *Phases of Fiction*, Donald Brace (of the publisher Harcourt, Brace) visited at Tavistock Square, where the huge Bell-and-Grant–painted panel covered part of a wall and Grant-designed fabrics adorned a sofa and window curtains.

Among the friends gathered to meet Brace, the discussion turned to the fiction of D. H. Lawrence and the issue of class. Richard Kennedy, the "soft duckling boy" (D, 3:195) at the Hogarth Press, captured in squiggly lines (Figure 50) Virginia's astonishment as she rolled her shag cigarette, at something Desmond MacCarthy said. Kennedy remembered:

> Mrs W held the stage and talked about working class writers being under a disadvantage, like women, as writers. Roger Fry thought it was easier to be a painter than a writer if you were working class, as a painter did not have to have an ear for subtle social distinctions as a writer did.[10]

Here again, the inevitable distinction between writing and painting surfaced in brilliant conversation, although Roger's comments about writing were newly naturalistic.[11] Virginia found herself depressed by Brace and the talk, but she consoled herself: "Fundamentally I am the happiest woman in all W.C.1. [Bloomsbury]. The happiest wife, the happiest writer, the most liked inhabitant, so I say, in Tavistock Square." Then George Duckworth called to ask her to lunch with a French couple whose taste was so bad they "preferred Brighton to Penshurst."[12]

Notwithstanding her increasing commitment to feminist issues, Virginia's "wild ideas" for "The Moths" were unusually subjective and detached. In the combined study and book-storage room at Tavistock Square, she considered organizing the book around a "lamp & a flower pot"; there should be no story, just "islands of light" and "mind thinking," and the narrative consciousness "should have no name." Yet in such an abstract design, some device for unifying the book had to be found (D, 3:229). Virginia thus resurrected an idea she first considered for *To the Lighthouse* (D, 3:34): "Could one not get the waves to be heard all through?" (D, 3:236). Planning to do so, Virginia renamed the book *The Waves*.

Given the number of people who descended on Vanessa in Cassis, Virginia claimed that she and Roger felt "some scruple about coming. However, its a scruple we seem to be getting over" (VWL, 4:58). At the beginning of June 1929, the Woolfs traveled by train to join Vanessa and Duncan in Cassis. Roger met them "en passante" in Paris, where Virginia may have told him about her "mystical eyeless" book. He wrote to Vanessa that his delightful visit with Virginia provoked a dream "in which I learned the title of an entirely unknown work of [Virginia's]" (RF/VB, 3 and 9 June 1929). Vanessa told Clive that she had "spent a week of chatter with the Woolves." Because of their rather "gloomy account" of Roger's health, she was glad she had insisted that he "go to a reputable nerve doctor" (VB/CB, 12 and ? June 1929).

French food, landscape, views of the Mediterranean, and astonishing heat all offered a welcome escape from London, but when Virginia talked "with people who have never heard of me & think me old, uglier than

Nessa, & in every way inferior to her" and when Vanessa bid for "almost overpowering supremacy" in her conversation about her "elder son," old rivalries resurfaced. Virginia countered: "I made £2,000 out of Orlando & can bring Leonard here & buy a house if I want." A home in France promised "a complete change; as buying a house so often does" (D, 3:232), but Vanessa was her old dismissive self: "The Woolves house here will be a very queer little object—like all their houses" (VB/CB, ? June 1929). With the newly affluent Virginia, Vanessa was a sadly humble self: "I am a failure as a painter compared with you, & cant do more than pay for my models." Virginia summarized, "And so we go on; over the depths of our childhood" (D, 3:232–33). One sister's success still, irrationally, meant the other's failure.

After her return to London, facing corrections of *A Room of One's Own*, Virginia felt less of a success, and the Woolfs extended a weekend at Monk's House because she had a headache. Depressed because she needed "to write more succinctly," Virginia mused, "What a born melancholiac I am! The only way I keep afloat is by working." For "directly I stop working I feel that I am sinking down, down." Ironically, sinking offered some opportunity to "reach the truth," but that truth was decidedly nihilistic: "There is nothing—nothing for any of us. Work, reading, writing are all disguises; & relations with people. Yes, even having children would be useless" (D, 3:235). Such a grim notion of truth partly explains Virginia's compulsive need to work, for without occupation and diversion, she would have to face unbearable emptiness.

During the summer of 1929, she was often depressed with headaches, not only because of Vita's infidelity, not only because of her sense of nothingness, but also because the telephone seemed "strung to my arm & anybody could jerk me who liked" (D, 3:238). Against such interruptions, the Woolfs frequently retreated to the isolation of Monk's House. Vanessa offered little sympathy: "If you run a press & are a celebrity too you can expect nothing else" (VB/RF, 8 October 1929).

Despite being troubled by mysterious ailments, Roger continued to paint, write, and lecture at a frenetic pace, and his celebrity resulted in an honorary degree from the University of Aberdeen. Roger was contemplating buying a painting by Renoir, although Clive maintained that "his flat is already no better than an old curiosity shop, with a Cezanne propped up behind the wash-stand and a Seurat in the W.C." (RF/VB, ? July 1929; CB/VB, 7 July 1929). Virginia's purchases were considerably more practical; rather than a house in Cassis, Virginia ordered an oil stove for Monk's

House. Then the Woolfs bought cottages nearby, in hopes of living without servants in residence. They were "very extravagant, for the first time in our lives." As Virginia saw it, they were "watering the earth with money." They bought furniture and crockery; their oil stove, on which Virginia imagined cooking adventurous dishes, arrived; and construction began on a new upstairs sitting room for them both, a greenhouse for Leonard, and a bedroom for Virginia (literally a room of her own without access to the rest of Monk's House). These purchases made Virginia feel "freer, more independent—& all one's life is a struggle for freedom" (D, 3:257).

Although Virginia sometimes worried that her money did not compare with Vanessa's children (D, 3:241), she no longer sentimentalized children. When Angelica went off to boarding school, Vanessa took a studio alone in London. Virginia gave Angelica a dress allowance about this time (Garnett, *Deceived with Kindness*, 111). As she became closer to her niece, she realized that Vanessa was "going back, perhaps rather sadly to the life she would have liked best of all once, to be a painter on her own" (D, 3:255).

While Virginia worked slowly and deliberately on "Waves or Moths or whatever it is to be called" (D, 3:259), *A Room of One's Own* was published on 24 October 1929. Except for the story "A Society," *A Room of One's Own* was Virginia's only extended political writing since the beginning of her Postimpressionist phase. This lively foray into "assiduous truth," favorably reviewed even by Desmond MacCarthy, sold remarkably well. Meanwhile, the stock market crash in America does not seem to have affected the Woolfs directly, but the resulting depression in England, as well as in the United States, set the stage for the horrors of World War II.

From time to time for Virginia, memories of past losses welled up, as when she thought of Thoby dying "fighting something alone. But then I had the devil to fight, & now nothing" (D, 3:260). Although she dreamed of death and in her dream feared insanity (D, 3:264), in her waking life Virginia assumed that her bouts with the devils of madness were over. Reviewing her major life choices (defined in terms of the houses she had left behind), Virginia thought of herself as "an old struggler after my fashion—not so valiant I daresay as Nessa, but tenacious too & bold" (D, 3:267). In fact, by this time she was the braver sister, as she continually embarked on new artistic and political ventures.

Virginia remembered her father now with considerable affection as he "lugged home [Hakluyt] for me—I think of it with some sentiment—father tramping over the Library with his little girl sitting at HPG in mind. He must have been 65; I 15 or 16 then" (D, 3:271). She also recognized, "Had I married Lytton I should never have written anything. . . . He checks & inhibits in the most curious way. L. may be severe; but he stimulates" (D, 3:273). Rereading Thoby's letters, Virginia considered him "that queer ghost. I think of death sometimes as the end of an excursion which I went on when he died" (D, 3:275). That she envisioned beginning her excursion at Thoby's death suggests that had he lived, his presence might have been as inhibiting as her father's or mother's. Certainly, facing his death without a recurrence of the horrors had been enabling for Virginia.

The Woolfs "crushed visitors—Morgan, Roger, Adrian"—at the end of 1929 and managed to be alone at Monk's House, with Virginia baking bread (D, 3:275). They were interrupted at Christmas only by the Keyneses and Vita, who now had become a casual friend. Vanessa had decorated the fireplace tiles in Virginia's new bedroom with a lighthouse and scenes reminiscent of *To the Lighthouse* and the tiles in the new upstairs sitting room with a basket of flowers. Somehow there were two extra tiles, which the Woolfs laid on the upstairs hearth in an intolerable and "absurd arrangement" that the Bells, nevertheless, did not comment on when they visited (Bell, *Elders and Betters*, 121).

Tacked onto the back of the house, without plumbing or internal access to the rest of the house, Virginia's new bedroom cost convenience but bought privacy, brightness, and two lovely views of the downs. Should they live there for most of the year, Virginia suspected that the views would become less special. That Christmas season, however, they were so glorious and soothing that Virginia thought of crusading, as Leonard did, to protect the downs against development (D, 3:281, 274).

When Angelica visited Monk's House and other topics of conversation failed, Virginia often tried to get her niece ("Pixerina") to agree that Monk's House was superior to Charleston (Garnett, *Deceived with Kindness*, 111). Despite such tactics, Vanessa did approve of the Woolfs' household improvements, although she stopped short of praising Virginia's taste. She told Clive that Monk's House

> has become a regular country house & is hardly any longer a queer poky jumbled cottage. The new rooms have large windows which look over the water meadows. Also they've made a very nice little courtyard of shaven grass with a pool in the middle of it owing to Roger's advice. (VB/CB, 4 March 1930)

Hibernating at Monk's House, Virginia wondered whether a habitual "denigration of human nature & adoration of solitude is suspect" (D, 3:281).

The Woolfs spent two weeks in Rodmell that Christmas and New Year's, walking, gardening, visiting, and, of course, writing. When they returned to London, Virginia paid for her solitude in Sussex with much socializing in elegant company. Then she paid for the socializing by having to convalesce at Monk's House (VB/CB, 4 March 1930). Back in London, Virginia refused an offer of £2,000 to write a life of Boswell and concluded: "I have actually paid for the power to go to Rodmell & only think of The Waves by refusing this offer" (D, 3:295).

Despite the toll that age was taking on him, Roger Fry remained an enthusiast. Virginia's description of him at a 1930 New Year's costume party in Vanessa's London studio for Angelica Bell (then twelve) demonstrates how that enthusiasm flowed into improbable channels. With guests dressing like characters in *Alice's Adventures in Wonderland* and *Through the Looking-Glass*, Roger as the White Knight was "a masterpiece, having called out all the resource and ingenuity of Woolworths stores" (VWL, 4:128–29). Vanessa wrote to Clive about another expression of Roger's enthusiasm: She had seen Roger "again in the character of the white knight, practically carrying among other things a Giotto under his arm. He has discovered (belonging to an American) the original portrait of Dante painted by Giotto. Of course if it existed it would be discovered by Roger." His boundless enthusiasm spilled over at a "Bloomsbury Sunday evening" at Tavistock Square (LWD) with the "Frys and Woolves" talking of beginning a new paper for discussiing of art and literature. Roger was "full of hope,"[13] but Virginia was mostly derisive (VWL, 4:133, 141), even though the idea of an inexpensive broadsheet, perhaps the *Hogarth News*, did intrigue her (D, 3:292–93).

Asked to write the foreword to an early 1930 exhibition of Vanessa's paintings, Virginia returned to Postimpressionist aesthetics in distinguishing painting from literature: "No stories are told; no insinuations are made. . . . If portraits there are, they are pictures of flesh which happens from its texture or its modelling to be aesthetically on an equality with the China pot or the chrysanthemum." She pointed out the pictures' impersonality, in that from them, unlike novels, one could tell nothing about the artist. These pictures, she insisted, contained "no words." Then Virginia questioned her own formalist values: "But is morality to be found there?" (*Recent Painting*).

Virginia's pronouncement about the paintings' wordlessness carries unintended ironic overtones, for at that time Vanessa had silently submitted to Duncan's plan to bring his young lover to Charleston while she was in London. Vanessa was resigning herself to losing Duncan to the young man from New York, but she told no one. Even to Duncan, after self-effacingly attributing her exhibition's success to his exhibition poster and Virginia's foreword, Vanessa avoided the topic of his infidelity and her pain (VBL, 350–51). When she finally expressed her fears and resentments, Duncan became so angry that Vanessa apologized. She painted furiously, often in Roger's steadfast company, but she kept her distress away from Virginia, Clive, and Roger.[14]

Meanwhile, Virginia's formalist foreword prompted Walter Sickert to ask her to write about him, and even the Royal Academy to ask her to "lecture on Art." Virginia declined the lecture invitation, maintaining, as of old, that the writer was out of place in the painter's "sublime silent fish-world" (VWL, 4:142). Sickert's proposition was more tempting, however. About this time, Virginia had a visit from George Duckworth (now Sir George, thanks to political and aristocratic connections). Impressed by Virginia's celebrity and shocked by Vanessa's nude paintings, he proved to be as sentimental, conventional, and self-complacent as ever, although now harmless. Still he preserved, Virginia recognized, "a grain or two of what is me—my unknown past; my self; so that if George died, I should feel something of myself buried" (D, 3:293–94).

A *Room of One's Own* brought a letter of praise to Virginia from Dame Ethel Smyth, a composer and former suffragette. Spending much of mid-February on the sofa at Tavistock Square, incapacitated by influenza and saddened by another friend's death, Virginia delayed a first visit from the admiring Ethel until 20 February 1930. Then Virginia poetically declared that Ethel "descended upon me like a wolf on the fold in purple and gold" (VWL, 4:146). Ashamed that she took so much longer to recover from being sick than Leonard did, Virginia followed Adrian's hypothesis (D, 2:287, 277) and blamed her poor health on "my father's philosophy and the Dictionary of National Biography" (VWL, 4:145). That facetious remark to Ethel began a long series of confidences from Virginia about her past. Posing as Ethel's helpless captive, Virginia wrote to Quentin, "An old woman of seventy one has fallen in love with me. It is at once hideous and horrid and melancholy-sad. It is like being caught by a giant crab" (VWL, 4:171). Clive heard from Roger that both Vanessa and Virginia "have been swallowed whole by Ethel Smythe [*sic*]" (CB/VB, 9 June 1930).

But even Ethel could not pry Virginia away from *The Waves*. She recorded 28 March 1930 as being a "day of intoxication when I said Children are nothing to this. . . . felt the pressure of the form—the splendour & the greatness." That day Virginia had tea at her London studio with Angelica and Vanessa, who was still enduring Duncan's infidelity, which Virginia still did not know about. She walked home through busy London streets, which even in their turmoil gave her "the greatest rest" (D, 3:298). By 23 April, in Rodmell for an Easter break, Virginia felt that she had "turned the corner" with *The Waves*. At last, she could "see the last lap straight ahead" (D, 3:301).

Virginia found Roger at sixty-three looking "like a ravaged scavenger," and yet his house was "alive with a living hand in it" that transcended decor (D, 3:284). The old flirtatious game she had played with Vanessa and Clive so long ago now resumed in her relations with Ethel and Vita. Twice to Ethel, she described a visit with Vita as lying in Vita's "adulterous sheets" (VWL, 4:185, 192). Virginia also told Ethel that if she were ill, she would as willingly go to Ethel as to Vita, for she was "diverse enough to want Vita and Ethel and Leonard and Vanessa and oh some other people too. But jealousy is not a very bad fault is it?" (VWL, 4:199). She confessed to Ethel having "physical feeling" for a man only twice. Titillation was aesthetic, not sexual: "I cannot get my sense of unity and coherency and all that makes me wish to write the Lighthouse etc. unless I am perpetually stimulated. . . . There must be this fanning and drumming—of course I get it tremendously from Leonard—but differently" (VWL, 4:200). Later Virginia was pleased that Vita was jealous, even of Ethel, "that old sea-monster encrusted with barnacles" (VWL, 4:247).

The two-faced way of conceptualizing her work that Virginia later termed "switching from assiduous truth to wild ideas" (D, 5:159) became increasingly pronounced by 1930. While working on *The Waves*, she wrote an introduction to a book of essays by working women that had been collected by Margaret Llewelyn Davis. During the summer of 1930, retreating to Monk's House, Virginia continued work on *The Waves*, which was "resolving itself (I am at page 100) into a series of dramatic soliloquies. The thing is to keep them running homogeneously in & out, in the rhythm of the waves." She respected herself for this book, "even though it exhibits my congenital faults" (D, 3:312). Presumably she considered her faults to be detachment, "unrounded" characters, and a lack of drama.

In early 1930, the Hogarth Press's success finally released Leonard from the literary editorship of the *Nation and Athenaeum*. The Woolfs were "very prosperous" (D, 3:305). Virginia, who had had to sell her jewelry to pay for her 1915 medical expenses, who had "never had a pound extra; a comfortable bed, or a chair that did not want stuffing," was amazed at her own ease in buying four new armchairs (D, 3:315–16).

The Woolfs spent the summer of 1930 alternating between London and Rodmell. Then toward the end of the summer, without prior warning, Virginia fainted at Monk's House. She had entertained Vita and her children for breakfast, Ethel later in the morning, and the Keyneses for tea. While showing them around the new greenhouse, according to Clive, she "fell down in a swoon." Maynard and his chauffeur helped carry her into the house, "whereupon the Keynes[es] and their chauffeur made off. As Virginia had a second and worse attack a few minutes later this conduct is generally condemned" (CB/Lytton Strachey, 6 September 1930). Virginia concluded, "This brush with death was instructive & odd" (D, 3:315). In fact, it may have suggested, as Quentin Bell proposes, an ending for *The Waves* (Bell, 2:154). It certainly suggested the precariousness of her health.

Unaware of Duncan's affair with the young New Yorker, Virginia romantically associated Charlestonian bohemianism with the rhythm and unconventionality of her new novel: "This rhythm (I say I am writing The Waves to a rhythm not to a plot) is in harmony with the painters" (D, 3:316). She further opposed rhythm and plot: "Though the rhythmical is more natural to me than the narrative, it is completely opposed to the tradition of fiction and I am casting about all the time for some rope to throw to the reader" (VWL, 4:204).

The fall began with an odd journey into the past. Still a sentimentalist, George Duckworth requested one more meeting with Virginia and Vanessa, "his only women relatives" (D, 3:316), before his imminent death. When Vanessa held a luncheon party at Charleston, George gave Virginia, as a "reward," a picture of their Pattle great-grandparents with their daughters, including Maria Pattle Jackson (VWL, 4:220, 224), the grandmother of both George and Virginia. That afternoon, George took a photograph of the Duckworths, Bells, and Woolfs that included, in one of the rare incidences since their childhood, Vanessa and Virginia in the same picture (Figure 51). It showed them, like Leonard and Duncan, if not Clive, both looking considerably healthier, handsomer, and more lively than Lady Margaret Duckworth. (The photo seems to validate Virginia's suspicion that "the delights of snobbishness somewhat fail in later life" [D, 3:316]).[15]

Work on *The Waves* continued to be taxing and discouraged Virginia, only slowly recovering from the last summer's brush with death, enough to threaten "to cut adrift—I will go to Roger in France—I will sit on pavements & drink coffee—I will see the Southern hills; I will dream" (D, 3:323). But instead, she kept working.

Listening to Beethoven on their gramophone in Tavistock Square that December, Virginia conceived the end of *The Waves*, in which she would merge "all the interjected passages into Bernard's final speech . . . thus making him absorb all those scenes" (D, 3:339). Ill with influenza, she meditated on the disadvantages and advantages of illness and gave copies of a new limited edition of "On Being Ill" to friends as a 1930 Christmas present. Virginia was sufficiently disillusioned with society dinners to contemplate burning her evening dress, although Clive's remark that she and Leonard were "provincial" may have deterred her resolve (D, 3:337). When Roger attacked Virginia's society friends—specifically Vita, Ethel, and Maynard—she found him no longer grumpily tolerant but savagely bitter. When he dined with the Woolfs in Tavistock Square just before Christmas, Roger complained about "poverty and the neglect of his art." Earlier, Virginia had taken "comfort" in prospects of a breakup between Roger and Helen, but now she concluded that his "mesalliance is souring him and Helen to wit" (VWL, 4:175, 264).

The Woolfs spent Christmas at Monk's House, where Virginia had to retire to her bed (LWD; VWL, 4:263). Because she was fighting off yet another fever, Virginia ended 1930 in her bedroom overlooking the frosty downs, elated over her accomplishment in *The Waves*: "a saturated, unchopped, completeness; changes of scene, of mood, of person, done without spilling a drop." As always, "it is this writing that gives me my proportions" (D, 3:343). She had conceptualized a way to end *The Waves*, but she still could not envision a way to follow through on her conception.

On a New Year's visit to Charleston, Virginia "fought, rather successfully with the usual depression. Is it their levity?—a sneer? But nothing so bad as usual" (D, 4:3). Her depression was aggravated by poor health. Back in London, walking with Leonard ten days later, she found herself in a "mood of tears . . . of pathos, for Leonard, for myself," feeling great "sorrow for people" (D, 4:6).

Unlike Virginia's fear in 1915 that both Woolfs were "despised," this mood seems to have been a recognition of various sadnesses and perhaps, regarding Leonard, of the anti-Semitism that still alienated their friends and sometimes even Virginia. Quentin Bell remembers being at the tea table with the Woolfs and Julian. Virginia might say, "Tell the Jew to pass the honey." Leonard would reply, "I shan't 'til you call me Leonard." Such remarks, according to Bell, were made jokingly as a sort of game, but

Leonard probably did not find them so innocent, as Virginia seems to have realized in a moment of "pathos."[16]

Like Mrs. Ramsay, "not without weariness," Virginia pulled her faculties together, assuring herself that in *The Waves*, she made "prose move—yes I swear—as prose has never moved before: from the chuckle & the babble to the rhapsody. Something new goes into my pot every morning—something thats never been got at before" (D, 4:4).

The Charlestonians' "levity" continued into the second generation. At her London studio, Vanessa held a masked party for which Julian Bell, now "likely to become a fellow of King's they say" (VWL, 4:269), had written what Roger called "a very smart satire of a mild kind on all of us and Angelica did lady Ottoline almost alarmingly well." Roger "appeared in my father's Lord Justice's wig and a horrible mask which I painted" (RF/CB, 22 January 1931). Virginia's account of the party was entertainingly manic (VWL, 4:281–82), but Vanessa, for once, wrote the better letter. She told Quentin that the Lord Justice's wig transformed its wearers (Roger and then Leonard) into "magnificent" figures (reminding Virginia of the aura of masculine power symbols). Virginia herself "came as Sappho I believe, at any rate a most voluptuous lady casting her eyes up to heaven." Vanessa's letter continued with a riotous account of the play, which ended with "an epilogue spoken by Ot (Angelica) in which any stray members of Bloomsbury who had escaped mockery in the play were given sharp raps" (VBL, 357–58). At twelve, Angelica had such well-developed impersonating skills that she seemed destined for a stage career. Virginia continued to give a special allowance to her niece, whom she now described as a "whirlwind of beauty and destruction" who would probably marry the Prince of Wales (VWL, 4:309).

Nearing the end of *The Waves*, knowing that "the didactive demonstrative style conflicts with the dramatic" (D, 4:6), Virginia turned from rhythms to realities, wild ideas to assiduous truths. Fed up with both male egoism and female acquiescence, she envisioned a novel about the "sexual life of women" as a "sequel" to *A Room of One's Own*, to be called "Professions for Women."[17] Virginia gave a speech by that title on 21 January 1931 at a meeting organized by Philippa Strachey and the London National Society for Women's Service. She asserted that women keep absorbing values not their own, but she cautioned them against anger and advised humor instead.

Virginia also instructed her audience to kill "The Angel in the House." This angel was

> intensely sympathetic. She was immensely charming. She was utterly unselfish. She excelled in the difficult arts of family life. She sacrificed herself daily. If there was chicken, she took the leg; if there was a draught she sat in it. . . . Above all—I need not say it—she was pure. Her purity was supposed to be her chief beauty—her blushes, her great grace. In those days—the last of Queen Victoria—every house had its angel. (CE, 2:285)

Of course, in Virginia's mother the Stephen house had held the angel par excellence. In 1931, even more clearly than when writing *To the Lighthouse*, Virginia seems to have recognized that Julia would have prevented women, especially Vanessa and Virginia, from taking up professions. In the original version of the speech, Virginia wrote that when she had tried to write, there had been an angel "in the house with me." Despite being "a dream, a phantom," it insisted that a real relationship "between men and women was then unattainable." This angel had a "special hatred for writers" because unlike painters and musicians, writers convey "charac[ters,] morality, human relations." Thus the angel tried to guide Virginia away from writing anything displeasing to men, threatening to the upper-class Victorian family, or revealing about the body. According to Virginia, "though I flatter myself that I did kill her in the end, the struggle was severe" (*Pargiters*, xxx–xxxii). She killed not Julia, however, but the ghost of Julia's conventional value system.

Writing a book about "the sexual life of women" promised to be "Lord how exciting!" (D, 4:6). In a fury of attention to this not-so-adorable world, Virginia began to "gather quotations and facts about women, men, law, sexuality, sports, religion, the Church, science, education, economics, politics, and social mores, taken from a wide variety" of sources (Silver, *Virginia Woolf's Reading Notebooks*, 22). Within a year, in three scrapbooks, she had "collected enough powder to blow up St Pauls" (D, 4:77). But first, Virginia "screwed [her] brain" to rhythms rather than realities, and on 7 February 1931, she wrote:

> I must record, heaven be praised, the end of The Waves. I wrote the words O Death fifteen minutes ago, having reeled across the last ten pages with some moments of such intensity & intoxication that I seemed only to stumble after my own voice, or almost, after some sort of speaker (as when I was mad). . . . I have been sitting these 15 minutes in a state of glory, & calm, & some

> tears, thinking of Thoby & if I could write Julian Thoby Stephen 1881–1906 on the first page. I suppose not. How physical the sense of triumph & relief is! . . . I mean that I have netted that fin in the waste of waters which appeared to me over the marshes out of my window at Rodmell when I was coming to an end of To the Lighthouse. (D, 4:10)

Even though her own inspired voice flew ahead of her at a frightening pace, Virginia could follow and transcribe it, whereas back in 1904 and 1915, when deluded voices had raced in her head, she had been helpless to conquer or counter them. Like her character Bernard, who saw the "bare visual impression" of a fin and recognized that "visual impressions often communicate" (W, 189), Virginia apprehended the end of her novel both verbally and visually. She "tossed aside all the images & symbols which I had prepared" (D, 4:10) and netted the illusive fin of truth and artistic achievement.

By January 1931, Duncan's infidelity was a subject of Bloomsbury gossip. To Quentin, Virginia expressed joking disbelief (VWL, 4:276), although to Vanessa she seems to have said nothing. When Vanessa criticized her newly curled hairstyle, Virginia worked to control "bottomless despair." She brooded over "my gnats perception of filamentary relations. With Nessa for example. I think she plumes & prides herself: I think she exists self sufficient: I think her beauty is praised; I think she does not want me." She understood that her own mood could be "the back wash of The Waves" (D, 4:11, 12, 13), the downside of her former elation, but she apparently did not consider how much Vanessa's crabbiness could have been the "backwash" of emotional agony. Virginia complained about "the egoism of the males," and Vanessa complained about Virginia: "Even Julian was blamed for talking of poetry to the Woolves. . . . But perhaps one ought not to expect her to listen." Vanessa figured that when John Lehmann began working at the Hogarth Press in early 1931, "that will be the end of him poor young man."[18]

When Helen Anrep was out of town, the sisters found Roger "more visible" (VB/CB, 23 January 1931). Vanessa still was dismissive about his painting, claiming that he was so pursued by the aristocracy "& then he has articles & lectures & wireless & obituary notices to do by the dozen. . . . It's no wonder he can't paint" (VB/CB, 18 February 1931). Roger thought that such a perspective only revealed prejudice against intellectual artists. He regretted that a novelist or poet could theorize about his art but that "a painter who knows and says what he thinks at once becomes a person lacking in inspiration" (RF/Charles and Marie Mauron, ? February or March 1931).

Stung by criticism of his forty-year retrospective show, Roger probably also was troubled by the aesthetic perversity of his favorite woman of letters. For about fifteen years, Virginia had been judging painting and literature on formalist grounds, and in 1930 she had explicitly praised Vanessa's paintings for having "no words." In 1931, however, she complimented Roger's exhibition for offering "all the elements of an absorbing novel laid before me." She told him she especially liked his portraits, "What a character monger you are!" and wished that he had asked her to write a preface, "but then it would have run to 6 volumes of small print" (VWL, 4:295–96). In "Solid Objects," she had exposed the dangers of disinterested contemplation of art; now she disparaged stories in pictures but also claimed to see pictures as novels. This contradiction may simply illustrate Virginia's often perverse tendency to adopt opposing opinions according to her audience. But perhaps it reveals her binary way of thinking or shows a developing interest in "literary" values, for Virginia was beginning to suspect that the Sickert paintings that Vanessa pronounced "idiotic" (VBL, 364) were actually quite "witty" (VWL, 4:336).

In March 1931, Ethel insisted on taking Virginia to Lady Rosebery's elegant party in honor of her (Ethel Smyth). Virginia anticipated "fashion and art" to "join hands"; instead, art was exhibited for fashion's credit. Virginia felt "pinioned," a term she usually reserved for aggressive men, by Ethel. She resented being "dragged to that awful Exhibition of insincerity and inanity against my will," just as she had been "dragged [to parties] by my half-brothers against my will." Virginia's angry letters to Ethel illustrate how feeling helpless could dredge up old horrors and revive old depressive suicidal impulses, so that she came home and told Leonard, as she wrote to Ethel, "'If you weren't here, I should kill myself—so much do I suffer'" (VWL, 4:297–98, 302).

Despite all her sparring with Arnold Bennett about materialism and characterization, Virginia and Leonard attended his funeral. Leonard (at fifty) and Virginia (at forty-nine, Julia's age when she died) considered their own deaths and decided to be buried together behind Monk's House (D, 4:24). On a lighter note, Virginia could now "write criticism fearlessly" because her assertiveness in *A Room of One's Own* had empowered author as well as reader. Her revisions of *The Waves* proceeded "like sweeping over an entire canvas with a wet brush" (D, 4:25). In May, however, an ominous specter from the past reemerged, for no apparent reason, in a devastating headache. While she was recovering, Virginia recognized that "if it were not for the divine goodness of L. how many times I should be thinking of death." Alone with Leonard at Monk's House, she realized that "this is happiness," partly because of "Nessa Roger & Clive being away so that I am not hauled about & ruffled" by the usual entanglements of art and affection (D, 4:27).

Even though Virginia and Vanessa had long laughed about his old age, Roger still considered it a distant prospect, something that might "happen to me suddenly," but only in "a few years" (RF/VB, 11 June 1931). In 1931, with Charles and Marie Mauron, he purchased Le Mas, a house in Saint-Rémy-de-Provence, planning to spend years of painting and theorizing there. But both Roger and Lytton Strachey were unwell. Lytton had cancer, and Roger suffered from various mysterious ailments. He traveled to Royat, a spa town where he bathed in carbolic acid to cure his sciatica (Spalding, *Roger Fry*, 272). From Royat, Roger speculated to Virginia about Italy and the difference between "abominable" Roman and "superb" Italian art, blaming poor Roman art on the prominence of literature there. Although he had "just written a history of painting and sculpture throughout the ages in which I've explained everything," he claimed ignorance. Hesitantly, like "Sargent offering to exchange a sketch with Velasquez," Roger requested both a letter and a copy of *The Waves* when it appeared (RFL, 2:660, 659).

With Leonard having written on the "future of British Broadcasting" and receiving offers to travel, write, and broadcast—"floating on the tide of celebrity"—Virginia was pleased not to be jealous (D, 4:34). During much of the spring of 1931, she had attended to rhythms rather than to realities for her final revision of *The Waves*, but she was often interrupted by the prospect of turning to realities. She finished the third retyping of *The Waves* and then sent it to a professional typist. On 17 July, she presented the typescript to Leonard, whose assessment, she knew, would be honest. She waited nervously, counting the book's faults but rationalizing that "anyhow I had a shot at my vision & if its not a catch, its a cast in the right direction" (D, 4:36). Two days later, she recorded with "jubilation": "'It is a masterpiece' said L. coming out to my lodge this morning. 'And the best of your books.'" She also noted that he had questioned whether the ordinary reader could get through the first hundred pages of *The Waves*, but nevertheless she was triumphant (D, 4:36), feeling "like a girl with an engagement ring" (VWL, 4:357).

As soon as *The Waves* and its rhythms were behind her, Virginia got an idea for an amusing foray into fact. As she later explained, "I was so tired after the Waves, that I lay in the garden and read the Browning love letters, and the figure of their dog made me laugh so I couldn't resist making him a life. I wanted to play a joke on Lytton—it was to parody him" (VWL, 5:161–62). With Lytton ill, however, this mock-heroic parody was less funny, and as a "biography" of Elizabeth Barrett Browning's dog Flush, it raised serious issues about the art of biography. Virginia began writing "Flush of a morning, half seriously to ease my brain, knotted by all that last screw of the Waves" (D, 4:37).

Headaches plagued Virginia off and on during the summer of 1931 (LWD). Then rereading *The Waves* at Monk's House in August called up the emotions she had felt writing it (D, 4:39). To ameliorate such moods, she merely switched them: "It is a good idea I think to write biographies; to make them use my powers of representation reality accuracy; & to use my novels simply to express the general, the poetic" (D, 4:40). Awaiting publication of the most extreme embodiment of her vision, Virginia had too much at stake, however, to remain calm. Once again, she became "acutely depressed," worried that her exquisite "sensibility verg[ed] on insanity." Although she had "tried to speak the truth, bombastic as the remark sounds," she was "trembling under the sense of complete failure." Then John Lehmann told her that he loved the book, for she somehow maintained "the speed of prose & the intensity of poetry" (D, 4:43–44). After such praise, Virginia's sails filled again. She responded exuberantly by beginning "A Letter to a Young Poet," addressed to John, which was later published as the eighth in the Hogarth Letters series. But when *The Waves* was described as only a "poem," the wind again went out of Virginia's sails, and she longed to hear that her book was "solid & means something" (D, 4:45).

Having installed a gardener in one of the nearby cottages, Leonard was making elaborate additions to the grounds behind Monk's House. As a result, by the early 1930s, the upper garden contained a fish pond surrounded by brick walks, a vegetable garden, a potting shed, several greenhouses, beehives, and pear and apple trees. Beyond one of the walks, Leonard created a lawn for bowling and archery with a round pond and, behind a hedge, vegetable gardens and a fruit orchard. Pampas grass inside a last wall marked the outer perimeter, and tall trees served as a rookery. The produce from these gardens was ample and good enough that each week when the Woolfs were in London, the gardener sent a supply of fruits and vegetables to them.[19] Virginia baked bread and made preserves from the fruit in their garden. However ambivalent she had first felt about Monk's House, now these surroundings were restorative. Virginia reminded herself "again—how happy I am: how calm, for the moment how sweet life is with L. here, in its regularity & order & the garden & the room at night & music & my walks & writing easily & interestedly." Not even a visit from George and Margaret Duckworth fazed her (D, 4:44).

In 1908, Virginia had written:

> attain a different kind of beauty [from painting], achieve a symmetry by means of infinite discords, showing all the traces of the minds passage through the world; & achieve in the end, some kind of whole made of shivering fragments; to me this seems the natural process; the flight of the mind. Do they [writing and painting] really reach the same thing? (PA, 393)

At that time, Virginia had believed the aims of literature and painting were different, but in *The Waves*, published on 8 October 1931, she most successfully made them the same. In this novel, a series of poetic interludes treat the sea and sky from predawn when they are indistinguishable, through the full coloring and glare of midday, to dusk when they are again indistinguishable. Each of these interludes is followed by interwoven soliloquies of six named characters carried from childhood to old age. By alternating italics and roman type in the interludes and the sections, which repeat and redefine the central moments in each character's life, Virginia distinguished the sections visually.

In the fourth section, the friends meet to say good-bye to Percival, a Thoby-like character, who is leaving for India. Bernard's description of their reunion owes much to the Cubist painters' evidence that perspective alters "reality":

> We have come together (from the North, from the South, from Susan's farm, from Louis's house of business) to make one thing, not enduring—for what endures?—but seen by many eyes simultaneously. There is a red carnation in that vase. A single flower as we sat here waiting, but now a seven-sided flower, many petalled, red, puce, purple-shaded, stiff with silver-tinted leaves—a whole flower to which every eye brings its own contribution. (W, 127)

The Waves shows how that contribution can work both positively and negatively. Like the young Virginia Stephen, Rhoda perceives a puddle as a yawning chasm that she cannot cross (W, 158–59). She also feels acute terror when words lose their meaning. Bernard fears the moment when "no echo comes when I speak, no varied words." This alienation is "more truly death than the death of friends, than the death of youth." It creates "a world without a self, I said." Deserted by self, the world in turn "withered. It was like the eclipse when the sun went out and left the earth, flourishing in full summer foliage, withered, brittle, false." But just as after an eclipse of the sun, the world miraculously returns (W, 284–87). Such changes in the phenomena of experience all emanate from the consciousnesses of the individual characters, except Percival, who only is observed.

In the fifth section, the characters react (each in her or his own way) to the news of Percival's pointless death. Images of completion and fulfillment in the sixth section are followed by images of entrapment in the seventh. In the eighth section, the characters meet again, and the "red carnation that stood in the vase on the table of the restaurant when we dined together with Percival is become a six-sided flower; made of six lives" (W, 229). Bernard, the writer, narrates the final section, the ninth, which, by explicitly repeating motifs from the first section, "encloses the whole." He wonders about the separation between himself and his friends, but in articulating the voice of the "human collective," he becomes indistinguishable from them (as the sea and sky are again indistinguishable and the waves break onto the shore).[20]

The entire novel is structured by a formal design, with Percival's departure at its center and later sections repeating earlier ones to suggest the parallels between the beginning and the end of life. It ends with Bernard exclaiming: "Against you I will fling myself, unvanquished and unyielding, O Death!" And then: "*The waves broke on the shore*" (W, 297). As nonmimetic as Postimpressionist painting, *The Waves* abstracts characters from their everyday contexts to achieve a "whole made of shivering fragments." Just as Bernard overcomes separation from his friends, so Virginia finally achieved a whole that aesthetically and emotionally overcame the separateness and loss that she had felt since infancy. Rather than give up writing for painting, as she had threatened to do in 1904, Virginia had by 1931 made her writing supremely painterly, thereby blurring the differences between verbal and visual art.

The reviews were favorable, if surprising. Virginia knew that her book was "not what they say. Odd, that they (The Times) shd. praise my characters when I meant to have none" (D, 4:47). At the same time, Leonard's *After the Deluge*, which considered the structure of society before and after the American Revolution, received fairly hostile reviews (D. Wilson, *Leonard Woolf*, 217–19). For a change, he was the more depressed Woolf, although by 30 October Virginia could record in her diary, "Happily that morbidity of L.'s is over" (D, 4:51).

After a weekend with "the Woolves," Vanessa wrote to Clive, "I didn't dare to mention The Waves. . . . I can't imagine what the ordinary reader can possibly make of it . . . it will be ticklish work saying anything about it." She told Clive that Roger claimed to have discovered the medium used by the Old Masters and was "almost as full of it as Virginia is of the Waves." Even though Virginia had written her most complete embodiment of the Postimpressionist vision, by 1931 Roger doubted that vision. Igor Anrep, Helen's son, remembers Roger shaking his head and saying, "I don't think it's quite come off." He thought that Virginia was "feeling her way toward pure poetry in an abstract weaving of words but hadn't achieved what she was trying to do." Virginia was "frightfully touchy" about criticism, but Roger could not tell an aesthetic lie and so prepared to discuss *The Waves* frankly with her.[21]

Vanessa must have felt similarly uneasy, but then her reading experience was so powerful that she grasped Virginia's accomplishment: "I have been for the last 3 days completely submerged in The Waves—& am left rather gasping, out of breath, choking, half-drowned as you might expect . . . over come by the beauty." She found reading this book "quite as real an experience as having a baby or anything else." Partly her reaction was personal: "You *have* found the 'lullaby capable of singing [Thoby] to rest'"; but it was also aesthetic, for Virginia had mastered the art of transmuting personal emotions into impersonal art: "Even though I know its only because of your art that I am so moved. I think you have made one's human feelings into something less personal." Vanessa's painting helped her "understand what you're about" (VBL, 367–68).

For her entire life, Virginia had been waiting for aesthetic approval from Vanessa. She exclaimed, "Nobody except Leonard matters to me as you matter, and nothing would ever make up for it if you didn't like what I did. So its an amazing relief—I always feel I'm writing more for you than for anybody." She also asked, "You didn't think it sentimental, did you, about Thoby?" (VWL, 4:390). E. M. Forster's letter saying that *The Waves* was a classic gave her "reason to think that I shall be right to go on along this very lonely path" (D, 4:52). And Roger's friend Goldsworthy Lowes

Dickinson, an Apostle and Fellow at King's College, sent her three typed pages and then another page of praise, stating that she did not write science's antithesis but "science made alive" (GLD/VW, 23 October and 13 November 1931). Virginia responded by explaining how the "six characters were supposed to be one." Expressing her earliest desires, she asserted that "all I can do is to make an artistic whole" (VWL, 4:397). Emboldened by many endorsements, Virginia could be true to her holistic vision: "Oh yes, between 50 & 60 I think I shall write out some very singular books, if I live. I mean I think I am about to embody, at last, the exact shapes my brain holds" (D, 4:53).

Roger's objection to abstraction in *The Waves* coincided with his ongoing determination to reconcile formalist with "literary" criteria. The extent of his retrenchment is conveyed by his title for a 1931 Cambridge lecture, "The Literary Element in Painting." A contemporary witness remembered that Roger distinguished between painting that "pleased, if at all, by such elements as spatial relationship, and painting that pleased, if at all, by its story or message." Roger illustrated the second sort of painting with Sir Edward Poynter's *Faithful unto Death*, saying, "The painting had no formal merit: its fame must be due to its 'message.'" Toward the end of the lecture, Roger defended himself for his stress on form, but he also "conceded that, in his reaction against public insensibility to [form], he had gone too far." He then remarked that he was "thinking of making amends" with a book on "Rembrandt as a dramatist." However "off the cuff" that last remark might have been, Roger's plans to write about the "Shakespeare" of painting and Virginia's plans to write didactic factual feminist nonfiction suggest a parallel reformation.[22] With this shift in emphasis, the charge that *The Waves* was "unreal" (VWL, 4:402) hit a nerve, and Virginia avoided both Roger and Lytton, sensing that they did not like her latest book (D, 4:53).

Headaches bothered Virginia toward the end of the year, leaving her a semi-invalid. Nevertheless, she was pleased to have the novelist Elizabeth Bowen to tea at Tavistock Square, about the same time that Leonard was discussing English colonialism with Mahatma Gandhi (LWD). By this time, Lytton Strachey was barely hanging onto life. The Woolfs made their usual Christmas retreat to Monk's House. While Vanessa was visiting the senior Bells with Clive, Virginia, although not well herself, repeatedly telephoned there to report Lytton's condition (LWD). She and Leonard pondered "the feeling of age coming over us: & the hardship of losing friends" (D, 4:55). The topic of death darkened lunch with the Keyneses on Christmas Day, as both Virginia and Maynard stated that they expected to die before their spouses. The conversation then lightened a bit when they supposed that Lydia and Leonard could marry and hence combine the dogs from both households that also were enjoying the Christmas visit (D, 4:56).

Virginia ended the year working on journal articles, having discovered that after this illness she, for once, did not feel that "if I dont write I shall whizz into extinction like an electric globe fused." Two presses were petitioning to reprint her books, and the sales of both Woolfs' works were high (D, 4:57). Anticipating Lytton's death, early in 1932 Virginia wrote to Vita, "I wish one's friends were immortal" (VWL, 5:3). "Can we count on another 20 years?" She planned the remainder of her writing career: four "novels: Waves, I mean"; one feminist treatise; and one account of all English literature (D, 4:63).

The Woolfs returned to London on 10 January 1932. Hearing that Lytton was much better, they went to a fancy dress party for Angelica, only to receive the news at the party that Lytton had died, news that oddly replicated the ending of *Mrs. Dalloway*. Virginia concluded that "much better" had really meant "much weaker." Remembering Vanessa's saying that Lytton "is the first of the people one has known since one was grown up to die," Virginia felt numb (D, 4:64–65). To one of his sisters, she reminisced that "Lytton came to me when Thoby died" (VWL, 5:8). His death saddened, but did not destabilize, Virginia. She worked through "the stages of Lytton's death" and concluded that "for us fame has no existence." But the absence of a funeral ceremony and the press's "flimsy" treatment of Lytton distressed and angered her (D, 4:72,74).

Virginia might have taken heart by reading in her diary the entries of some twelve years before: "Greetings! my dear ghost; & take heed that *I* dont think 50 a very great age. Several good books can be written still; and here's the bricks [in her diary] for a fine one" (D, 2:24). But now, mourning Lytton's death, Virginia did not mention her fiftieth birthday. She found Roger "rather sunk & aged after influenza" and Vanessa and Duncan, whose relationship had settled into something of a mother–son balance, "almost too severely malicious" (D, 4:67, 68) about Roger. When she considered publishing Lytton's letters, she expressed an affinity unlike their formalist phase: "Roger & I love to have facts poured over us." Both liked portraiture, and both were highly conscious of the need to keep contact with the "adorable" world. Her record of an evening's conversation, however, revived the old conflict between brush and pen in Roger's protest that there was "no public fund for helping painters—only writers" (D, 4:71, 72).

About three weeks after Lytton's death, Virginia and Leonard heard Roger lecture on French art at the Queen's Hall. Virginia described Roger's lecture visually:

> Roger rather cadaverous in white waistcoat. A vast sheet. Pictures passing. He takes his stick. Gets into trouble with the lanternist. Is completely at his ease. Elucidates unravels with fascinating ease & subtlety this quality & that: investigates (with his stick) opposing diagonals: emphasises the immediate & instantaneous in French art. (D, 4:76)

In his lecture, Roger voiced regret that ever since Horace wrote about the similarities of poetry and painting, people had assumed that painting, like literature, must deal with august, grandiose subjects (RF/Marie Mauron, ? December 1931; *Characteristics of French Art*, 54). Of course, both Roger and Virginia had argued and shown that painting and literature could reflect the most humble, small subjects.

At an after-lecture supper party in Vanessa's studio, the Bloomsbury friends did not talk about art but about the historical Jesus and "about a new life of Jesus, which offers two proofs of his existence—witness Roger's intellectual vitality after speaking to the Queen's Hall for 2 hours—really excited about the reality of Jesus" (VB/CB, 17 February 1932; D, 4:77). Virginia still worried that "with Lytton there was no mark to say This is over" (D, 4:78).

On 29 February 1932, Virginia Woolf received an invitation to deliver the Clark Lectures at Cambridge for the next year. Although she was greatly flattered, she turned down the invitation:

> This, I suppose, is the first time a woman has been asked; & so it is a great honour—think of me, the uneducated child reading books in my room at 22 H.P.G.—now advanced to this glory. But I shall refuse: because how could I write 6 lectures, to be delivered in full term, without giving up a year to criticism; without becoming a functionary; without sealing my lips when it comes to tilting at Universities. . . . And I am pleased; & still more pleased that I wont do it; & like to think that father [who, in 1883, had been Cambridge's first Clark lecturer] would have

> blushed with pleasure could I have told him 30 years ago, that his daughter—my poor little Ginny—was to be asked to succeed him: the sort of compliment he would have liked. (D, 4:79)

Despite the certainty of that immediate resolution, "the devil whispered, all of a sudden, that I have six lectures written in Phases of Fiction; & could furbish them up & deliver the Clark lectures, & win the esteem of my sex, with a few weeks work" (D, 4:79). Virginia was sorely tempted: "Such is the perversity of my mind . . . swarming with ideas for lectures"; for the moment, her refusal seemed "lazy & cowardly" (D, 4:80). She boasted to Clive of the invitation, even though "as Desmond was once a Clark lecturer, the honour is not overwhelming." The temptation notwithstanding, Virginia stuck to her refusal. She also gossiped that "Nessa predicts what we all secretly desire," an end to Roger's relation with Helen Anrep (VWL, 5:27).

The young man for whom Duncan had almost left Vanessa in 1930 again upset the equilibrium of her domestic life. Duncan had set up something of a love nest with him at Brighton. He tried to arouse maternal feelings in Vanessa toward the young man, but Vanessa could not be so easily placated.[23] Besides, her maternal instincts were otherwise engaged. Although Vanessa and Virginia had expected Julian Bell to become a Fellow of King's College (VWL, 4:269), he failed to win a fellowship. Vanessa wrote to him with disconcerting ambivalence both that his interests were too broad for him to be an academic and that he could try again for the fellowship. Virginia's account of his "vociferating" about pure poetry suggests that Julian had mouthed Roger's theories rather thoughtlessly, but Vanessa faulted the Woolfs for any criticism of Julian (VB/JB, 20 February 1932; VB/CB, 8 February 1932).

On 10 March 1932, the Woolfs visited Dora Carrington to console her over the death of Lytton. When Carrington burst into tears, Virginia took her into her arms. But when she lamented, "There is nothing left for me to do," Virginia found consolation difficult. As the Woolfs were leaving, Virginia and Carrington kissed and warmly waved good-bye, but the next morning Carrington shot herself and died a slow painful death (D, 5:82–83).

Both Leonard and Virginia felt guilty about the ineffectuality of their consolation; thereafter, what Leonard called their "mausoleum talks" intensified. Virginia felt that Lytton's memory had been tainted by Carrington's suicide: "I sometimes dislike him for it. He absorbed her[,] made her kill herself." The Woolfs responded to Lytton's death with a longing for new experiences and old friends: "We're both in the mood for ventures after this morbid time; so much talk of death; & there death is of course" (D, 4:83–84). To Ethel, Virginia wondered whether she should have said more to Carrington "in praise of life." But she "couldn't lie to her" (VWL, 5:38).

She and Leonard renewed their canceled 1927 plans for an excursion to Greece with Roger, this time with his sister Margery (called Ha), rather than Helen Anrep. Virginia expected trouble, whose only advantage would be their amusing Vanessa: "Its all likely to end in bugs, quarrels, playing chess, disputing about expeditions and so and on and so on—but I shall feel this all makes bones for your pot: that is still the cauldron in which my life is brewed" (VWL, 5:44–45). The holiday began on 15 April and lasted nearly a month.

In Venice, a church service that Virginia attended with Roger evoked her longing for some magical wholeness (D, 4:90). To Vanessa, who wanted to hear about quarrels and imagined Leonard "obligated to play chess & you probe the soul of Ha," Virginia did not mention spiritual longing (VB/VW, 19 April 1932). Rather, she amused her sister with an account of their aesthetic competitiveness: Margery bore

> the brunt of aesthetic criticism. I've heard her even contradict Roger about Bellini, and she always has a feeling about a sky or a pillar or the use of bald heads in design which turns the blade off the poor ignorant Wolves. I did however, attack Titian and Leonard says he made a point about a diagonal—anyhow, it doesnt matter, as Roger is urbanity itself. (VWL, 5:50)

This trip brought the past into the present. In Athens at the Parthenon, remembering the trip in 1906, Virginia found that "my own ghost met me, the girl of 23, with all her life to come." But she signed herself aged fifty at the hotel, "the flourish in the face of death" (D, 4:90–91). Greece itself was like "England in the time of Chaucer" (D, 4:92). Renewed friendships sparked old discussions of "Roger's theories of art" (D, 4:93). But the travelers were not so hardy as they had been in their younger days. Roger reported that "poor Virginia has been terribly scorched by the sun," so she devised "a kind of silk hood" to protect herself (RFL, 2:670). She also devised an improbable metaphor as protection from the evidence of Roger's mortality: "Some part of Roger's inside is coming through, so that he cant sit or stand—but this makes little difference. Its merely a question of going behind a hedge now and then with a button-hook" (VWL, 5:54).

To Vanessa, Virginia expressed suspicion of talk and talkers such as the Frys (VWL, 5:56). Roger concluded that

> Virginia in particular doesn't seem to want to talk as well. I think she gets immense pleasure from just having experiences. Only when Vanessa's grand letter from Cassis came denouncing the human race ("mice are easier to get rid of") Virginia launched out about how she would write to Vanessa about the Frys. (RFL, 2:670)

Vanessa was pleased to get Roger's accounts, for "Virginia's letters though very amusing are not altogether to be trusted" (VB/RF, 13 May 1932). On receiving a charming letter from Angelica, Virginia was pleased that her niece had summed "up all the gifts as well as graces" (VWL, 4:57). The spiritual longing that Virginia had experienced in Venice blossomed in Greece: "All that is in me of stunted and deformed religion flowered under this hot sensuality. . . . Why, we almost wept, we pagans" (VWL, 4:59). Roger brought local peasants into interaction with the Woolfs, and that plus intimacy with an old friend and an alien culture made this "the best holiday these many years" (D, 4:95). Leaving the Frys in Greece, the Woolfs journeyed homeward elegantly on the Orient Express and were in England by 15 May.

After the elation of the trip, Virginia experienced a period of deep depression, partly due to the deaths of Lytton and Carrington, partly to bickering with John Lehmann at the press, partly to an "odious quarrel" with Vita's cousin (VWL, 4:67), partly to a review faulting her reliance on "feminine charm" (D, 4:101 and note), and partly to the number of people who visited, interrupting her work schedule and tiring her. Virginia felt "terror at night of things generally wrong in the universe." After Carrington's suicide, she had seen "all the violence & unreason crossing in the air: ourselves small." Only writing could bring "order" into her universe (D, 4:102–3) and offer what she had called a "narrow bridge of art" over a chaotic abyss.[24]

Despite his casual infidelities, Duncan acknowledged the necessity of Vanessa to his existence. Her tolerance of his behavior continued to contrast sharply with her exasperation with Virginia and callousness toward Adrian, who, in a depression, expected attention from Vanessa that she was unprepared to offer (VB/CB, 27 June 1932). Virginia was more sympathetic, for his "suicidal face at night gave me the nightmare" (D, 4:118). In addition, Adrian's condition and her own depression perhaps seemed ominous reminders of Stephen "madness."

On 7 July, the day her *A Letter to a Young Poet* was published, Virginia fainted in a London restaurant and remembered a "curious sensation. Feeling it come on: sitting still & fading out," helpless before the sensation. On the same night, she sent Vanessa a check for £100 and Leonard split £100 between his mother and brother. Intended to "release [Vanessa] from worry," Virginia's wanted to do "solid good things" (D, 4:114–15), a desire heightened by her latest "brush" with death, by her need to be free of guilt, and by her conviction that art and humanity were one's only defenses against unreason and nothingness. Then in mid-July, Virginia wrote a very angry letter to Ethel Smyth, who expected her to spend her time with Ethel's society friends rather than visit people like Roger (who had just had an operation), the Stracheys, and Thomas Hardy's widow, who all were "good to me years ago when I was a gawky tongue tied impossible girl" (VWL, 5:76–77).

One evening, Virginia found herself "sleeping over a promising novel" after reading *The Life of Joseph Wright*, by his wife Lizzie Wright. Virginia's pleasure in that biography came from Lizzie Wright's respect for Joseph and Joseph's respect for her and for his working-class mother, who "went charring" to pay for his education. In the Wrights' relationship, Virginia found "odd how rare it is to meet people who say things that we ourselves could have said. Their attitude to life much our own" (D, 4:115–16). This marriage of equals offered a focus for her new feminist novel (Leaska, Introduction to *The Pargiters*, xiii–xiv, xv).

Despite her fainting spell, Virginia fended off depression emotionally and financially by cultivating immunity from the "range of darts," by letting the reconciled "Nessa & D. go to Paris without envy," by feeling "no one's thinking of me" (D, 4:117), and by germinating plans for a new and very different novel.

Virginia's recollection of fainting once again that summer at Monk's House strangely replicated the terms of her aesthetic endeavor: "I lay presiding, like a flickering light, like a most solicitous mother, over the shattered splintered fragments of my body."[25] The task was to reintegrate those fragments in the same way that she had integrated the shivering fragments of her Postimpressionist novels. Again, although "not without weariness," Virginia did so by working. She shelved the idea of a new novel, corrected proofs of another *Common Reader* collection of her essays, and composed tentative versions of *Flush*, her "biography" of the Barretts' dog (VWL, 5:83).

When Goldsworthy Lowes Dickinson died, Virginia remembered, "Goldie had some mystic belief." She was grateful for his praise of *The Waves*; she was depressed that "people will go on dying until we die, Leonard said. Lytton, Carrington, Goldie"; but she was determined not to "let friends lapse" (D, 4:120). Virginia wrote to Roger (then sixty-five), "How I hate my friends dying! Please live to be a thousand. You cant think what you mean to us all" (VWL, 5:87). Virginia also wrote to Vita, praising her poem "The Bull," mentioning her jealousy of Vita's "new loves," but admitting that Vita loved women "physically, I mean, better, oftener, more carnally than me." To Vita, Virginia maintained that her "one heroism" was to let Leonard coddle her ("Four Hidden Letters," 29).

In September, Virginia was asked by the *Times* to write about her father on the centenary of his birth. She intended to decline the invitation (D, 4:123), but with her head "full of him," Virginia decided to accept (VWL, 5:100). The renewed interest in Leslie led the Stephen siblings to consider publishing their father's *Mausoleum Book*. Virginia had doubts, but Vanessa was more cavalier: "I dont myself care much either way as I suppose I should have neither much trouble nor much profit" (VB/VW, 15 September 1932).

While Leslie had a centennial, Virginia was about to have published a full-length study of her works. She had bravely instructed Winifred Holtby to treat her work "with the candour and impartiality applied by critics to the writings of the dead" (*Virginia Woolf*, 8). Despite giving Holtby such an intellectual carte blanche, Virginia worried that the accompanying illustration would reveal her to the world as "a plain dowdy old woman."[26] Against her "ill joined web of nerves," the countryside rested Virginia in the summer of 1932. Even though some of the beautiful downs were being ravaged by limestone mining, the Sussex landscape suggested a "holiness" that "will go on after I'm dead" (D, 4:124). Despite recurring heart trouble, Virginia was determined to defy death by limiting her visits and visitors and vowing to do nothing she disliked (D, 4:129).

Sometime in this period, the widowed Lettice Ramsay, then Julian Bell's lover, took several photographs of Virginia with her niece at Monk's House (Figure 52). Feeling so close to Angelica, ruminating about what would survive her, Virginia had a disconcerting dream in which Angelica, too, died (D, 4:124). Vanessa, who had assumed that women were never honest, found her daughter growing into a good and honest companion: "Its almost a new discovery to me, as Virginia cant be called honest whatever else she may be." Vanessa further insisted that "real intimacy is impossible without" honesty (VB/RF, 2 October 1932). That belief explains why Virginia had lost her close relationship with her sister so long ago. Virginia had been dishonest to flirt with Clive; thus, in Vanessa's value system, "real intimacy" had become impossible. From then on, however much Virginia might have touched Vanessa's heart with *To the Lighthouse* or *The Waves* or her letters, her sister never trusted her again.

Although Virginia still defined art as "being rid of all preaching" (D, 4:126), she began work on a didactic essay that so preoccupied her she hardly thought about the publication of *The Common Reader: Second Series* (D, 4:128). Then she "entirely remodeled" her feminist "essay" to encompass her "promising novel" idea of the past summer. Now she planned an "essay-novel" alternating art and ethics, vision and fact, entitled "the Pargiters—& its to take in everything, sex, education, life &c." (D, 4:129). A *pargeter* is a plasterer or whitewasher, thus, figuratively, one who flatters or covers up.[27] Virginia used the term to suggest the results of Victorian repression and denial (Leaska, Introduction to *The Pargiters*, xiii–xvi). Rather to her surprise, she discovered that "after abstaining from the novel of fact all these years—since 1919—& N[ight]. & D[ay]. Indeed, I find myself infinitely delighting in facts for a change" (D, 4:129).

On 18 December 1932, the Woolfs entertained Roger and Helen for dinner at Tavistock Square and then went to Clive's in Gordon Square, where Virginia encountered "the first of the little pricks which will be so lavishly provided when The Pargiters comes out." Although she knew that it would be bombarded with more antifeminist "pricks," Virginia was pleased that she had "secured the outline & fixed a shape for the rest." This essay-novel seemed so significant that she felt she "mustn't take risks crossing the road, till the book is done" (D, 4:132). She had written sixty thousand words of alternating narrative and exposition, five chapters, and six commentaries on her chapters.

In her essays, Virginia considered the conventions that impeded her characters and even those that impeded her. For example, when describing a man exposing himself, "the three dots used after the sentence, 'He unbuttoned his clothes . . . ' testify, a convention, supported by law, which forbids, whether rightly or wrongly, any plain description of the sight that Rose [Pargiter], in common with many other little girls [like Vanessa and Virginia Stephen], saw." Virginia explained that Milly Pargiter did not learn painting at the Slade School of Art, where she would have had to paint from the nude, so she sketched flowers instead. Virginia protested the power that money gave Captain Pargiter, especially because he supported expensive educations for his sons, but not for his daughters. She was even moderately explicit about the process of male sexual sublimation and about female sexual repression (*Pargiters*, 51, 29–30, 31, 81–82, 109).

Virginia's unique dual accomplishment, however—alternating narrative with commentary—covered the Pargiter family only in the 1880s, thereby promising to make the book insufferably long. She had spent "a very fruitful varied & I think successful autumn—thanks partly to my tired heart," which excused her from socializing and allowed her to "impose terms" on the many visitors who sought her out. Work on a factual saga of a family over the years interposed with essays on mostly feminist issues had been as enthralling as her Postimpressionist experiments: "I have never lived in such a race, such a dream, such a violent impulsion & compulsion—scarcely seeing anything but the Pargiters."

At the end of the year, Virginia returned to revising *Flush*, depressed over its being both "too slight & too serious" and yet determined not to "pay for it with the usual black despair." Tied to such a "slight" work, she longed to feel her "sails blow out" as they did when she worked on the magisterial *Pargiters*. Virginia ended the year remembering Goethe's *Faust* and reflecting, "If one does not lie back & sum up & say to the moment, this very moment, stay you are so fair, what will be one's gain, dying?" She concluded, "I am now going in, to see L. & say stay this moment" (D, 4:133–35).

In early January 1933, the Woolfs went to another fancy dress party given by Vanessa, this time for Angelica's fourteenth birthday. Virginia "dressed up as Queen Victoria on her wedding night and fell into the arms of the Prince Consort [Leonard]" (VWL, 5:145). Despite the costumes, the unpretentiousness of the party pleased her: "Then there's Rosamond Lehmann, Roger Fry and myself all sitting on the studio floor and eating ham and chicken" (VWL, 5:148).

Later in January, after a ballet at Sadlers Wells for which Vanessa had designed the set and costumes, the Woolfs went to another party at Vanessa's studio. Even though they had had the Frys to dinner less than a month before, Virginia wondered whether Roger was "older, less volatile?" She "maliciously observed" that Roger "cross-questioned [her] about his lectures," which were published by Chatto & Windus as *Characteristics of French Art*. Despite sharing a sense of the inappropriateness of the grandiose in art, Virginia disliked being cross-examined. Even though Roger's "insistent egotism has its charm," she observed, by comparison, "I never ask him to read my essays" (D, 4:144).

In early 1933, George Gissing's son accused Virginia, in print, of disregarding facts in her introduction to his father's *By the Ionian Sea* (D, 4:150). This exchange, like Vita's mother's irrational response to *Orlando*, must have heightened Virginia's awareness of the dangers of dealing imaginatively with actual people's lives. Fortunately, Elizabeth Barrett Browning's dog had no relations who could complain about her treatment of their ancestor in *Flush*.

In *The Pargiters*, Virginia could "take liberties with the representational form which I could not dare when I wrote Night & day." Her excitement about this rechanneling of her creativity impeded her progress on *Flush*, although she did finish the final typescript in mid-January 1933, and "having bent my mind for 5 weeks sternly this way, I must unbend them the other." Recognizing the mind's need for change, Virginia was apprehensive about the didactic. She could even see "the shape of pure poetry beckoning me" after she finished her novel of fact (D, 4:142, 145).

But now the prospect of continuing the alternating pattern of narratives and essays from the 1880s to the present seemed unmanageable. In fact, Virginia was so desperate about the prospects of ever finishing the unwieldy huge book that by 2 February 1933 she made a drastic excision. She jettisoned all "the interchapters—compacting them in the text" (D, 4:146). Her theories, thereby, were to become implied, not stated. On 17 February, she was "launched again in The Pargiters, in this blank season of the year—Nessa at Charleston, Clive in Jamaica, Roger in Tangier—Vita in America—which of my friends is left." Even in that "blank season," the Woolfs seem to have been very active socially, although Virginia did not write in her diary for about a month (D, 4:147 and note).

Always thinking visually, Roger found that the Moroccans had "developed their sensuality far beyond Europeans," for they knew the "grammar of colour" (RF/VB, 18 February 1933). He wrote to Virginia from Alicante, "Well I'm accustomed by now to Jehovah's little ways and don't expect much else." He was referring to the weather, but he might have had in mind the Slade Professorship, which had been refused him in 1904, 1910, and 1927. Finally, in March 1933, Roger Fry was appointed Slade Professor of Fine Art at Cambridge University. The professorship would necessitate his being in England two out of three terms, and it also would force him "to work out some of my ideas more fully" (RF/VW, 26 March 1933; RF/VB, 24 March 1933).

At the same time, while Virginia was describing Elvira Pargiter's refusal to accept honors offered by a corrupt society, the University of Manchester offered Virginia a doctorate of letters: "What an odd coincidence! that real life should provide precisely the situation I was writing about! I hardly know which I am, or where: Virginia or Elvira; in the Pargiters or outside" (D, 4:148). Virginia turned down the doctorate and wondered how Roger could be a professor "without disgrace" (VWL, 5:180). Quentin Bell speculates that Roger's professorship revived Virginia's old jealousy of her brothers' Cambridge educations (Bell, 2:173). Also, Virginia may have felt resentful because despite her great need for approbation, she had turned down the Clark Lectures at Cambridge and an honorary degree from Manchester, but the antiestablishment Roger had accepted an honorary degree from Aberdeen and a Slade Professorship at Cambridge.

Whatever their differences about such honors, Virginia and Roger were again working along similar aesthetic tracks. Virginia's plans for her new novel were

> bold & adventurous. I want to give the whole of the present society—nothing less: facts, as well as the vision. . . . It should aim to immense breadth & immense intensity . . . millions of ideas but no preaching—history, politics, feminism, art, literature—in short a summing up of all I know, feel, laugh at, despise, like, admire hate & so on.

The Pargiters should somehow combine "The Waves going on simultaneously with Night & day" (D, 4:151–52). In a major lecture given in Brussels, Roger proposed a similar combination. In "La Nature double de la peinture," he argued that the best paintings appeal because of the coherence of their plastic and literary elements.[28] Between vision and fact, plastic and literary, Roger and Virginia now saw congruence rather than division in the best art.

In late April 1933, Virginia recorded the warning that the Woolfs heard from the prominent conductor Bruno Walter, who had been forced to leave Germany after Hitler's rise to power in January of that year. Walter had denounced Hitler for an intolerance that was poisoning Germany. Accordingly, Walter suggested that the civilized world not fight but essentially boycott Germany. Virginia was impressed by the intensity of his convictions but rather skeptical that the horrors could so bad (D, 4:153).

The Woolfs spent most of May driving through Italy in a new car. Virginia concluded that Vanessa did not "mind me so much in writing as in fact." Her problem continued to be that "Nessa never writes to me, or thinks of me" (VWL, 5:160, 161). She addressed one letter to Vanessa "Vile Wretch!" and went on to protest: "I wrote you 2 long letters: one you ignore."[29] After their holiday, Virginia felt that her brain was "extinct." She was depressed not only by others' absences but also by Ethel Smyth's presence, described in an image of aggressive masculinity: "Its like being a snail and having your brain cracked by a thrush—hammer, hammer, hammer." In this mood, *The Pargiters* now seemed like an "empty snail shell" (D, 4:160, 161).

Roger faced the sort of catastrophe that would have devastated other people: He had been working for about fifteen years on translations of Mallarmé's poems, the only copies of which were in a suitcase that was stolen from a Paris train station, "with all my most precious belongings," including his inaugural Slade Lecture. Calmly, however, he rewrote his lecture and, with the help of Charles Mauron, "reconstructed" his translations (RFL, 2:681). He remained fascinated with Mallarmé, "almost the purest poet that ever was in the same sort of way as Cézanne was, in the end, the purest of painters." Their arts were "purified of everything accidental and external to be a direct communication" of feeling (RF/VB, 20 June 1933, and 30 May 1934). Whereas Mallarmé and Cézanne communicated feelings, Virginia now saw imagination as the "picture making power" (D, 4:176).

After Vita returned from America, her visit to Tavistock Square only caused Virginia to wonder why she had painted her lips, "unskilfully" at that (D, 4:168). Virginia wrote to Quentin Bell, who was ill with whooping cough: "I wish I were sitting by my own fish pond with my own nephew writing indecent and vulgar lives of the living," as they had done for Christmas entertainments so many times (VWL, 5:207).

The Woolfs returned to Monk's House in late July 1933, and Virginia and Roger experimented further with the art of autobiography, both giving Memoir Club papers at a "country club" meeting at Charleston in September (RF/VB, 13 September 1933; VB/RF, 17 August 1933). Roger read a memoir of his childhood, and Virginia's now lost memoir was on the proposals of marriage she received after Vanessa's marriage. Exploring her novel of fact in more detail, Virginia considered another attempt at "a biographical fantasy—an experiment in biography?" (D, 4:180). Meditating on death late that September at Monk's House, Virginia decided to rename *The Pargiters* "Here & Now." Even without the interchapters, it appeared to be "the longest of my little brood" (D, 4:176).

Although *Flush*, Virginia's latest biographical fantasy, published on 5 October 1933, promised to be a great popular success, Virginia again feared that she would be thought of as a mere "ladylike prattler" (D, 4:181). After David Garnett praised *Flush* in the *New Statesman* on 6 October, Virginia's letter of appreciation revealed a recurring doubt: "Oh lord how does any one pretend to be a biographer?" She went on to answer, "Yes, the last paragraph as originally written was simply Queen Victoria dying all over again—Flush remembered his entire past in Lyttons best manner; but I cut it out, when he was not there to see the joke" (VWL, 5:232). Other praise for *Flush* was reassuring, but when the *Granta* at Cambridge saw in her most recent works "the death of a potentially great writer," Virginia felt once again helplessly exposed, but she bravely determined: "I will go on adventuring, changing, opening my mind & my eyes, refusing to be stamped & stereotyped. The thing is to free ones self" (D, 4:186 note, 187).

Freeing herself to some extent meant defying all those to whom she had deferred. In 1931, Virginia had considered Sickert's paintings "witty" (D, 4:336), but in 1932, she had heard them deemed "very bad" (VWL, 5:46). Nonetheless, Virginia had long admired Sickert's paintings, even though she knew that Vanessa, Clive, and Roger thought her judgment was deficient in this matter.

Then in November 1933, Virginia saw a retrospective exhibition of Sickert's paintings that confirmed her judgment. Perhaps encouraged by Vanessa, Virginia wrote to Sickert telling him how much she admired his work. He replied, challenging what he mistakenly called "Bloomsbury clichés" such as the alienation of writers from painting. Sickert also urged Virginia to write about his painting.[30] Virginia thus went back to the exhibition "with a view to writing" (D, 4:190). She tested this artistic venture with Quentin, then aged twenty-three, asking him whether she could treat Sickert's paintings as novels. Knowing that Roger and Clive were critical of Sickert, she asked for Quentin's opinion (VWL, 5:253–54). No doubt, Virginia wanted to show how much she valued her nephew's opinion, but that very day (26 November 1933) she had written, without Quentin's advice, a version of "Portrait of Walter Sickert," which begins, "'I am a literary painter, so are all painters of any excellence'" (Berg 66B0244).

When Virginia and Leonard had Roger and Helen to dinner in

November (D, 4:191), Virginia seems to have made mischief by showing Roger her Sickert piece. He was furious: "'Well,' he said to me, 'you wouldn't find any literature in my paintings'—this referred to my essay on Sickert. 'What should I find?' I asked" (VWL, 5:256). Even though Virginia was implying that Roger's painting had no content, it was increasingly concentrating on representative scenes of enduring value, especially landscapes and church interiors, whereas her new novel was firmly grounded in English social and political history.[31]

On 15 December 1933, Clive had Virginia and Walter Sickert to dinner. She described the event to Quentin: "[Clive] primed us with wine and turkey; cigars and brandy; in consequence we all kissed each other, and I am committed to write and write and write about Sickert's books—he says they are not pictures" (VWL, 5:262). In her diary, she noted that Sickert told "great jokes about Roger" (D, 4:194), who was giving a "cascade" of lectures in Cambridge, Brussels, Edinburgh, and Queen's Hall in London, where he anticipated "rousing a mob" against him for his criticism of Turner and other English painters (RF/Charles Mauron, 17 December 1933).

The Woolfs spent three weeks at Monk's House, seeing Vita and her children and the Keyneses on Christmas Day and the Charlestonians for much of this time. Virginia was so "divinely happy & pressed with ideas" in Sussex that her diary failed to say "a word of farewell to the year" or to record "the walks I had ever so far into the downs; or the reading—Marvell of an evening, & the usual trash." When they returned to London, in mid-January 1934, she again began writing in her diary (D, 4:199).

In January 1934, tired from the "Ethel Smyth Festival" that belatedly honored the accomplishments of this female composer, from finally parting with her cook, and perhaps from still smarting over the wrangle over her essay on Sickert, Virginia did not go to hear "old Roger" lecturing. She still hoped that he "could scrape his back of all Russian barnacles"—Helen Anrep and her children (VWL, 5:272, 273). Virginia gave herself a headache writing the raid scene in her novel, by recalling the dinner party that the Woolfs had held in their cellar back in 1918 (D, 4:200–201). Vanessa painted another portrait of Virginia, maybe because she was rather "hard up" for money. To help her out, Virginia proposed "some caricatures" to be "issued between us" (D, 4:200), and Vanessa painted a caricature of Roger (Figure 53), using thick paint and rough brush strokes. But because Virginia was already feeling guilty about listening to Sickert's jokes about Roger and because Vanessa's exhibition at the Lefevre Gallery, opening on

7 March 1934, made her the money she needed, the sisters abandoned their caricature project.

As if to compensate for all that she had said about Sickert, Virginia argued in the foreword to the catalog of Vanessa's exhibition, as she had in 1930, that Vanessa's paintings told no stories and offered "only silence." In honoring Vanessa's art, she also disparaged her own: "Words talk such nonsense that it is best to silence them."[32] In March, Sickert read a draft of Virginia's essay on him and came to thank her. She began to find Sickert's praise suspicious, however: He liked her writing only because "it praises him to the skies." She also told Quentin that Sickert was "bitter though against all Rogers and Clives I imagine; says they dont know a picture from a triangle" (VWL, 5:282).

Over the past two years, Virginia's friendship with Elizabeth Bowen had blossomed enough for the Woolfs, on a trip to Ireland, to visit the novelist and her husband in late April 1934. After Bowen's Court, Kildorrery, "pompous & pretentious & imitative & ruined—a great barrack of grey stone," Virginia found "how ramshackle & half squalid the Irish life is, how empty & poverty stricken." At a hotel, Leonard found the first *Times* they had seen in nearly a week and read that George Duckworth, who had thought himself dying nearly four years before, had just died. Virginia summoned up some rather fond memories (D, 4:210, 211) and wrote to Vanessa, "Now suppose this had happened 30 years ago, it would have seemed very odd to take it so calmly." Recalling a memoir of George that Vanessa had written, however, vanquished Virginia's fond memories (VWL, 5:299).

After the trip, Virginia contracted influenza. Although she recovered sufficiently to attend *Figaro* at the first Glyndebourne Opera Festival in Sussex on 8 June, she was too sick to go to see T. S. Eliot's *The Rock* at Sadler's Wells. Reading it anyway, she felt the "dogmatism too full in the face," a reaction echoed by Roger, who left the performance "in a rage." Despite admitting an "anti-religious bias" (VWL, 5:315), Virginia, like Roger, was chiefly offended by Eliot's use of art to advance religion, thereby making art "kinetic" rather than "static."

Virginia had been rather intrigued the year before by Maynard Keynes's theory that their generation "had the best of both worlds. We destroyed Xty [Christianity] & yet had its benefits." In other words, they had, as Leonard said the Jews had, "great morality but no religion." Julian's generation, however, lacked not only religion but also morality, tradition, and discipline (D, 4:208).

Still regarding the absence of religion as a loss, Virginia was nevertheless outraged by Ethel's "Xtianity and egotism" because it was aggressive, aligned with the masculine point of view. Virginia was even more offended at Ethel's blaming Leonard for Virginia's lack of religiousness. Virginia, who had considered reading the New Testament for the first time in 1933 (D, 4:187), now had an ideal of religion that was basically feminine and that Leonard represented: "My Jew has more religion in one toe nail—more human love, in one hair"—than, presumably, Ethel and her High Church friends (VWL, 5:321).

In the summer of 1934, the Woolfs advertised for a "cook-general" for whom they would provide a small cottage. They hired Louie Mayer, who vowed to run Monk's House according to a daily timetable so detailed that she concluded the Woolfs "really loved time" (Noble, *Recollections of Virginia Woolf*, 187–98). Whereas Louie's presence afforded the Woolfs considerable comfort, Duncan's hypochondria was upsetting Vanessa. After "a most depressing view of family life at Charleston," Virginia was distressed over Vanessa's "passive submission" to Duncan's demands (D, 4:239). Virginia seems to have seen Roger only once that summer, at Clive's, when Roger refused to confirm a Whistler attribution. Clive amused Virginia by claiming, in his self-promoting manner, that Roger "had been duped over a Whistler once, when Clive was in the right" (D, 4:226). After that meeting, Roger departed for France, where at Saint-Rémy he worked with Charles Mauron to finish the reconstituted Mallarmé translations (RF/VB, 3 August 1934).

On 30 June 1934, Hitler shocked the world by murdering and executing rival members of the Nazi Party and other Germans whom he distrusted. In all, some twelve hundred Germans were summarily eliminated. Reading the news, like "an act in a play," numbed and terrified Virginia (D, 4:223 and note). No longer was she skeptical about the horror of Hitler's "ideal of human life" (VWL, 5:313), and consequently, she felt even more committed to her novel of fact and the social issues it raised. Writing another "mystical eyeless book" began to seem as ethically irresponsible as collecting solid objects; thus Virginia decided to break "the mould made by The Waves" (D, 4:233). Again, she was tempted to write a biography. By 2 September, she thought she had never "been more excited over a book than I am writing the end of . . . [her novel was once again nameless]" (D, 4:241).

From Saint-Rémy, Roger returned to the dreaded Royat for another attempt at a cure. Even he found "it hard to enjoy painting nature when she's as repulsive as she is here" (RF/VB, 8 August 1934). On 3 August, he broke the silence precipitated by the Sickert episode by begging Virginia for a letter for "the poor exile who sits in darkness" (RFL, 2:692). Then on the last day of the month, he wrote to her:

> If we go on writing and painting till 92 we shall be dreadfully in the way of the young. Would you like to be the "grand old woman of letters."[33] There's never been one has there? it wld. be a sublime position. You could say exactly what you liked & no one cld. dare to differ.
>
> But I am disgusted not to have talked to you for such ages. I shall relapse into barbaric stupidity. In fact I'm so far there that I can't write to you. . . . I want some excitement. My book on the National Gallery progresses slowly about half done I calculate. It'll be a book which no one can read through but I like writing it. I'm putting the old Masters through their paces as though they were contemporaries. Now you see that softening of the brain has set in from want of intellectual conversation. I know you'll say it's my fault for not coming to Rodmell but indeed I couldn't fit it in. Well you must let me have an evening as soon as ever I get back in October. [Actually he returned in September.] Yrs. ever Roger. (RF/VW, 31 August 1934)

If Virginia answered either of Roger's August letters, her replies have not survived. She no doubt would have answered the second, with its invitation to begin literary discussions as of old; she would have explained her new commmitment to combine vision and fact; and she would have told him about her elation over having "finished" her novel of fact and social relevance. She did not, however, because Roger Fry died suddenly on 9 September 1934, shortly after returning to London.

In one of her shifts from "wild ideas" to "assiduous truths," Virginia had been tempted to write a "great historical picture" of living persons, but her stock of living friends was fast dwindling. In these last six years, Virginia had survived several rather terrifying reminders of her own mortality, but death had put down his paw on her friends instead. Even though she had begged Roger to live to be a thousand, he had died at sixty-four, and at fifty-two Virginia took this loss badly.

11
An Odd Posthumous Friendship

1934–1938

When news of Roger's death arrived, Vanessa was visiting at Monk's House. She took the telephone call, delivered the message, and then fainted. Back at Charleston, the usually stoic Vanessa was so distressed that she cloistered herself in her room, where her daughter heard her "howling in anguish." Not yet sixteen, Angelica found herself disoriented by this first encounter with death, but she did not dare venture into her mother's room, fearing "the creature who made those noises could not be my mother" (Garnett, *Deceived with Kindness*, 104). The day after Roger's death, the Woolfs drove Vanessa to London to see Margery Fry and Helen Anrep (LWD).

Virginia wrote to Ethel of their "awful blow—Roger Fry's death—You'll have seen—It is terrible for Nessa. . . . He was to me the most heavenly of men—so I know you'll understand my dumb mood.—so rich so infinitely gifted—and oh how we've talked and talked—for 20 years now" (VWL, 5:330). She assessed her feelings: "I feel dazed: very wooden." Leonard told her that "women cry," but Virginia did not "know why I cry—

mostly with Nessa." She felt "too stupid to write anything" and thought "the poverty of life now is what comes to me. a thin brackish veil over everything. . . . The substance gone out of everything" (D, 4:242). She found herself comparing her present despair with her reaction to Julia's death: "I remember turning aside at mother's bed, when she had died, & Stella took us in, to laugh, secretly, at nurse crying. She's pretending, I said: aged 13. & was afraid I was not feeling enough. So now" (D, 4:242). In 1934 at Roger's death, as in 1895 at Julia's, Virginia felt numb.

Roger Fry's funeral was held on 13 September 1934. Passages from Milton's *Comus*, Fry's *Transformations*, and Spinoza's Proposition LXVII were printed on the program. There was no eulogy. Virginia appreciated the ceremony's "wordlessness" and reflected on Roger's living "with such variety & generosity & curiosity." She felt herself "jaded; cant write." Needing to hold onto old friends, she told Desmond MacCarthy, "Dont die yet" (D, 4:243). The Woolfs returned to Monk's House, where they played bowls in the evenings as a diversion and took Vanessa blackberry picking (LWD).

In an obituary, E. M. Forster named as Roger's chief virtue a complete rejection of all authority, a rejection that led others to question both values and experiences ("Roger Fry," 38–39). Virginia valued Roger in those terms, too, but her sense of loss was more personal. Roger's dying just as he had begun his exciting and ambitious lectures at Cambridge forced her once again to confront her own mortality and "the vainness of this perpetual fight, with our brains & loving each other against the other thing: if Roger could die" (D, 4:244).

The Bloomsbury friends had hoped that with their brains and their hearts, their work and their friendship, they would do something to make the world more civilized. But for Virginia, that assumption paled in the face of death. She felt "stupid and depressed and dull. I hate my friends dying. Roger was so full of his lectures and his plans, and now its all over" (VWL, 5:332). Only toward the end of the month could she settle down to work again and congratulate herself: "The last words of the nameless book were written 10 minutes ago; quite calmly too. 900 pages." Even the weather seemed to share her brief exhilaration (D, 4:245). Although these were hardly the "last words" of her novel of fact, as she found out later, claiming that she had finished helped her banish the demons of helplessness and loss. But this sense of accomplishment soon faded, and the memory haunted her: "When Nessa came across the terrace—how I hear that cry Hes dead" (D, 4:246).

Guilt for all that Virginia had not told Roger began to nag at her. She wrote to Roger's daughter, Pamela, that she had "felt since his death how little one gave him—how much I wished I had told him what he meant to me" (VWL, 5:335). Virginia's sense of giving too little and accepting too much echoed her comments when her father died in 1904: "I never did enough for him all those years" (VWL, 1:130) and "if one could only tell him how one cared" (VWL, 1:133).[1] Despite needing to assuage such feelings, she suspected, as she had at her mother's death, that in fact she felt too little.

Roger Fry's death marked the twenty-first anniversary of Virginia's nearly successful suicide attempt in 1913. We do not know whether she considered that conjunction of dates, but we do know that she felt guilty: because he was the giver, because she had not told him what he meant to her, because he was dead just when his life was taking splendid shape and she, who had more than once wanted to die, was alive. And because Virginia replayed both the numbness she had felt at her mother's death and the guilt she had felt at her father's, Roger's death unsettled her much more than Katherine Mansfield's or Lytton Strachey's had.

Back in London in early October, Virginia's sense of guilt was further exacerbated. Before Roger had been dead a month, her article on Sickert, entitled "A Conversation About Art," was printed in the *Yale Review*; then on 25 October, the Hogarth Press published it (slightly revised) as *Walter Sickert: A Conversation*. After attending classes taught by Sickert in the 1890s, Roger had objected to the lack of formal interest in Sickert's painting. Indeed, in Roger's projected portrait of Bloomsbury, Sickert was to be looking in the window unable to comprehend the Bloomsbury spirit. Publicly, Sickert had attacked Roger's avant-garde tastes and privately, as Virginia testified, had ridiculed him (VWL, 5:282).

For years, Sickert had flirted with Vanessa, and ever since Vanessa had arranged a private meeting for him with Virginia in 1927, he had taken every opportunity to flatter Virginia. Challenging Virginia to write, Sickert had primed her with the lines about his being a "literary painter" because her stature in the service of his ideas would offer a powerful refutation of Roger's theories about non-"literary" painting.[2] Now her laudatory Sickert piece suggested that she had been manipulated into siding with Sickert against Roger's ghost. Espousing "literary" values, Virginia called Sickert a "true poet" with "any number of stories and three-volume novels" in his painting.[3] By concluding that he was "probably the best painter now living in England" (CE, 2:244), Virginia indirectly insulted Vanessa Bell, Duncan Grant, and the ghost of Roger Fry.[4]

The timing of the article's publication only heightened the insult, especially when Roger's posthumous "The Toilet" was published almost simultaneously with Virginia's "Sickert." Roger's essay clarified for a general audience the issues she had heard him lecture on more than a decade eariler. In "The Toilet," Roger examines a Rembrandt nude in terms of its "impassioned contemplation" of "natural forms." Then he describes his pleasure in the painting's title, *Bathsheba*, which evokes the biblical story of King David's adulterous lust for the beautiful Bathsheba. The "psychological" appeal of "literary" association adds to the painting's "plastic and pictorial appeal," since they are harmoniously or "operatically" integrated. Roger's balanced modification of the extreme implications of formalism upstages *Walter Sickert: A Conversation*, which simply jettisons them.

Virginia regretted her betrayal of the aesthetics that she and Roger had promulgated in a joint campaign in the 1920s. Virginia maintained that she wrote "at command of the old tyrant himself" (VWL, 5:340) chiefly because Sickert was old and "on bad times financially" (VWL, 5:343, 314). She called the essay a "little joke about Sickert" and claimed that she was "ashamed now to explain it" (VWL, 5:354). Virginia did not often admit to being ashamed, but when she did, she worked very hard to compensate. But Roger was dead, and so she had no ready way to compensate for betraying his friendship and denying his lessons in aesthetics. As a consequence, she felt so guilty that she later wrote that she "blamed" herself when Roger died (D, 5:104).

There is no rational reason that a woman of fifty-two would have blamed herself for the death by natural causes of a sixty-eight-year-old man, especially one who had been "poor old Roger" almost since they first met and who had been in fragile health for much of the past decade. Although she suspected that the doctors had mismanaged Roger's care, Virginia assured herself that "he was in such a worn out state that he could not have lived long, and might have died at any moment" (VWL, 5:332). Actually, according to Igor Anrep, Helen's son, even though he had a painful intestinal problem, Roger could still "drive all day in France, see many churches, and spend five or six hours in a gallery without turning a hair." He was "hardly ready to die." In fact, he would have lived had it not been for his doctor's mistakes. Anrep, a retired cardiologist, explained that Fry had fallen and broken his hip. His own doctor and others they knew were out of town on a bank holiday weekend. An unknown conservative doctor at the Royal Free Hospital put him in a huge hip cast instead of following the more up-to-date method of putting a pin in his hip. The cast then caused an obstruction of the bowel. When an intern realized what was happening, he tried to remove the cast—even having a terrible fight (which Helen observed) with the nurse, who did not want her scissors ruined on the cast—but Roger Fry died. Clearly, the doctor at the Royal Free Hospital was to blame, and equally clearly, Virginia Woolf was not.[5]

Almost immediately after his death, Roger's friends began speculating about a likely biographer for him. Virginia had publicly attacked the practice of biography in *Orlando*, *The Waves*, and, most recently, her essay on Sickert. There she hypothesized that Sickert's portraits were truer than biographies and criticized "the three or four hundred pages of compromise, evasion, understatement, overstatement, irrelevance and downright falsehood which we call biography" (CE, 2:236). Morgan Forster's biography of Goldsworthy Lowes Dickinson was a current example of such evasion and dishonesty. Privately, Virginia condemned it as a "quite futile" tribute that failed to mention the condition that shaped Dickinson's life—his homosexuality (D, 4:247).

Suddenly, Virginia herself "got tempted into making notes for biography" (D, 4:252). But because she was depressed about Wyndham Lewis's attack on her fiction and about Roger's death—which was "worse than Lytton's"—she found it impossible to write (D, 4:251, 253). When the Woolfs dined with Helen at the Bloomsbury apartment that she had shared with Roger, Virginia for a moment imagined he was with them. Trying to placate Roger's ghost, she supposed her "little flurry" over Sickert to be a failure (D, 4:257).

Virginia's impulse to make "notes for biography" could have been a theoretical or practical follow-up of "The New Biography," in which she had insisted on biography's potential as an art form. But soon that impulse opened the option of herself becoming Roger Fry's biographer. Such a possibility would indeed placate Roger's ghost. It would assuage loss and alleviate guilt. It would offer diversion, a means of rotating her crops. It would constitute artistic vindication by proving her reality gift. And it would not entail great effort, for "Julian [Bell would] collect all facts, make a skeleton; I to sum & compose." Offering Julian a writing opportunity in the service of Roger seemed guaranteed both to divert and to gratify Vanessa. But Margery Fry herself might want to write about Roger. Virginia considered compiling a collection of essays by Margery and Roger's friends.[6] Editing such a collection, however, would offer none of the advantages of writing a biography herself.

While Virginia was steeling herself "to tackle re-reading & re-writing" the massive *Pargiters* (D, 4:261), Helen Anrep visited on 8 November. Both she and Margery Fry wanted Virginia to write Roger's biography. Now the psychological and artistic advantages of "writing R.'s life" made that possibility "uppermost" in her consciousness. She became tremendously excited about this "splendid, difficult chance—better than trying to find a subject—that is, if I *am* free (D, 4:260). Then Margery Fry and Vanessa came to tea at Tavistock Square on 18 November (LWD). Margery at last proposed that Virginia write a "study" of her brother, but Virginia replied that such a book would be "unreadable." When Margery assured her that she could be "quite free," Virginia declared that she would also "have to say something about his life." The "something" she had in mind was no doubt adultery, especially as Roger had lived for nine years with Helen Anrep while still being married to Helen Fry, who was still living in an asylum. Making the matter even more complicated, Helen had not divorced her husband, Boris, when she left him for Roger. Confronted with the prospect of such revelations about Roger's life, Margery hedged and asked Virginia to be "careful" not to offend Roger's aged mother and his other spinster sisters (Figure 54). (The grouping in Figure 54 suggests how central Roger was to their lives.) Despite the unpromising exchange with Margery and the silent pressure of Vanessa's desire, Virginia agreed to go through Roger's letters and papers while revising *The Pargiters* (D, 4:262).

Virginia and Leonard were invited to a private showing, on 25 November 1934, of Man Ray's photographs in Bedford Square. There they met the wealthy and cultured Argentinian Victoria Ocampo, who courted Virginia's friendship for a time. Virginia reluctantly agreed to Man Ray's request to photograph her. The next evening, the Woolfs had Vanessa, who was "lonely without Roger," to dinner. Virginia's efforts "to comfort Nessa" failed against "the religion & superstition of motherhood" when they told her that Julian's poems were not ready for them to publish. "Ruffl[ing] like a formidable hen," Vanessa was "reasonable but cold." Virginia regretted "exacerbating ourselves, instead of consoling," but blamed the friction on Vanessa's "maternal partiality" and her own "discreditable" reaction to it. It was "not a nice evening" (D, 4:264).

On the next day, 27 November, Virginia sat for her photograph. When Man Ray insisted that she wear lipstick, Virginia objected, but he argued that it would not show but would technically enhance the picture.

Then, after he put the lipstick on Virginia's mouth, she forgot to remove it—perhaps she liked the effect. The resulting studio poses of a short-haired, be-lipsticked Virginia are powerful pictures that Leonard thought the best ever taken of her (Richardson, *Bloomsbury Iconography*, 299–300; Ray, *Self-Portrait*, 189; LWL, 546). One (Figure 55) echoes the poses and the facial angularity of Figures 11 and 17; however, the earlier pictures show a wounded and defensive Virginia, whereas Man Ray's shots show a poised, handsome, self-confident woman.

In early December, anguishing over another friend's approaching death, Virginia found herself unable to "feel any more at the moment—not after Roger" (D, 4:265). Three months after his death, Virginia still was tempted to write something about him, although she intended something less grandiose than a "whole big life" (VWL, 5:352). Vita's portrait appeared on the cover of the Christmas 1934 issue of *Bookman* where Virginia must have seen it and mourned the loss of romance with her androgynous friend (Figure 56).[7] Meanwhile, Victoria Ocampo was negotiating with Virginia about Spanish translations of her works, and Victoria sent Virginia orchids and roses for Christmas. In an odd resurrection of the *paragone*, Virginia asked Vita whether she knew that Potto had "taken up art, to cure his heart?" She also told Vita about Victoria's orchids, "in the hope of annoying you" (VWL, 5:358, 359).

The Woolfs' annual retreat to Monk's House included visits with the Keyneses and Harold Nicolson, just back from a visit in New Jersey with Charles and Anne Lindbergh (VWL, 1:359, note). Three quiet weeks at Rodmell at Christmastime and the beginning of 1935 also saw a rehearsal of Virginia's revised comic play *Freshwater* (a satire of, in particular, high-minded Victorian notions of art) at Charleston.

Virginia acquired another room of her own about this time. Her old writing lodge was demolished and replaced by a lodge across the garden behind Monk's House that had few amenities except a heater, "open doors in front," and a view of the downs (D, 4:265). Leonard also used this building to store fruit in an "apple loft," reached by a rickety outside stairs, but it principally served Virginia's need to write in solitude (Trekkie Ritchie Parsons to me, ? February 1995), although the end of 1934 was almost too wet for Virginia to get to her new lodge.

Virginia's last diary entry of 1934 asked: "And Roger dead. And am I to write about him?" (D, 4:267). Apparently she believed she was. She began her 1935 diary planning to finish *The Pargiters* by July, to have it published in October, to read Roger's papers in August, to write "Despised" (which later became *Three Guineas*), and then to write a biography of Roger in 1936 (D, 4:271). Virginia's sense of emptiness after Roger's death may have made her decide in early 1935 to illuminate a "dark spot" in her education by reading St. Paul, the Acts of the Apostles, and Ernest Renan's studies of the origins of Christianity. Her creative juices seemed to be bubbling again (D, 4:271).

In an increasingly ominous political climate, Leonard wrote *Quack, Quack*, a warning against neoauthoritarianism and irrationalism in fascism, in Nazism, and, with such philosophers as Bergson, even in philosophy (D. Wilson, *Leonard Woolf*, 225–27). Virginia made plans to write "Despised," a prescient exposure of connections between fascism and misogyny (D, 4:273 and note).

The Woolfs and their friends got away from death and grim political realities with "an unbuttoned laughing evening." What was left of old Bloomsbury, along with new and younger friends, met on 18 January 1935 in Vanessa's London studio for a riotous performance of *Freshwater*, in honor of Angelica's sixteenth birthday. This play, which Virginia had first conceptualized in 1919, concerns her great-aunt Julia Margaret Cameron and the high-minded company she kept, especially Alfred Tennyson and George Watts. This absurdist drama spoofs Watts's motto the "Utmost for the Highest" in art.[8] Ellen Terry, the young actress whom Watts married in his old age—chiefly, it seems, to procure her services as a model—deserts him for a passionate young man. After her desertion and after the Camerons depart with their coffins for India, Tennyson and Watts are left "alone with our art" (FW, 51). The play ends as Queen Victoria offers them society's consolations of a peerage and the Order of Merit (FW, 53).

Playing Ellen Terry, Angelica was "ravishing" but "too grown up" for Virginia's taste. All along, Leonard had been opposed to Angelica's casual upbringing, and Virginia, too, began to have misgivings about Vanessa's indulgence of her daughter (Garnett, *Deceived with Kindness*, 108). Even while she laughed at the highly moral notion of art endorsed by Tennyson, Watts, and Julia Margaret Cameron, Virginia wondered whether Bloomsbury aestheticism had been too pure or detached from moral responsibilities. While writing *The Pargiters*, planning "Despised" and a biography, and reading the New Testament, Virginia now "rather relish[ed]," in contrast to Angelica, Adrian's daughters, in whose Cambridge educations Virginia had played no part. She supposed that the Stephen girls would probably become activists and perhaps make a difference in the practical world (D, 4:274).

Their dead friends were "our ghosts now," and Virginia became increasingly loyal to Roger's ghost (D, 4:275). Virginia accused Ethel of compounding "'intelligence' with destructive criticism." By contrast, Roger, "the most intelligent of my friends was profusely, ridiculously, perpetually creative: couldnt see 2 matches without making them into a boat. That was the secret of his charm and genius" (VWL, 5:366). Her defensiveness may have been related to a diary entry written on the same day as that letter: "Yes, I ought to have explained why I wrote the Sickert" (D, 4:275). She had not had the chance to rationalize to Roger, as she did to others, that the Sickert piece had been a "joke" or an act of compassion for, or coercion by, an old man. Then, when she and Vanessa dined together in London, she reminisced, "Oh dear, how the ghost of Roger haunted us! . . . An extraordinary sense of him: of wishing for him; of vacancy" (D, 4:279).

Virginia had long ago abandoned her experimental plans to make *The Pargiters* into a novel-essay.[9] But even without the essay portions and even though she thought that she had finished it the preceding September, rewriting *The Pargiters* seemed interminable. Feeling a "certain emptiness," Virginia also was dreading Vanessa's expected departure, "the defection of Vita; Roger's death; & no-one springing up to take their place; & a certain general slackening of letters & fame, owing to my writing nothing" (D, 4:287). Virginia compensated for her loneliness with mischief making, telling Ethel to write to Vita and then telling Vita that she "did my best to persuade her not to write."[10]

Although the sisters had drawn closer when Virginia comforted Vanessa over the loss of Roger, Vanessa still resented Virginia's criticism of her children. Not only had the Woolfs rejected Julian's poems for the Hogarth Press, but Virginia also had trouble praising Quentin's painting. She exhibited "constraint with Nessa about her sons art—as usual: am I too critical? Why this difficulty in praising?" (D, 4:288, 289). While Virginia resisted Vanessa's pressure to praise her children, she also coped with criticism of herself and Bloomsbury in general. She considered, "If I write about Roger, I shall include a note, a sarcastic note, on the Bloomsbury baiters."[11] But even though she and Leonard suggested that Morgan Forster write a "comic guide to Bloomsbury" (D, 4:289), she resisted the idea of baiting the baiters.

When the Charlestonians went to Italy in the late spring of 1935, Vanessa took along all of Roger's letters to her to reread before passing them on to Virginia, because, as she told Julian, "some might easily be indiscreet." Certainly, some were "very intimate," but she felt that she could trust Virginia not to publish those. With now all-too-typical self-denigration, Vanessa wrote to Virginia, "I suppose you're swallowed up by admirers and won't have any time for old crones, and I have nothing to offer in return" (VBL, 388, 387).

Writing Roger's biography would satisfy Virginia's "great desire" for him (D, 4:290), just as her 1908 "Reminiscences" and her 1928 *Orlando* had compensated for the losses of Vanessa and Vita. But she worried whether she "could write about Roger" frankly (VWL, 5:378). Whereas Leonard honored his friendship with Roger simply and efficiently by being a financial executor of his estate, Virginia had not yet done anything.[12]

When James Strachey asked Virginia to write a character study of Lytton as an introduction to a biography, she wondered how far the ideal of frankness could extend. Could a biographer mention "buggery?" (D, 4:296). At least if she wrote about the heterosexual Roger, she would not face that problem.[13] A gathering at Clive's in Gordon Square was "slightly empty as usual—but why? Without Roger, I suppose" (D, 4:297). Julian Bell began readying Roger's translations of Mallarmé's works for the Hogarth Press, but Margery Fry's meddling was reportedly creating a "prickly time" for him. For Virginia, Margery's silence was equally prickly, as she needed a firm commitment but had "not heard a word" from her (VWL, 5:387).

While Margery left Virginia dangling, Morgan Forster insulted her. First hinting that the officers of the London Library might ask Virginia to become a member of the Library Committee, Morgan then explained that her father—after working, before the turn of the century, with Alice Green, widow of the eminent historian J. R. Green—had pronounced women "impossible." (Actually, by 1900 Leslie thought better of widow Green.) Morgan's point was that women were still thought not suitable to serve on the committee. Virginia was furious: "The veil of the temple . . . was to be raised, & as an exception [a woman] was to be allowed to enter in. But what about my civilisation? For 2,000 years we [women] have done things without being paid for doing them. You cant bribe me now" (D, 4:297 and note, 298). Virginia's fury sprouted into ideas for a book, called "Despised," on the "outsider" position that women played in English civilization, a position that encouraged militarism.

Even though Virginia considered Hitler to be "very frightening," the Woolfs had yet to buy gas masks because no one took seriously the threat of war (D, 4:304). Then, when the Woolfs traveled to Italy in May, driving through Germany, they saw signs of anti-Semitism and Hitler's growing power, which now could not be ignored. Seeing banners proclaiming, "'The Jew is our enemy' [or] 'There is no place for Jews in ——,'" Virginia and Leonard found their nerves frayed and their fear and anger building (D, 4:311).

Before leaving on her trip, Virginia had hoped to be "done with" her rivalry with Vanessa (D, 4:301), but it was too deeply planted to be easily uprooted. When Virginia and Leonard reached Aix-en-Provence, she remembered the "old triumphant Vanessa of early married days": "How she would bear off in full sail with Roger Clive & me attendant" (D, 4:314). Suitor then, she now was rival again: "Segonzac thinks Nessa the best painter in England, much better than Duncan. I will not be jealous, but isnt it odd—thinking of gifts in her? I mean when she has everything else." If Virginia's logic was faulty in thinking that because Vanessa had children, she did not need success, Virginia was justified in thinking that Vanessa was cruel never to write to her (D, 4:322, 328).

While Virginia was jealous of Vanessa's fame and resentful of the one-sidedness of their correspondence, Vanessa wrote to Clive making fun of the Woolfs' reluctance to try new things, their attempting (she claimed) to drive through all of Italy in one day, Virginia's reluctance to eat (a complaint not voiced in twenty years), and Leonard's desire to return to England and his dog Pinka (VB/CB, 26 May 1935). When the Woolfs arrived back at Monk's House, however, they found that Pinka had died the day before. Virginia felt "something of our play private life" had gone with their dog (D, 4:318). With Roger and now Pinka dead, fascism gaining power, Vanessa irritated, the public hostile, and more revisions of the gargantuan *Pargiters* ahead of her, depression threatened Virginia.

She confided to Ethel that she was "in the cavernous recesses (excuse this language) because Roger is dead (I never minded any death of a friend half so much: its like coming into a room and expecting all the violins and trumpets and hearing a mouse squeak)" (VWL, 5:399). While Virginia longed to write something to compensate for the loss of Roger and for her denial of his influence, she still resented the Fry family's "counting on me to write a life" (D, 4:323). Although Virginia had made a tentative commitment to her, Margery had yet to make a corresponding commitment to Virginia's autonomy.

To compound her problems, *The Pargiters* would not be finished soon, and the Stracheys resented her writing about Roger. Nevertheless, in those increasingly troubled times, Virginia returned to the issue that Leonardo da Vinci had confronted so long ago: Which art is more true, more significant? Having confirmed that her art could be as pure and antimimetic as a Postimpressionist painting, Virginia needed to prove also that her art could affect the world of experience. Anticipating both a respite and a challenge, Virginia denied herself a clear look at the threats looming ahead. She valued Roger for his creativity and his iconoclasm, for the very factors that had signaled his alienation from his family. Loving Roger for his unconventionality while having to please the straitlaced Frys put Virginia in a bind. She began to feel trapped by Margery's assumptions and overwhelmed by the documents being gathered for her. Privately planning to write the biography, she nevertheless publicly maintained that she wished others would contribute (VWL, 5:402). But the more time she invested in reading Roger's papers, the less chance there was that she could withdraw or share the responsibility.[14]

In Rome, Vanessa reread the letters between Roger and herself. Then, through friends in the diplomatic corps, she had the letters sent to Virginia. Vanessa's commitment to truth forced her to send "almost everything, however personal about me," including examples of her frequent irritation with Virginia and the details of her affair with Roger, such as his drawing of her breast from the side as she lay in bed (Figure 18), but Vanessa instructed Virginia not to let "anyone else see them" and to "read the whole lot straight through when you begin" (VB/VW, 25 and 14 June 1935). After dining in Paris with various artists who "talked a good deal about Roger," Clive suggested to Vanessa that there "ought to be a chapter on Roger in France by a French painter" (CB/VB, 2 June 1935). Instead, Vanessa wanted Virginia alone to write.

Back in London that summer, one of many flare-ups between Leonard and a servant caused Virginia to think about her husband and the class system. His severity resulted from "not being a gentleman partly: uneasiness in the presence of the lower classes" and "his desire, I suppose, to dominate. Love of power"—odd characteristics in one who opposed political dominance and power. If the desire to dominate entered their relationship, Virginia could easily "get up and curse him," but he apparently rarely tried to dominate her. Nonetheless, she was afraid that other people noticed the disparity between his devotion to justice, good deeds, and equity and his tyranny over servants. Virginia vowed, "All this I shall tell him again," which she must have done, for Leonard became "contrite" about the affair, but still "Nessa does not write" (D, 4:326, 328).

In early July, the Woolfs paid "£18—dear me" (D, 4:328) for a black-and-white cocker spaniel they named Sally, who "at once [fell] passionately in love with Leonard" (VWL, 5:409). Then on 12 July 1935, in a "wave of old sentimentality" (VWL, 5:409), Virginia opened the Roger Fry Memorial Exhibition at the Bristol Museum and Art Gallery. There she announced that Roger's great achievement was teaching those "outside" the art world, like herself, how to "enjoy looking at pictures" (CE, 4:88). This tribute, despite "Roger's face on the canvas, smiling at me," was a miserable experience (D, 4:330–32). That is, the "horrors" of giving the lecture resulted from her suspicion that Margery would "disapprove of anything I write. Yet I feel she means to hold me to it" (VWL, 5:416). The trip to Bristol was slightly redeemed by two nights with Leonard in country hotels (D, 4:330), but Virginia still felt trapped in a no-win situation, obligated to write yet unable to please Margery.[15]

Virginia also was unable to begin, for *The Pargiters* remained her first obligation. She finished a "wild retyping" the week after the Bristol exhibition and found the book now down from 900 "to 740 pages." She decided to shape, cut, and add "spaces of silence, & poetry & contrast." Virginia set herself the task of "typing out again at the rate, if possible, of 100 pages a week, this impossible eternal book." By late summer, she finally was sending pages to a professional typist.

Then one July evening, Julian Bell came to Tavistock Square to tell his aunt that he had accepted a three-year term as a professor of English in China. Virginia understood Julian's decision as a need to be political, not "merely a poet, a writer." But she would miss him, for when he returned, "he will be thirty & I 56 alas" (D, 4:332).

Fleeing from her "eternal" book to Monk's House later that summer, Virginia summoned her "indomitable courage" (D, 4:335) to begin the biography of Roger. The complexity of sorting documents "rather dazed" her (VWL, 5:426) when in addition to "my 3 large boxes of Fry," there seemed to be a "whole room full more, I believe" (VWL, 5:427). Virginia speculated about the bipolarity of her brain: "Half my brain dries completely; but I've only to turn over, & there's the other half, I think, ready, quite happily to write a little article" (D, 4:338). While continuing to work in her lodge on her novel, now rechristened *The Years* (D, 4:337), she rested the imaginative side of her brain by dropping "a few facts into [it]" from her "three boxes of Roger" (D, 4:341).

She explained her dilemma to Julian Bell, now in China: "Rogers letters are fascinating; an awful mix; the family ones very stiff; the travel ones rather dull; but always some flash of interest; and some to Basil Williams extremely amusing. But I cant think how to deal with it—or whether to deal with it" (VWL, 5:433). She also confided her dismay to Julian: "His love letters are prolific; he must have had a love every new year; and most of them are foreigners. So I am plodding away, when the light fails, and I can no longer write my long dull novel, And now the Stracheys want me to write about Lytton" (VWL, 5:436).

However ideal the arrangement of writing *Three Guineas* (the former "Despised" or "On Being Despised") in the mornings and taking notes for *Roger Fry* between tea and dinner (D, 4:346), Virginia still sometimes maintained that she had not committed herself to writing a biography (VWL, 5:435). But by 27 October 1935, she was encouraged to think she could finish *The Years* by January and then "dash off" *Three Guineas* and be ready to "do Roger" in the summer of 1936. Visiting the hilly London suburb of Highgate with Leonard, she remembered "thats where Roger was born & saw the poppy" that he had described in a Memoir Club paper. She thus decided to begin her book with that scene and confidently assured herself, "That book shapes itself" (D, 4:348).

A year after Roger's death, Virginia was planning a decidedly, even aggressively, experimental marriage of granite and rainbow. Her biography "might be a series of portraits of places—peasants: pictures/landscapes." It would be ordered "as his character dictates: not in order of time." It would heal the *paragone* by positing the "alliance between the arts" and explaining Roger's "aesthetic feelings" (MHP/B17a). But to Angelica Bell a month later, Virginia expressed a sense of entrapment: "And I have to write about Roger. At least Nessa wants me to." Virginia felt the coherent selves that biographies project are untrue because "people are all over the place" (VWL, 5:445). Indeed, in *The Waves*, Virginia has Bernard express that very "distrust [of] coherency" (Graham, *Virginia Woolf's "The Waves,"* 381–82). Although her scattered 1935 notes about Roger's various characteristics did avoid artificial unities, her desultory approach threatened to jeopardize coherence altogether.

Virginia looked forward to finishing her novel and being free to write *Three Guineas* with "wild excitement" and considerable anger and to "go on accumulating Roger." *Three Guineas* and *Roger Fry* fell on the "realities" end of the artistic spectrum, as did *The Years*. But in revising her novel, Virginia began to apply some of the techniques she had developed in her Postimpressionist fiction, finding it "rather like writing The Waves—these last scenes" (D, 4:321).

As she attempted to circumvent the constrictions of biography, Virginia encountered people who could not imagine anything beyond them. For example, Robert Bridges's widow told her that Roger had asked his father's permission to violate Quaker prohibitions and play games on Sunday (D, 4:355). Virginia must have been wondering how any art could grow out of such arid depths. Other facts were more engaging, however: "And now for 30 mins of Roger's letters to Helen [Anrep]—that vast sparkling dust heap, the best so far; but how to dig out? how to represent?" The problem was creating a hybrid out of both experimentation and representation. Virginia was unsure how to manage but decided to "read & read & wait on the moment of illumination" (D, 4:355). The letters made her like Helen Anrep "better and better" (VWL, 5:450). Virginia thought "of love, & L. & me; & the different lives." Despite her wanting to be free of obligations, she "read & read & the packets hardly lessen" (D, 4:356).

Julian and Roger were oddly "joined together" for both sisters. Vanessa felt Julian's departure for China was "like some dim reflection of [Roger's] death" (VBL, 402). After coaxing, Virginia received a long letter from Julian about his recollections of Roger.[16] On 1 December 1935, she thanked him for giving "the feeling of [Roger] extremely well" and confided that being privy to such knowledge posed new problems. She told Julian "be discreet; for [Roger] says very sharp things about Clive . . . about his bagging Rogers ideas; his lack of understanding of art; his reverting to the Bell type and so on" (VWL, 5:447, 449). Later she concluded that "Clive did pilfer a good deal without acknowledgement from Roger" (VWL, 6:20). For her part, Vanessa still felt Roger's absence "constantly and long[ed] for his help and criticism" (VB/JB, 13 December 1935).

Despite complaints, the prospect of writing an experimental biography continued to excite Virginia. A 16 December diary entry recorded a complete plan for the book, to begin with Le Mas, Roger's home since 1931 in Saint-Rémy-de-Provence:

> Why not begin at the end with Le Mas: a whole day; & then work backwards: give the elements in combination in action, first; & then trace them—give specimen days, all through his life.
>
> Le Mas: the mosquitoes &c. his cooking: the colour, the martins: the French novel: freedom—cast back to childhood. Quote. Then Cambridge, then America: then us. Then the end. (D, 4:358)

Virginia's excitement over that experimental plan was both aesthetic and personal. She wrote to Julian, "How I miss Roger on Sundays for then there was always a substance, not mere froth in the talk. Yes I wish Roger were here more and more" (VWL, 5:452). Explaining that she was trying to "master" Roger's aesthetics, Virginia summarized a point that she had made herself in "Phases of Fiction": "What he says is if you cut off your practical senses, the aesthetic [senses] then work" (VWL, 5:455).

Despite having "finished" *The Pargiters* just after Roger's death in September 1934, Virginia found herself toward the end of 1935 trying "to read Roger in the lapses of finishing my book; but it won't finish; its like some snake thats been half run over but always pops its head up" (VWL, 5:448). She was often "so congested I cant even copy out Roger." Yet she copied out an amazing, even excessive amount, for she distrusted her memory: "I'm almost extinct . . . that is my brain."[17] Nevertheless, she ended 1935 on a happier note: Having "just put the last words to The Years," "reading Roger I become haunted by him. What an odd posthumous friendship—in some ways more intimate than any I had in life. The things I guessed are now revealed; & the actual voice gone. Clive Quentin, Nessa Duncan" (D, 4:360, 361). Apparently, her sampling of the letters revealed much that Virginia had only guessed about Vanessa's sexual relations with Clive, Roger, and Duncan and her almost loverlike devotion to her sons. Later Virginia was disturbed to read in Vanessa's letters numerous caustic remarks about herself. But in 1935, reading Roger's letters about struggles and triumphs that his "actual voice" had not shared with her established a posthumous intimacy with him.

With seemingly interminable revisions on *The Years* still before her, although she had written the "last words" in late September 1934 and again in late December 1935 and planned to be really finished in February, Virginia began 1936 "sick of fiction" and doubtful that biography would be less arduous (VWL, 6:7). She also had such a bad headache that she spent the first three days of the new year in bed at Monk's House. The challenge of composing an artistic biography thrilled her, but obstructions, "the Frys—etc etc," inhibited her (VWL, 6:9–10). The Woolfs returned to London on 8 January (LWD). On 20 January, King George V died, and his funeral procession was to pass through Tavistock Square. Virginia expected that only those residents with keys to the fenced-in garden would get into the square, but instead democracy "swarmed through; leapt the chain, climbed the trees" to see the coffin and the new king, Edward VIII, walking behind it with his brothers (D, 5:12 and note; VBL, 404). Juxtaposed against such pomp, however, was the specter of a female beggar at the Woolfs' door. Virginia reflected that she had never seen "unhappiness, poverty so tangible." As a person of substantial capital, albeit a socialist, she "felt its our fault" (D, 5:19).

Partly in deference to Clive's straitlaced parents but also in violation of her reputation for truthfulness, Vanessa had never told Angelica the facts of her parentage. In addition, Angelica was so sexually ignorant that Julian felt that he must tell his mother to explain human sexuality to her daughter. Vanessa's inept explanation of sex suggests that the woman who, in her youth, had used words such as *cunt* and *fuck* so freely had become, in middle age, afraid of even the language of the body. (Conversely, Virginia was now writing bawdy letters to Ethel and even Quentin.) Later, Angelica talked about her resentment of her mother's reticence and the repression that had presented both her true parents and the Woolfs to her as "asexual if not virginal" (Garnett, *Deceived with Kindness*, 65, 121–23; VBL, 406–7). Whether or not they were asexual, the Woolfs managed that February—although they were in London—to refuse invitations and make "a space & quiet thats rather favourable to private fun" (D, 5:13).

In early 1936, Virginia kept her various friends and correspondents at bay, working desperately but in increasingly poor health, to finish *The Years*. By means of dated chapters, she interwove a fabric of historical and cultural allusions into the saga of the Pargiter family over the course of some fifty years. This large family experiences death, war, the beginnings of the dissolution of the patriarchy, the rise of the feminist movement, the ominous threats of another war, and various class and economic issues. Fearing that her work had swung too far toward the ethical and away from the aesthetic pole, Virginia worked to redress the balance. As she revised, she introduced more Postimpressionist techniques into her novel of fact. Although much of the novel's weight was still carried by dialogue, she broke up speeches by "skipping, & parenthesis" and balancing different points of view (D, 4:291). Virginia constructed a "reverberative structure" of echoes and reechoes of gestures, objects, colors, and sounds, including the wood pigeons "crooning in the tree tops. Take two coos, Taffy; take two coos, Taffy" (Y, 115).[18] And she added details of weather and scene and gathered all the diverse characters and themes into a final chorus.

Julian Bell mailed his "memoir" of Roger from China in March 1936, hoping that the Woolfs would publish it in their Hogarth Letters series. Vanessa immediately read the memoir and passed on the "moving and beautiful" tribute to Helen Anrep (VBL, 408). After about a month, when Vanessa "at last disgorged your Roger," Virginia wrote to Julian that she was "so bemused with headache" she had to send out the memoir to be typed before she could read it. She also registered to Julian her alarm about Angelica's "passion for clothes . . . like a craze for drink" and spoke of her plans to "solicit the muse of biography" (VWL, 6:33). Seven months earlier, she had planned to finish reading all the Fry letters in six months (VWL, 5:428–29). Now she announced that she had "extracted the whole lot" of Fry correspondence, except for Charles Mauron's letters (VWL, 6:33). Typically, however, she overestimated the task behind and underestimated the task ahead in order to spur herself on.

Many of Virginia's metaphors for helplessness—a hare hypnotized in the glare of oncoming headlights (D, 5:64), for example—could describe not only her situation as she tried desperately to finish but also Europe's under the shadow of Hitler (VWL, 6:19). Because "Hitler has broken his word again," it became "odd, how near the guns have got to our private life again." With Hitler marching on the Rhine, Virginia felt like a "doomed mouse, nibbling at my daily page" (D, 5:16, 17). Such images of personal and political helplessness threatened the dissolution of the private self that she had so successfully maintained for twenty years. Virginia continued to be uncertain about and depressed over *The Years*, thinking of it as "feeble twaddle"; noting that "I have never suffered, since The Voyage Out, such acute despair on re-reading, as this time,"; yet at other times pronouncing it "my best book" (D, 5:8, 17). Devastated by Roger's death and unable to finish *The Years*, Virginia recorded an ominous suspicion: "I must very nearly verge on insanity I think" (D, 5:20). With old horrors replenishing themselves, she still managed to send *The Years* to the printer a portion at a time. The Woolfs then retreated to Monk's House, where Virginia could gather her strength to fight off her demons. As Leonard began to read the galleys, Virginia felt undermined by the "tepidity in his verdict so far," and she vowed never to write another novel (D, 5:22).

These excerpts are from Virginia's last four entries before she put away her diary in April 1936. For the first time in more than twenty years, Virginia's depression brought her close to a serious breakdown. She was, in Keats's words, "half in love with easeful death"; she was, in her own words, like the sailor who "thinks how, had the ship sunk, he would have whirled round and round and found rest on the floor of the sea" (TL, 84). Fortunately, she was able to find rest in the country. Walking, playing bowls, resting, and seeing few people other than Leonard, Virginia could moderate her mood swings so that they did not splinter her into "shivering fragments."

After two months in the country, Leonard took Virginia back to Cornwall. There they "crept into the garden of Talland House and in the dusk Virginia peered through the ground-floor windows to see the ghosts of her childhood" (LWA, 2:300). Viewing the now-mythologized center of childhood delights and taking a restful holiday relieved the headaches that had tormented her. But when more proofs of *The Years* arrived, Virginia could not bear even to look at them for more than two months (D, 5:23–24; VWL, 6:26, note).

Although Virginia did rouse herself to pen a few letters, she did not write in her diary at all between 9 April and 11 June. She sent Julian a long letter telling him that his memoir was being typed and urging him not to write a novel because "there ought to be a scrambling together of mediums now. The old are too rigid." She suggested that perhaps Julian would be the one to "explode the old forms and make a new one," although, of course, she herself had already made such an explosion. She let him know that she had tried a "little sketch" of Roger, "but I can't see my way and must do my proofs first" (VW/JB, MFS, 186). Vanessa told Julian that she avoided "asking her many questions [about his memoir] as I know we shall get against each other in the queer way we always do about you" (VBL, 413).

Indeed, Virginia was now in such a desperate state that she confessed to Ethel: "Never trust a letter of mine not to exaggerate thats written after a night lying awake looking at a bottle of chloral and saying no, no, no, you shall not take it." Virginia thought it was "odd" that sleeplessness so frightened her that it invited suicide (VWL, 6:44), but she knew from experience that insomnia could signal—or precipitate—the onset of a manic-depressive episode. Back in London on 11 June, she acknowledged how serious this last episode had been: "At last after 2 months dismal & worse, almost catastrophic illness—never been so near the precipice to my own feeling since 1913—I'm again on top." She luxuriated in "the divine joy of being mistress of my mind again!" (D, 5:24).

Victorious over depression, Virginia's brain was "teeming with books I want to write—Roger, Lytton, [a sequel to] Room of ones Own etc etc—and I can only just manage one wretched 3/4 hour proof correcting [of *The Years*]. Thats my plaint" (VWL, 6:44). The difficulties of finishing *The Years* had precipitated and world politics had contributed to that condition, but unrealistic expectations about *Roger Fry* aggravated it: "My folly was I would do all the Fry papers in between times up till Easter: made 3 stout volumes of extracts" (VWL, 6:44). Whatever those volumes were, they hardly dented the task before her.[19] But they were the last of *Roger Fry* for some time.

Virginia was feeling what she called "new emotions": "Humility" and "impersonal joy" may have felt new, but "literary despair" had threatened her all her life (D, 5:25) and had not been suppressed this time. Unlike Vanessa, Virginia and Leonard found Julian's memoir of Fry artistically and politically flawed. Julian was wordy, disorganized, and so determined to make "Fry a kind of patron saint of the Left" that he ended up contradicting himself (Stansky and Abrahams, *Journey to the Frontier*, 282). On 28 June, Virginia faced a task that might well have caused a relapse, when she chose aesthetic and logical standards over family loyalties. She wrote to Julian to tell him that the Hogarth Press had decided not to publish his memoir as is. She also told him about her "sleeplessness, and the usual old pain." She talked of political pressures, a writers' congress in "defense of culture" in the face of fascism (which she did not attend), Leonard's activism, and her own efforts: "What can I do but Write? Hadn't I better go on writing—even by the light of the last combustion?" (MFS, 189).

Given the turn her work had taken toward ethical commitment, her frequent clippings illustrating the absurdity of masculine militarist values, and Julian's leftist leanings, Virginia could attack the painters for their aesthetic disengagement:

> There they sit, looking at pinks and yellows, and when Europe blazes all they do is to screw their eyes up and complain of a temporary glare in the foreground. Unfortunately, politics get between me and fiction. I feel I must write something when this book is over—something vaguely political; doubtless worthless, certainly useless.

In this revocation of the formalism that she and Roger had championed, she maintained, "I'm all in favour of biography instead of fiction." In between those personal and political comments, Virginia tactfully explained to Julian that he had not "mastered the colloquial style," that his memoir was too long, and that it lacked "personality" (MFS, 188–91).

When the Woolfs rejected Julian's poems, Vanessa had "ruffle[d] like a formidable hen" (D, 4:264). But when they rejected Julian's memoir, Virginia risked ruffling more than feathers. Vanessa wrote to Julian, "Those blessed Woolves won't publish your Roger letter. They're very tiresome" (VBL, 418). Later Vanessa passed on the theory, acquired from David Garnett, that Virginia rejected Julian's memoir because she "can't face any other writer of any real merit" (VBL, 424). Vanessa laughed to Julian about her well-developed "powers of repulsion" in keeping neighbors and guests away (VBL, 417). Indeed, she kept away almost everyone from her childhood and even once fended off George Duckworth by pretending to be her own maid and insisting on the telephone, "I am afraid Mrs Bell is out."[20] She might well have used those powers on Bunny Garnett, who was becoming reinvolved with the Charleston group. He was especially interested in Angelica, who by this time was studying in London to be an actress. But because she took Angelica's sexual innocence for granted, Vanessa hardly noticed his interest in her daughter.

Trying to delay for as long as she could the repercussions of the rejection, Virginia held onto her explanatory letter for a month while she wrote two more letters to Julian. She mailed the three together. In the other two letters, she spoke of gossip and politics: Leonard was "concocting policies with the Labour party; Margery Fry is on the frizzle; and in a moment I shall turn on the wireless and listen to the latest massacres in Spain. Oh my dear Julian, you're lucky to be feasting your eyes on the rice fields." Nevertheless, it was peaceful at Monk's House, with "only the hum of a plane to remind us now and then of the future" (MFS, 192, 194). In the two later letters, Virginia did not mention to Julian the Woolfs' rejection of his memoir of Roger.

Before he received the package of three letters, Julian had heard from Vanessa that the Woolfs were not going to print his memoir. He did not reply to Virginia, but to his mother he responded in terms he might have borrowed from Vanessa: "I do think Virginia is exceedingly tiresome. She wrote me a letter saying she thought my Roger work needed rewriting etc. I really don't believe it. I don't think I could say it any better, anyway not at present. But since they won't have it, there's an end of them."[21] Presumably, he meant an end of them in his affections. Virginia's (and Leonard's) uncompromising publishing standards had now alienated both Vanessa and Julian. Meanwhile, Helen Anrep expected that "surely surely [the Woolfs] must be impressed" by a memoir that would have "delighted" Roger. Hearing otherwise, she too faulted the Woolfs' judgment (HA/VB, n.d.; HA/JB, n.d.).

In June 1936, Virginia had claimed to be mistress of her mind, but her victory over depression was short lived. Between June and October, she wrote few letters and again stopped keeping her diary. Despite country walks and games of bowls (LWD), the summer was "filled with an unending nightmare" (LWA, 2:300). Fighting off the ravages of manic-depression, Virginia remembered this time as "the worst summer in my life, but at the same time the most illuminating" (D, 5:67). Her scrapbooks continued to document the horrors of war, male aggression, and the oppression of women. (For example, she clipped out of the *Sunday Times* of 13 September 1936 the text of Hitler's speech on the different places in society that men and women should occupy [Abel, *Virginia Woolf*, 91].)

By the early fall, Virginia had regrouped her forces. She was determined to regain her health and to deal with what Anne Olivier Bell calls the "incubus of her unfinished proofs" of *The Years* (D, 5:25, note). Virginia made severe cuts in the proofs, which in some instances improved the text. For example, she deleted a speech about the equality of the sexes but left in the example of the happy, egalitarian marriage of Maggie and Renny. Most of her prunings, however, compromised the novel's political implications, as they muted the force of militant ideas, especially that of her earlier attacks on militarism and the doctrine of female chastity. Virginia had intended to explore the effects of childhood sexual trauma on Rose Pargiter, and she had also let Rose's cousins speculate about her lesbianism. But in the final version, even when Rose finds herself unable to tell her cousins about the man who exposed himself to her when she was a child, Virginia substituted the vague "she had lived in many places, felt many passions, and done many things" for the explicit references to lesbianism (Y, 166–67). She also deleted explicit cause-and-effect associations between sexual ignorance and sexual coldness. Thus, even while writing to expose the effects of repression, Virginia showed herself to be in print as repressed as her characters. In the final version of *The Years*, Virginia also deleted a statement that was personally revealing: "Probably people who have been bullied when they are young, find ways of protecting themselves. Is that the origin of art he asked himself: . . . making yourself immune by making an image?" (Radin, *Virginia Woolf's "The Years,"* 95). All her life, Virginia had attempted to immunize herself against separation and loss by immersing herself in her writing, but she took out that bit of information from *The Years*.

Virginia's extensive revisions of the proofs had been "like a long childbirth. Think of that summer, every morning a headache, & forcing myself into that room in my nightgown; & lying down after a page: & always with the certainty of failure" (D, 5:31–32). She wrote in Monk's

House, not in her lodge, so that she might rest frequently. Given rest, isolation, and Leonard's support, Virginia managed to suffer through this depression. Indeed, she made herself finish *The Years* partly so that she could resume her "odd posthumous friendship" with Roger, but "how to make a life of him" still eluded her (VWL, 6:77). She had one imaginative plan for making a life: "The method should be to find out what his qualities were and proceed to illustrate them by events. To be very free with sequence of facts."[22]

On 30 October 1936, Virginia began writing in her diary again, but her proofs of *The Years* remained a debilitating incubus. She was desperate enough to ask a question that went to the core of her being: "Can I still 'write'? That is the question, you see" (D, 5:26). She feared that she had deceived herself into thinking she still could. And so she conceived a solution almost as self-defeating as her 1904 proposition to abandon writing and take up painting. On 3 November, Virginia decided to unburden herself of the incubus of a failed novel. The proofs were so bad that she decided to carry them from her study in the packing room at the back of 52 Tavistock Square "like a dead cat, to L. & tell him to burn them unread. This I did. And a weight fell off my shoulders." Relieved, Virginia walked by Coram's Fields, along Holborn, and through Gray's Inn. She felt "no longer Virginia, the genius, but only a perfectly insignificant yet content—shall I call it spirit? a body? And very tired. Very old. But at [the] same time content to go on these 100 years with Leonard." At peace with her failure, she began to make rational plans to pay for the abortive costs of her destroyed proofs and to begin earning money as of old, by reviewing (D, 5:29).

Leonard, of course, did not burn the proofs but read them. Knowing this to be "a difficult and dangerous task," he faced an excruciating choice: If the book was as bad as Virginia claimed, he would have to sacrifice his integrity (on which Virginia counted) to praise it. Or he could offer his usual forthright judgment and risk precipitating, given her shaky health, a breakdown.

Leonard decided to take his first option: "To Virginia I praised the book more than I should have done if she had been well" (LWA, 2:301). Fortunately, Virginia trusted his judgment: "Miracles will never cease—L. actually likes The Years!" Suddenly he "put down his proof & said he thought it extraordinarily good—as good as any of them." Leonard finished the long novel in tears, perhaps as much for Virginia as for her novel. Nevertheless, he told her, "It is 'a most remarkable book—he *likes* it better than The Waves.' & has not a spark of doubt that it must be published." For

Virginia, "the moment of relief was divine" (D, 5:28, 30). Leonard's approval was a ballast against depression: "I can only cling to L's verdict" (D, 5:31), which included reservations only about the book's length.

Virginia revised again, cutting "in the ruthless and drastic way . . . two enormous chunks" and making further deletions on almost every page (LWA, 2:302). The deleted "chunks" convey Eleanor Pargiter's recognition that she had not acknowledged the force of evil and that much of her charity work merely satisfied a lust for power over the poor. Without these epiphanies and various political theories, *The Years* became less radical but also less explicable. For example, after Virginia cut out all explicit references to women's bodies and passions, the damage done to women by sexual prohibitions is such a muted theme that it might be missed by a reader not already sensitized to this issue.

Virginia was obeying the aesthetic imperative that she and Roger had promulgated against didacticism in art. But she also was succumbing to a conventional prohibition against speaking out about female sexuality. The chorus bringing everyone and every theme together therefore seems to diffuse rather than to focus. The last couple getting out of a taxi is supposed to represent a relationship of equals, an image retrieved from *A Room of One's Own*, as Eleanor watches them and says "There," but only the woman's "tweed travelling suit" suggests the couple's equality. Certainly, Virginia's terminology does not, for he is a "man" and she is a "girl."[23]

When Virginia did not hear from Julian, she dreamed about him. Then later in November, Vanessa had a nightmare that Julian was dead (VBL, 430). As if to assuage her conscience about not publishing his memoir, Virginia wrote to him that she thought *The Years* hardly worth publishing, that she was doing so only on Leonard's advice, and that she liked his work on Roger's Mallarmé translations (VWL, 4:84).

Losing Julian's affections depressed Virginia. Aging depressed her. Sexism depressed her. But she had enough energy to dash off the Memoir Club paper "Am I a Snob?" after visiting Lady Sibyl Colefax's Argyll House and finding her elegant furnishings being readied for auction. After Roger's death in 1934, Virginia had defended him against Ethel Smyth and the snob world (VWL, 5:354). Roger's attacks on English snobbism and his irritation with her for keeping company with aristocrats also belatedly catalyzed this memoir, in which Virginia admits hating "being badly dressed; but I hate buying clothes."[24]

One evening in November 1936, Virginia enjoyed discussing psychoanalysis (and perhaps Melanie Klein) with Adrian and Karin but found depressing a discussion with the analysts Alix and James Strachey two nights later (D, 5:32, 33). She lectured herself, "I will not be low" (D, 5:36). Finally, at the end of November, Virginia could congratulate "that terribly depressed woman, myself" for bringing off the seemingly interminable task of composing *The Years* (D, 5:39).

Virginia confronted the ghost of her own past when a package containing two typescript volumes of her adolescent letters to Violet Dickinson arrived unexpectedly. Thanking Violet for transcribing some 350 letters, Virginia could "barely bring" herself to look at her "childish scribbles" (VWL, 6:89). Had she done so, they would have confirmed Klein's theories about longing and buoyed Virginia's spirits with their testimony to her early genius.

Nonetheless, Virginia's spirits were buoyed or at least distracted by news of the abdication of King Edward VIII. Virginia theorized in the same way that she had described the 1910 shift in human existence: "Things—empires, hierarchies—moralities—will never be the same again" (D, 5:40). Anne Olivier Bell points out that Virginia's "perpetual interest in [such] wholly external and non-threatening matters" makes her diaries accessible to nonliterary readers (D, 5:vii); it also made diary writing restorative. Virginia ended the medically expensive and emotionally draining 1936 by sending off the "stinging nettle" of galleys of *The Years* and noted "the absolute necessity for me of work."

Another necessary condition for keeping herself whole was "always to be after something," so she resolved

> now I am not going to think can I write? I am going to sink into unselfconsciousness & work: at Gibbon first: then a few little articles for America; then { Roger / 3 Guineas } Which of the 2 comes first, how to dovetail, I dont know. . . . At least I feel myself possessed of skill enough to go on with. No emptiness. (D, 5:44)

Simpler, more urgent, and psychologically compensating for the political muting of *The Years*, *Three Guineas* took precedence over *Roger Fry*.

On Christmas Day, the Woolfs had lunch with the Keyneses at Tilton. They spent New Year's Eve at Charleston (LWD). Apparently, they did not talk about Julian's memoir, nor did Julian write about it. While still at Monk's House for the Christmas holiday, Virginia at last heard from him and replied, "to say how sorry I am—indeed angry with myself" for hurting his feelings (MFS, 196). Later, she tried to explain the rejection to Vanessa but succeeded only in turning "Nessa to steel by talking about J.'s essay on Roger—a most curious transformation: as if some tigress lay in a cave, growling" (D, 5:51).

In early 1937, Virginia and Leonard drove to Brighton to visit Elizabeth Robins, who had shared her recollections of Julia Stephen with Virginia in 1928 and had just submitted a book about herself and her brother to the Hogarth Press. She then lived with Dr. Octavia Wilberforce "in one of those rounded houses in Brighton–Montpelier Crescent." Although Virginia described the interior as "a solid, clean, rather unsophisticated house," the Grand Regency exterior of the 1840s Montpelier Crescent was truly imposing. Octavia Wilberforce was a distant relation by marriage not blood, but Virginia felt an "I think mythical, relationship" with her after this visit (D, 5:48–49 and note).

A descendant of the famous abolitionist, herself a pioneering female physician, and the manager of a Sussex farm with its own dairy that also served as a rest home for professional women, Octavia Wilberforce exemplified activism and public commitment. While others accused Virginia of "lyrical emptiness," Octavia gratified her by praising her understanding of people in *A Room of One's Own*. This visit gave Virginia a "flushed & exuberant feeling." They "broached many scraps of memories," including Elizabeth Robins's highly charged memory of hearing Leslie Stephen speak about Julia after her death in 1895 without mentioning her name (D, 5:49–50 and note).

In January 1937, Virginia sank "once more in the happy tumultuous dream: that is to say began 3 Guineas this morning, & cant stop thinking it." She prepared herself for the publication of *The Years*: "I must plate myself against that sinking in mud" (D, 5:52). According to Vanessa, Charles Mauron had encouraged Virginia to write openly about Roger Fry, "regardless of the family, who will he says really thoroughly enjoy being upset and scandalised." Vanessa hoped that her sister would follow Mauron's advice (VBL, 429), but openness was easier for Vanessa to suggest than for Virginia to achieve. At the beginning of 1937, she claimed to be "fingering Fry papers nervously; and feel once more the various horrors and delight of book making. Only how does one square the relatives? How does one euphemise 20 different mistresses? But Roger every day turns out more miraculous" (VWL, 6:104).

Facing the publication of *The Years*, finding Vanessa "in one of her entirely submerged moods," and suddenly worried about Leonard's health,

Virginia could have slipped back into the near despair that she had suffered in 1936. But when Leonard's symptoms were "almost over," the Woolfs drove to Rodmell "in that odd relieved state which seems as much physical as mental; as if one's body could unfurl." Virginia's image is particularly apt for the holistic synthesis of mind and body, left and right brain, that she so desired all her life. It was "as if another space of life had been granted us." In this mood, she claimed to have developed the rhinoceros skin that at the age of fifteen she had so regretted "one has not got!" (PA, 132). Now criticism of *The Years* would feel only like the critic's "tickling a rhinoceros with a feather" (D, 5:55). Her confidence spilled over into a collaborative project in which she was to describe and Vanessa to illustrate a series of "Faces and Voices," to be published by the Hogarth Press.[25]

As the publication of *The Years* approached, however, Virginia's confidence waned. With Leonard out for lunch on 1 March, Virginia found herself "very cold: impotent: & terrified. As if I were exposed on a high ledge in full light. Very lonely." Publication was two weeks away; menopause had unsettled her; and "Nessa has Quentin & dont want me. Very useless." Excluded from Vanessa's emotional circle of children and lovers, Virginia dreaded "a roar of laughter at my expense" and was "powerless to ward it off." She could not "unfurl my mind & apply it calmly & unconsciously to a book," yet she knew "I must go on doing this dance on hot bricks till I die" (D, 5:63).

Publication of *The Years* brought the "dazzle of that head lamp on my poor little rabbits body . . . dazed in the middle of the road." Virginia's temperature began to vacillate, as it had during her breakdown twenty-two years earlier; however, she knew that if she could make metaphors, she was not helpless (LWD; D, 5:64, 65). Being still "fitted out for another 2 books—3 Gs & Roger," Virginia became more rhinoceros than rabbit and determined to "hold myself aloof" (D, 5:65).

Julian Bell left China early, planning to enlist in the International Brigade in Spain. Horrified, Vanessa insisted that he return home and discuss his plans. On the weekend that reviews of *The Years* were anticipated, the Woolfs dined at Charleston, where Virginia found Julian "grown a man" and "on the defensive." Vanessa explained that Julian would go to Spain if he did not get a job in England. Perhaps Leonard felt that Vanessa was pressuring him to hire Julian; at any rate, he was irritated with the "intense 'self centredness' of my family." Privately, Virginia acknowledged "some truth" in Leonard's observation that "Nessa only cares for her children," but she did not allow herself to dwell on that shattering truth (D, 5:68).

Virginia's dread of the "fatal" publication day turned out, given the early reviews of *The Years*, to be unfounded. When later reviews were more judgmental, however, Virginia concluded "that odious rice pudding of a book is what I thought it—a dank failure" (D, 5:66, 67, 75). Headaches again began to torment her (LWD).

In spite of her suspicions and the deletions that had weakened it, *The Years* encapsulates a sad but slightly hopeful saga of English mores for women and men between the Victorian past and the present. More accessible than most of Virginia's fiction, it sold better than any of her other books, so money seemed assured for her "last lap" (D, 5:79, 68). Virginia had sent Louie Mayer to cooking school in Brighton so that she might cook more elegant fare; now Virginia could promise her "cook-general" the kitchen renovations that would simplify entertaining at Monk's House (Mayer, in Noble, *Recollections of Virginia Woolf*, 191–92, 194).

Expecting her last lap to last only another ten years, Virginia became curious about the "mystic survival" in which the aged Janet Case believed. Invited to broadcast on the BBC, Virginia "got my pecker up & read with ease & emotion." Against that image of male potency, Virginia remembered walking alone after the broadcast through the cold London streets and imagining that no one had listened anyway (D, 5:79, 68, 76, 83).

Planning a holiday in France would have made Virginia "wholly content" if Julian Bell, apparently having found no work in England, had not decided to enlist in Spain. To placate Vanessa, Julian agreed not to fight but instead to volunteer as an ambulance driver for Spanish Medical Aid (D, 5:83). At a dinner party at Tavistock Square in early May, Julian bitterly blamed his lack of a profession on Bloomsbury's notions of education. Virginia maintained that she had wanted him to go into law. "Yes, but you didn't insist upon it to my mother, he remarked, rather forcibly" (D, 5:86). Even though Leonard had openly objected to Vanessa's indulgent ways with her children, Virginia had endorsed her sister's child-rearing methods, only to be accused by her nephew of failing him. This no-win exchange must have reminded Virginia of how little influence she had had on Angelica's upbringing.

After spending much of May in France while *The Years* topped the best-seller lists and she was "most intelligently (& highly) praised by Faulkner in America,"[26] Virginia returned to work on *Three Guineas*, pouring into it much of the anger that had fueled her scrapbook collecting and that she had suppressed from *The Years*.[27] On 1 June, she did some private assessing: "Were I another person, I would say to myself, Please write criticism; biography; invent a new form for both; also write some completely unformal fiction: short: & poetry." She planned to "work hard at 3 Gs for a month—June: then begin reading & rereading my Roger notes" (D, 5:91). She seems to have thought that she could review old notes for two months and then "write Roger" (D, 5:91) according to a "new form" she would invent.

By 12 June 1937, she ceased procrastinating and broached the "reading & rereading" of her notes. Confronting the boxes of Fry letters that filled her lodge, Virginia could not imagine "how anyone writes a real life." She seems to have given up on her experimental plans, for she wrote to Janet Case that an imaginary biography "wouldn't so much bother me. But oh, the dates, the quotations!" (VWL, 5:135). Then when Janet died on 15 July 1937, Virginia considered how Janet had been "anchored in some private faith" and recollected "how great a visionary part she has played in my life" as an independent woman of learning (D, 5:103).

On 16 July, the Woolfs entertained a meeting of the Rodmell Labour Party, with Quentin Bell as their dinner guest. On 19 July, they returned to Tavistock Square. Then the next day, news reached them of Julian Bell's death while driving an ambulance in Spain.[28] The Woolfs met Vanessa at her studio (D, 5:104), and then Leonard drove down to Charleston to fetch Quentin to be with his distraught mother (LWD). Virginia found it difficult to record her reactions, even in her diary, to the death of her nephew; the news had been "like a blow on the head: a shrivelling up." The only comfort was "being there with Nessa Duncan, Quentin & Angelica" and losing her feeling of being outside the circle of Vanessa's loves (D, 5:104–5).

Continuing an odd pattern of associations, she compared Julian's death with Roger's: "When Roger died I noticed: & blamed myself: yet it was a great relief I think. Here there was no relief" (D, 5:104). Presumably, whereas Roger's death at sixty-three could be rationalized, Julian's at twenty-nine could not. Virginia had blamed herself for not acknowledging Roger's influence. But she might have felt guilty about telling Julian that he was lucky to be looking at Chinese rice paddies far away from the massacres in Spain. She did feel guilty for not having stood up to Vanessa and insisted on a rigorous education for him. And she felt excruciating guilt for having rejected his memoir of Roger.

Near the end of July, the Woolfs installed Vanessa at Charleston, and themselves at Monk's House. Back in her lodge, with a "huge box of Roger's articles" still to be read, Virginia sensed Julian "stalk[ing] beside me, in many different shapes" (D, 5:107). Every afternoon, Leonard drove Virginia from Monk's House to Charleston to comfort the distraught Vanessa, who wrote to Vita on 16 August, "I cannot ever say how Virginia has helped me. Perhaps some day, not now, you will be able to tell her it's true" (VBL, 439). Later Vanessa remembered lying about "in an unreal state" of grief with Virginia's voice alone keeping "life going" for her sister (VBL, 475). But Vanessa failed to realize how much Virginia needed to hear that message of gratitude. Virginia wrote to her sister on 17 August, "I rather think I'm more nearly attached to you than sisters should be. Why is it I never stop thinking of you?" (VWL, 6:158).

When she did not work, it seemed to Virginia that "nothingness begins." Rather than confront nothingness, she dovetailed her writing assignments, planning to finish *Three Guineas*, "& so to Roger this autumn." She felt now "entirely & for ever my own mistress" (D, 5:105). She and Leonard entertained Angelica and Quentin, playing bowls in the afternoons, trying to "beat up talk" about anything but Julian. But the strain made the two of them "quarrelsome" in this "unbecoming stage of sorrow." While considering writing a well-paid article for *Cosmopolitan*, Virginia could hear the "familiar tune" of her motivation: "Nessa's children; my envy of them, leading to work" and tried to keep herself from thinking about Vanessa (D, 5:106, 107).

In late September, Virginia remarked how "happy & rounded off" her and Leonard's lives would have been if "Julian had not died." Then Vita conveyed the message that Vanessa could not voice: "that I have 'helped' her more than she can say." Virginia was "profoundly" touched to be, for once, the more needed sister (D, 5:111, 112). Because Julian's death was "unnatural," Virginia could not "make it fit in anywhere." Death meant only "emptiness: the sight of Nessa bleeding: how we watch: nothing to be done" (D, 5:113). Despite such despair, Virginia did do two things: She wrote *Three Guineas* to expose the patriarchal values that fueled war and hence had killed Julian, and she tried to comfort Vanessa.[29]

Virginia had long ago learned to mitigate her own grief with work. Now she forced herself "to drive ahead with the book" (D, 5:113) and applied this cure to Vanessa as well, giving her the Helen Anrep, "old Lady Fry," and Marie Mauron correspondences to summarize (VWL, 6:152–53, 159, 163). She hoped that Vanessa would be intrigued by Roger's exchanges with women and that copying the letters would distract her.[30]

Virginia's therapy seems to have partially succeeded. Having other women's lives opened to her that summer, Vanessa at last decided to open her own. She took Angelica aside and told her that she was not Clive Bell's daughter but Duncan Grant's. She seems to have felt "that she owed this gesture to Julian's memory" and needed to "unburden herself of the lie we had all been living under for the last seventeen years." But then neither Vanessa nor Duncan spoke further to Angelica about her parentage. Angelica later angrily concluded that rather than have two fathers, Vanessa's prevarication had denied her any and that her mother had spoiled her to compensate for her own guilt (Garnett, *Deceived with Kindness*, 134–36). But having had the frank talk that Julian's ghost urged on her, Vanessa felt cleansed of that guilt. By September, she could write to Virginia, "I really need not be visited like an invalid now" (VBL, 440), but Virginia continued to comfort and to "shower benefactions regardless" on Vanessa (VB/VW, 17 September 1937). In these low times, Virginia wondered what she would do if Leonard or Vanessa died and how she might use her work to assuage such grief. She consoled herself with Roger's example of trusting life without "cut & dried rules" (D, 5:109–10).

Virginia still claimed to be ambivalent about the Fry biography: "How can one write the truth about friends whose families are alive? And Roger was the most scornful of untruths of any man" (VWL, 6:169), but her energies went into *Three Guineas*, which she "finished" on 12 October 1937. After the strain of *The Years*: "Oh how violently I have been galloping through these mornings! . . . I have deserved this gallop" (D, 5:112). She raced through *Three Guineas* because she did not think of it as art. Instead, it was an outlet for her intense anger at her culture's misogyny; it set out "kinetically" to change society.

Three Guineas protests against patriarchy's militarism, its dismissal of women, and its disregard for feminine values. On a more private level, this answer to the question, How in your opinion are we to prevent war? protested Julian's death. In addition, Virginia exposed the function of dress that announces power: The robes, wigs, ribbons, chains, hoods, and gowns that "advertise the social, professional, or intellectual standing of the wearer . . . your dress fulfills the same function as the tickets in a grocer's shop" (TG, 3, 19, 20). She would give several guineas to help "the daughters of educated men to enter the professions."[31]

First, Virginia suggests that the current college system, with its vested associations with power, be burned to the ground, and she would donate one guinea to rebuild the college. The second guinea would be given for a noncompetitive education in tolerance and justice for both men and women. It would offer the best means to prevent war, ensure against

Monk's House. (Private collection) [35]

Lytton Strachey and Virginia Woolf at Garsington, June 1923. (Courtesy of Sandra Lummis Fine Art) [36]

Virginia and Leonard Woolf at Studland, September 1923. (Private collection) [37]

Richard Kennedy, the packing room at Tavistock Square, from *A Boy at the Hogarth Press*. (Courtesy and permission of Mrs. Olive M. Kennedy) [38]

Angelica Bell with Duncan Grant and Roger Fry, 1926. (Courtesy of the Tate Gallery Archive, Vanessa Bell Photographic Collection, Q48) [39]

Angelica Bell, Clive Bell, Stephen Tomlin, and Lytton Strachey at Charleston, ca. 1926. (Courtesy of the National Portrait Gallery, London) [40]

Leonard and Virginia Woolf in Hyde Park, June 1925. (Private collection) [41]

Duncan Grant's studio at Charleston, n.d. (Courtesy of the Charleston Trust) [42]

Roger Fry, Virginia Woolf at a Memoir Club meeting, with Mary Hutchinson in the background, n.d. (Fry Papers. Courtesy of Annabel Cole, Betty Tabor, and King's College, Cambridge) [44]

The garden room at Charleston, n.d. (Courtesy of the Charleston Trust) [43]

Raphael, *Bridgewater Madonna*, ca. 1508–1509, possibly as early as 1506. (In the collection of the Duke of Sutherland, courtesy of the National Gallery of Scotland) [45]

Julia Stephen at Talland House, with Adrian (covering his face) and Virginia in the background, ca. 1892. (Leslie Stephen's Photograph Album. Courtesy of the Mortimer Rare Book Room, Smith College) [47]

Roger Fry, "translation" of the *Bridgewater Madonna*, 1917. (Collection of the author) [46]

Virginia Woolf, ca. 1927. (Courtesy of the Mortimer Rare Book Room, Smith College) [48]

*L*eonard and Virginia Woolf beside their Singer automobile, Cassis, France, April 1928. (Courtesy of the Tate Gallery Archive, Vanessa Bell Photographic Collection, album 5, p. 5) [49]

*R*ichard Kennedy, conversation at Tavistock Square, with Desmond MacCarthy, Roger Fry, Donald Brace, Richard Kennedy, Leonard Woolf, and Virginia Woolf, 1929, from A *Boy at the Hogarth Press*. (Courtesy and permission of Mrs. Olive M. Kennedy) [50]

A family reunion, Charleston, 1 October 1930. Standing: Henry Duckworth, Duncan Grant, Julian Bell, and Leonard Woolf; sitting: Virginia Woolf, Lady Margaret Duckworth, Clive Bell, and Vanessa Bell. Photograph apparently was taken by George Duckworth. (Courtesy of the Tate Gallery Archive, Vanessa Bell Photographic Collection, A16) [51]

Angelica Bell and Virginia Woolf in a photograph taken by Lettice Ramsay, 1932. (Courtesy of the National Portrait Gallery, London) [52]

Vanessa Bell, caricature of Roger Fry, 1934. (Photograph courtesy of Richard Shone) [53]

Roger Fry with five of his sisters: clockwise from him are Ruth, Agnes, Margery, Joan, and Isabel, ca. 1930. (The eldest sister, Mariabella or Mab, died in 1920.) (Courtesy of the Tate Gallery Archive, Vanessa Bell Photographic Collection, album 5, p. 35) [54]

Man Ray, photograph of Virginia Woolf, November 1934. (Courtesy of the National Portrait Gallery, London) [55]

Vita Sackville-West, ca. 1934. (Courtesy of Nigel Nicolson) [56]

Virginia Woolf at Ray Strachey's Mud House, 1938. (Permission and courtesy of Barbara Strachey Halpern) [57]

Leonard Woolf at Ray Strachey's Mud House, 1938. (Permission and courtesy of Barbara Strachey Halpern) [58]

Roger Fry, sketch of a well-dressed woman, in a letter to Vanessa Bell, 1911. (Courtesy of the Tate Gallery Archive, 8010.5.596) [59]

Gisèle Freund, photograph of Virginia Woolf, 24 June 1939. (Photo Researchers)
[60]

Gisèle Freund, photograph of Virginia Woolf, 24 June 1939. (Photo Researchers) [61]

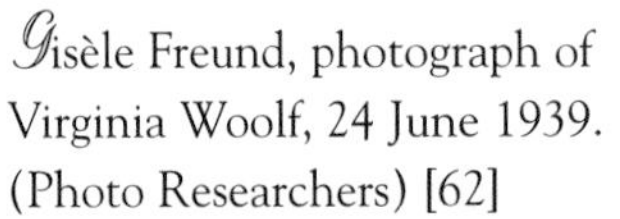

Gisèle Freund, photograph of Virginia Woolf, 24 June 1939. (Photo Researchers) [62]

*G*isèle Freund, photograph of Leonard and Virginia Woolf, 24 June 1939. (Photo Researchers) [63]

*V*anessa Bell, *Leonard Woolf*, 1940. (Courtesy of the National Portrait Gallery, London) [64]

Montpelier Crescent, Brighton, including Octavia Wilberforce's house, number 24; an inverted memory of Brunswick Terrace, Brighton, including Maria Jackson's house, number 5; and the capitals shared by both rows of Grand Regency Brighton houses. (Photographs by Stephen Ford, Brighton. Computer manipulation of images by Herbert Goodman, Department of Art, Louisiana State University) [65]

Stephen Tomlin and Charlette Hewer, sculpted heads of Virginia and Leonard Woolf, Monk's House Garden. (Photographs by the author. Computer manipulation of images by Herbert Goodman, Department of Art, Louisiana State University) [66]

discrimination, and encourage the "daughters of educated men" to maintain their ties with the redefined goals of "poverty, chastity, derision and freedom from unreal loyalties" (TG, 79). So that she would not assume the sort of authority she deplored, Virginia's persona would give the third guinea without conditions (TG, 144). Virginia's scrapbooks provided her with materials for more than forty pages of notes on examples of patronizing and insulting treatment of women and the horrors of fascism and war. If she could not say such things in her novel, she could voice them in this bold polemic. After years of suppressing and tempering her anger, Virginia was liberated from the residue of her "tea table" training by writing *Three Guineas*. Then having let her fury erupt, she could rotate her crops again with her public conscience clear.

Virginia's private conscience may not have been so sanguine. When she visited Adrian's daughter Ann Stephen at Newnham College, Cambridge, Virginia told the young women there "how lucky we were to go to the university and have a proper education" (Noble, *Recollections of Virginia Woolf*, 24). But Vanessa viewed formal education as a waste of time, and so when Angelica went to school, Vanessa insisted that she be allowed to skip requirements, not take exams, or quit when the work was difficult. Angelica later came to believe that Virginia "probably agreed with Leonard about my education" (Garnett, *Deceived with Kindness*, 77, 78, 86, 117, 114). Nevertheless, even when she took on the entire educational and political establishment, Virginia did not take on her sister.

Shifting from her activist writing to the artistic approach that she and Roger had long believed to be essential to civilization, a new scheme for alternating two kinds of work occurred to Virginia on 19 October 1937. She "saw the form of a new novel." Virginia was determined to "keep the idea at the back of my mind for a year or two, while I do Roger &c" (D, 5:114, 115). This reference is the germination of her last novel, *Between the Acts*. For two and a half years, this experiment in fiction subverted Virginia's plans to write an imaginative biography that itself would marry "granite and rainbow."

One day in October, Virginia and Leonard "walked round the [Tavistock] square love making—after 25 years cant bear to be separate." Virginia felt it "an enormous pleasure, being wanted: a wife. And our marriage so complete." With her happiness "radiant still under my skin," Virginia and Leonard walked again "to see Roger's Highgate birthplace," and Leonard suggested relieving themselves of much of the burden of the Hogarth Press by letting young intellectuals (John Lehmann, Christopher

Isherwood, W. H. Auden, and Stephen Spender) run the press as a cooperative (D, 5:115, 116). But then, after Christmas at Monk's House, the Woolfs' happiness was interrupted when Leonard became ill. Thus they ended 1937 back in London, with Leonard in bed waiting for the results of X rays, their fiscal accounts stable but their emotional accounts in jeopardy (D, 5:122, note).

In Virginia's first diary entry of 1938, she vowed to "force" herself to begin "this cursed year." Leonard faced a possible prostate operation; Vanessa was away at Cassis; the final polishing of *Three Guineas* was still ahead of her; almost the entire daunting task of researching and writing *Roger Fry* was still to come; and Hitler was on the march. To keep from worrying about Leonard, Virginia immersed herself in work. Then words failed her when she tried to describe her relief at learning that Leonard would not need an operation (D, 5:125–26). To relieve the stress on them both, the Woolfs at last relinquished part of their responsibility for the press. John Lehmann bought Virginia's share and became a partner with Leonard in the Hogarth Press.[32]

At Christmas, Virginia had sent Vanessa a substantial check. Once again, Vanessa maintained her dignity by protesting,

> However many more stories you get accepted in America no more of such ill gotten gains are to find their way to me. I see that you'll cut up so crusty if I throw this cheque into the fire that for this once I refrain—but its the last time.

Vanessa was glad, however, to hear that Leonard was well (VB/VW, 13 January 1938). On Julian's birthday, Vanessa herself finally acknowledged to Virginia her need for her: "You do know really don't you how much you help me—I cant show it & I feel so stupid & such a wet blanket often but I couldn't get on at all if it weren't for you." She ended the letter, "It's stupid to mind dates isnt it—this one I cant help thinking of. Yrs VB."[33] On the same day, Leonard "gravely" approved the final version of *Three Guineas*, which Virginia now assertively thought had "more practical value" than her novels (D, 5:127).

At last able to leave Charleston for her London studio, Vanessa felt propelled out of artistic detachment as well. She also wrote a third letter to Virginia, offering to help Leonard "in the political world"; for "when things have got to such a pass, I dont see how the least politically minded can keep out of it. I can read and write and even typewrite & have no pride."

She urged Virginia to go to America, as "the whole continent worships you" (VBL, 444).

As the terrible events of the late 1930s unfolded, Virginia worried that "when the tiger, ie Hitler, has digested his dinner [Austria] he will pounce again." She had intended not to let herself "be submerged in Roger quite so completely as in 3 Gs." However, given world events, becoming "submerged" could be a "remedy; an anodyne" (D, 5:132) against both the loss of Julian Bell and the threat of fascism. Consequently, she embarked on her life of Roger with determination, if not delight. After putting aside the preliminary work for almost all of 1936, deferring it while she wrote *Three Guineas*, and returning to it only sporadically in late 1937, with boxes of innumerable unread documents still before her, Virginia set out to compose a "whole and manysided picture" of Roger Fry (VWL, 6:10).

But as soon as Virginia had gotten "two pages of Roger written," she began writing the novel (D, 5:132–33 and note) that became her major distraction from the biography. As *Pointz Hall* (which was retitled *Between the Acts*) increasingly absorbed her creative faculties, often only the left side of her brain was available to work on what now seemed by comparison the "donkey work" of biography (D, 5:133). At first, Roger's memoirs covered his early life in such vivid and succinct prose that Virginia needed merely to paraphrase or quote it. But she then faced a conflict between her desire to "cut loose from facts" and the need for accuracy "contradicting my theories" (D, 5:138). She began to resolve that dilemma by channeling one impulse into "the airy world of Poyntz Hall" and the other into "the solid world of Roger" (D, 5:141).

For the first time in her career, that solid world confronted Virginia Woolf with something that might seem as filthy to the English public as Joyce's prose had long ago seemed to her: Roger's school memoir described in horrified detail the brutality of his headmaster, who levied out punishments by sadistically beating miscreant boys. Because Roger was the "head boy," he had to observe these beatings. Then, observing himself, he found that

> my reaction to all this was morbid. I do not know what complications and repressions lay behind it but their connection with sex was suddenly revealed to me one day when I went back to my room after assisting at an execution by my having an erection, so far as I can remember it was the first I ever had. It was a great surprise to me. I had not even the faintest idea of the function of the

> organ whose behaviour so surprised me for all ideas of sex had been deeply repressed in me in my unremembered past. (FP)

Roger's ironic tone was calculated to delight the Memoir Club, but the general public might be less amused. When Virginia was considering "what to leave out in the School fragment," she was referring to how much to include or to omit in Roger's horrified discovery of a connection between sadism and sex (D, 5:134).

Death continued to stalk the Woolfs' friends. Virginia wrote an obituary tribute for Ottoline Morrell, but the death of Ka Cox Arnold-Forster was more unsettling. Like Madge Vaughan, Ka had become, after marriage, a conventional, public-spirited English matron. Virginia had drifted away not because she found Ka increasingly dull but because Ka had seen her ravaged by apparent madness. After that, Virginia had never been at ease with her. Now she sounded a familiar theme in regretting that she could not feel enough "for the deaths of ones friends!" (D, 5:143).

At about the same time, the Woolfs spent a weekend with Lytton's brother Oliver Strachey and his wife, Ray. An activist in the women's movement, Ray found *Three Guineas* "simply *perfect*" (D, 5:149, note), but her daughter Barbara was not so entranced with Virginia in person. At twenty-six, Barbara saw Virginia, thirty years her senior, as a "mischief maker," an intellectual snob with a "cruel tongue." Barbara spent most of the weekend arguing with Virginia and accusing her of a snobbery every bit as offensive as the social variety. Virginia defended herself by saying that she and her friends believed in the "vital importance of good [frank?] emotional relations." But Virginia seems to have been depressed and wearied by this attack from a young person. Indeed, a photograph that Barbara took at the time (Figure 57) shows a Virginia unlike the self-possessed person in Figure 55, who instead is rather dowdily dressed, her hair once again in an old-fashioned twist, and looking defensive. Leonard, who stayed out of the argument, looks gloomy and tired (Figure 58), Barbara remembers, like "a benevolent bloodhound" (to me, 23 March 1995).

Barbara's argument seems to have confirmed Virginia's sense that she got "no thanks: no enthusiasm from the young for whom I toiled" (D, 5:149). Nevertheless, she was generally happy in the spring of 1938, free to write what she pleased, "afraid of nothing," and proud that the little hand press that she and Leonard had worked on, on the dining room table, had grown into an impressive financial asset. Awaiting the publication of *Three Guineas*, Virginia could "sit calm as a toad in an oak at the centre of the storm . . . & stodge away at Roger" (D, 5:146–47). Perhaps because "its a fact I want to communicate rather than a poem," this was "the mildest childbirth I have ever had." She now saw *The Years* and *Three Guineas* as a single book written over the course of "six years floundering, striving, much agony, some ecstasy." Julian had been a presence behind these exposés of destructive masculine values. But whereas Virginia had funneled her private anger and grief into cathartic public protest, Vanessa kept her grief bottled up inside her (D, 5:148).

Roger's memoirs carried Virginia swiftly through his early years. When she had finished them, however, she faced the enormity of the task that she had set herself. The notes she had taken sporadically since 1935 were, she discovered, hopelessly impressionistic and incomplete. Consequently, during the last half of June 1938, when the Woolfs drove through northern England and Scotland on a postpublication holiday that should have been "a freshener before Roger" (D, 5:149), Virginia took along some of Roger's voluminous correspondence (VWL, 6:233). She wrote to Vanessa with trepidation about the job ahead of her: "Writing lives is the devil. I shiver at the thought of Roger" (VWL, 6:245).

Apparently desperate to keep the Fry materials under control, Virginia adopted the method of working chronologically and doing necessary reading only when it fit the time frame about which she was writing. This method granted her more control and eased the strain of deciphering old notes, but a conversation with Marjorie Strachey raised again the issue of whether a biography could be written at all. Virginia needed to "lay my mind out in pigeon holes" and "doggedly" continue writing "till I meet him myself—1909—& then attempt something more fictitious" (D, 5:155). These plans for a more "fictitious" treatment of the years after she met Roger suggest that Virginia was contemplating a modernist exercise in point of view, but she was "stuck in a bristle of dates" (D, 5:156) when she most wanted to switch "from assiduous truth to wild ideas" (D, 5:159).

Virginia marked the first anniversary of Julian's death but refrained from mentioning it to Vanessa, who still encased her grief in stony silence. Meanwhile, Vanessa received a letter from Adrian showing how little—despite his psychoanalytic training—he had transcended the dynamics of the Stephen nursery. He complained,

> In my beastly profession I have to be careful about my reputation & if it gets about that one's family thinks one a "looney" suffering from "black madness" it is liable to put patients off

> coming to one. I think I already suffer a bit as things are even from Virginia's reputation for madness. People mostly feel that they want to be treated by some one who is pretty sane.[34]

Juxtaposed against that example of insensitivity was Virginia's great sympathy for and generosity to Vanessa. The Woolfs tried to mollify Vanessa and Julian's ghost by publishing *Julian Bell: Essays, Poems and Letters*. Echoing the Woolfs' earlier reservations, John Lehmann wanted to omit Julian's memoir of Fry, but Vanessa insisted on including it, and this time the Woolfs did not object (VB/VW, 1 July 1938).

They made their usual summer retreat to Monk's House, where they saw much of the Charlestonians, occasionally including Bunny Garnett (LWD). But there was no retreat from world events. Hitler's military buildup signaled "the complete ruin not only of civilisation, in Europe, but of our last lap. Quentin conscripted &c" (D, 5:162). Julian's death was still an irrational absurdity: "How could he ever become a ghost[?]" (D, 5:164). Against these troubles, the issue of the efficacy of language seemed decided in the negative, "as if articles mattered!" (D, 5:165).

When Vita patronizingly wired Virginia to say that her son Benedict Nicolson would be glad to "read and annotate" Roger Fry's articles for her, Virginia replied to Ben that she found the articles so important she must read them herself (VWL, 6:264). Although she was grateful for private information from those who knew Roger, Virginia clearly was not prepared to trust theoretical aesthetics to anyone except herself, despite the residue of her sister's and now apparently Vita's notion that she knew nothing about art.

The Woolfs were adding a library and balcony to the upper floor of Monk's House, and these improvements offered a distraction from the threat of war in Europe and the donkey work of making one book out of Roger's "2 lives [:] enough for 6 books: emotion & art." Another distraction was "taking a gallop in fiction" (D, 5:159). On 7 August 1938, Virginia described her biography to Ethel Smyth as a "barren nightmare" (VWL, 6:262), but according to her diary entry for that day, she was looking forward to writing about "the Post I[mpressionists]. & ourselves" (D, 5:160). Virginia still expected that after Roger met "ourselves," she could stop transcribing letters. This "change of method" would introduce her own memories and allow her to be free with the facts. She could finish writing the book by Christmas 1938 and then revise it in 1939 (D, 5:160). In response to negative reactions to *Three Guineas*, especially its pacifism,

Virginia assured herself that "no one can bully me." She could write "a thorough good book—i.e. Roger," but soon her *Roger* seemed "too detailed & flat" (D, 5:163, 165).

When Bob Trevelyan's wife, Bessie, sent to Virginia recollections of Helen Fry, Virginia commented on both "the dread of insanity" and Helen's "curious individuality" (VWL, 6:268). In 1896, when Lady Fry had objected to Roger's courting Helen, he had assured her that there was no insanity or even a tendency toward it in Helen's family. Then he responded to his mother's queries about Helen's family's class and wealth. Virginia copied this whole letter several times, presumably because it revealed so much about Lady Fry's strict and conventional standards and Roger's filial deference as he attempted to justify his love within his mother's value system.[35] Since Helen began to suffer from mental instability less than two years after her marriage and had to be committed in 1910, there is a terrible irony in Lady Fry's inquiry. Despite returning almost obsessively to that letter again and again in her notes, Virginia did not explore that irony in her book.

Under the threat of fascism, Virginia thought 1938 might as well be "3rd Aug 1914" [the beginning of World War I] and asked herself, "What would war mean? Darkness, strain: I suppose conceivably death. And all the horror of friends: & Quentin."[36] Yet somehow she did not "feel that the crisis is real—not so real as Roger in 1910 at Gordon Square, about which I've just been writing" (D, 5:166, 167). Even with Hitler threatening Czechoslovakia, Virginia's posthumous friendship with Roger continued to be an antidote. It now reassured her that art was real rather than irrelevant: "I'm thinking of Roger not of Hitler—how I bless Roger, & wish I could tell him so, for giving me himself to think of—what a help he remains—in this welter of unreality" (D, 5:167). Indeed, the very tediousness of her research validated her devotion (D, 5:173).

Virginia's letter of thanks to Bob Trevelyan for offering more Fry correspondence indicates that she intended to use other voices, including her own, as a counterpoint to Roger's (VWL, 6:271). In seeking those other voices, Virginia found herself "easy and intimate" talking with Morgan Forster about Roger's relationship with Goldie Dickinson (D, 5:168). Presumably their "intimate" talk referred to the homoeroticism of Roger's early relationship with Goldie, which ceased when, as Virginia put it, an older woman "undertook to educate [Roger] in the art of love" (RF, 94). But if Virginia and Morgan broached such a delicate topic, she did not write about it, any more than Forster did in his biography of Dickinson.

Writing the biography continued as Virginia's counterbalance to Hitler (D, 5:170). Although *Three Guineas* was explicitly antifascist, she saw *Roger Fry* as the match she could strike against darkness, for Roger's example proved the efficacy of art and humane values. For her, the process of writing was, as Anne Olivier Bell writes, "a safety-valve for despair" (D, 5:vii) when Czechoslovakia was "sacrificed. War staved off for one year" (D, 5:170). Being "chained to my RF. who is, in himself, magnificent" (VWL, 6:272), Virginia had both the safety valve of work and the reassurance of human grandeur. Even her dreams were affected:

> Dreamt of Julian one night: how he came back: I implored him not to go to Spain. He promised. Then I saw his wounds. Dreamt of Roger last night. How he had not died. I praised [Roger's book on] Cezanne. And told him how I admired his writing. Exactly the old relationship. Perhaps easier to get this in dreams, because one has dreamt away the fact of his death, to which I woke as L. came in. (D, 5:172)

Virginia still planned a major change in the style and tone of her biography after Roger became part of the Bloomsbury Group (D, 5:160, 164). She was reading the letters between Roger and Vanessa from the years 1910 to 1916 and making more discoveries about their sexual relationship. He had sent her a drawing (Figure 59) of an "outrageously seductive lady" to prove that women are "physically attractive *when* they dress well" (RF/VB, 1911?). Furthermore, Vanessa had pronounced "Woolf rather an idiot" to think that Clive really meant to start an affair with Virginia—with a seductive nude drawing of herself (Figure 29) (VB/RF, 7 April 1913).

In 1915, Vanessa had written to Roger several times about the likelihood of Virginia's being certified and confined to an asylum. Caustic remarks about both "Woolves" run throughout Vanessa's letters.[37] Reading and seeing such materials must have dredged up Virginia's "clothes complex," her conviction that the visual arts were aligned with the body and against writing and herself, and her fears that her sister secretly hated her. Virginia wondered how not to discuss Roger's affair with Vanessa (D, 5:172–73). She must also have wondered whether Vanessa's current gratitude really had uprooted her old resentments of and hostility toward Virginia.

As in 1914, a sense of helpless despair over the survival of civilized values, of art, and of themselves sapped the energies of the remaining

Bloomsbury friends. Feeling a "sense of horror & calamity coming nearer & nearer" as she drove through France, Vanessa asked whether "Leonard & Maynard still think war unlikely." She concluded, "As for looking ahead 6 weeks how can we possibly do it nowadays?" (VB/VW, 30 and 26 September and early October 1938).

In late September 1938, Virginia's mental juxtaposition of Hitler and Roger became physical. When she was reading in the basement of the London Library about Roger's first Postimpressionist exhibition, an "old sweeper gently dusting" told her they should "try on our masks." Virginia talked with him about saving books from bombing, and then the man dusted under Virginia's chair. Afterward, she walked past the majestic state buildings and "looked in at the Nat Gall. being warned by a sober loud speaker to get my gas mask as I walked down Pall Mall." Sheltered in the National Gallery from ominous warnings outside, Virginia "looked at Renoir, Cezanne &c: tried to see through Roger's eyes" (D, 5:174–75). Although no bombs fell on that occasion, men were piling sandbags and boarding up shopwindows in preparation for an attack.

A few days later, Prime Minister Neville Chamberlain announced Britain's appeasement of Hitler. Virginia tried to be consoled: "Peace for our life time: why not try to believe it?" (D, 5:179). Apparently, though, no one believed it, as everyone who could leave London was instructed to do so. The Woolfs and John Lehmann made emergency plans for moving the press, and in a torrential rain, with the road packed with evacuees, Virginia and Leonard drove from London to Sussex, Virginia carrying "Roger's letters to [Vanessa]" (VWL, 6:277). On 3 October, Virginia wrote Vanessa a long gossipy letter that included the news that she was "trying to describe the first PIP show [1910] without success" (VWL, 6:279). Vanessa replied from Cassis that her family was so "enraptured" by that letter that she had to read it aloud twice; she bargained: "If you want compliments for your pen please write again" (VBL, 448, 451).

Virginia was facing various biographical dilemmas, still including the question of how to deal with Roger's love affair with Vanessa. She claimed her work so far to be unreadable,

> but Roger himself is so magnificent, I'm so in love with him; and see dimly such a masterpiece that cant be painted, that on I go. Also, reading his books one after another I realize that he's the only great critic that ever lived. For instance Cezanne—a miracle. French Art—another. Why did one only read his little arti-

> cles? So, as you see, on I go; and grumble; and sweat; and sometimes get so hot in the head I roll in the cabbage bed. Do give me some views; how to deal with love so that we're not all blushing. (VWL, 6:285)

These comments add a new element to Virginia's lifelong rivalry with Vanessa, that in some way being in love with Roger (even posthumously) and making his life her artifact (as she had done with Vita in *Orlando*) claimed him as hers, not her sister's. Virginia also claimed, as Vanessa could not, an enduring relationship with a supportive man. Happy in her "divine loneliness" with him and excited over her new room, she told Vanessa that she was about to call Leonard in for tea, with her homemade bread and his honey, when she saw him on a ladder picking apples "where he looked so beautiful my heart stood still with pride that he ever married me" (VWL, 5:286).

After the Woolfs left Sussex for London that fall, a new complication entered Virginia's posthumous relationship with Roger Fry. Helen Anrep dined with Virginia on 19 October and told her that she had seriously overdrawn her bank account. Impulsively, Virginia lent Helen £150 to cover the overdraft. She reflected, "It is right to give her the money, as she's doing without a servant. I mean if one cared for R. this is a way to show it—better than buying clothes or another room." This loan, as Anne Olivier Bell writes, "was to prove a source of disproportionate vexation" to Virginia (D, 5:181 and note).

Having urged Virginia to write about him, Vanessa was curious about "what you [are] saying about Roger." She agreed with Virginia's judgment about *Cézanne* and *French Art* and of Roger as "the only great critic who ever lived—of painting, anyhow"; besides, the other critics were "too boring to read." Then Vanessa's frankness touched Virginia's weak spot: "I hope you wont mind making us all blush . . . the only important thing is to tell the truth for the sake of the younger generation" (VBL, 450). In her spirited reply, Virginia told Vanessa about a visit from Margery Fry and her sister "deaf Agnes," who "disapproves of any truth told about the Frys. But she was a charming apparition . . . walled still in the 19th century."[38] Virginia continued the letter with information about Helen Anrep's financial troubles. She did not tell Vanessa that she had already lent Helen a substantial amount; instead, she asked whether she ought to help (VWL, 6:295–96). Vanessa replied by suggesting that Leonard, who had helped her with money matters, make a budget for the desperate Helen (VB/VW,

27 or 28 October 1938). Virginia's response had been more generous but, in the long run, was much more troubling.

In that long letter of 24 October 1938, Virginia also asked Vanessa for factual information about the murals (Figures 20 and 21) that Roger and Duncan and others had painted at the Borough Polytechnic in 1911. She described her visit to Hampton Court to see Roger's Mantegna restoration. There, "thanks [to] Roger, I'm now seeing in chairs pictures tapestries a remote world of inexplicable significance." She went on to register a Fry-like perspective:

> I think the art of painting is the art for ones old age. I respect it more and more. I adore its severity; its bareness from impurity. All books are now rank with the slimy seaweed of politics; mouldy and mildewed. I wish I could settle to pure fiction; indeed had to rush headlong into a novel [*Pointz Hall*] as a relief; but am now back at Roger and the compromise of biography again. (VWL, 6:294)

Virginia's gifts to Vanessa included this compliment to painting, the memorial volume of Julian's writings (which Vanessa found too painful to read), and a guaranteed £50 a year for two years for Angelica's education, on which Virginia imposed no terms (VB/VW, 6 November 1938).

A 14 November 1938 diary entry clearly associates Virginia's loan to Helen with her biography and both with her debt to Roger: "But my brain racked by biography wont describe life; or analyse the lending emotions. Some were very happy. I mean, paying the debt to Roger. Then the reaction. Irritation" (D, 5:186). Now she felt that she was writing about Roger only to please the important women in his life: "I could go on lending money; earning money in order to lend money; writing books in order to please Nessa & Ha & Helen" (D, 5:186). Rather than let her "irritation about Helen" rankle, however, Virginia—once again her own best therapist—reminded herself that "pessimism can be routed by getting into the flow: creative writing. A passage in Bio[graph]y. came right . . . & I'm floated" (D, 5:189).

Probably Virginia was not referring to her Fry biography but to the essay "The Art of Biography," published on 16 November 1938. It had been eleven years since, in "The New Biography," she had written about marrying granite and rainbow and over a year since she had thought about inventing new forms for criticism and biography (D, 5:91). Laboring in 1938 over writing an actual biography, her hopes for inventing a new form were dimming, for the "novelist is free; the biographer is tied" or "bound by facts" (CE, 4:222–26).

Old doubts dogged Virginia so much that she could "hardly see how to make a book out of [Roger]; for he was in touch with every kind of idea, and emotion and human being; and left a huge glow." Once again, she distrusted the art of the pen, fearing that Roger's glow was "impossible to make visible in ink" (VWL, 6:305). After "rather a debauched Sunday evening at Clive's," Virginia pondered the difference between the gossip and bickering there and the substantive conversation that Roger and Lytton had fostered (D, 5:191). On 19 December, she reminded herself of her accomplishments: She "began, about April 1st, Roger: whom I have brought to the year 1919." She had written essays and a story, and there was talk of her being awarded the Order of Merit (D, 5:193). She quickly refused another honorary doctorate (D, 5:206) because, as *Three Guineas* made clear, accepting such honors meant acceding to an authoritarian system founded on discrimination and militarism.

On 20 December, with chains on their tires (D, 5:197) for the "deep snow," the Woolfs drove to Rodmell for a stay of nearly four weeks at Monk's House. On 22 December, Leonard "unfroze pipes." On Christmas Eve, there was another "heavy snowstorm," and the news reached them that Jack Hills had died (LWD). Virginia remembered him as "my poor old brother in law." What with power failures, frozen pipes, and snow, Vita's gift of paté "practically saved our lives" (VWL, 6:307). In still very cold weather, the Woolfs drove to Tilton for Christmas tea with the Keyneses and to Charleston for Christmas dinner. On 29 December, Leonard finished the second volume of his *After the Deluge* and then, when it warmed up, went out to prune trees.

After traveling to Monk's House, Virginia made no more diary entries, and after Christmas she wrote no more letters in 1938. Instead, huddling by the heater in her lodge, she devoted her writing energies to *Roger Fry*. She intended, first of all, simply to finish the biography, but she also wanted to establish her fix on the "reality gift" and shed the stigma of "lyrical emptiness." If she managed to do so, her "odd posthumous friendship" with Roger Fry could result in a biography that was both artful and truthful. Hoping that it might be a bright match she could light against the oncoming darkness, Virginia settled down to what Leonard later described as an exhaustive, relentless "grind" or "push" to finish *Roger Fry* (interview with Trekkie Ritchie Parsons, 25 July 1993).

12
And I Shall Be Abused

1939–1941

After the post-Christmas thaw, the beginning of 1939 seemed almost June-like, but Virginia spent little time enjoying the weather because for her, "the dominant theme is work: Roger." Her early 1939 work on Roger's life between 1922 and 1924 seems to have been largely preparatory. She had "brought my brain to the state of an old washerwomans flannel over Roger," but she still had "ever so many packets of R.'s letters" yet to read (D, 5:197). After the Woolfs transplanted themselves to London in mid-January, Virginia paused over a disturbing passage in Delacroix's journal that again demonstrated that painters could not praise their art without denigrating hers (D, 5:199).

Virginia was angered by a sniping review of Julian Bell's now published writings that equated his various equivocations and inconsistencies with all Bloomsbury. She also was irritated to hear that Helen Anrep's son Igor had bought a car. What Virginia "owed" Roger seemed to reach an absurd limit when she wondered whether he bought the car "out of my £150?" (D, 5:199). Then Vanessa aggravated her sister by telling her that

Helen would "never repay you" (D, 5:205–6). *Harper's Bazaar*, however, did pay Virginia $600 for the story "Lappin and Lapinova," in which a couple's happy marriage depends on—as the Woolfs's marriage may have in early difficult times—a private set of animal identities. When the husband destroys their fantasy, as neither Woolf did, "that was the end of that marriage" (CSF, 268). Then the *Atlantic Monthly* bought her article on biography, and Virginia decided to regard her recent earnings as overpayments for her loan to Helen (D, 5:200, 201). But her emotional accounting did not balance so positively.

After enforced dormancy, Virginia's creativity blossomed in *Pointz Hall*, in which she invented "poems (in metre)" outlining the history of English literature. The poems were written in "the prose lyric vein, which, as I agree with Roger, I overdo." But rather than overdo, in *Pointz Hall*, Virginia balanced lyricism with satire. She now remembered Roger's reservations about the "Time Passes" section of *To the Lighthouse* as "the best criticism I've had for a long time" and reminded herself that "certainly I owed Roger £150" (D, 5:200). On the eve of her fifty-seventh birthday, Virginia was content to ride on a bus with Leonard to see a show of spring plants and describe to him "my new 'novel.'" Confident about their future, the two Woolfs "planned the books we shd. write, if we could live another 30 years" (D, 5:202).

The Woolfs tried to help a Jewish couple from Austria who had fled to England with only "a bundle of old pictures" that turned out to be fakes (D, 5:201, 207). To Ethel, Virginia complained, "Even I have to write letters, try to be 'kind.'" She concluded that "politics are coming much too close" (VWL, 6:311). After socializing with aristocrats and famous theorists, including Sigmund Freud (now a refugee in London) and Marie Stopes, Virginia felt like an outsider. She spent a "sensible" day walking by the Tower down to the warehouse district beside the Thames. The bitter cold wind reminded her of the "refugees from Barcelona walking 40 miles, one with a baby in a parcel." Then, finding Pepys's church, she felt comforted by the literary associations in even this slippery, "rat haunted, riverine place" (D, 5:203–4). Worried about her loss of fame, Virginia consoled herself: "There are no limits unless one submits" (D, 5:204–5).

Virginia's knowledge of Leonardo da Vinci is demonstrated in a lifetime of questioning, as he did, which art was more true, which more significant, and also in her futile efforts to persuade Vita in 1939 to write a biography of Leonardo for the Hogarth Press ("Four Hidden Letters,"

30–31). Even while she encouraged Vita, Virginia's own biography of an artist threatened to choke off her creativity, whereas *Pointz Hall* released it. By the end of February, Virginia was so upset that she literally needed to "stamp" out her obsessions walking the London streets so that she could "be sane tomorrow" in an insane world in which "Franco was recognised. And Julian killed for this." In early March, Virginia refused another honorary doctorate, insisting that such honors were merely "sops"—illusions of being insiders thrown to outsiders to placate them (Wilberforce, *Autobiography*, 174; D, 5:205, 206).

Virginia's moment of "mild gratification" at finishing a sketch of Roger was undermined by doubts: "Why am I so impulsive? Why am I so old, so ugly so —[?]" Then the worst doubt surfaced: "—& cant write" (D, 5:207–8). In such a mood, after meeting Melanie Klein and having her to dinner, Virginia realized that Klein possessed "some submerged . . . subtlety" that was "like an undertow: menacing" (D, 5:209), perhaps because her theories exposed Virginia's deepest longings.

Having been scolded and prodded by Vanessa—"if I did not start writing my memoirs I should soon be too old"—on 18 April 1939, Virginia Woolf began digging up her nineteenth-century recollections in an insightful memoir entitled "A Sketch of the Past." Even though she was "sick of writing Roger's life" (MB, 64), Roger's own memoir, focusing on a red poppy, surely inspired Virginia's beginning with her "first memory . . . of red and purple flowers on a black ground—my mother's dress."[1] And Roger's memoir of the overtly sadistic and covertly sexual beatings that he had had to witness at school may finally have released Virginia to unearth the memory of Gerald Duckworth's lifting the little Virginia onto a hall shelf and exploring her private parts (MB, 69). Last, Melanie Klein's theories seem to have informed Virginia's treatment of rapture and loss in "A Sketch of the Past."

Writing "A Sketch of the Past" may have been a "holiday from Roger" (MB, 75), but it also was a refinement of the same process she used in *Roger Fry*. Realizing that writing *To the Lighthouse* had been a kind of covert self-analysis (MB, 81), Virginia now made her analysis overt. She wove her ambivalence about writing a biography into the fabric of *Roger Fry* and similarly incorporated the date of composition (a "platform" [MB, 75]) and her present circumstances into the narrative of her past. In commenting on the art of memoir writing, Virginia made her "Sketch" into a "meta-memoir" (Broughton, "Virginia Woolf's Meta-Memoir"). But by uncovering "a pattern hid behind the cotton wool" (MB, 73) in her life, she worried that there might be a similar pattern in Roger's life that was not accessible to her.

At Monk's House from 6 to 24 April, Virginia was sick briefly. As usual, she was reading widely, partly to inspire the pageant scenes in her novel, partly to prepare herself for writing a book on English literature, and partly to stave off the "refrigeration of old age." Bravely she vowed "to flout all preconceived theories." Performing her own self-analysis, Virginia found that a noncerebral activity like cooking could be an antidote to the depression that dogged her (D, 5:214, 215).

At the beginning of the year, Virginia had been writing about Fry's life in the early 1920s, but then she circled back. By 13 April, she had, thanks to Roger's memoirs, finished "my first 40 pages," and by 26 April she had "done" 100 pages. Yet two days later, back at Tavistock Square, she was working on Roger's life before the twentieth century. She still planned somehow to "be free in August" (D, 5:214, 215, 216). But such an unrealistic expectation, like others she had used to spur herself, only created a sense of impossibility and futility.[2] Still, she refused Ben Nicolson's assistance, suggesting instead that he take on an independent project of editing Roger's essays for publication (VWL, 6:330).

Virginia increasingly used metaphors of oppression to describe writing *Roger Fry*: "What a grind it is; & I suppose of little interest except to six or seven people. And I shall be abused" (D, 5:216). Nevertheless, "like a worm under a stone [I] try to lift the weight of Roger. The stone is then firmly stamped down" (VWL, 6:333). Although in January the Woolfs had planned thirty years more of writing books, in April they thought about death and wondered who would most mind losing the other (D, 5:216).

Because the adjoining houses were being torn down, the noise from Southampton Row traffic along Woburn Place[3] intruded into the peaceful Tavistock Square, and the Woolfs began negotiations to buy the lease to a house in stately Mecklenburgh Square, across Woburn Place, beyond Brunswick Square, and next to Coram's Field, where the Foundling Hospital had been. Virginia longed for 37 Mecklenburgh Square, an "oh so quiet house, where I could sleep anywhere" (D, 5:218).

When the painter Mark Gertler dined with the Woolfs at 52 Tavistock, he opened old wounds by comparing literature's supposed vulgarity with the "integrity of painting." Recalling her mother's death, Virginia then mused whether only a painter like Cézanne could convey the integrity of Julia Stephen's personality (MB, 85). That thought prompted her to send Vanessa for her sixtieth birthday a check to spend "*simply and solely on Models*" (VWL, 6:333). Vanessa first flippantly resisted the money by advising Virginia to "think of Helen. She'll want another 100 or 2 before long I feel sure" (VBL, 454).

Then Vanessa offered the sort of compliment that Virginia always craved. Virginia had sent a rewritten story to Vanessa in May 1939 (CSF, 310, note). In the revised version (not published in Virginia's lifetime), a searchlight looking for enemy planes reminds a woman of a story about her great-grandfather and his telescope. This intertwining of past and present struck Vanessa as "lovely, only too full of suggestions for pictures almost." Vanessa wanted to complement the story with an analogous painting—although not a mere illustration—but she feared that she "should come a cropper by comparison" with Virginia's writing (VBL, 454). Overlooking Vanessa's self-demeaning comment, Virginia ecstatically responded that only Vanessa's praise could fill her with such joy. Vanessa was again the lost mother figure, "the source of all joy and succulence" from whom Virginia wanted love and so offered money for models and requested a painting. Vanessa's scolding for "your wicked custom of sending me checks" did not deter Virginia's generosity (VW/VB, 5 June 1939; MFS, 200, 201; VB/VW, ? ? 1939).

Vanessa now had cause to regret her former laissez-faire attitude toward Bunny Garnett's lechery. This man—who had been Duncan's lover, who had almost raped Angelica's doctor after Angelica's birth in 1918, and who had then vowed to marry Angelica someday—had now begun an affair with her in London, even while his own wife was dying in the country. Without models of emotional fidelity, Angelica thought that it was "perfectly natural for him to make love to me while at home he had a wife and children."[4] When Angelica herself became sick, with a kidney infection, Bunny was the one to check her into a nursing home. Vanessa was so outraged she was unable to "talk calmly and say what I meant," so she wrote to Bunny, telling him that Angelica must recuperate at Charleston. She wrote "as truthfully as I can" but did not identify the source of her distress, which surely was the affair itself as much as Bunny's assuming a position of authority (VBL, 457–59). Virginia knew "all the old anxieties [were] rampant" (D, 5:220) for Vanessa, but she thought their cause was Angelica's illness.

When the Woolfs traveled to Brittany and Normandy in June 1939, Virginia let Ben Nicolson use her studio and go through Roger Fry's papers with an eye to reprinting his essays. The Woolfs returned to London to find that Ben had kept Virginia's key and "spread my studio with MSS all laid on the floor" (VWL, 6:342 and note).[5] They also found that they had been able to get the Mecklenburgh Square house and so had to find a renter for 52 Tavistock Square (D, 5:219–20). Alternating writing about Stella Duckworth in her memoir and about the Postimpressionist exhibitions in Fry's biography, Virginia sought ways to "toss & lighten" the oppressive biography and "drive thro' to the end" (D, 5:220, 222). Trying to assure herself that the Bloomsbury Group "indented" themselves on the world, Virginia worried that her confidence might be only an illusion (D, 5:222).

A visit from Victoria Ocampo one June day turned out to be a pretense for introducing Virginia to another photographer, Gisèle Freund, who brought along her portfolio of pictures of writers and asked to photograph Virginia. Although Virginia earlier had twice refused Freund's overtures, she was reluctant to be rude in her own house. Feeling trapped, she agreed to sit the next day (VWL, 6:342–43). For the two-hour sitting, Virginia and Freund chose several outfits for Virginia. First (Figure 60) she wore a plaid blouse with a dark jacket and skirt and smiled at the camera from a sofa covered, like the curtains, with different versions of Duncan Grant's "Grapes" pattern. However much Virginia claimed that the spaniel Sally loved only Leonard, the dog sat adoringly at Virginia's feet for that session. Virginia wore an eyelet blouse for photographs that show two different sides of her personality: In one (Figure 61), she looks right into the camera; in the other (Figure 62), taken in front of the Bell and Grant panel of a vase of flowers, she stares away from the camera, as her mother had done when photographed. In Figure 62, Virginia both adopts her mother's favorite pose in profile and relies on her own emblem, a book, to distinguish herself, as she did in Figure 9, about forty-seven years earlier. In several other pictures taken that day, she wore a dark dress with a frill at the neck, including one with Leonard and Sally (Figure 63) in front of a Bell and Grant panel.

These superb photographs reveal a sometimes severe, sometimes wistful, sometimes playful Virginia. And they show why Leonard would call her the most beautiful of women. Whatever irritation she felt, clearly Virginia cooperated with Freund. Two months later, however, after hearing that Freund planned to show the pictures—even though Virginia had "made it a condition she shouldn't"—Virginia now saw herself a victim of gate crashing and Freund a "devil woman." If she had indeed imposed a condition against showing the pictures, Virginia had a right to be angry. Both in being a willing model and in posing as a victim of a "devil woman" and of "treacherous vermin" (VWL, 6:351), however, Virginia once again showed herself adopting different poses for different audiences.[6]

When Leonard's mother fell and broke two ribs, Virginia concluded that Mrs. Woolf was one of those old women who "have the immortality of the vampire." But when Mrs. Woolf died on 2 July at the age of eighty-nine,[7] Virginia found "regret for that spirited old lady, whom it was such a bore to visit." She channeled her regret into a private cathartic tribute to the woman whom she and Leonard had not included among their wedding guests some twenty-seven years earlier (D, 5:222–25).

Leonard hypothesized about the barbarity not only of fascism but also of capitalist excess and communist repression. His publisher, however, thought that these associations in *Barbarians at the Gate* were so controversial that he delayed its publication (D, 5:220 and note). Leonard's subsequent gloom provoked a "grim thought" in Virginia: "Wh. of these rooms shall I die in?" But she fended off her depression by remembering the words from Spinoza read at Roger's funeral: "A free man thinks of death least of all things; and his wisdom is a meditation not of death but of life" (D, 5:226 and note).

Virginia's regret over her debt to Roger and her desire to prove her "reality gift" continued to trouble her, leaving her angry about the drudgery of writing a biography. Writing "the new Omega" chapter of 1913, Virginia could only hope that she "might still pull [Roger's life] through" (D, 5:225). Amid preparations to move their London quarters, and despite the threat of war, the Woolfs retreated from a busy London social life to Rodmell on 25 July. At Monk's House, a quarrel about a new greenhouse and the need to take out one of Leonard's fruit trees "because of me" soon made Virginia feel "headache; guilt; remorse." Later, "L.'s adroitness in fathering the guilt on me" depressed Virginia even more. Nevertheless, she was "so happy in our reconciliation. 'Do you ever think me beautiful now?' 'The most beautiful of women'" (D, 5:227–28).

In writing biography, details, "old bits of bones," and even permissions throttled experimentation. And Virginia's unsystematic process of composition further hampered her progress. Some three weeks after writing about the Omega Workshop, Virginia needed to "copy out Roger on J.A.S.," or John Addington Symonds, whom Roger had known in Venice in 1891. She also needed to "tackle" Symonds's youngest and last surviving daughter, Katherine Furse.[8] In an 1891 letter to Goldie Dickinson, Roger had written: "Symonds has been awfully good to me—taking me out in his gondola and so on. He is the most pornographic person I ever met but not the least nasty. He also delights greatly in discussing paederastia which is almost a special subject with him and he has become most confidential to me over certain passages of his life."[9]

After pondering such salacious topics, Virginia decided to ask Katherine Furse for permission to quote at least some of Roger's letter. Earlier, she had speculated about "the law of Libel" that brooded over the biographer, "with its dark wings and hooked beak" (CE, 2:236). In 1939, she realized that "all books now seem to me surrounded by a circle of invisible censors" (D, 5:229). Katherine replied to her generously: "You are welcome to use anything with regard to J.A.S. in your book about Roger Fry. To my mind, the more said the better." In another letter, Virginia approached the issue more specifically, to which Katherine replied: "You know I leave you absolutely free to use Pornographic or not as you like. . . . Do what seems best from Roger Fry's end." But then she must have had second thoughts. Virginia assured Katherine that the open acknowledgment of homosexuality would not "damage your character—rather the other way." However, she also conceded, "I'll consider pornographic when I re-read" (CS, 424, 423). Apparently, Virginia's own self-censor prevented her from discussing the issue of pederasty. Roger had volunteered to Symonds, "I will also send my Essay on Sex, if you like." Virginia made no reference to pederasty or the presumably lost "Essay on Sex" (KF/VW, 31 July and 26 October 1939; RF/JAS, 2 and 13 November 1891).

The Woolfs drove to Charleston on the last Sunday in July, finding "Bunny there; Angelica moody; conversation however well beaten up." With Hitler's ambitions growing, "rumours of war" were inevitable (D, 5:228). On two Thursdays in August, the Woolfs drove up from Rodmell to London to supervise the wearisome job (made even more onerous because workers were needed for the army [LWL, 244]) of moving first the press and then their household goods from Tavistock Square to the Georgian row house on Mecklenburgh Square.[10]

Back at Monk's House, tired of "the word filing and fitting that my life of Roger means" (MB, 98), Virginia retreated to her memoir and *Pointz Hall*. "All in favour of the wild, the experimental" (D, 5:228), Virginia was including in her novel a pageant, which would offer a quick overview of all English literature. Her plot, such as it was, would include the gentry's reactions to the pageant performed by the villagers. In late July 1939, she "spun off" the heroine's speech for the pageant's mock Restoration drama sequence. In this parody of a reconciliation scene, the heroine hides herself until her true love seems about to kill himself, "like the Duke in the story book!" She then, of course, reveals herself, the two embrace, and they foil a plot to rob her of her inheritance (BA, 138, 141).

Writing this comic bit of mimicry provided a rest "from Roger" (D, 5:228). Whereas earlier she had been upset by what Roger left out of his letters to his mother, now she was stymied by what he put in (VWL, 6:352). His early letters to Vanessa, including the sketch of her breast seen from the side (Figure 18), revealed so much about their passionate intimacy that Virginia was both touched and appalled at the prospect of writing about this and other affairs. Another problem, as the editors of the letters note, was that new Fry materials "kept pouring in" (VWL, 6:356, note).

Hardly "free" of Roger in August, Virginia filled in a small hand-sewn book with pages alternately outlining the last halves of *Pointz Hall* and *Roger Fry*.[11] This alternation suggests how much the two books in tandem expressed Woolf's writerly pendulum swing between rhythms and realities. The *Pointz Hall* pages hint at an ending with the "man & woman of Eden" and queries about what the pageant taught. One comment in the novel section could also have come from the biography section: "Some say art doesn't teach." The biography section (running from the end of World War I to Roger's death) establishes that even as late as August 1939, Virginia's plans for treating the last sixteen years of Roger's life were not definite. Her "possible scheme for last part of Roger" ends, "Ah then to Keep moving & give the sense that Death was an accident." The scheme's vagueness, combined with the goals of speed and finishing by Christmas, put Virginia in another fix. If she treated those last sixteen years as thoroughly as she had the first fifty-two, she might be at work on this drudgery for years to come. A scribbled note in this little book indicates near despair: "I stop. I am not sure. I have got as far as I can: & may indeed change my mind."

There were other causes of despair. In her diary, Virginia describes the threat of fascism and her fears of death and aging. On "this possibly last night of peace," living in the Sussex countryside seemed like "being on a small island" (D, 5:231). Once when she called Leonard to come in to listen to Hitler on the radio, he refused, saying that the irises he was planting would be flowering long after Hitler was gone (LWL, 249, note). The Woolfs were "privately so content. Bliss day after day. So happy cooking dinner, reading, playing bowls. No feeling of patriotism. How to go on, through war?—thats the question" (D, 5:231). But after Hitler invaded Poland, "writing, & re-writing one sentence of Roger" seemed a stay against chaos. More and more anxious to finish, however, Virginia found her biography "the worst of all my life's experiences." She forced herself to write, but as always, force was "the dullest of experiences" (D, 5:233, 234).

Feeling spiritually vulnerable, Virginia wondered "if we had a church? The relief of having some common outside interest or belief. If it were a belief." But that relief was unavailable, for she could not believe in any transcendent power. With Britain at last at war with Germany, Virginia's global despair became similar to her youthful bouts with depression, in that "all creative power is cut off." But once again, she conquered her despair by reminding herself that "any idea is more real than any amount of war misery. . . . And the only contribution one can make" (D, 5:235). Leonard's contribution was a second volume of *After the Deluge*, an attempt to understand the causes of war. Believing "its so little one can do," Virginia offered to write for the *New Statesman and Nation* (VW/Raymond Mortimer, 10 September 1939, MHP). Her "Roger" seemed "hopeless. Yet if one cant write . . . one may as well kill oneself" (D, 5:239).

In another notebook written that fall, dealing with Roger from after the war to the Slade Professorship he received in 1933, Virginia noted, "I do my best to retire from all benevolence, but it is very difficult & I am very weak."[12] The last page of the book contains only four poignant words: "I do my best." Writing *Roger Fry* was a benevolence for Roger, for Vanessa, for Margery, and even (along with the loan of £150) for Helen. Longing to "retire from all benevolence," Virginia's "I am very weak" and "I do my best," more than any of the facetious remarks she made in her letters, show just how emotionally and physically draining the task of writing *Roger Fry* had become.

Virginia's diary kept returning to external threats: "Poland being conquered, & then—we shall be attended to." With Poland "gobbled up," it seemed that "civilization has shrunk. The Amenities are wilting." They had to stint on "paper, sugar, butter, [and were] buying little hoards of matches." Now with a gasoline shortage, Leonard and Virginia could just as well be back at Asheham on bicycles in 1915. Again they began to monitor their income and to work as journalists, for a drastic decline in Virginia's earnings jeopardized her prospects of an independent old age.[13]

Ironically, even though writing Roger's biography could serve as a way to confirm that Virginia was not broken, it was also the chore that threatened to break her. Writing principally served as a benevolence calculated finally, ultimately, to secure her sister's love and gratitude. Vanessa made an exchange, a portrait of Leonard (Figure 64; D, 5:238). But Virginia's portrait of Roger required much more time and trouble. She vowed, "I will have the whole book typed & in Nessa's hands by Xmas—by force" (D, 5:241).

On 24 September 1939, Virginia noted, "Freud is dead, the stop press says. Only these little facts interrupt the monotonous boom of the war." Country living too was monotonous; she would "like to be rubbing my back against London" (D, 5:238). On 6 October, Virginia mentioned that she had succeeded in "copying out again the whole of Roger," but she still had to "hold the Roger fort" (D, 5:240, 241). The "whole of Roger" seems to have been the sections that took him up to 1919. Her method was to circle back over the already completed whole, spreading and filling and then revising again by increments. As Virginia revised her descriptions of the Postimpressionist exhibitions, she depersonalized her narrative and adopted the persona of an authoritative aesthetician. About this time, in his introduction to Roger's *Last Lectures*, Sir Kenneth Clark enlarged the market for her biography by proclaiming, "In so far as taste can be changed by one man, it was changed by Roger Fry." Both Virginia and Clark aimed to show how Fry's "doctrine of detachment can survive in a world of violence" (Fry, *Last Lectures*, ix, v).

The Woolfs spent a week in mid-October back in London at their new Mecklenburgh Square house. Visiting with Beatrice Webb, now over eighty, Virginia found herself once again opposing what Roger had called an ethical rather than an artistic approach to life. In her diary, she wrote that she had been too "harassed & distracted" to have any impression of the visit (D, 5:242). But a fragment of writing that mixes parts of *Between the Acts* with descriptions of "London at War" and the Woolfs' visit with Mrs. Webb shows Virginia reclaiming those impressions: Mrs. Webb looked "like the veins of a leaf when the pulp has been eaten away. One can almost see through her nose. But the old brain still hums." The old woman allotted ten minutes each to Leonard and Virginia and told Virginia she was "thankful for hard V[ictoria]n. training in morality. I said we were moral in fighting that morality" (MHP/A.20). Somehow the Bloomsburian victory over Victorian proprieties seemed less an achievement when in the newspapers, on the radio, and in conversation, "you never escape the war." A dreaded "trap feeling" seemed to be closing around Virginia. When the Woolfs went back to Rodmell, however, "the world rises out of dark squalor into this divine natural peace," which soothed Virginia for a time. By late October, she had decided to summarize the last nine or so years of Roger's life in one chapter called "Transformations" (D, 5:242, 243).

About this time, Virginia made another attempt to marry "granite and rainbow" in an experimental biography. In a typewritten essay entitled "Roger Fry; a series of impressions," she explained that she had tried "to make a portrait of Roger Fry almost as a novelist might make a character in fiction." She insisted that she had "not aimed at accuracy of date or fact" and explained that she and Roger Fry had planned to collaborate by making him "the subject in a biographical experiment." She outlined their ideas:

> He was to talk about himself, and I was to take notes and write a life in which fiction was to be allowed full play—the idea being that it was only by having full liberty to invent and create that a true life could be written. Unhappily the sittings never took place; I have nothing but memory to depend upon; so that the portrait is merely an impressionist sketch written in the hope that it may help a more serious biographer.

Virginia acknowledged something of her identity and reactions to the post-1910 Roger Fry and his early version of the *paragone*: "Literature was affected by the same disease as painting." She showed Roger pointing out how authors still relied on "useless adjectives" and "merely literary phrases," just as nineteenth-century painters had played on the ready-made associations with their subjects.[14]

The lack of chronology in Virginia's "impressions" suggests that she was so weary she seriously considered giving up the role of conscientious biographer and instead writing an impressionistic summary of the last years of Roger's life. Doing so, however, would be admitting defeat, so she plowed on. By 9 November 1939, she was "nearing the end of my trouble with Roger, doing once more, the last pages" (D, 5:245). Despite her compromise with the muse of biography, Virginia had not completely abandoned her craft, although she still feared "R. a failure—& what a grind" (D, 5:248). Her haphazard way of writing, for example, of Roger's death in 1934 and then of his marriage in 1896 (D, 5:245, 250) only made the grind more tedious, the ending more elusive.

The previous summer, Virginia had written a rather imbalanced attack on "reviewing" that had been rejected by the *Atlantic* in September (D, 5:235) but was then published by the Woolfs as a Hogarth Sixpenny pamphlet. Leonard took the remarkable step of printing a disclaiming "Note," which Virginia had first viewed as an attack (VWL, 6:336).

He explained that he agreed with many of her arguments but found "doubtful" her conclusion that literature would be improved if reviewing were abolished (V. Woolf, *Reviewing*, 27). Perhaps Leonard was only trying to disarm the critics by making an obvious point first. He may have thought that being impersonal, his disclaimer did not violate whatever agreement he and Virginia had made in 1915. Virginia, however, felt "stung with my pamphlet," presumably by Leonard. But she was gratified that John Lehmann saw *Reviewing* as part of her "Outsider Campaign." While she was being publicly stung, Leonard's controversial *Barbarians at the Gate*, after a delay, came "out and [was] noticed at length" (D, 5:248), another cause for resentment.

The biography of Roger Fry needed to be tightened and also expanded. Virginia's unsystematic composition had jumped from her treatment of the Omega Workshop and the war to Fry in the mid-1920s, leaving a gap of more than five years in her coverage. Around the beginning of December 1939, as a quick way both to expand and to vitalize, she seems to have dipped back into the correspondence. From this, she created a new penultimate chapter for the years between 1919 and 1925, largely by making a pastiche of Roger's letters to Vanessa from this period.[15] In one of these letters, Virginia found a clarification about reading Freud that Roger appended to a 1919 letter to Vanessa: "Virginia's anal and you're erotic."[16]

Virginia might have granted the appropriateness of describing Vanessa as "erotic." But to find out why she was "anal," she seems to have read Freud for the first time, with genuine curiosity. Virginia discovered associations between the anal and the repressive (Freud, *Civilization*, 67, note). In fact, one way she tried to defy her own repressive nature was to include Roger's reference (in the school memoir) to his first erection. She then asked the Keyneses (at their 1939 Christmas visit) "about Roger. 'Can I mention erection?' I asked." Maynard rather pompously advised against it, but Virginia defied "public school" proprieties by including the reference (D, 5:256).

On Christmas Day, after bicycling with Leonard to Charleston in a fog, Virginia had to admit that she did not have a completed *Roger Fry*, as she had vowed she would, to give to Vanessa (D, 5:252 and note). In addition, needing to cut her allowance for Angelica's drama lessons intensified her sense of failure and guilt (VWL, 6:377). Although Virginia and Leonard were in good health, recurring words like *force*, *grind*, *no letters*, *no echo*, and *abused* suggest that psychologically, Virginia was troubled.

Leonard went skating on the last day of 1940, and in the terrible cold that began the new year, various services were cut off. The temperature was well below zero, the snow hard and white, and the street "like glass." Nonetheless, Virginia's fertile brain was hatching new projects: "All the little cuckoos shoving the old bird—Roger—out of the nest" (D, 5:255). The Woolfs traveled up to London briefly in mid-January, with Virginia carrying sixty pages of *Roger Fry* to be typed, but then she returned to work on the Omega chapter (D, 5:258). Although she implied to Margery that the whole book was being typed, apparently only the first chapters were ready.

Back in Sussex, Virginia analyzed the workings of her mind, offering a sort of "brain graph" of the way she had recovered from *Roger Fry* with a wild sprouting of ideas for a lecture. With manic exuberance, Virginia "syllabled" her "Leaning Tower" speech "in the bath, on walks" until the "fever" passed and she could return to *Roger Fry*. She wondered, "Why this sudden pressure on the brain? Its uncontrollable." She reflected that at Mecklenburgh Square she had made up some of *Pointz Hall*, a "new combination of the raw & the lyrical." Continuing her brain graph, she decided that "2 years of Roger may have filled the cistern" (D, 5:258, 259). Despite the war news, she created a positive and healthy attitude toward her life at this time (D, 5:260).

On 26 January 1940, Virginia was trying to tighten her last summary chapter of Roger's biography, "Transformations." Her "moments of despair . . . have given way as they so often do to ecstasy" (D, 5:260–61, 262). The Woolfs gave up going to London, and Virginia hoped "somehow to meet" Vanessa (VWL, 6:380). But the cold, an out-of-order telephone, and another quarrel kept the sisters apart. It was so cold that Virginia's ink froze. She walked to Lewes over icy stubble and came home "by the short cut; which was painful" (D, 5:262). She had not heard from Charleston.

At the end of January 1940, Virginia decided to "concentrate & sweat out R.'s last Transformations." Then the sisters met on 2 February outside a grocer's in Lewes, "so the frost of isolation is over" (D, 5:262). Throughout her diaries and most explicitly in *Mrs. Dalloway*, Virginia had written about the joys of walking in London. Now trapped by the cold in Rodmell, she found it

> odd how often I think with what is love I suppose of the City: of the walk to the Tower: that is my England; I mean, if a bomb destroyed one of those little alleys with the brass bound curtains & the river smell & the old woman reading I should feel—well, what the patriots feel. (D, 5:263)

On 1 February 1940, Virginia wrote to Ethel about being snowbound at Monk's House: "Never was there such a medieval winter. The electricity broke down. We cooked over the fire, remained unwashed, slept in stockings and mufflers." During this "frozen pause," she confronted a "long last grind at R.F." She disowned her own artistry, calling her biography "only a piece of cabinet making," but she also could brag, "I've learnt a carpenter's trick or two" (VWL, 6:381). By 9 February, she had "got my teeth I think firm into the last Transformations. . . . I cant help thinking I've caught a good deal of that iridescent man in my oh so laborious butterfly net" (D, 5:266). By 11 February, she exclaimed, "The authentic glow of finishing a book is on me." Immediately her thoughts raced back to plans for the "Leaning Tower" lecture she was scheduled to give in March. Even though there was no news from Charleston, Virginia was so absorbed in her lecture plans that for once she felt indifferent (D, 5:266–67).

In mid-February 1940, the Woolfs drove to the city for a brief stay. Meeting Margery Fry, however, who was off to France, made Virginia deplore her own apoliticism. At a dinner party, during talk about civilization, she found "all the gents. against me" when she "flung some rather crazy theories into the air." A shopping excursion on which she bought new outfits and was persuaded against her better judgment to buy a blue striped coat did not help. Virginia "suffered from [her] clothes complex acutely" and concluded, "Of course I looked a shaggy dowdy old woman" (D, 5:267–69).

Back in Sussex, on 23 February, Virginia sent the typescript of *Roger Fry* to Margery Fry, "like a small boy showing up an exercise" (VWL, 6:383). Virginia left in the word *erection* but backed away from her earlier defiance by writing in the margin, "Omit?" Margery predictably wanted the reference to be omitted, but in deleting it and thereby obscuring the sense of Roger's memoir, Virginia showed herself as anal as Roger had claimed she was.[17] Soon, however, feelings of being trapped by convention intensified her sense of failure and vulnerability. Then the psychological became physical as influenza debilitated her and the lecture that had so tempted her imaginings had to be postponed.

As it turned out, Virginia had to fear Margery's reaction less than Leonard's.

> L. gave me a very severe lecture. . . . It was like being pecked by a very hard strong beak. The more he pecked the deeper, as always happens. At last he was almost angry that I'd chosen

> "what seems to me the wrong method. Its merely anal, not history. Austere repression. In fact dull to the outsider. All those dead quotations." (D, 5:271)

(Virginia's use of the word *anal* to paraphrase Leonard suggests how much she, after reading Freud, had taken to heart Roger's description "Virginia's anal.")

Virginia interrupted her detached analysis of Leonard when Vanessa arrived at Monk's House and disagreed with him about *Roger Fry*. Virginia was reassured by Vanessa's, Duncan's, and Margery's approval. But the most gratifying and exhilarating response was a heartfelt letter from Vanessa (D, 5:271–72). In another of those curious associations of Roger and Julian, in a letter dated "Charleston Wednesday [13 March 1940] midnight," Vanessa wrote (without salutation): "Since Julian died I havent been able to think of Roger—now you have brought him back to me. Although I cannot help crying, I cant thank you enough" (VBL, 461). Keeping her reserve, she signed the letter only "VB," but her words were truly generous and grateful.

After all the years of trying to win affection and to earn her sister's approval, Virginia had made a gift prompted by love. By doing so, she had at last touched Vanessa's hidden inner self.[18] As Virginia told Vanessa, "I never wrote a word without thinking of you and Julian and I have so longed to do something that you'd both like" (VWL, 6:385). Virginia defined her achievement: "& Lord to have given Nessa back her Roger, lost since Julian died" (D, 5:272). All the associations, from both Vanessa and Virginia, of Roger and Julian confirm that Virginia's first aim was regaining Vanessa's love and (at a long-submerged level) her lost mother's. Virginia could be satisfied, for she had proved both her art and her reality gift and, it seemed, secured Vanessa's love.

Virginia spent much of March in bed in Leonard's room, making the corrections that Margery requested. As she recovered from bronchitis and influenza, with London increasingly dangerous, Virginia enjoyed the spring in the country. This was her first spring entirely in Sussex since she had been "ill at Asheham—1914—& that had its holiness in spite of the depression." Dreaming of a "poet-prose book," she vowed to "relish the Monday & the Tuesday & dont take on the guilt of selfishness feeling: for in Gods name I've done my share, with pen & talk, for the human race. . . . I owe nobody nothing" (D, 5:276).

Despite influenza and, as she described it to Ethel, "a brain like old macaroni" (VWL, 6:387), Virginia finished her revisions of *Roger Fry* in

April, highly conscious of the journey beyond formalism that both she and Roger had taken. When she wrote to George Bernard Shaw to let him know that she had "lifted" some paragraphs from his letter about Roger and inserted them in her proofs, she was signaling an antiformalist end of those recurring arguments about Shaw. Even though she had written in the little notebook that art does not teach, Virginia assured Shaw, "I should have been a worser woman without Bernard Shaw." She explained her initial coldness to him as a remnant of the past: "I was set against all great men, having been liberally fed on them in my father's house" (CS, 428).

By 1940, Virginia was pondering in earnest the writer's responsibilities to society. The ideas that had possessed her like a fever were ready in April for a lecture to the Workers Educational Association at Brighton. Virginia asserted that "with help from Dr. Freud," younger "leaning tower" (as opposed to ivory tower) writers have done a great deal "to free us from nineteenth-century suppressions," which she equated with repressions. Again she defined writing in opposition to painting, for the writer's subject is not "so simple as a painter's model." Instead, the writer has to deal with the whole of "human life." That life had been apolitical in the past but could hardly be so for a new generation of writers. Insisting on literature's new responsiveness to political realities, Virginia envisioned a "stronger, a more varied literature in the classless and towerless [postwar] society of the future" (CE, 2:178, 162, 179). Thus writing would achieve a range and depth that painting could not match.

In the audience that day was Dr. Octavia Wilberforce, who responded not to the cheering words about literature's relevance to a future classless society but, with her clinical eye, to Virginia's thinness (Wilberforce, *Autobiography*, 161).

After the lecture, the Woolfs went to London for a few brief days of intense socializing. Burrowing back into the past again, Virginia found her perspective on her parents to have been enlarged: "My mother, I was thinking had 2 characters." Now she could "see father from the 2 angles. As a child condemning; as a woman of 58 understanding—I shd say tolerating." The death of her eminent cousin H. A. L. Fisher[19] reminded her how much she shared with Leslie Stephen his outspoken irritation at Fisher's priggishness (D, 5:281). Virginia acknowledged that in herself, not in Vanessa, who felt only anger, "rage alternated with love," for her father. Thanks to Freud, Virginia had recently discovered that "this violently disturbing conflict of love and hate is a common feeling; and is called ambivalence" (MB, 108).

With war now the "back curtain of the mind" (D, 5:279), family life too had its back curtain. Angelica, aged twenty-one, had gone to live with the now widowed Bunny Garnett for two months in a cottage in Yorkshire. On 6 May 1940, Vanessa came to Monk's House to break the "astonishing" news (D, 5:282). She excused her previous silence by maintaining that the affair was kept silent while Bunny's wife was still alive. Vanessa found Virginia not "as much upset as I had expected. She wasn't exactly pleased!" (VBL, 464).

In fact, Virginia was more displeased than she let on. Long ago, she had found Bunny to be an unfortunate drain on Vanessa's physical energies (although she may not have known exactly how much he taxed her emotional reserves). In 1916, both Woolfs had resented Bunny's breaking into Asheham, and Virginia had particularly resented his stealing her *Oxford Book of Poetry*. Now he had taken her dear niece. Despite all her theories about professions for women, Virginia had joked that Angelica would marry the Prince of Wales (VWL, 4:309). Julian had faulted Virginia for not insisting to Vanessa that he be sent to law school. And she faulted herself for not insisting that Angelica's education be rigorous enough to offer her a choice of professions. Her terrible nightmare about Angelica's death (D, 4:124) now seemed to be coming true metaphorically as Angelica sacrificed whatever theatrical prospects she had for a man whom Virginia found loathsome. She could only hope that Angelica "may tire of that rusty surly slow old dog with his amorous ways & his primitive mind." She may have felt guilty; certainly she felt "oddly old: even to me comes the emptiness that Nessa feels, as I can guess." Angelica's affair meant "Julian's death renewed," for now there was "no Angelica to hoard gossip for" (D, 5:282, 283).

As the Allied troops withdrew from Norway, the Woolfs considered what might be the consequences for them of Hitler's plan to invade England and incarcerate Jews and writers such as themselves (D, 5:283). Rather than endure that horror, Leonard and Virginia agreed to commit suicide together, to asphyxiate themselves in the garage. Hitler and Roger continued to be juxtaposed in Virginia's mind: "Though L. says he has petrol in the garage for suicide shd. Hitler win, we go on. . . . So intense are my feelings (about Roger): yet the circumference (the war) seems to make a hoop round them." But two days later, she changed her mind: "No, I dont want the garage to see the end of me." She thereby countered the threat of fascism with work: "Thinking is my fighting" (D, 5:284, 285).

For herself, in the country "alone with L., ones certainly happy." Here she experienced the physical pleasure of feeling "the wind-blown state of ones body in the open air." After its long fermentation, *Pointz Hall* began "bubbling" (D, 5:290). Nevertheless, as she checked the proofs for *Roger Fry*, Virginia was painfully aware that the future might mean that "we [would be] in concentration camps, or taking sleeping draughts." The Woolfs were not being paranoid, only realistic: Heinrich Himmler's list for immediate arrest after the invasion of England included the names "Woolf, Leonard, Schriftsteller" (author) and "Woolf, Virginia, Schriftstellerin" (author) (LWL, 164). Virginia was well aware that "capitulation will mean all Jews to be given up. Concentration camps. So to our Garage. Thats behind correcting Roger, playing bowls." That knowledge was also behind the Woolfs' various survival strategies of canning gooseberries, blackberries, and vegetables and bottling honey. Furthermore, Virginia had the sense that "the writing 'I,' has vanished. No audience. No echo. Thats part of one's death" (D, 5:292–93).

With such horrors lurking below the precipice on which their existence was precariously balanced, Virginia forced herself to plod along at pedestrian tasks. By 10 June 1940, she had sent off the proofs of "my Roger." She tried to assess her accomplishment, admitting that "there are patches of anal; too much quotation; sometimes its cramped & poky" (D, 5:293). She compared her "tethered" repression in *Roger Fry* with Ethel Smyth's autobiographical "free handed and profound revelation," her concealment with Ethel's self-confession (VWL, 6:404).

Carrying "A Sketch of the Past" beyond the memories of Stella written the year before, Virginia wrote appreciatively of Jack Hill's "wholesomely" attempting to overcome her repression and ignorance by teaching her nearly forty years earlier about "sex in the life of the ordinary man" (MB, 104).

Italy declared war on Britain and France, and soon there was more "black news." On the day that Virginia and Leonard accompanied Vita on a visit to the Elizabethan Penshurst Place, Paris fell to the Nazis. When the "French stopped fighting," Virginia wondered, "Whats to become of me?" and put morphine in her pocket (D, 5:294, 297). She seems to have been outwardly calmer than Leonard, who was so distraught over the fall of Paris that friends "seriously thought he might collapse" (LWL, xiii), but both Woolfs kept their terror out of their letters. While she worked in her lodge on *Pointz Hall*, Virginia wondered, "If this is my last lap, oughtn't I to read Shakespeare? . . . oughtn't I to finish off P.H.: oughtn't I to finish something by way of an end?" (D, 5:298). On 27 June 1940, she reassessed the situation:

> Further, the war—our waiting while the knives sharpen for the operation—has taken away the outer wall of security. No echo comes back. I have no surroundings. I have so little sense of a public that I forget about Roger coming or not coming out. . . . We pour to the edge of a precipice . . . [her ellipses] & then? I cant conceive that there will be a 27th June 1941. (D, 5:299)

Roger Fry was published on 25 July. Virginia ruminated:

> What a curious relation is mine with Roger at this moment—I who have given him a kind of shape after his death— Was he like that? I feel very much in his presence at the moment: as if I were intimately connected with him; as if we together had given birth to this vision of him: a child born of us. Yet he had no power to alter it. And yet for some years it will represent him. (D, 5:305)

Here Virginia acknowledged how much writing this biography had been—to borrow a phrase from William Faulkner—a "shape to fill a lack." As with Vanessa, Julia, Vita, and others, she had appropriated Roger's life to her own ends. Aware that a biography is not a life but a necessary abstraction of it, she was pleased with her intimate relation with Roger and also with the "shape" she had made of his life, the child born of her posthumous love affair with him.

After the publication of *Roger Fry*, Virginia expected "my family will accept this book in complete silence. 2 years hard work. But I shall hear eno' from 'the public'. Why dont we praise, or blame, each other?" (D, 5:307). By "my family" she seems to have meant the Stephens, or herself and Vanessa, but in this instance, she had certainly had praise from Vanessa (not to mention blame from Leonard) for *Roger Fry*. But Virginia's remarks seem less related to this biography than to all those years when she had been complimenting, teasing, testing, buying paintings, giving presents, and writing books to gain Vanessa's approval while writing partly to prove how independent she was of her sister's approval or disapproval.

At the evenings' diversions on the lawn behind Monk's House, looking over the peaceful Ouse valley to the downs, Virginia associated losing at bowls with England's submission to an expected German invasion. Although such fears are omnipresent in her diary, in her letters Virginia made light of them. To others, she sometimes seemed exhilarated by the war. On the train from Sussex to London, she once invented a wild fantasy about a "perfectly innocent nun" in their carriage being a "Nazi spy in disguise," much to Leonard's amusement (Lehmann, in Noble, *Recollections of Virginia Woolf*, 53). Thinking of the difference between the public and the private self, she noted that each of her books "accumulates a little of the fictitious V.W. whom I carry like a mask about the world" (D, 5:307). That mask remained the entertaining, effervescent, healthily manic self that invented fantasies and entertained others to cover her nearly despairing, too often depressive self.

Clive Bell wrote that he did not feel as others did a "disconcerting break between the first part [of *Roger Fry*] (which ends in 1910 I take it) and the second." Although he liked her treatment of a "delicate situation," Clive's praise was nonetheless double-edged and sparse.[20] Virginia consoled herself that these negative responses tended mostly to be attacks on Roger rather than on her (VWL, 6:410–11). The third printing of *Roger Fry* was ordered. But Virginia found herself depressed by one direct attack. In a full-page review in *The Listener* (15 August), Arthur Waley attacked Virginia for omitting the intimate details of Fry's life (D, 5:311). His objection revived her fear that she was indeed "anal" and had committed the very error to which she had objected long ago in Forster's biography of Dickinson, but Desmond MacCarthy's extended review in the *Sunday Times* helped allay her worries. She claimed that she respected herself for suppressing herself and instead letting Roger paint himself (VWL, 6:417–18).

As a German invasion appeared more and more likely, England began incarcerating refugees, including the Austrian Jew who the Woolfs had tried to help the year before. Writing on his behalf, Virginia, with the help of Lady Oxford, secured his release (D, 5:302–3 note, 326 note). She had complained that politics were coming too close; now war was. On the day the invasion was anticipated, bombs fell "very close" to Monk's House, permanently destroying the illusion of peaceful security there. Virginia and Leonard lay facedown under a tree with their hands covering their heads. The sound was like "someone sawing in the air just above us." Expecting to be hit, Virginia consoled herself that "we shall be broken together." She thought "of nothingness—flatness" (D, 5:311).

Three days later, more planes swooped down just over Monk's House. Virginia "looked at the plane, like a minnow at a roaring shark. . . . The closest shave so far" (D, 5:312). She had observed part of a complex German maneuver to destroy, Anne Olivier Bell explains, "key R.A.F. Fighter Command airfields in the south-east." But the maneuver failed when British guns damaged or destroyed the German planes. Virginia had observed part of the "most momentous day in the Battle of Britain," forestalling the German invasion.[21] Still there were bomb scares, such as in London when Virginia was in the London Library. Back in Sussex, a German plane flew over and then crashed near Lewes. Virginia mused that "it wd have been a peaceful matter of fact death to be popped off on the terrace playing bowls this very fine cool sunny August evening." Then the "feeling of pressure, danger horror" settled in (D, 5:313).

The war and Roger Fry continued to be played off against each other. Maynard Keynes managed to lower Virginia's spirits a bit by asking something like, having finished the official life, why did she "not write the real life for the Memoir Club?" Such a reaction to her labor and craftsmanship was, she thought, "morose & savage" (D, 5:314). Perhaps still smarting from Waley's accusations (and haunted by "Virginia's anal"), Virginia now tended to apologize for her "experiment in self suppression" (VWL, 6:417, 423).

Given that at one time she had planned to place herself as a character and a voice in the biography (as she had in "Impressions"), suppressing herself affected the book in fundamental ways. So did her inhibitions, although she blamed them on "clouds of Quakers buzzing round me all the time I wrote" (VWL, 6:428). However much a "compromise" this biography was, Virginia had nonetheless followed an artistic, symphonic design, but she had been burdened by the need for accuracy and discreteness. Around Roger she had "sometimes felt overpowered, and so uneasy," but now she could reveal what she had long denied: "Nobody—none of my friends—made such a difference to my life as he did. And yet, writing about him, one had to keep that under."[22]

Meanwhile, Virginia's fear that she would be abused because of her biography of Roger Fry materialized in a most improbable exchange. On 6 August, Ben Nicolson wrote to Virginia an extremely patronizing attack on Roger. Virginia was outraged that Ben said Roger had shut himself off from

> "disagreeable actualities and allowed the spirit of Nazism to grow." Lord, I thought to myself, Roger shut himself out from

> disagreeable actualities did he? Roger who faced insanity, death and every sort of disagreeable—what can Ben mean? Are Ben and I facing actualities because we're listening to bombs dropping on other people?

She compared Ben's privileged existence with the troubles that Roger had faced. Then Virginia quoted Ben Nicolson's high-minded notion of the intellectual's obligation to fight stupidity and untruth and asked: "Who on earth, I thought, did that job more incessantly and successfully than Roger Fry?" Virginia suggested that if anyone could be accused of a failure to "check Nazism," Nicolson's own influential family members were more to blame than Roger Fry. She faulted herself for Ben's obtuseness. In a handwritten postscript, she insisted that she did not blame Ben for the current state of the world (in the way that he had blamed Fry and Bloomsbury) (VWL, 6:413–15).

Ben's reply, dated 19 August 1940, only made Virginia more defensive about Roger, about Bloomsbury, and about class prejudices.[23] Her much revised response may be characterized as a defense of the social consciousness of Bloomsbury, especially herself, Leonard, Maynard, Lytton, Duncan, and Roger, who had "done their very best to make humanity in the mass appreciate what they knew and saw" (VWL, 6:419, note). In particular, Roger "could fill the Queens Hall with two thousand listeners from all classes when he lectured" (VWL, 6:420). Although Virginia ended up mailing a milder revision of this letter, her vehement defense suggests how much her "odd posthumous friendship" had intensified her admiration for and identification with Roger (VWL, 6:421–22). That defense echoes her sense of social responsibility (as in the "Leaning Tower" talk) and of Roger's contribution to society (MFS, 202).

While roaming the countryside south of Rodmell, Virginia observed men "excavating gun emplacements in the bank. They look like little swarms of busy ants, as I walk." Preparations for war were "like the raising of the gallows tree, for an execution now expected in a week or fortnight." Virginia longed to "expand & soar" into the fantastic fun of *Pointz Hall*, but the intense "concentration—a screw"—of writing it sometimes seemed to be only an exhausting part of the general oppression (D, 5:310, 311).

By September 1940, the invasion of "our majestic city" once again seemed inevitable (D, 5:317). Vanessa's London studio was bombed (VBL, 473), as was 52 Tavistock Square. When bombs dropped on Mecklenburgh Square, the Woolfs tried to get to their house but were turned back in an atmosphere of chaos and impending doom. Then a time bomb in Mecklenburgh Square exploded, rendering their new home uninhabitable. Virginia sometimes thought about "violent death" (D, 5:319 and note, 320). But needing courage, she forged ahead writing *Pointz Hall*. Resisting the "sense of invasion" and the "feeling of pressure, danger horror," she made herself write, still believing that thinking was her sole defense (D, 5:319).

Angry that "Nessa has told those d——d Anreps they can come to the [Rodmell] cottage for a fortnight" and helpless before the threat of invasion, Virginia turned back to St. Ives in her memoir. She and Vanessa once again became "cold & distant, after our wrangle" over "Helen & the 2 oafs" (D, 5:323, 324). Helen Anrep seems to have brought out the best and worst of Virginia's personality: She appreciated Helen more after reading Roger's letters and lending her the £150 (still unrepaid), but she also treated her dismissively and even nastily, characterizing her children as oafs and her as having a "little artists back kitchen smell."[24] Virginia could dismiss Helen, but she could not lose Vanessa. Even with reduced finances, she settled their quarrel by buying a statue and giving Vanessa a refrigerator (D, 5:326).

Also in September 1940, the Woolfs for the first time in twenty-three years severed "their physical bond with the press." They put their large treadle press in storage in Rodmell, and John Lehmann had to move the Hogarth's editorial and staff operations north of London to Letchworth (Willis, *Leonard and Virginia Woolf as Publishers*, 3).

With bombs dropping nearby, Virginia showed courage and perseverance: "Said to L.: I dont want to die yet," but she made herself imagine what death by bombing would be like (D, 6:326–27). She continued to work on *Pointz Hall* and "A Sketch of the Past" and to formulate a book on English literature: "I shall thread a necklace through English life & lit." Despite the new defense installations, Virginia still found peace in the downs, pastures, and marsh around the River Ouse. Even though happiness seemed almost "treasonable," Virginia used an image of wholeness ("the Globe rounds again") to convey the completeness of her life. She was terrified "of passive acquiescence" and thus vowed to "intensify." Still living in the country was "all so heavenly free & easy—L. & I alone" (D, 5:328, 329).

Along with the quickening pace of dangers and disasters, however, the aggravations multiplied. Dame Katherine Furse, who had earlier given Virginia permission to quote freely and honestly Roger's words about her father, now wrote to say, "I note you could not resist [using] 'Pornographic' to which I still take exception. I think you wrote that you would not use it."[25] John Pierpont Morgan, son of J. P. Morgan, also took offense at sentences that Virginia had quoted from Roger's memoirs about the senior Morgan and his mistresses (VWL, 6:438). And a woman named Rachel Dyce Sharp complained about the "repulsive" passage about flogging at school that Virginia had partly expurgated but still used. In any case, Virginia considered Sharp to be "whirling raving mad" (D, 5:334 and note; VWL, 6:458 and note).

On 20 October, the Woolfs went to London to survey the damage in Mecklenburgh and Tavistock Squares. At Tavistock, the huge panels that Vanessa and Duncan had painted when the Woolfs moved in were "suspended over the rubble" (VWL, 6:449), exposed to the gawking eyes of Londoners. Virginia could "just see a piece of my studio wall standing: otherwise rubble where I wrote so many books. Open air where we sat so many nights, gave so many parties." From the rubble at Mecklenburgh Square, she salvaged her diaries, plates made by Vanessa, and glasses by Duncan. Virginia did not despair and even felt (or pretended to feel) relief at losing possessions. It was said that you could hear a bomb whistle before it hit you; on 23 October, playing bowls at Monk's House, Virginia heard the whistle of a bomb, but it exploded "over the field path" (D, 5:331–32, 333).

On 7 November 1940, Virginia had the immense pleasure of saying no to Morgan Forster's belated suggestion that he nominate her to serve on the London Library Committee: "This was a nice little finish to a meeting with EMF years ago in the L. L. He sniffed about women on Cttee. One of these days I'll refuse I said silently. And now I have" (D, 5:337). When a "bomb burst the Banks [of the Ouse]," rather than be depressed, Virginia claimed to be elated: "We are so lovely—all sea, right up to the gate" (VWL, 6:446). She took pleasure in walking with Leonard "in top boots & trousers through the flood." The hilly town of Lewes "looks like a harbour—like a French town spreading its skirts round a bay" (D, 5:338–39).

Although reading her "Leaning Tower" essay in print was rather like opening a "raw wound," Virginia determined to be a "mental specialist now. I will enjoy every single day." The rhythm in the last chapter of *Pointz Hall* "became so obsessive that I heard it, perhaps used it, in every sentence I spoke" (D, 5:338, 339). She "rotated her crops" between *Pointz Hall* and either "A Sketch of the Past" (now recollecting Leslie Stephen) or essays entitled "Anon" and "the Reader" for the book on English literature.

Virginia could be proud of her biographical professionalism: "Have I not conveyed Roger from one end of life to the other?" (MB, 136). She even began to feel "a little triumphant." Even though it had been "written at intervals when the pressure was at its highest, during the drudgery of Roger," *Pointz Hall* embodied "an interesting attempt in a new method. I think its more quintessential than the others." Her literature book, variously entitled *Reading at Random* or *Turning the Page*, was now serving as her "daily drugery" or "fact book," so that she could "brew some moments of high pressure. I think of taking my mountain top—that persistent vision—as a starting point."[26] Virginia claimed that "it wont matter" if her imagination could not brew a new concoction from the mountaintop image, but clearly a failure of her imagination would be devastating (D, 5:340–41).

In her memoir, Virginia pointed out that although describing scenes was her "natural way of marking the past," her relationship with Vanessa had been "too deep for 'scenes.'"[27] As Virginia half-admitted, her relationship with her sister was derived from that longed-for oneness she had sought to restore all her life. After Angelica appeared with Bunny at a Memoir Club meeting, producing "a sense of strain" (D, 5:314), Virginia made peace by writing to Angelica in October 1940. On 9 November, the Woolfs invited Vanessa to lunch (D, 5:338), and on 15 November, they had lunch and tea at Charleston (LWD).

On 22 November 1940, Virginia and Leonard had lunch with Angelica and Bunny in the farmhouse they had rented that fall near Charleston. Virginia found Bunny "surly from the start," and Leonard agreed. "A.'s position, with B. as her mentor, struck us both as almost grotesque—a distortion: a dream; for how can she endure Bottom. And when will she wake?" (D, 5:341). Leonard was recording air raids everyday (LWD). The United States seemed poised to hand over England to Hitler without intervening, and a cloud of national doom hung over "our little scene." Four days spent moving their salvaged London possessions to quarters in Rodmell so wearied Virginia that she wished "Hitler had obliterated all our books tables carpets & pictures" (D, 5:343).

At the end of November 1940, beset by demands from all sides, Virginia felt a seismic shudder that momentarily wrenched her psyche and allowed old demons to resurface. She saw "vampires. Leeches" after her: "Anyone with 500 a year & education, is at once sucked by the leeches. Put L. & me into Rodmell pool & we are sucked—sucked—sucked." Parasites visited her: "Leech Octavia asks to come. B[unny]. & A[ngelica]. tomorrow" (D, 5:342). But by the next day, when Virginia and Leonard entertained the couple for lunch and tea (LWD), Virginia was no longer seeing leeches. Her internal tectonic plate had meshed again, but nonetheless, a fault line had been marked. Seventeen days later, it split open again and exposed a vision of John Lehmann as "that large, rather pretentious livid bellied shark." Privately, Virginia fulminated over Lehmann's interest in the press: "Must I spend my last years feeding his double row of teeth?" (D, 5:343).

In December, Octavia Wilberforce visited when Monk's House was at its least tidy, with "books and wine bottles and cases in process of being unpacked" from their London quarters. Concerned that both Woolfs looked like "waifs," Octavia brought them Jersey milk and Devonshire cream, for which Virginia insisted they should give her apples or books (Wilberforce, *Autobiography*, 164, 167; VWL, 6:450, 454).

In her memoir about her parents, Virginia was touched by "how beautiful they were, those old people" in their untroubled, uninvolved, nonintrospective existences (D, 5:345, 347). As she argued, "Roger Fry said that civilisation means awareness; [Leslie Stephen] was uncivilised in his extreme unawareness" (MB, 146). Although she earlier had been "set against" great men, now Virginia was nostalgic. Echoing Roger on the subject of genius, she ended her memoir, "I cannot remember ever to have felt greatness since I was a child."[28]

An unexamined life nurtured the confidence behind old-fashioned "greatness," and in 1940 this was an option available only to the insulated or dull witted. Virginia's introspection turned on her country, person, and art. On 23 December, when Octavia brought cream to the Woolfs, she found Virginia "inside with hands worse than icicles." Virginia said that she felt "swept away" by her father's love letters to her mother. She had gotten so involved in her memories that "'poor Leonard is tired out by my interest in my family and all it brings back'" (Wilberforce, *Autobiography*, 166). On Christmas Eve, after having lunch with Helen Anrep in a Sussex farmhouse, Virginia mused how "England consoles & warms one, in these deep hollows, where the past stands almost stagnant" (D, 5:346). Christmas morning brought the extravagant gift of two pounds of butter from Vita, which Virginia and Leonard, "economising with a duck" rather than a turkey, enjoyed as a rare delicacy (VWL, 6:454). Apparently, they heard nothing from the Charlestonians.

Toward the year's end, Virginia noted that she detested the "hardness of old age—I feel it. I rasp. I'm tart," but she found strength in her "growing detachment from the hierarchy, the patriarchy." She steadied herself with the knowledge that "I am I," which was the "only justification for my writing & living." Because of the severe rationing in Britain, she even found herself making up imaginary meals (D, 5:347). Leonard spent the end of the year walking, writing, marketing, moving furniture, and oiling the now-stored printing press (LWD).

On the last day of the year, on a sheet of paper salvaged from their bombed London house, which Octavia described as "quite clean" but Virginia saw as soiled by "dirt on it—bombs," Virginia told her that *Roger Fry* was "more or less an experiment in self-suppression." She theorized that Roger's "mastery" resulted from never analyzing, as a writer must, "character, but always art." Virginia confessed that she felt little mastery: "I've lost all power over words, cant do a thing with them." She could not even "make my hand cease to tremble" (VWL, 6:456). As the "trap feeling" intensified, Virginia even asked Octavia what sort of writing she should try. Octavia would not presume to plant ideas in Virginia's mind but assured her that "every true genius has to lie fallow for a time while the seed germinates—as you know perfectly well."[29] Indeed, Virginia knew that, but she was afraid that her imagination had lain fallow too long to nurture any new writing.

As the new year began, with the countryside glittering with ice and snow and the war drawing closer, Virginia recalled Walter de la Mare's "look your last on all things lovely" (D, 5:351). The Woolfs had Angelica to tea on 6 January 1941 and to lunch on 17 and 21 January (LWD). Perhaps in an attempt to remind her of the possible career as an actress she was sacrificing to her liaison with Bunny, Virginia had her tell theatrical anecdotes to the Rodmell Women's Institute. When James Joyce—"about a fortnight younger than I am"—died, Virginia remembered Harriet Weaver's approaching the Woolfs about publishing *Ulysses*: "Roger I think sent her." She also remembered reading the "indecent" novel "with spasms of wonder, of discovery, & then again with long lapses of intense boredom" (D, 5:353).

In London on 13 January, Virginia walked through the part of London she had called "my England." She "wandered in the desolate ruins of my old squares: gashed; dismantled; the old red bricks all white powder . . . all that completeness ravished & demolished." Seeing the "globed wholeness" of her England shattered into fragments, Virginia felt her own life also starting to splinter. For once, she "decided to eat gluttonously" (D, 5:353), perhaps in reaction to rationing, perhaps to buffer herself against the destruction of "my England," or perhaps to rebuild her own precarious completeness.

About this time, Leonard began worrying about Virginia's psychological health and conferred with Octavia, who perhaps had an illusory sense—based on their "remote cousinship"—that she understood Virginia.[30] Octavia, whom Leonard described as "large, strong, solid, slow growing, completely reliable, like an English oak" (LWA, 2:427), thereupon began surreptiously observing Virginia (LWA, 2:431). Once she entertained Virginia with the story of the residents from the Montpelier Crescent area of Brighton meeting to organize a bomb watch. When the chairman insisted that ladies should not watch after midnight, Octavia lost her temper and stood up and protested that "at this stage of the world's history we were all in equal danger, and that there should be any sort of sex distinction seemed to me utter nonsense" (Wilberforce, *Autobiography*, 168, 169, 170).

Vanessa, too, maintained that "female eyes are as good as male ones" (VBL, 469). As a campaigner for equality between the sexes, Virginia must have approved her sister's sense of equity and her doctor's "healing the sick by day, and controlling the fires by night" (VWL, 6:479). As one who was herself exempted from fire watching in Rodmell—although Leonard was a regular fire watcher (LWD)—Virginia felt her weaknesses exposed by Octavia's tale of strength and assertiveness.

In a letter to Virginia, Lord David Cecil, nephew of her old friend Nelly Cecil, mentioned that his Oxford lectures had helped distract him from the war. He saw no reason to "contemplate events we cannot alleviate" (LWL, 251, note). But Virginia defensively and irrationally interpreted the note as a "silly sneer at Lytton & Mrs Woolf, withdrawing from life to cultivate their art in quiet," and wrote a blasting retort (D, 5:352 and note). Virginia did, however, sense how much her reaction to the Cecil letter had been a paranoid distortion, perhaps connected to her reaction to war, for when next she saw Octavia, she told her about the exchange, explaining, "I was raving mad for 4 years during the last War."[31]

Virginia had a "damp, perhaps rather strained, visit to Charleston," where she heard that "Adrian has almost died of pneumonia." Vanessa retreated into her monolithic self, "apprehensive, on guard, when I spoke of Angelica's dirt" (D, 5:354). Presumably Virginia did not openly refer to Angelica's affair with Bunny as her "dirt," but she must have implied as much. Or perhaps the schism that opened when she thought of John

Lehmann, Octavia Wilberforce, and Angelica and Bunny as "leeches" momentarily opened again, and she really did refer to Angelica's affair as "dirt." Just as when Virginia had rejected Julian's poetry and memoir, Vanessa wrapped the maternal mantle around herself and her children. Although she herself felt betrayed by and estranged from Angelica and Bunny (VBL, 481), when Virginia criticized, Vanessa closed the valves of her affection. Virginia made various overtures, only to receive minimal replies: "Do you find painting gets slower? Yes. One can do more. And money? Never think of it. And Helen? She does nothing. I like being alone. How can one do nothing?" After this conversation, Monk's House was especially cheerful by contrast (D, 5:354).

When *Harper's Bazaar* rejected a story after "clamouring for" anything from her, Virginia fought the curse of depression by cleaning out her kitchen (VWL, 6:463), swearing that this "trough of despair" would not "engulf" her this time (D, 5:354). But living "without a future" and with "our noses pressed to a closed door" (D, 5:355), Virginia began to feel as oppressed and caged as she had in 1904. By 7 February 1941, Helen Anrep finally repaid £25 of the £150 that Virginia had lent her in October 1938. Even though the repayment was minimal, Virginia liked Helen better for the effort (D, 5:355).

On 11 February, the Woolfs visited in Cambridge with Lytton's sister Pernel Strachey at Newnham College and with Dadie Rylands at King's College, an outing Leonard said that "Virginia seemed to enjoy" (LWA, 2:431). Virginia contrasted the Cambridge atmosphere with that at Letchworth, where, in another of her distorted images, she described the Hogarth Press the next day as operated by "slaves chained to their typewriters" (D, 5:356).

In February, Virginia entertained Elizabeth Bowen and arranged for Vita to give a lecture on Persia (VWL, 6:462 and note). At Virginia's request, Vita brought her books on Elizabethan lives to use as sources for her book designed to explain literature "from our common standpoint, to painters," and to move "from criticism to biography" (Woolf, "'Anon' and 'The Reader,'" 370, 373).

If Virginia was taking tranquilizers to sleep, she only compounded her problems (or tried to) by asking the writer Enid Bagnold, who was having lunch with Virginia and Vita, for amphetamines (Bagnold, in Stape, *Virginia Woolf*, 21). Then, when Vanessa came to tea, Vita apparently brought up the topic of Angelica's affair with Bunny. Explaining to Harold that Bunny had once had an affair with Duncan, Vita joked: "As Duncan is

Angelica's father, it seems to add incest to sodomy. My joke; not theirs, though no doubt they have made it" (Glendinning, *Vita*, 312). Perhaps they recognized the irony, but neither Vanessa nor Virginia could laugh about Angelica's affair.

Virginia teased Ethel about her "far away lover . . . a doctor, a cousin, a Wilberforce, who lives in Brighton and . . . sends me a pot of cream weekly." She asked Ethel, "Did I tell you I'm reading the whole of English literature through? By the time I've reached Shakespeare the bombs will be falling. So I've arranged a very nice last scene: reading Shakespeare, having forgotten my gas mask, I shall fade far away" (VWL, 6:465, 466). Despite the jocular tone, this "nice last scene" conveys both a mood of doom and the need for an aesthetically proper end.

When in a *Sunday Times* review of "The Leaning Tower," Desmond MacCarthy objected to Virginia's identifying with working men, she protested with intense vigor: "I never sat on top of a tower! Compare my wretched little £150 education with yours, with Lytton's, with Leonard's." She assured him, "My tower was a mere toadstool, about six inches high." Despite her protest and the cold that "reduced my hand to such a frozen claw that I type," Virginia assured Desmond that she still read him with warmth (VWL, 6:467–68 and note). The next week, she again apologized for typing because her trembling hand was "like the cramped claw of an aged fowl" (VWL, 6:471).

Despite all these tensions, on 26 February 1941, Virginia finished *Pointz Hall* (later renamed *Between the Acts*) and gave the typescript to Leonard to read. Then she retraced a dialogue that she had overheard from "common little tarts" in a ladies' lavatory at the Sussex Grill in Brighton where Virginia had sat "behind a thin door, p——ing as quietly as I could." Virginia also recorded bits of another conversation that she had overheard in a tea shop. Again the fault lines in her psychic plate opened as Virginia wondered, "Where does the money come to feed these fat white slugs? Brighton a love corner for slugs. The powdered the pampered the mildly improper." Examining her mood, Virginia asked herself: "Shall I ever write again one of those sentences that gives me intense pleasure?" (D, 5:356–57).

While Leonard was reading her typescript, Virginia tried to make use of her ladies' lavatory observation. Called "The Watering Place," the sketch sees a divided world with the lavatory door separating the need to obey nature with the desire to improve it, but it does so in a nastily manic manner. It depicts people as "shells" of real men and women. The sound of the tide coming in and out is obscured by the sound of a toilet flushing or a woman urinating—the "great gush of water from the next compartment" of the lavatory. The tide withdraws, "and there are the fish again, smelling very strong of some queer fishy smell that seems to permeate the whole watering place" (CSF, 291–92).

This bit of writing contrasts sharply with Virginia's just-completed novel. In *Between the Acts*, there is no more plot than in *Jacob's Room*. The landed gentry and their guests at Pointz Hall, a country home, watch a pageant, written and directed by a Miss La Trobe. Virginia's graphic sense of typography prompted her to interpolate the text of the play into her novel. The frame narrative connects the present with England's historical and even prehistorical past. Then the interplay between frame and pageant and nature itself raises various questions about the interplay between experience and imagination. The emphasis on the reader's reactions and on the need for an aesthetic sharing between the author and the reader reflects Virginia's acute sense that her writing must elicit an echo. The entire notion of representation is parodied as the actors turn on the audience with mirrors. Miss La Trobe concludes, "Reality [is] too strong" (BA, 179).

Just as *To the Lighthouse* incorporated aesthetic issues into its text, *Between the Acts* uses what we might term *reader response* issues. Both the presentation and the novel itself are formed out of "orts, scraps and fragments like ourselves," which are part of a whole endangered by airplanes and bombs in the June 1939 setting (BA, 188). Because the "play hung in the sky of the mind," it showed how art serves as a defense against the threat of war (BA, 212). The allusions in the novel thus unite the public and the private.[32] Virginia treated religion and sexual desire more positively here than ever before, and her revised conclusion unites also the aesthetic and the social, the poetic and the sensual, since the primal lovemaking between Isa and Giles may be the new play that Miss La Trobe imagines: "Then the curtain rose. They spoke" (BA, 219). After its many revisions, this ending, as Susan Kenney argues, represents an artistic victory over despair ("Two Endings," 281).

During most of the 1920s, Virginia's art had sought to surpass nature with formal inventiveness. During most of the 1930s, Virginia's art had tried to reflect and affect the world of human experience. Finally, in *Between the Acts* she did both. Her last novel "enclosed the whole" by being both as formally designed as a canvas and as politically relevant as a tract. Virginia Woolf thereby proved that her writing could be both aesthetic and relevant, artistically contained and provocative of more than disinterested contemplation. Despite the exhilaration with which Virginia had begun it, when she finished this novel, the weakness and exhaustion that she had held at bay for so long began to sweep over her, threatening to return her—after twenty-six years—to a state of shivering fragmentation.

Virginia dated as 1 March 1941 a draft of the mountaintop story that she had envisioned the preceding fall. This tale, called "The Symbol," as Susan Dick points out, clearly associates with death the snow, the mountain, and a letter writer's inability to find words. On the same March day, Virginia wrote to Ethel that she felt "that this is the worst stage of the war." She had told Leonard, "We have no future." But whereas he was energized by the fire watching and other civil defenses, she was suspended and trapped. She felt encumbered by chores, village life, and intrusions on her time. But she continued to pay her debt to Roger by asking Octavia Wilberforce to find Helen Anrep an apartment in Brighton, which seemed to Virginia a more appropriate location for Helen than Rodmell. And she asked Vita to find hay for Octavia's cows, which were, she observed, "so thin you could put a safety pin through them. You see, I'm no good at practical affairs: I'm fished out of my element and lie gasping on the ground" (VWL, 6:475).

Thanking Vita for her gift of a firelighter, Virginia instructed her, "You must stop. You cant add anything to fire. You see the poetic fitness of ending there." She imagined Vita's orchard beginning to dapple as one "of the sights I shall see on my death bed" (VWL, 6:476). Against so much awareness of endings (as in a "nice little finish" or a "nice last scene") and death, Virginia seems to have tried spring cleaning, although as she told Ethel, "I'd no notion, having always a servant, of the horror of dirt" (VWL, 6:478). Such passages suggest that she felt overwhelmed by quotidian details and chores. As always, Virginia tried to conquer that feeling through writing, asking Octavia whether she might write a "sketch" of her. Octavia first thought that she meant literally to draw her, but Virginia intended to "sketch" her in words, as she had Roger Fry, for she saw Octavia as a "paintable" subject "that composes well" (VWL, 6:477).

In the fall and winter of 1939/1940, when she had written probably the last fifteen years of *Roger Fry* under great pressure, Virginia had confessed to herself, "I do my best," "I retire from all benevolence," and "I am very weak." Nevertheless, she had grown a rhinoceros skin against air raids, property loss, friends' deaths, a declining income and reputation, and the vagaries of Vanessa's affections. Virginia also had persevered by writing journalism and her brilliant last memoir and novel. But these efforts had also weakened her resilience to "pressure, danger horror" (D, 5:313). Virginia may also have heard that Adrian had been delirious.[33] And she might have feared, as she had so long ago, that madness did, after all, run in the family; if so, her own mental health could slip again as well.

That March, the flowers Leonard had planted rather than listen to Hitler on the radio began to bloom (LWL, 249 and note). On 7 March, Virginia analyzed to Leonard "my London Library complex" and repelled "that sudden terror" over Morgan Forster's antifeminism. The next day, Virginia was annoyed at having to follow Octavia's and Leonard's advice not to accept an invitation to go with Clive to meet the publisher Hamish Hamilton (Wilberforce, *Autobiography*, 174, 175). And she was haunted by memories of the ravaged Piccadilly and Oxford Street.

Even though she took care of one worry only to have another appear, Virginia was determined to "conquer this mood." She vowed:

> No: I intend no introspection. I mark Henry James's sentence: Observe perpetually. Observe the oncome of age. Observe greed. Observe my own despondency. By that means it becomes serviceable. Or so I hope. I insist upon spending this time to the best advantage. I will go down with my colours flying. . . . Occupation is essential. (D, 5:357–58)

Although Virginia's 8 March 1941 entry echoes a lifetime of resolutions to rise above despondency, it is unusually solipsistic. Rather than work, self-observation would sustain. And even though Virginia wished to keep her "colours flying," she contemplated going down with them.

Before Virginia wrote that entry, Leonard had that day given a speech, "Common Sense in History," at Brighton, and Rachel Dyce Sharp had presented Virginia with a bunch of violets. The next day, this same woman—the one who had complained about "repulsive" passages in *Roger Fry*—wrote Virginia a four-page letter praising Leonard for the assurance that after this war Hitler would be "irrelevant": "We shall forget about our hate in our plans for a better world." She insisted, "It is the thinkers like Mr Woolf who have brought this about." She and Virginia exchanged letters about visiting each other's gardens, letters that reduced Virginia to the trivial, especially when she had written repulsive material while Leonard had delivered a speech about saving the world (RDS/VW, 9 and 12 March 1941).

Because Vanessa still seemed to be "on guard" after Virginia's talk—in whatever terms—about Angelica's "dirt" (D, 5:354), Virginia tried to placate her sister by inviting the lovers to visit. On 11 March, Leonard drove into Lewes to do the marketing and brought back Angelica and Bunny for tea. That evening, when Leonard went out to check the haystacks set on fire by incendiary bombs (LWD), Virginia brooded about her niece.

When Octavia visited Monk's House on 12 March, Virginia seems to have used the pretext of "painting" Octavia to seek her counsel about her general depression. Octavia later wrote that Virginia told her she was "feeling desperate—depressed to the lowest depths, had just finished a story. Always felt like this—but especially useless just now. The Village wouldn't even allow her to firewatch—could do nothing—whereas *my* life." Octavia assured Virginia of the enormous value of her writing, but Virginia rejoined, "Yes, but people *need* you. You're doing something worthwhile, practical."

Virginia also told Octavia about the deaths of Julia and Stella and said that Leslie thereafter "rather went to pieces." He made "'too great emotional claims upon us and that I think has accounted for many of the wrong things in my life. I never remember any enjoyment of my body.'" When Octavia asked her what she meant, Virginia explained that she had not played games out of doors or in the woods.[34] Actually, she had played bowls and cricket (Figure 5), swum, hiked, skated, and roamed the hills above the seacoast, chasing butterflies. Virginia thus seems to have been referring to the period following her mother's death when her father lapsed into self-pity and petulance and the "healthy outdoor" summers at St. Ives ended.

Octavia saw Virginia as a "rather hauntingly fearful but brilliant mind. She said she envied me my touch with reality." Whereas Octavia described herself as forward looking, she saw Virginia as a "backward-looking spirit" beset by losses and regrets, expectations and insecurities. Octavia knew that Virginia had "lost hold" during the last war and thought that she was afraid of doing so again. The good doctor hoped her steady influence could be "*tranquillizing*."

Unable to trust her own creativity, in 1914 Virginia had offered to type for Lytton and Violet; in a similar situation in 1941, she offered to catalog books for Octavia. She said she envied Octavia her "touch with reality" and also asked her about "drink in women. Was it curable?" She pretended to be asking for "a friend—gifted poetically," but Octavia "thought I knew whom she meant." If Virginia meant herself, she was following a familiar ploy among manic-depressives who, as Thomas Caramagno explains, try to medicate themselves with alcohol or drugs, only to make their depression worse and increase the risk of suicide (*Flight of the Mind*, 62). Octavia suspected that Virginia was frightened that the war would reactivate her experience of losing hold, as World War I had.

On 14 March 1941, Virginia and Leonard drove to Lewes and caught the train for London. They met with John Lehmann in Mecklenburgh Square and then had lunch with him near the Houses of Parliament. Virginia seemed to John to be "unusually tense and nervous, her hand shaking now and then, though she talked absolutely clearly and collectedly." After Leonard revealed that Virginia had just completed a novel, she

> began now rather confusedly, to say that it was no good at all, couldn't be published, must be scrapped. Very gently, but with great determination, Leonard rebuked and contradicted her, and said to me and to her that he thought it was one of the best things she had written.

John begged to read it for himself (Noble, *Recollections of Virginia Woolf*, 55–56). Assuming that it would be published, and without consulting either Woolf, Lehmann added *Between the Acts* to the Hogarth Press's advertisement in the spring books issue of the *New Statesman* and *Nation*.

On 16 March, despite suffering from severe anxiety and depression, Virginia wrote an obligatory letter of thanks for Ruth Fry's praise of *Roger Fry*. The letter shows a calm Virginia Woolf, perfectly in control of herself (VWL, 6:479–80). The self-control in that letter notwithstanding, manic delusions had now begun to inhabit Virginia's brain. She heard voices, as she had after her father's death. Seas of horror threatened to become tidal waves. On 18 March, Leonard noted in his diary that Virginia was "not well," and he remembered in his autobiography that in "the next week I became more and more alarmed."[35]

By 20 March, Virginia protested to John Lehmann that her "so-called novel" was "too slight and sketchy" to be published. But because Leonard disagreed, she sent Lehmann the typescript and asked him to "give your casting vote." She also warned him, "Meanwhile dont take any steps" about publishing *Between the Acts*. Back in 1924, Virginia had thought that tedious work for the Hogarth Press could prevent her brooding and had schemed, "If I can't write, I can make other people write: I can build up a business" (D, 2:308). Now, in 1941, doubting her creative talents, she asked for the mundane and self-effacing task of reading other people's manuscripts for Lehmann and the Hogarth Press (VWL, 6:482 and note).

Vanessa came to tea at Monk's House on the same day, Thursday, 20 March 1941. Virginia had last written in her diary on 8 March when she had vowed to observe her own despondence and at least to go down "with my colours flying" (D, 5:357–58). She did not record Vanessa's visit. Our only records are from Leonard, who wrote, "Van [to] tea" (LWD), and Vanessa, who later wrote to Vita that Virginia had "talked to me about herself" (VBL, 475). Apparently, Vanessa shrank away from Virginia's talk "about herself." Later, Vanessa wrote to Jane Bussy, daughter of Simon and Dorothy, that she had no idea until then "that anything was the matter at all. Then she talked to me, and so did L., and I saw that she was in a state when rest and food were essential—as of course had often been the case before." Perhaps on that visit, Leonard told Vanessa that he "could not get [Virginia] to rest" (VBL, 477, 278).

On 21 March, Virginia wrote to her old friend Nelly Cecil that she was glad Nelly liked Leonard's book *The War for Peace*. Virginia's low self-esteem and final rejection of formalist art also appear in this letter: "It seemed to me the only kind of thing worth writing now. Do you find you can read the novelists? I cant." She told of being bombed in London and raided in Sussex and stated, "I cant help wishing the invasion would come. Its this standing about in a dentist's waiting room that I hate" (VWL, 6:483).

Virginia had recently received a letter from a childhood friend, the newly widowed Lady Tweedsmuir. On 21 March, she wrote a reply, calling forth fond memories of Lady Tweedsmuir's parents and telling about being "bombed out of London" and leading "a rather vegetable existence here, surrounded by the melancholy relics of our half destroyed furniture. All this afternoon I've been trying to arrange some of my father's old books." Despite the war, she hoped they could meet sometime in London, but she did not mail that letter (VWL, 6:483 and 484, note).

When Octavia Wilberforce came to tea in the late afternoon of 21 March, Leonard told her that he thought "Virginia [was] on the verge of danger" (LWA, 2:434). Virginia was still proposing to write a portrait of Octavia, but she announced that both *Orlando* and *Roger Fry* were failures and that she could not imagine "how I'd do you." Virginia despaired: "I can't write. I've lost the art," just as she had after her mother's death.

These remarks ostensibly refer to a projected biography of Octavia, but they also reflect her worries about *Between the Acts*. From her earliest days, Virginia's art had fulfilled "the instinct of self preservation" (Woolf, "'Anon,' and 'The Reader,'" 403), but now she was convinced that her art would instead lead to self-immolation. With her brain turned to macaroni and her inspiration to dust, Virginia feared herself incapable of revising her novel. But she also feared that publication of the book in its current form would sacrifice her to a torrent of critical abuse that she could not withstand.

Not only was Virginia fearful that she had lost her writing gift, but she doubted the value of writing altogether when Octavia was "doing more useful work, helping things on." Virginia complained of being "buried down here—I've not the stimulation of seeing people. I can't settle to it [writing]." She also must have hypothesized about madness in her family, a theory of "blood thicker than water" that Octavia refuted (incorrectly) with the label "balderdash." Virginia told her that she had even taken "to scrubbing floors when she couldn't write." If she had lost the art of writing, the *paragone* not only was irrelevant but also seemed to be over.[36]

The dangerous depression that Leonard and Octavia observed that Friday appears to have been precipitated by a letter from Vanessa that Virginia received that afternoon.[37] The letter has no salutation; it begins without even Virginia's name. Vanessa instructed: "You must be sensible, which means you must accept the fact that Leonard & I can judge better than you can. Its true I havent seen very much of you lately, but I have often thought you looked very tired." Therefore, Vanessa argued that Virginia should be

> only too glad to rest a little. You're in the state when one never admits what's the matter, but you must not go and get ill just now.
>
> What shall we do when we're invaded if you are a helpless invalid—what should I have done all these last 3 years if you hadnt been able to keep me alive and cheerful. You dont know how much I depend on you. . . . Both Leonard and I have always had reputations for sense and honesty so you must believe in us. Yours VB. I shall ring up sometime and find out what is happening. (VBL, 474)

Despite its testimony to Vanessa's need for Virginia, the letter is remarkably insensitive. Certainly it contributed to Virginia's sense of incompetence beside Vanessa and Leonard, asking her to believe in her sister and husband as if they were superiors.[38] After Virginia had spent months consoling Vanessa over Julian's death and years of drudgery in order to "give Nessa back her Roger," in this letter Vanessa considered Virginia to be worth only a phone call "sometime," not a visit; not a salutation to "Beloved William" or even "My Billy"; not even a proper signature. If Virginia had felt helpless and guilty, Vanessa's callous letter can only have made her feel worse.[39]

Back in 1881, Leslie Stephen had written to Julia that after a little coaxing, Vanessa, almost two then, agreed to send a kiss to her mother, but she seems to have had "odd little misgivings in her little mind as to doing anything demonstrative." In the late 1930s, Vanessa showed that her character had not changed in nearly sixty years, for she confessed to Vita that when Virginia is "demonstrative, I always shrink away" (VWL, 6:xv). Virginia must have sensed what Angelica wrote years later: That at Charleston, her aunt Virginia was someone "who could be teased and laughed at, sometimes to her face, but mostly behind her back."[40] No matter that she had paid her debt to Roger, no matter that she had written Postimpressionist fiction, no matter that she had bought paintings and decorations from Duncan and Vanessa, no matter that her writing had stimulated Vanessa's painting, no matter that she had written *Roger Fry* for her sister: Vanessa's letters to Roger and others and now to Virginia show that in the end she could love only Duncan, her children, and her painting. Virginia would never be admitted inside that closed circle.

Perhaps Vanessa's talk of "what shall we do when we're invaded if you are a helpless invalid" reminded Virginia that Leonard, who had loved her for thirty years, would be the one who would have to care for her if she became ill again. And if there were an invasion, Leonard would have enough problems without Virginia's adding to them. While Vanessa had kept her distance, Leonard had remained by her side, and in the country in the last year, Virginia had rediscovered the contentment of being alone with him. Now, feeling helpless in her art, acutely conscious of Vanessa's alienation, guilty over past and future troubles she had caused and might cause Leonard, Virginia was in the throes of the fatal concatenation of demons against which she had spent twenty-six years protecting herself.

On Sunday, 23 March 1941, Virginia replied to Vanessa's letter with heavy irony: "Dearest, You cant think how I loved your letter." She could have "loved" Vanessa's letter only if it released her, as it seems to have done, for she continued by explaining why she planned to kill herself

(VWL, 6:485). She was "certain now that I am going mad again." She echoes what Vanessa had said repeatedly in 1913 and 1915: "It is just as it was the first time, I am always hearing voices, and I know I shant get over it now." Virginia devoted a paragraph to Leonard's "astonishingly good" devotion to her. Then she ended: "I can hardly think clearly any more. If I could I would tell you what you and the children have meant to me. I think you know. I have fought against it, but I cant any longer. Virginia" (VWL, 6:485).

Probably about the time that she wrote her suicide letter to Vanessa, Virginia also wrote, I believe, another letter to Leonard that repeats such phrases as "I want to tell you" and "All I want to say" but does so in an even shakier hand. (For my justification for reordering these letters and events, see Appendix D.)

In her undated letter to Leonard, Virginia acknowledges that her illness means that

> I shall never get over this: and I am wasting your life. All I want to say is that until this disease came on we were perfectly happy. It was all due to you. No one could have been so good as you have been, from the very first day till now. Everyone knows that.[41]

An odd postscript to this suicide letter, written along the left margin of the page, tidies up a final remnant of the Fry biography: "You will find Roger's letters to the Maurons in the writing table drawer in the Lodge. Will you destroy all my papers." The reference to Roger's correspondence with the Maurons shows how in control of details and responsibilities Virginia was as she prepared for suicide "by way of an end."

The evening before Virginia wrote the suicide letter to Vanessa, John Lehmann received the typescript of *Between the Acts* along with her 20 March letter saying that the novel could not be published "as it stands" yet asking him to give the "casting vote" (VWL, 6:482, 486 note). Lehmann read *Between the Acts* that Saturday night, 22 March, and brooded over it during his hours of guard duty:

> The first thing I had noticed was that the typing—her own typing—and the spelling were more eccentric, more irregular than in any typescript of hers I had seen before. Each page was splashed with corrections, in a way that suggested that the hand

> that made them had been governed by a high-voltage electric current. The second impression I had, was that the book had a quite shattering and absolutely original imaginative power, pushing her poetry—I have to call it poetry—to the extreme limits of the communicable. It was obviously in some ways unfinished, but I was deeply moved, disturbed, thrilled in the face of a revolutionary work of art.

On Sunday, 23 March, Lehmann telegraphed Virginia, praising her novel. Knowing that the announcement of its publication was about to appear, he also wrote her a letter insisting that *Between the Acts* "simply must be published" (Noble, *Recollections of Virginia Woolf*, 56–57). When she received Lehmann's telegram, Virginia answered with a coherent but brief, unsigned recommendation about manuscripts. Her only personal comment was "But my head is very stupid at the moment." She forgot to sign the letter (VWL, 6:484–85). Virginia said nothing about her novel or, of course, about the suicide letter or letters she may already have written that day.

Even though Leonard had stayed up until 4:00 A.M. fire watching (LWD), on that Sunday, trying to keep Virginia distracted, he took her to see a Mrs. Chavasse in Rodmell. Virginia wrote a suicide letter to Vanessa sometime that day and perhaps that evening wrote the one to Leonard; then she folded the two letters together, possibly with the letter to Vanessa outside, and hid them, probably in her writing-table drawer along with Roger's letters. Although she had felt herself perched beside an abyss, Virginia allowed herself a last chance to wither gracefully into old age and natural death.

Despite having hidden the letters, by now Virginia was too upset for grace and calm. On Monday, 24 March, Leonard thought her "sl[ightly] better," perhaps because she wrote in her diary for the first time in sixteen days. Most of what she wrote, however, sounds distinctly disturbed. She described Mrs. Chavasse as having a "nose like the Duke of Wellington & great horse teeth & cold prominent eyes." Sunday had been a windy day, with "all pulp removed" from life. "This windy corner. And Nessa is at Brighton, & I am imagining how it wd be if we could infuse souls" (D, 5:359). The Brighton reference associates, I believe, Vanessa with "Brighton a love corner for slugs." The infusion of souls seems most certainly to have recalled the emotions first articulated in 1895 by the author of the *Hyde Park Gate News*: the desire to look into her best friend's mind and the fear that she would find herself hated. Reading Vanessa's letters to

Roger and her last letter to Virginia had offered Virginia just such a disturbing infusion of souls.

In the same diary entry, Virginia tried to rise above despondency with more pleasant thoughts, such as letters from friends that she enjoyed, but that she could not "tackle" answering, and "Octavia's story. Could I englobe it somehow? English youth in 1900." But Virginia no longer felt capable of fulfilling her "insatiable desire to write something before I die" (D, 3:167). She had tried three very fragmentary sketches of Octavia, but with two suicide letters waiting in her drawer, Virginia could no longer englobe the past or envision the future (Dick, "'The Writing "I," Has Vanished,'" 143). Her diary ends: "L. is doing the rhododendrons . . . [her ellipses]" (D, 5:359).

All of Virginia's last three major works were retrospective: Her biography and memoir incorporated the present into the past, and her novel incorporated the past into the present. Together, these works rounded out Virginia Woolf's life and art, establishing that her art could embrace facts, psychology, aesthetics, and politics while remaining free of tedious referentiality. She had proved not only that writing could encompass more than painting could, as she said in "The Leaning Tower," but also that her writing could so in such dazzling new ways. Nonetheless, David Cecil, Ben Nicolson, Desmond MacCarthy, and others considered her Postimpressionist art too dissociated from everyday exigencies. She therefore was convinced that nothing she could write—not *The Years*, not *Three Guineas*, not *Roger Fry*, not *Between the Acts*—could dispel that misconception. Probably on Tuesday, 25 March, Virginia received Lehmann's letter arguing that *Between the Acts* "must be published." Although she had instructed him "don't take any steps," possibly on the same day, the step he had taken appeared in print.[42] The advertisement begins:

THE
HOGARTH
PRESS
announce for publication
Spring and Summer 1941
VIRGINIA WOOLF
A new novel
Between the Acts

Virginia might have exclaimed with pride over this last novel, as she had more than forty-one years before over her description of a sunset, "well may an Artist despair" (PA, 156). Instead, feeling that the advertisement now compelled the publication of her "so-called novel" and fearing that its publication would expose her to public abuse, Virginia despaired.

Virginia took out the two letters. The rather cryptic letter to Leonard, with its hasty postscript, must have seemed inadequate to express her love, gratitude, and concern, so she wrote a new one. This much fuller "Tuesday" letter begins, "Dearest, I feel certain that I am going mad again: I feel we cant go through another of those terrible times. And I shant recover this time." Part of her proof was her very shaky handwriting, "You see I cant even write this properly. I cant read." She repeated her pleasure at their happiness and her fear that she was "spoiling your life." She assured him, "If anybody could have saved me it would have been you. Everything has gone from me but the certainty of your goodness" (VWL, 6:481). Virginia apparently stuffed the rejected "Leonard" letter back into a drawer (as its smudges indicate) and put the Sunday letter for Vanessa and the new, more expansive Tuesday letter for Leonard into matching blue envelopes labeled simply "Vanessa" and "Leonard."

Probably she placed the letters where they could be found and then went "for a walk in the water-meadows in pouring rain." Leonard had been to Lewes to the market (LWD). On his return, he went out to find her in the rain and saw her trudging "back across the meadows soaking wet, looking ill and shaken. She said that she had slipped and fallen into one of the dykes." Walking back with her in the mud and rain, Leonard had "an automatic feeling of desperate uneasiness," but only later did he realize that Virginia may have attempted suicide on Tuesday, 25 March (LWA, 2:433–34). If she had tried to drown herself, something—the will to live? love of Leonard? of Vanessa? desire to write her book on English literature? hope that her creative powers would return?—had held her back. If she had taken out her suicide letters, upon returning she retrieved them, leaving the original rejected suicide letter to Leonard back in the dusty drawer and putting the envelopes addressed to Leonard and Vanessa into a cleaner spot.

Hoping to cheer her up and distract her, Leonard seems to have taken a bathed and warmed Virginia to another tea with Angelica and Bunny late that afternoon. When they arrived, Angelica realized that Virginia "was under the weather, and she made this tremendous demand for love that she was in the habit of making rather." Angelica explained that Virginia would often carry on: "'But, Angelica, don't you love me, don't you adore me, you hate me, you know you don't like me at all'—this sort of way of going on." Angelica remembered that Virginia "did it particularly on that day, and I was particularly cold and undemonstrative." Angelica further explained that Virginia begged "with more than her usual insistence" for attention, but Angelica, following the pattern she had learned from her mother, "reacted with more than my usual impatience."[43] So rather than cheer up Virginia, the visit must have made her even more depressed, for it would have confirmed her fear that Angelica, like Vanessa, was "lost to me, lost to me." That evening, Leonard made another attempt to calm Virginia by turning on a soothing Brahms trio on the BBC (LWD).

Leonard later wrote to Vita that Virginia "was terrified that she was going mad again. It was, I suppose, the strain of the war & finishing her book & she could not rest or eat. . . . She has been through hell" (LWL, 250). By Wednesday, 26 March, Leonard recognized that "Virginia's mental condition was more serious than it had ever been since those terrible days in August 1913 which led to her complete breakdown and attempt to kill herself." But he knew that the "mere hint of pressure, even a statement of the truth might be enough to drive her over the verge into suicide" (LWA, 2:434). Leonard called Octavia to say that he was frightened about Virginia and needed help. Herself ill with influenza, Octavia nevertheless agreed to see Virginia on 27 March.

The precipitous prepublication advertisement for *Between the Acts* was on the newsstands by 27 March. Included in that same (29 March) issue of *New Statesman & Nation*, in the regular "London Diary" column, written by a "CRITIC" who was clearly Leonard, were two paragraphs on the bombing of London.[44] Describing a bombed-out house, Leonard lamented, "There they stood, in rows, the familiar volumes that belong to a period when one could still lecture about the past and theorise about the future. On the top, seventy volumes of Voltaire in faded calf; the whole lot only cost me fifteen shillings." He also noted that whatever Bloomsbury "was once, it is gone now." He concluded that the destruction of books mattered less than the destruction of minds and maintained that in England "one may still read Voltaire and even Marx and Mill and *The New Statesman and Nation*."

That reminder of loss and futility, combined with the pressure put on her by the announcement of her forthcoming novel, reopened the fault line in Virginia's psyche and allowed old demons to emerge once again. That Thursday, Virginia typed a letter to John Lehmann, returning her notes and the manuscripts that she had been reading. With the physical and mental destruction of Nazism threatening civilization, with political writing humankind's only hope, Virginia concluded that her novel was "too silly and trivial" to publish (VWL, 6:486).[45]

After typing her apologetic letter to Lehmann, Virginia agreed, "against her will" (LWL, 251), to be driven to Brighton for a consultation with Octavia Wilberforce. For Octavia, the experience was "rocky going," a "battle of—not wits, but *minds*." Virginia was fidgety and "resistive," protesting her examination "like a petulant child!"[46] Desperately, Virginia begged Octavia not to prescribe a rest cure and instead to assure Leonard that she was well. Octavia could not offer that assurance but could promise to do nothing unreasonable. Gradually, Virginia became more forthcoming. She reflected that she had "'been so *very* happy with Leonard' with much feeling and warmth in her face."

Virginia admitted that she felt especially useless, given the machine guns firing in the distance. She confessed her obsessive fear that with the war, "her madness would recur and render her unable to write." Octavia believed that Virginia was "haunted by her father," later noting that Leonard agreed. Virginia detested the "hardness of old age" in others and feared it in herself. Perhaps she imagined herself becoming like her father in his last years, self-pitying, irascible, convinced of failure. In addition, in her memoir and diaries she had come to remember her father before he became petulant and tyrannical and may have been pursued by the idea that her long-nourished anger with him had been unfair.

Despite being sick herself, Octavia tried to calm Virginia by walking with her around Montpelier Crescent. With their half-columns, topped by Corinthian and Ionic capitals, the houses in the crescent might have reminded Virginia of her grandmother Maria Jackson's house on Brunswick Terrace. Octavia's home was on a hill above the center of Brighton, however, and Maria's home was on the seafront, but both were built by the same architect in the same Grand Regency style (Figure 65) (Carder, *Encyclopedia of Brighton*).

With the handsome houses on one side and the peaceful park, shaped like a half-moon, on the other, Virginia walked along and asked whether Octavia herself ever had "black moods" or was depressed. Octavia seems to have tried to evade the question, whereupon Virginia angrily accused Octavia of being a "damned fool" because even though she meant well, she did not know much about Virginia. Before the elegant facades of the homes that may have recalled her many childhood exiles in Brighton, Virginia announced, "I *won't* rest," maintaining that she knew more about herself than Octavia possibly could.

Octavia replied that Virginia had only her own perspective, whereas as a doctor, she had several. She argued that just as a painter sees "far more than others do because he knows the architecture, the anatomy of the body," so does a doctor. That reference to the supremacy of the painter's knowledge must have dredged up old insecurities. And Octavia's analysis that the "signs and symptoms of overreaction of [the] autonomous negative nervous system are outstanding in you" would hardly have comforted Virginia. Despite Octavia's viral infection and Virginia's severe depression, the two women braved the cold wind blowing up from the sea to walk around the crescent arguing and cajoling. A giant white willow tree offered a canopy of shade but no protection from the wind or the persistence of Octavia's argument.[47]

Finally, Octavia outlined a bargain in which Virginia would agree to rest, be easier on herself, do housework, "watch the birds, do a bit of embroidery, design a new chair cover," and eat carbohydrates. Octavia stated she hoped that a recovered Virginia might write the "greatest masterpiece of this century." Talk of another masterpiece, however, only increased Virginia's sense of inadequacy. Beset by demons, she could not have been comforted to find how alien her experiences were to those of the sensible, activist woman doctor.

After taking her patient back to the office, Octavia went to the front room to consult with Leonard. She told him that Virginia needed to be secluded in a nursing home. Although she had more compassion, Octavia repeated the advice of the doctors of Virginia's youth: "No writing or criticism for a month. She has been too much nurtured on books. She never gets away from them. Let her be rationed and then she'll come good again. *If* she'll collaborate." While Leonard and Octavia conferred, a German bomber flew over their heads, down the line of the street toward the sea. They heard the crash of exploding bombs but were too preoccupied with their dilemma even to notice them, although Leonard remembered the bomber with a shock as he drove Virginia back to Rodmell (LWA, 2:434–35).

Remembering 1913, when the prospect of incarceration had precipitated a suicide attempt, Leonard felt that he had to listen to Virginia, not Octavia. Although voices were again plaguing Virginia, Leonard expected his care and her courage to bring her back once again from the precipice. Octavia promised to visit in a day or two and observe Virginia's progress. That evening, as Leonard later told Octavia, Virginia was "cheerful and quite different." The Woolfs listened to Beethoven's "Appassionata"

Sonata on the BBC (LWD), and Virginia seems to have kept her bargain with Octavia by working on a piece of embroidery.[48] Leonard felt reassured. He knew that Virginia had often reclaimed herself from despair, as one sets a watch to ticking again or "as a sailor not without weariness sees the wind fill his sail and yet hardly wants to be off again and thinks how, had the ship sunk, he would have whirled round and round and found rest on the floor of the sea" (TL, 78). Even after Virginia's death, Leonard "passionately" insisted that "'she could have . . . recovered as she had done from the previous attacks.'"[49]

Despite the more cheerful evening, Virginia's "hairy devils" must have returned in the night to howl in her ears, as they had thirty years before, that she was "a failure—childless—insane too, no writer" anymore (VWL, 1:466). Perhaps the devils counseled the nihilism that she had expressed nearly twelve years before: "There is nothing—nothing for any of us. Work, reading, writing are all disguises; & relations with People."[50] Without the illusion that the "narrow bridge of art" could lift her over the abyss of nothingness, with both art and affection, she must have believed, lost to her, Virginia could no longer resist her own devils.[51]

Even though Octavia had told Virginia that "there's nobody in England I want more to help than you" and had urged her to call "night or day if you feel things are hellish," Virginia did not call. Her suicide letters were ready. She did not reach out even to Leonard, fearing, "We pour to the edge of a precipice . . . & then? I cant conceive that there will be a 27th June 1941" (D, 5:299). And for Virginia Woolf, there was not.

On the morning of 28 March 1941, Leonard mailed Virginia's last letter to Lehmann, with a cover letter explaining that she was on the verge of a nervous breakdown. Then, before he went out to the garden, Leonard asked Louie Mayer, who was dusting his study, to give Virginia a duster. This request seemed "very strange" to Louie, but Leonard was trying to follow Octavia's suggestion to help distract Virginia from her worries. Leonard knew that Virginia would not have tolerated trained nurses and that "one would only have made it impossible and intolerable to her if one attempted the same kind of perpetual surveillance by oneself." Virginia dusted a bit, and then put down the duster and went upstairs. Later Louie saw her go out to her lodge.

Determined to escape her devils, Virginia retrieved the undated letter to Leonard from her drawer, leaving it on her desk, and pocketed the two envelopes labeled "Leonard" and "Vanessa." Leonard seems to have followed Virginia to her lodge and got her to return to the house. Louie remembered that she had returned after "a few minutes." When Leonard went back to the garden, Virginia took the two blue envelopes out of her pocket and placed them where they would be found. Louie then recalled that she "put on her [fur] coat, took her walking-stick and went quickly up the garden to the top gate." She also wore, according to Vita's later account, "big gum boots (which she seldom did because she hated them)."[52]

Leonard must have been working at the other end of the garden, for Virginia walked across the bowling lawn unobserved. She passed along the fence by two elm trees and let herself out at the top gate. With huge black rooks cawing in the tall trees above, Virginia set out toward the river valley. She walked across the meadows, buffeted by the wind from the sea, until she reached the River Ouse, put stones in her pocket, left her walking stick on the bank, walked into the water, and sank into a tidal current, hoping to find "rest on the floor of the sea."

When Leonard came in for lunch, Virginia was gone. He ran down to the river and found her stick on the bank. There was no sign of Virginia. He and Louie, with the gardener and a policeman, searched for her until dark.

Then Leonard drove to Charleston with the news of Virginia's disappearance. He simply recorded in his diary, "Van wept," but otherwise she and he remained, as Vanessa said of him, "amazingly self-controlled" (VBL, 474). On 29 March, Leonard again walked down to the river, but there still was no sign of Virginia (LWD). Vanessa expected him to "simply plunge into work," as he did. She regretted that the possibility of Virginia's committing suicide had "never occurred to me," even on her last visit with Virginia (VBL, 475). When Vita visited her on 7 April, Vanessa found it difficult to "believe Virginia wasn't there too talking to us for I think I've hardly ever seen you without her."[53]

Vanessa wrote to Vita, "I cannot bear to think of [Leonard] alone in that house, though I know he would not stand being anywhere else" (VBL, 476), but Leonard saw Vanessa (several times), Octavia, Clive, the Keyneses, and other friends while taking respite from grief in his work and his garden (LWD). On 18 April, he received the news that Virginia's body had been found by picnickers.[54] Leonard identified Virginia's body that afternoon and attended the inquest on 19 April. He described these experiences to Vanessa as "more horrible than all the rest" (VBL, 476).

As Vanessa explained to Vita, Leonard "arranged for the cremation in Brighton [on 21 April] and didn't want me to go, so I didn't" (VBL, 476). Only when he told her about his plans for Virginia's burial did Vanessa see

Leonard "break down completely" (VBL, 479). He buried her ashes under the great elm that the Woolfs called "Virginia," on the edge of the bowling lawn overlooking the field and the water meadows behind Monk's House. On the chalk-bound flint wall facing the formal pond, Leonard had Stephen Tomlin's head of Virginia, cast in lead, mounted with the words from *The Waves* that Virginia had spent her life, until its end, asserting: "Against you I will fling myself, unvanquished and unyielding, O Death!"

Virginia had written in her "Tuesday" letter to Leonard that she feared "going mad again. I feel we can't go through another of those terrible times," but her suicide was almost universally misreported as an inability to face the "terrible times" of war. Then the wife of the bishop of Lincoln wrote to the *Sunday Times* that many people had lost everything but, unlike Virginia Woolf, continued to fight nobly "for God against the devil." Leonard was furious but wrote a calm response explaining that Virginia "took her life, not because she could not 'carry on,' but because she thought she was going mad again and would not this time recover." Leonard could take consolation only in thinking "how amused Virginia would have been by the extraordinary things people write to me about her."[55]

Losing Virginia made both Leonard and Vanessa reach out to others. Vanessa wrote to David Garnett on the day after Virginia's body was found that she wished "to forget these painful months of estrangement" (VBL, 481). He and Angelica married in 1942 without inviting either Vanessa or Duncan to the wedding.

Vanessa lived on for twenty years, during which time she seemed to need "to freeze some part so herself." She and Duncan continued their Charleston decorations, their painting, and their travels. Duncan's affairs no longer threatened Vanessa, but she increasingly retreated into herself.[56] Quentin Bell remembers that Angelica, who had "suffered most" from Vanessa's devotion, "was the first to find the antidote to the poison that infected the sad years after Julian's death. She presented Vanessa with four grandchildren who did a great deal to make her later years tolerably happy" (*Elders and Betters*, 56).

Leonard Woolf lived for twenty-eight years after Virginia's death. In the 1950s, he began writing his remarkable five-volume autobiography, which, in its meandering chronology, owes much to Virginia's "A Sketch of the Past." He continued his political activities at an astonishing rate, refusing a titled honor in 1966. Meanwhile, he answered every attack on

Virginia's work (and there were many) with a stern rebuttal. Although he came to feel that his political activities had made little difference to the progress of civilization, Leonard took comfort in tending his garden, in sorting out the "inextricable confusion" of Virginia's papers,[57] in preserving and managing her literary estate, and in an abiding companionship with Trekkie Parsons. Before his death, "his greatest pleasure came when Quentin Bell visited and read passages from the draft manuscript of his biography of Virginia."[58]

Leonard Woolf died on 14 August 1969 at the age of eighty-eight. His ashes were buried near Virginia's under the other great elm which the two of them had called "Leonard." Charlotte Hewer's head of Leonard Woolf, also cast in lead, was placed down the wall from Tomlin's head of Virginia. The paired sculptures (Figure 66) convey the sensitivity and bravery that both Woolfs displayed throughout their lives. Together these sculpted heads seem to stare across the peaceful water meadows of the Ouse valley, to the downs that Virginia and Leonard had so loved, and beyond.

Appendix A
Virginia's Childhood and Her Grandmother's Letters

The letters from Maria Jackson to her daughter Julia Stephen are, in many ways, appalling. Among other reasons, they are a travesty of the language, not because Maria did not bother with rules of capitalization and punctuation (she did not), but because she did not bother to think much about the words she used. Julia is "my own dear Lamb," "my own," "dearest," and other combinations of the same formula. All of Julia's children are "my sweet," "my pretty," "my dear," "my darling." And all of her letters end, with slight variations, with this formula: "Love to my dear Leslie and kisses to my darlings yr devoted old mother."

In the thousands of letters that Maria wrote to her dearest daughter, there is scarcely a word or phrase to suggest that this abuser of the language would have, among her grandchildren, a future prose stylist. Maria Jackson's language is full of clichés, and, not surprisingly, she cannot describe a scene adequately. For example, after her husband died, Maria was looking for another lodging. "Little Ginia" came in, looking "very radiant," to tell her about a house that she had spotted. Maria Jackson "went afterwards with Ginia to please her to Powys Grove feeling since it s[hou]ld not suit me of course it did not a nice little villa house like the M Villas."[1] It is unfortunate for us that Maria did not describe her granddaughter's disappointment at failing presumably the first of many house-hunting ventures.

Many of Maria's letters are, furthermore, both syntactically incoherent and boring. Of the four pages of a folded note, Maria Jackson typically used three to tell what time her bath was scheduled for; whether the temperature could be warmer; how her bowels were working; when she drank mineral water, beef tea, solvolatile, or brandy and water; when she had had a bilious attack; whether her legs had been rubbed; what sort of sleep she had had the night before; why she did (or did not) go out in her chair; and that Julia "would be glad to know" she would be "pricked" with morphine after a particularly poor night. Typical of her insensitivity to verbal resonance is "God bless you my precious—My Bowels are quite comfy now."

Through her own words, Maria Jackson becomes a caricature of upper-middle-class Victorian respectability and indulgence. Even though a contemporary witness considered her to be a "great invalid" as well as a "great reader," her letters attest only to the former (Hill, *Julia Margaret Cameron*, 68). Among public events, she notices mostly deaths of contemporaries such as Browning. Eight months after the fact, Maria lamented: "What a sad end of the Crown Prince—it shows how necessary it is to be cautious when there is the slightest affection of those organs."[2] Yet she had untold household servants caring for her house and rubbing, bathing, feeding, carrying, and medicating her. She was too "gouty" to walk. Her many trips to take the waters at Bath were aimed—futilely, it seems—at mobilizing her enough to learn how to use crutches. She tried being hypnotized, but no hypnotist could override Maria Jackson's own enjoyment of dependence.

Because Maria got virtually no exercise, one might suppose that she would consider the harmful effects of immobility and overeating. On the contrary, Maria advised Julia, "I wish you could be stuffed as well as you stuff [others]. As for me I stuff quite as much as my old digestive organs will submit to." Such statements suggest that even though Figure 2 shows Maria Jackson as still a thin woman, twenty more years of stuffing would have contributed to her immobility. She was convinced that illness was brought about by thinness—if it was not introduced by a lower-class governess or a "shoe black" employed in the garden.

Maria Jackson did, nevertheless, care about her grandchildren. (They spent enormous amounts of time with her and Dr. Jackson.) And Maria adored her daughter Julia. When Julia sent her mother flowers from St. Ives, Maria replied that they made her "see you stand before me in yr. lovely maidenhood far more lovely than" Julia appeared in Edward Burne-Jones's famous painting *The Annunciation*. Not surprisingly, Maria's letters were full of worries about Julia's being too thin, taxing her strength, getting a chill, not wearing combinations (long underwear), not wearing enough petticoats, going out in the night air, traveling in four wheelers, and taking the U.G. (Underground) where a man might sit beside her. Again and again, Maria found it hard to part with Julia's "sweet sweet face," yet she continued to exhaust her daughter's energy by depending on her devotion.

Maria Jackson's letters are superb Victoriana. But hers was not the Victoriana of those men and women who believed in progress so fervently that they challenged (with mixed results) the boundaries of knowledge, class, and nation. Instead, hers was the Victoriana of self-righteous privilege. She had no compunctions about exploiting anyone, even her much beloved daughter (although after the fact she repeatedly regretted having taxed Julia's health). She showed little generosity of spirit to anyone outside her own family, yet she seems never to have doubted the immortality of her soul. Her values may be disturbing, but they are archaeologically very interesting, revealing that the same class that produced those Victorians of prodigious energy could also produce its own parasites.

Virginia Woolf had much to say about Maria Jackson's eccentric sister, the photographer Julia Margaret Cameron, but almost nothing to say about her own grandmother. When Julia Stephen was nursing Maria Jackson in her last illness, Virginia did refer to her grandmother as "the invalid of Hyde Park Gate" who would be "properly shocked" to know the "vulgar little song" with a chorus "ta ra ra bomdeay" that the Stephen children merrily sang under the same roof. Then in "Sketches of the Past," Virginia distinguished between her maternal great-aunts, who were "worldly in the thoroughgoing Victorian way," and her grandmother, who was "devout and spiritual" (MB, 88). Otherwise, for more than fifty years, Virginia said little or nothing about Maria Jackson.

On their miserable Brighton holiday in 1897, Virginia enlivened a boring Easter tea with her Fisher cousins by searching for letters from Maria Jackson to the Fishers (HPGN, 7 and 28 March 1892; PA, 76–77). No doubt Virginia wanted to laugh over Maria's typical news about dinner and drugs, baths and bowels, advice about petticoats and combinations, dangerous men on the Underground, and reliable gentlemen on the ferry. Julia kept thousands of letters from her mother, which Virginia read and tried to sort, as her scribbles—"Letters by Granny, 1881 @ AV's death," "Granny 1888," "Granny 1890," and "~~Granny~~ Mrs Jackson 1888–89"—on the backs of the envelopes, indicate.[3] To Virginia's great disappointment in Brighton in 1897, she found that her aunt Mary Fisher had kept none of Maria's letters. However unintentionally amusing, Maria's interminable letters no doubt reinforced Virginia's early commitment to write only intentionally entertaining letters.

Despite their banality, however, Maria's letters also offer considerable insight because, after three pages of hypochondriac complaints that provide no information about anyone or anything but herself, tucked away on the last page are occasional bits of information about her daughter and grandchildren that have helped me piece together heretofore unreported information about her granddaughter Virginia Stephen's first years.

Appendix B
On Manic-Depression

Manic-depression is a biochemical imbalance that now can be alleviated by the drug lithium. A predisposition toward manic-depression can be inherited but does not fully manifest itself until adolescence. Unlike those of psychosis and other psychological disorders, the symptoms of manic-depression occur only periodically. Then they may disappear for long periods of time, although the predisposition toward manic-depression remains. In fact, the principal fear of manic-depressives is that they—as sane, functional, and even superbly capable persons—can suddenly be taken over (seemingly possessed) by irrational and apparently uncontrollable moods. Frederick Goodwin and Kay Jamison explain that manic-depression "magnifies common human experiences to larger-than-life proportions. Among its symptoms are exaggerations of normal sadness and fatigue, joy and exuberance, sensuality and sexuality, irritability and rage, energy and creativity" (*Manic-Depressive Illness*, 3). In manic-depressive persons, normal delight in or discomfort with others or themselves can develop (especially under systemic stress) into major manic or depressive episodes, turned on either the external world or themselves.[1]

Being elated over one's successes or depressed over one's losses or failures is normal. Even exuberant, extravagant talk that entertains and delights others is healthy. But with manic personalities, that exuberance can accelerate into delusions and hallucinations.[2] Similarly, sadness over loss, frustration over helplessness, and guilt over meanness either contemplated or accomplished are normal and appropriate. Depressive personalities, however, become tortured by worries totally unconnected with their causes. Himself a manic-depressive, the novelist William Styron asserted that these worries can become a "veritable howling tempest in the brain" (*Darkness Visible*, 38).

This tempest is the result of a temporary dysfunction of the brain. Normally, the left brain apprehends sequentially, as when one reads a book, and the right brain apprehends holistically, as when one first sees a picture. Furthermore, the left hemisphere "is responsible for language and speech, and the right hemisphere directs skills related to visual and spatial processing" (Bloom and Lazarson, *Brain, Mind, and Behavior*, 282). Each hemisphere "translates" the other's apprehensions into its own language. In the mood swings of a manic-depressive, however, the hemispheres are out of sync and so can no longer communicate. The left brain, with its analytical and linguistic powers, becomes dissociated from the right brain, with its emotional and spatial sensitivities, and so perception spirals out of control.[3] Then, when normalcy is restored, the manic-depressive relishes the sensation of wholeness (Oakley and Eames, "Plurality of Consciousness," 227). Sensitive minds, like Virginia Woolf's, are especially attuned to the stable world around them and so are especially grateful for the restoration of harmony. For Virginia Woolf, making a whole or "englobing" her world signified reintegrating the verbal and the visual, left and right, her self and her world.

A good study for understanding both Virginia Woolf and this inherited illness is Thomas C. Caramagno's *The Flight of the Mind: Virginia Woolf's Art and Manic-Depressive Illness*. With the kind permission of Robert Langenfeld and *English Literature in Transition*, I quote the following from my review of Caramagno's *Flight of the Mind*:

> Woolf criticism has been plagued by reductionist approaches to her life and art. We have been told that because Woolf evidenced affective disorders, as do incest victims and others, then she must have been the victim of incest. We have been told that because Virginia railed against Leonard Woolf in her 1915

breakdown he was and continued to be an oppressor (never mind that she railed against her sister in her 1904 breakdown). We have been told that because Virginia was a brilliant and prolific writer, her "madness" was an invention of persons determined to belittle her achievement. We have been told that her nervous breakdowns were rational responses to an oppressive patriarchy. We have been told that Woolf's physical symptoms (headaches, influenza, etc.) were an expression of guilt, grief, and unresolved conflict. And we have been told that her fiction likewise was an expression of these feelings. Countering such pronouncements with logic or biographical evidence has had too little impact.

Thomas Caramagno's study offers a real possibility for shifting the balance toward objective analysis. Caramagno argues that literary scholars should abandon outdated Freudian models and stop pretending that biology does not affect the mind. He is devastating in exposing the absurdities in various reductionist interpetations such as those I have outlined.

Among manic-depressives, the speed, duration, and intensity of swings between wellness, depression, and mania varies in different circumstances. "Circumstances" is a pivotal term. Freudians see the circumstances of Virginia Woolf's life as cause for mental instability (if they acknowledge it existed). Caramagno establishes that the cause of her mood disorder was an inherited syndrome which could be triggered by physical and mental circumstances. Before drug therapy, bipolar mood swings could be by-passed or muted by avoiding situations that could trigger (though not cause) them. Thus Leonard (so maligned recently for the regime of nourishment and rest he set up for Virginia in 1913) was right. Even though she was bored and irritated by it, Virginia herself acknowledged that this regime, for which she took increasing responsibility, was enabling.

Given stability, though, the factor that most maintained Woolf's health over the years was her writing. She reflected "how nothing makes a whole unless I am writing." Along with other critics, before learning more about this illness, I had taken such remarks to mean simply that writing gave her control over herself and her environment. Writing did counter that sense of helplessness which was a major precipitator of the depressive side of her mood swings. Caramagno amplifies that theory by explaining how Woolf's fiction is neither a symptom nor a disguise but a "transformation" of the perceptual problems presented to her by manic-depression. Virginia Woolf sought an artistic fusion that integrated depressive and manic experiences or the sense of being either the object or alternately the subject. She neither discounted her "mad" feelings nor "fell back upon a Freudian explanation." As Caramagno says, she "turned the issue around and questioned all mental states—normal and abnormal, in herself and in others—and the unexamined assumptions about their integrity."

Caramagno sees Virginia Woolf progressively working through the phenomenon of having both sane and insane selves. He finds the turning point toward in-depth understanding in the 1926 diary entry she titled "A State of Mind." There she in effect duplicated the techniques of modern cognitive psychotherapy. Having separated herself from her manic and depressed selves, Virginia Woolf could locate those selves and their ways of perceiving in various characters and finally, like Bernard in *The Waves*, achieve harmony and balance in her own plurality. Without specific knowledge of the brain, but with an acute awareness of the effects of its splintering, she bravely, deliberately, projected her own mood swings onto her characters. Through her fiction she learned to understand her own bipolarity and "enclose the whole."

Appendix C
Leonardo da Vinci, Roger Fry, and Virginia Woolf

From the time when human beings first made art and, further, talked about it, two questions emerged that have continued to shape aesthetic discussions: How should (or should not) art relate to human experience? And which art most fully accomplishes (or negates) that relationship? At different times, art has been thought to have magic power; to give order to a chaotic world; to validate an existing order; to elucidate meanings beyond this world; to teach lessons; to entertain; to delight; to record; to express emotions, dreams, and truths hidden beneath the surface world; to civilize; to humanize; and (in the modern period) to exist beyond the relevance of or even relationship with our experience. The visual and verbal arts have often shared each era's assumptions about the function of art but nonetheless have quarreled over which accomplished that function better.

From Greek Classicism until the High Renaissance (and sometimes thereafter), these quarrels were masked by convenient but largely unexamined phrases proclaiming the unity of the arts. One such phrase was

Simonides' stating (and Plutarch's relating) that poetry was a speaking picture and pictures were mute poems. Another was Horace's *ut pictura poesis*—"as with pictures, so with poetry." Another was the notion of "sister arts," which were sometimes claimed to be sisters born at the same time. The currency of such formulations can be attributed to their wit and symmetry and their users' need to overcome what we might call aesthetic inferiority complexes. Disputants usually seized on such phrases to establish either that painting was as elevated and intellectual as writing or that writing was as realistic and "true" as painting.

Painters have been particularly defensive because the ancients excluded painting from the ranks of the liberal arts. Their assumption was that painting and sculpture were mechanical rather than mental arts and thus lesser than the arts of literature, rhetoric, dialectics, arithemetic, geometry, music, and astronomy. Then, in the Renaissance, when the Italians discovered and systematized the principles of perspective, the painters' mathematical sophistication defied their supposedly inferior status. Hence, in Raphael's fresco *The School of Athens*, the representatives of the seven liberal arts are gathered around Plato and Aristotle, with two painters (one being himself) advancing to join the group.

Whereas Raphael saw the painters' bid for inclusion as an expression of their rising status, his near-contemporary Leonardo da Vinci believed that painters already held superior status. Indeed, the illustrations for Leonardo's treatises visually proclaimed the mathematical and scientific sophistication of painting. Leonardo rationalized that painters had been excluded from the ranks of the highest arts only because they did not advertise their accomplishments in words. He argued that "painting is superior to poetry. But because those who practice it do not know how to justify it in words, it has remained for a long time without advocates" (12).[1] Painters may indeed have been mute and modest, but Leonardo was neither, writing eloquently and flamboyantly in defense of his art.

Leonardo used the ancient formulation "Painting is a poetry that is seen and not heard, and poetry is a painting which is heard and not seen" (17) to set up his argument. Given his premise that sight is a nobler sense than hearing, he contended that the art appealing to sight must therefore be the nobler art: "Painting makes its end result communicable to all the generations of the world, because it depends on the visual faculty" (9). He asked why people traveled to beautiful sites: "Why not accept the poet's description of those places and stay at home, escaping the excessive heat of the sun? Would this not be more useful to you and less fatiguing?" (29). His answer was that people need visual impressions that words cannot replicate.

Leonardo's argument privileged immediacy, for painting requires no interpreters and "at once satisfies mankind, no differently than do things produced by nature" (9). The poet would be worn out if he tried to "describe fully what the painter with his medium can represent at once" (24). Because literature takes place in time, its harmony cannot be fully understood: "The poet's creations are read over long intervals of time, and frequently they are not understood and need various commentaries, but rarely do the commentators [using words] understand what was in the poet's mind" (25). Painting, on the contrary, can be comprehended at once by everyone.

Leonardo maintained that the painter's creation is God-like, whereas the poet's is only man-like:

> Painting presents the works of nature to the senses with more truth and certainty than do words or letters. But letters present words with greater truth than does painting. Let us state that science to be more admirable which represents the works of nature, than that which represents the works of workers, that is, the works of men. Words, in the case of poetry and such creations are the work of men and are made known by means of the human tongue. (9)

Leonardo's argument was that God created nature, which in turn gave birth to painting. Accordingly, "painting [is] the grandchild of nature and related to God" (5). Words, though, offer only shadows of nature, and commentary offers only shadows of shadows (14).

The grandchild and shadow metaphors mask a fundamental ambiguity in what Leonardo was saying. As a direct descendant of nature, painting resembles or imitates nature. He advised, "Poet, make yourself an imitator also" (21). But when Leonardo compared painting and sculpture, imitation was no longer an adequate yardstick for measuring the superiority of painting, for sculpture is not only visual but arguably more mimetic or "lifelike" than painting. Faced with this comparison, Leonardo no longer privileged imitation. He then argued not from nature but from mind: "There is no comparison between the mental effort, amount of skill, and analysis required in painting and that required in sculpture" (33). In fact, painting "compels the mind of the painter to transform itself into the very mind of nature, to become an interpreter between nature and art" (41).

Suddenly being the "grandchild of nature" did not mean reproducing God's works but creating in the same way as God does. A painter can induce men "to love and to fall in love with a painting that does not represent any living woman" (22). (Here perhaps Leonardo was referring to his *Mona Lisa*.) If a writer claimed creative superiority, a painter could best him:

> If you say: I shall describe hell or paradise, and other delights or horrors, the painter will surpass you because he will place before you things that, although silent, will tell you of delights, or will terrify you and move your spirit to flight, for painting moves the senses more quickly than does poetry. (22)

Thus even though Leonardo sometimes privileged imitation for itself, at other times he preferred the mind's capacity to reshape or create on its own.

Presumably Leonardo intended to edit these notes into a coherent treatise that would have eliminated much of their redundancy and oversimplification, but it is hard to imagine that he could have avoided the fundamental inconsistency of his argument. This is the crux of the dilemma that has plagued aesthetic discussions throughout time: Is art to be valued for its relationship to the world or for its transformation of that world? Should the human mind replicate or reshape nature? Do the visual or the verbal arts better achieve the superior goal?[2]

Although these issues cannot be resolved, what I term a *paragone* has conveyed a sense of urgency in certain historical periods: the Renaissance (with Leonardo in particular),[3] the eighteenth century (with Gotthold Lessing and Edmund Burke in particular),[4] and the early modern period (with countless painters and writers, including the particularly telling instances of Roger Fry and Virginia Woolf).[5]

By the early twentieth century, relativity, alienation, and skepticism had undermined many of Leonardo's assumptions, principally that the artist re-creates in the way that God does, that art "mirrors" nature, and even that "nature" can be known. Modernism countered that because pen and brush could not really reproduce or represent a world, their signs were arbitrary. Given the arbitrariness of pen and paint, Roger Fry insisted that writing and painting be judged by values other than faithful representation and association. All his life, he sought to resolve the *paragone* with an aesthetic common to all the arts. Fry believed that the brush could lead the pen, and from Virginia Woolf especially, he learned that the brush could also follow the pen.

Although Fry was almost as aggressive an advocate for painting as Leonardo was, he did not argue with him on a theoretical level. And even though he turned Leonardo's arguments inside out, Fry's theory traveled on similar paths and stumbled on similar obstacles. In fact, Fry's early writing had the same problems as Leonardo's did. Fry praised paintings for both their dramatic "truth" and their formal designs that remade the "true" world. But by 1910, when Virginia Stephen first knew him, Fry had purged the "truth" factor from his aesthetic judgment. By then, he was contending that representative values were irrelevant to aesthetic judgment. Later, he qualified the extreme implications of that statement by claiming that he meant only that the representation of particulars was irrelevant to art.

At the same time, Fry's formalist theories were as influential in literary as in art criticism. Certainly, e e cummings studied them.[6] They perhaps influenced Joyce and Eliot, and they may have inspired the New Criticism, which dominated literary studies for some fifty years. Virginia Woolf both absorbed and modified them.

In the early 1920s, there is so much overlap between Virginia's and Roger's essays and reviews that their common aesthetic can be distilled into basic precepts. Except in their complementary pieces on Spain, in which Roger regretted not having the words to describe accurately what he saw and Virginia regretted not being able to paint what she saw, their aesthetic assumes that all art is an abstraction from life. Hence, each usually considered "realism" a naïve notion.

Roger and Virginia offered an alternative to "realism" that we might call *essentialism*: The "reality" to which both arts should be "true" was not the material surface of life but an essence beyond, behind, above, or around material existence. The smallest and most quotidian detail may therefore contain more essential life than the largest and apparently most grand. This argument also was expressive: Art expresses (Fry) spiritual and (Woolf) imaginative experiences in forms that are not imported but are created by those experiences. Thus both Roger and Virginia followed something of Leonardo's argument when he compared painting with sculpture, that the best artist is "an interpreter between nature and art" who can reshape or create essences not immediately comprehensible in nature.

In essentialist art, pattern (not an accurate rendering of the subject) can organize a painting. And pattern (not conventional ordering mechanisms such as plot) in a novel can be appreciated independently of subject, meaning, and such empathetic emotions as pity, pleasure, and fear. Like Leonardo when he compared painting and sculpture, Virginia and Roger believed that art is not *about* something else; it *is* something in its own right: autonomous, self-contained, controlled by its own inner harmony.

Leonardo, however, did not face the problem these modernists did. They saw art becoming purer and realized that pure or abstract art would appeal more to an aesthetic appreciation of structural design and harmony than to the emotions of everyday life. Thus both Virginia and Roger hesitated over the issue of how pure art should and could be. Virginia, in particular, as we see in this book, evolved in her art an imaginative amalgam of fact and vision, realism and essentialism.

Appendix D
A Redating of Virginia Woolf's Suicide Letters

The chronological order of Chapter 12 reflects my own redating of Virginia Woolf's suicide letters, based on both my reading of Leonard Woolf's autobiography and the documentary evidence.

Leonard wrote that there was "a note in my diary on 18 March that she was not well and in the next week I became more and more alarmed. I am not sure whether early in that week she did not unsuccessfully try to commit suicide" (LWA, 2:433). The editors of Virginia's letters believe that her three suicide letters were written "on the 18th, 23rd and 28th March respectively."[1] In addition, they assume that when Leonard wrote on 18 March that Virginia was not well, she tried to drown herself that very day. I disagree; I read Leonard's "in the next week" as referring to the week beginning (as English calendars do) on Monday, 24 March. Therefore, when he followed the reference to the "next week" with the phrase "early in *that* [emphasis added] week," I think that he meant the Monday or Tuesday of the week beginning on 24 March.

On Tuesday, 18 March, Leonard noted Virginia's illness, and on

Thursday, 20 March, Vanessa came to visit. Then probably on Friday or perhaps on Saturday, Virginia received Vanessa's letter, which I regard as a major provocation for Virginia's suicide. In fact, Virginia's reply was her suicide note to Vanessa, dated "Sunday," which would have to have been Sunday, 23 March. The editors of Virginia's letters think it unlikely that she would have written to Vanessa before Leonard. They probably are correct on that point but, I believe, incorrect in concluding therefore that the "Tuesday" letter to Leonard must date from the preceding Tuesday, 18 March.

I think that the editors are wrong by a week for the "Tuesday" letter and by three to five days for a short undated letter to Leonard. Virginia definitely wrote the letter to Vanessa on 23 March, and I believe that she wrote the undated one to Leonard about the same time, on Friday, Saturday, or, most likely, the same Sunday. The editors consider this undated letter "almost certainly the last time she used her pen," and they date it 28 March. Its brevity and haphazardness, however (she remembered as an afterthought and noted in the margin the location of the Mauron/Fry correspondence), suggest a first, not a final, version of her suicide letter to Leonard.

On 24 March, Leonard noted that Virginia was "slightly better" (LWA, 2:434). Then, in that same week, on Tuesday, 25 March (the week *after* the week of 18 March, as I read Leonard's recollections), Virginia wrote the more elaborate suicide letter (dated "Tuesday") to Leonard and may have unsuccessfully attempted suicide.

My evidence is empirical as well as interpretive. The editors tell us that the two letters that I—not they—believe she wrote about the same time, probably on Sunday, 23 March, were "written on paper 8 in. × 10 in. torn from the same pad. The 'Tuesday' letter to Leonard is written on slightly smaller paper, 6 ½ in. × 8 in., which increases the likelihood that it was written on a separate occasion" (VWL, 6:490). The same argument could, of course, establish that the two letters on identical paper were written at the same time.

Virginia also handled alike her left margins on those two letters, indenting them under her salutations of "Dearest" (to both Leonard and Vanessa). And the wording in the two letters that I believe were both written on Sunday is very similar ("All I want to say" and "I want to tell you," for example). Despite the similarities in phrasing, salutations, and positioning, however, the two letters do not seem to have been written in immediate sequence, for as Christopher Fletcher of the British Library confirmed, they "were written with different pens and ink and the hand varies quite noticeably between them: in the letter to Leonard the hand is markedly weaker than that in the one addressed to Vanessa."[2] Fletcher also pointed out that in the undated letter to Leonard,

> Virginia claims that "I cant write this even" whereas everything in the [letter] to Vanessa suggests it to have been written in a more composed state of mind. The letter to her sister begins in quite conventional terms before broaching the subject of her mental health, warmly thanking her for her recent letter, whereas that to Leonard betrays a more urgent, distressed and distracted tone from the start. Virginia, after all, decided that the letter to Vanessa was good enough to be sent [or left for her], whereas that to Leonard required re-drafting. (to me, 8 November 1995)

Hence the undated letter to Leonard was, as he wrote on its verso, "not the one left for me by V. I found it later on the writing block."

Likewise, the similarities in format suggest affinities between Virginia's "Sunday" letter to Vanessa and her undated letter to Leonard. But it is impossible to know whether Virginia wrote the shaky letter to Leonard before or after the more composed letter to Vanessa. Perhaps Virginia first—in an assertive, confidently ironic mood—wrote to her sister that she "loved" her letter and planned to kill herself. Then late in the day, perhaps after her visit into the village, Virginia tried to write an equally composed letter to Leonard but succeeded only in writing the shaky, undated one.

Or, in distress perhaps on Friday or Saturday but certainly by Sunday, she could have written the undated letter to Leonard; then later on Sunday, she might have pulled herself together to write with bravado the last of a lifetime of letters to Vanessa. Between the undated letter to Leonard and the "Sunday" letter to Vanessa, we cannot know which Virginia wrote first, but we can see that those two letters were closely related and that the undated letter to Leonard preceded the more fully developed "Tuesday" letter.

One more piece of evidence indicates a more concrete and crucial connection between the undated letter to Leonard and the "Sunday" letter to Vanessa: Not only were they written on identical notepaper, but they originally were folded together as one. First, Virginia made a central straight horizontal fold, then she made a second, slanting horizontal fold—which, of course, appears twice on both letters—and then she made a vertical fold, reducing each sheet to a thick packet whose height and width were one-eighth its flattened size.

Two identical pages each folded into one-eighth its original size might prove only that Virginia habitually folded notes into eighths. The second horizontal folds on both letters, however, slope upward at identical angles on the top halves of the sheets and downward at corresponding identical angles on the bottom halves.

The unique foldings of these letters, then, establishes that they must have been folded together at some point.[3] There still is a problem with this empirical evidence, however, for both are soiled along their outward creases, and the undated Leonard letter is more heavily soiled. Possibly the letter to Leonard was outside, but then how would the letter to Vanessa have become soiled at all? And how do we know that Leonard did not fold these two letters after Virginia's suicide?

I propose the following scenario: Virginia wrote the two letters around the same time and then folded them together, with the letter to Vanessa on the outside (hence the appropriateness of labeling it "Vanessa" on its back, where she could see the label, which she could not if it were the inside letter). In that case, the "Vanessa" letter would have collected dust for two days, between Sunday and Tuesday. Then on Tuesday, 25 March, Virginia decided to go ahead with her plans to kill herself. She removed her letters from their hiding place, found the one to Leonard to be inadequate, and decided to write another. She carefully redrafted her undated letter in a second, fuller letter to Leonard (dated "Tuesday") on smaller 8 x 6 ½ notepaper, using both sides and keeping her left margin flush with the salutation. Then she put the new letter to Leonard (which has no soiled marks on it) and the slightly soiled "Vanessa" letter in separate, clean, blue envelopes. Finally, she returned the original, rejected, letter to Leonard (folded just the same way as the "Vanessa" letter) to the less than pristine drawer.

If Virginia attempted suicide on that Tuesday, she may have put out the two envelopes, labeled "Leonard" and "Vanessa," then. But if she left out two letters that day, Leonard did not find them. Or she may have put them out and then retrieved them after an abortive suicide attempt and placed them in a clean place (perhaps between the pages of a book). Meanwhile, the original letter to Leonard was pushed to the back of a dirty drawer, where it gathered dust for three days. On 28 March, Virginia took the envelopes out of the book (or whatever) and probably put them in her pocket. Leonard said that he had visited her that morning and "found her writing on the block." But that does not mean that she was at that moment writing a suicide letter. Or perhaps she was only pretending to be writing, so he would not be suspicious of her activities. She then went into the house with Leonard. When he went back out into the garden, Louie, at his instructions, seems to have tried to get Virginia to do some dusting.

After a few minutes, Virginia may have taken the two letters, in their blue envelopes, from her pocket and placed them where they could be found. (On the verso of the "Tuesday" letter Leonard wrote, "This is the letter left for me on the table in the sitting room which I found at 1 on March 28." In his autobiography, however, he remembers finding this letter "on the sitting-room mantelpiece" [LWA, 2:435], where she may have left the letter to Vanessa. But because he wrote his autobiography many years later, Leonard probably confused the two.) Or Virginia may not have had the letters in her pocket but, after she stopped dusting, went back to the lodge, retrieved them, and returned to the house. At some point, she also took from her drawer the short, undated, fairly soiled letter to Leonard and left it out on her writing desk. Then Virginia went out to take her last walk.

Of course, Virginia may have separated the letters written on identical paper and placed the letter to Vanessa in one blue envelope and the fuller "Tuesday" letter to Leonard in another. Then, after her death, when these letters were found, Vanessa and Leonard could have folded them together. But the evidence suggests otherwise. The very different folds in the two letters to Leonard were present on 11 May 1941, when Leonard used their creases as margins for his notes on their backsides. Because he did not mark the letter to Vanessa, she apparently kept it with her.

When deposited in the British Library, the "Sunday" and "Tuesday" letters came in their blue envelopes, just as Virginia had left them. As Sally Brown pointed out to me, "The envelope addressed to Leonard has been carefully opened," whereas the one to Vanessa "has been ripped apart as though in great haste and apprehension." Their differing approaches to the envelopes, like their differing treatment of Virginia, make it unlikely that either Leonard or Vanessa would have taken the "Sunday" letter to her out of its clean envelope and folded it with the grimy undated early letter to Leonard, leaving the "Tuesday" letter in its clean envelope. If neither of them would have folded those letters together, then Virginia must have been folded them together when she first decided to commit suicide, probably on Sunday, 23 March.

Although this scenario is necessarily speculative, it is derived from enough evidence, I believe, to restructure correctly the sad chronology of Virginia Woolf's preparations for suicide.

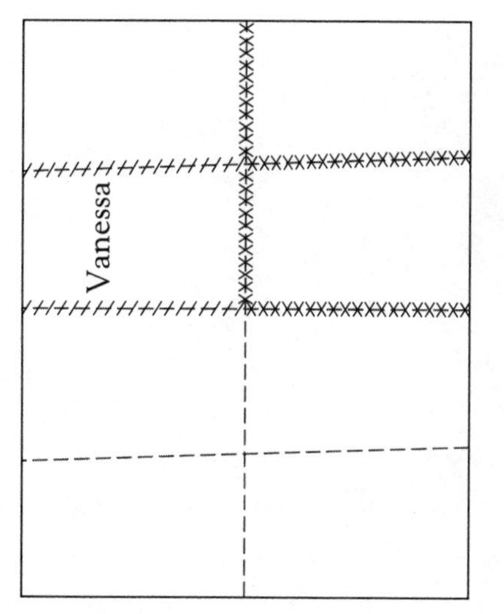

"Sunday" letter to Vanessa

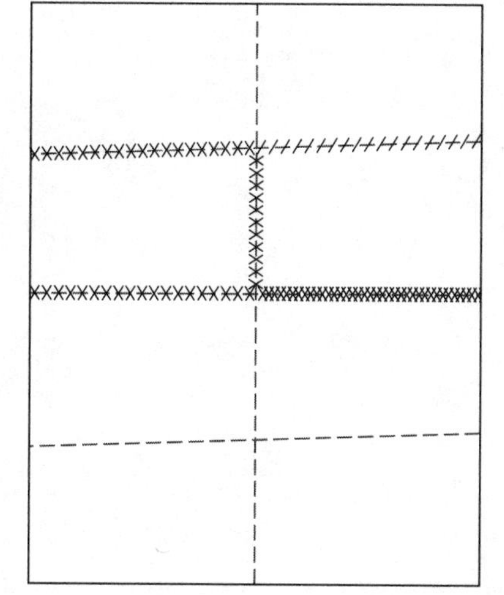

Undated letter to Leonard with P.S. about Mauron letters

//// = Very light soil
XXXX = Slight soil
XXXXX = Very heavy soil
– – = Fold lines
—— = Outside edge

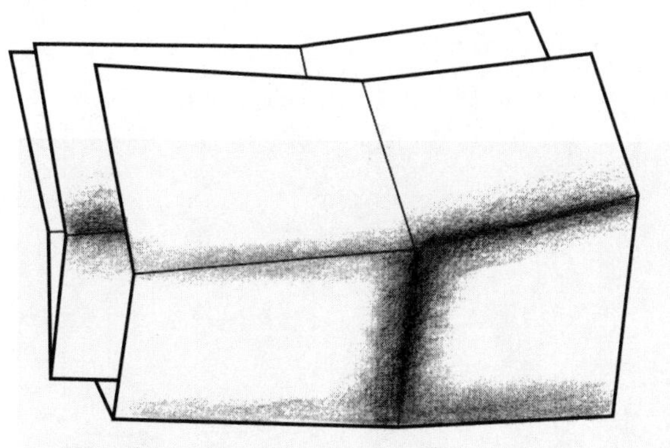

The folds and placement of the letters

Notes

Chapter 1

1. LS/JP[S], 19 July 1877. For an assessment of Stephen's eminence, see Annan, *Leslie Stephen*.
2. Marsh says that Prinsep's *Head of a Girl* is "possibly Julia Jackson" (*Pre-Raphaelite Women*, 29). After comparing it with Figure 1, I believe that it is indeed a drawing of Julia, even though the date Marsh gives as ?1875 must be nearly twenty years off, since Julia was born in 1846. The date for Figure 1 in the Stephen family photograph album is 1856, but Julia seems somewhat older. In Weaver, *Whisper of the Muse*, several of Julia Margaret Cameron's photographs taken on the lawn at Little Holland House include a large tree in the background; this is surely the same tree as the one the young Julia stands before in Figure 1. Her uncle, Lord Somers, was an amateur photographer and for some time vice president of the Photographic Society of London. If the picture was not taken by Julia Margaret Cameron, he seems to be the next most likely photographer.
3. For a description of Saxonbury, see Stephen, *Mausoleum Book*. For photographs of John Jackson, Maria Jackson, and Saxonbury, see Richardson, *Bloomsbury Iconography*, plate 3.

4. Sir Winston Churchill lived at 28 Hyde Park Gate into his nineties.
5. Vanessa was named after the second important woman in Jonathan Swift's life, the first being Stella. (Leslie Stephen wrote a life of Swift.) Virginia Woolf's reflections on Swift suggest that to her the names seemed almost self-fulfilling prophecies. See "Swift's Journal to Stella" (CE, 3:71–79).
6. LS/JPS, 4 January 1881?, 28 April and 29 September 1882, and 4 and 24 January, 14 April, and 4 January 1881. Maria Jackson was less pleased with Vanessa's early outspokenness, writing, "My sweet little V tell her angels are never stupid" (MJ/JPS, 17 June 1881).
7. MJ/JPS, 16 April and ? April 1881. Apparently a month or two beforehand, Bunch had ceased to be a live-in nurse.
8. LS/JRL, 28 January 1882; MJ/JPS, 10, 9, and 15 February 1882; LS/JPS, 9 April 1882. Julia may have ceased nursing Virginia, given Maria Jackson's instructions, as early as February and still have felt guilty about it in April when Leslie assured her of his approval. Or she may have nursed Virginia for about two months, in which case the April "bottling" would have been a recent step.
9. For more details, see Annan, *Leslie Stephen*, chap. 3.
10. MJ/JPS, ?, 9, 15, 10, and 25 February, 26 June, and 7 February 1882. In MJ/JPS, 9 February 1882, Maria Jackson tells of reading one of Yonge's books to Stella, perhaps the popular novel *The Daisy Chain*, a tale of generations growing up, marrying, and bearing children—in short, of "uneventful family life." See Battiscomb, *Charlotte Mary Yonge*.
11. In the mother–infant plot, Klein posited an alternative to Freud's oedipal plot. According to her, prenatal fusion with the mother is best approximated during nursing, but the baby also is conscious of the breast as withdrawn, empty, or, like the baby, vulnerable. As the child becomes more and more conscious of separation from its mother, it more desperately seeks to make whole, to achieve the lost oneness, and to overcome its angry urges. From loss and guilt emerges a desire to make reparation to the mother. Klein found these desires "the driving forces in all constructive activities and interests" (*Love, Guilt, and Reparation*, 295).
12. LS/JPS, 10 October 1883; Thomas Bewick (1753–1828), a wood engraver, was famous for *British Birds* (1797–1801); Stephen, *Mausoleum Book*, 85; LS/JPS, 9 April 1884.
13. Bicknell, Introduction to *Selected Letters of Leslie Stephen*.
14. Chodorow argues that "mothering" need not be done by the biological mother but by one stable and loving person (*Reproduction of Mothering*, esp. 68–80). Caramagno exposes flaws in Freudian interpretations that focus on the date of weaning as the solely relevant aspect of mothering (*Flight of the Mind*, 129–31).
15. Bollas explains that the mother's failure to provide a "facilitating environ-

ment, through prolonged absence or bad handling, can evoke ego collapse and precipitate psychic pain." Furthermore, "what was an actual process can be displaced into symbolic equations" with the support of a mother substitute. If there is no such substitute, the child will suffer considerable psychic pain (*Shadow of the Object*, 33, 14, 15). See also Chodorow, *Reproduction of Mothering*, 68.

16. Stella Duckworth/JPS, 13 April 1884 (Berg). Julia and Adrian were visiting her mother at Brighton. My transcription differs slightly from Dunn, who where I read "Pigswash" reads "Pugwash" (*Very Close Conspiracy*, 9). Leslie's record of Virginia's saying "dont go, Papa!" is also dated 13 April 1884. His earlier story of her saying "Kiss" to him at twenty months (10 October 1883) seems to show Virginia emerging from her speechless days. Bell says she did not talk until age three (1:22), perhaps as family legend had it, but these examples prove that she talked at age two.
17. Virginia says "we" persecuted Vanessa, but Vanessa, in her "Notes on Virginia's Childhood," makes it clear that the persecutor was Virginia alone (LWP, 2:6a). Nevertheless, Virginia's calling Vanessa "Saint" seems less "horrid" than Vanessa's calling Virginia "Goat." Neither sister provides an age for this "persecution."
18. In his *Mausoleum Book*, Leslie Stephen copied excerpts from Julia's letters to him, omitting salutations and closings. These excerpts testify to Julia's deep affection for him and her honesty and also that Julia consoled Leslie but did not offer him "a diet of verbal treacle." She could disagree with and criticize him (Bicknell, "Ramsays in Love," 6).
19. These have been gathered together and published in JDS. See Broughton, Review of *Julia Duckworth Stephen*. See also Stemerick, "Virginia Woolf and Julia Stephen," 51–80, and Marcus, "Virginia Woolf and Her Violin," 27–49, both in Ginsberg and Gottlieb, *Virginia Woolf*.
20. LS/JPS, 5 February 1887. Maria Jackson was pleased that Julia did not spoil Vanessa (MJ/JPS, 10 July 1881).
21. Elizabeth Steele, who edited the stories, explains that although "cleverness in girls is unusual in these stories," they convey "Julia's sneaking admiration for devilish ingenuity, especially as practiced by small boys" (JDS, 31,32).
22. LS/JPS, 29 January 1887. Bicknell tells me that in an 1887 letter, Leslie speaks of being able to walk twenty to thirty miles without becoming exhausted.
23. LS/JPS, 6 October 1886. In LS/JPS, 12 October 1886, Leslie Stephen refers to Laura's living with a Miss Searle, a fact previously unknown to other biographers.
24. LS/JPS, 1 April 1887; MJ/JPS, 13 February 1888. For Maria Jackson's rather incoherent narrative of the house-hunting venture, see Appendix A.
25. Virginia seems to locate this event at Talland House. The current owner

of the house told me that no such slab or shelf exists there now (Mrs. M. Hanson, July 1995). Also, because Talland House was so much more open than the other houses that figured in Virginia's childhood, this experience seems more likely to have happened at either 22 Hyde Park Gate or Brunswick Terrace, both of which had many levels, halls, and stairways.

26. Quentin Bell wrote, "If Gerald had broken the hymen in the course of his nasty investigations his victim would I imagine have felt something more than resentment and dislike; she might well have let out a howl that would have brought family and servants to the spot" (Review of *Virginia Woolf*, 4).
27. Anne Olivier Bell writes that given that the "degree of ignorance and the lack of information then available to young people on the bodily organs and functions was almost complete," it is "not very surprising that in the 1880's a fatherless upper-class Eton school-boy should be curious about the female anatomy, and should seek to satisfy that curiosity by examining an intimidated little sister." Olivier Bell further explains that there is no evidence for seeing, as DeSalvo (*Virginia Woolf*) does, this event as a matter of continual abuse (letter).
28. She later told Ethel Smyth about the experience (VWL, 6:460). There seems to be no evidence that she told Leonard Woolf about it. There was a now lost "subjective" Memoir Club presentation on 15 March 1920 that created "a kind of uncomfortable boredom on the part of the males" (D, 2:26). However subjective and distasteful, that memoir does not seem to have been a version of the Gerald Duckworth tale she wrote about in 1939 and 1941. Olivier Bell suggests that "it was her treatment rather than the recollections themselves which caused her discomfiture and determined her to write her next memoir *22 Hyde Part Gate* in a fearfully brilliant manner" (letter to me, 6 May 1991).
29. MJ/JPS, 28, 8, and 30 January and 8 February 1889. Bicknell tells me that the dates of their visit were 20 January to 15 February 1889.
30. LS/JPS, 29, 25, 27, 31, and 29 July and 2 August 1893. Leslie mentions the train ride on 25 July 1893 but says only that Vanessa declined food and said she was not uncomfortable.
31. For analyses of Stephen's work and character, see Rosenbaum, *Victorian Bloomsbury*, and Annan, *Leslie Stephen*.
32. Stephen was, however, too good a critic always to follow his own dictum. In his essay on Byron for the *Dictionary of National Biography*, for example, Stephen dismissed rumors of Byron's incestuous relations with his half sister, because evidence (which he does not identify) "proves this hideous story to be absolutely incredible." While acknowledging the poet's "strange infirmity of will" and "debasing life," Stephen also credited the strength of Byron's "intellectual activity" and the power of his poetry (*Dictionary of National Biography*, 1:140, 145, 143).

33. From an unpublished biographical study of Jonathan Swift by Sir Harold Williams. My thanks to John Irwin Fischer for providing me with these quotations.
34. Julia Stephen expected the servants to enjoy vicarious participation in the Stephen family, not to have family lives of their own. See ["The Servant Question"], in JDS, 248–52, esp. 252.
35. JPS/George Duckworth, n.d. (Berg). In his chapter "Genetics and the Stephen Family," Caramagno traces Virginia's inherited predisposition to manic-depression (*Flight of the Mind*, 97–113, 313–14). Annan calls Virginia "flesh of his flesh and bone of his bone" (*Leslie Stephen*, 113–145, esp. 94–95, 134–35). For further discussions of the similarities between Virginia and Leslie, see also Hill, "Virginia Woolf and Leslie Stephen," and Hyman, "Reflections in the Looking-Glass."
36. LWP, 2:6a. "Goat" may also, according to Elizabeth Heine, be a friendly term, not unlike calling children "kids."
37. LS/JPS, 22 July 1885, and 15 and 17 April 1887; MJ/JPS, 5 June 1887, and 13 October 1889. On 30 June 1890, for example, Julia was up all night nursing Thoby in some illness; on 1 July, she gave a party; and on 4 July, she was in Liverpool, seeing Leslie off on a trip to America to receive an honorary doctorate from Harvard University (MJ/JPS, 3 July 1890; John Bicknell to me, 11 February 1995).
38. Volumes 1, 2, and 5 are now in the British Library, ADD. MS. 70725. Quentin Bell offers a detailed account of this remarkable production (1:28–30).
39. Thoby and Virginia were listed as joint authors (Vanessa was the unlisted scribe) of the serialized installments of the *Hyde Park Gate News* called "A Cockney's Farming Experiences" and "The Experiences of a Paterfamilias." Quentin Bell believes that Thoby was the "dominant partner" in that coauthored task (*Virginia Woolf Miscellany* 38 [Spring 1992]: 2). Olivier Bell transcribed the sixty-nine surviving issues of the *Hyde Park Gate News* before its acquisition by the British Library. She writes, "My own view is that the greater part was actually written out by Vanessa." She thinks that Vanessa was sometimes the editor as well as the scribe, although not necessarily the author, and that the paper was often a collaborative effort, organized by Vanessa (letter to me, 17 February 1992).
40. HPGN, 7 December 1891, and 23 May 1892. The only illustrations are in the issues dated 25 December 1891 and 11 January 1892. These clumsy drawings, according to Olivier Bell and Quentin Bell, are by Thoby's slovenly hand (interview, 27 July 1993).
41. HPGN, 6 April 1891, and 4 April, 16 May, 3 October, and 8 August 1892. Although she was in private care with Miss Searle, Laura joined the family for

the summer outings to St. Ives until 1893 (LS/JPS, 12 October 1886, and 1 August 1893).

42. Psychoanalysts argue that the mothering process carries its own aesthetic and that the baby "takes in not only the contents of the mother's communications but also their form." Thus the aesthetic experience is "an existential recollection" of "the maternal aesthetic" (Bollas, *Shadow of the Object*, 33, 16, 34–35).
43. HPGN, 21 and 28 November and 19 December 1892. King mistakenly assumes that Adrian actually produced a rival paper (*Virginia Woolf*, 155).
44. HPGN, 21 November 1892, LS/CE Norton, Harvard Collection. Also quoted by Alan Bell, Introduction to Stephen, *Mausoleum Book*, xxviii. Of course, Leslie also may have used those terms in conversation, so it is impossible to know who borrowed from whom.

Chapter 2

1. LS/JP[S], 18 July 1877; JP[S]/LS, 19 July 1877; LS/JP[S], 19 July 1877. I am quoting from both the original letters, now housed in the Berg Collection, and the "Calendar of Correspondence" (British Library, ADD. MSS. 57922, 78). This "Calendar" is Stephen's summary of important portions of the correspondence, and the letters offer a strong defense of his position on education for women. The "Calendar," written after Julia's death, tends (by selection) to place more emphasis on his apologies than on his defenses. See Annan, *Leslie Stephen*, 120.
2. Annan suggests that Laura was probably the Stephen child who suffered most for being Leslie's daughter (*Leslie Stephen*, 122; for Annan's discussion of Julia's distaste for female education and female careers, see 119–21).
3. HPGN, 13 June 1892. The accepted notion of Julia as "perfect" mother persists into our own time. Annan writes, "She had the gift of combining genuine tenderness with firmness towards children so that they always knew where they were and what was expected of them, and their father boasted that she never had to punish them. She could heal a child's wound before it could fester, read thoughts before they were uttered, and her sympathy was like that of the touch of a butterfly, delicate and remote" (*Leslie Stephen*, 102).
4. MB, 80. Even Julia's admirers could censure her for her conventionality. For example, see *Letters of James Russell Lowell*, 2:348, and *Letters of George Meredith*, 2:426–27.
5. JPS/GD, 18 March 18?? (Berg). Since Thoby is absent, this letter might date from 1892, when he went off to school. Virginia was ten then.
6. JDS, 123. For a girl's natural rejection of her mother, see Chodorow, *Reproduc-*

tion of Mothering, 121–29. For the results of having this psychic weaning process radically interrupted, see Suttie, *Origins of Love and Hate*, 87.

7. The friend was James Payn, who had increased the *Cornhill*'s sales by lowering its literary standards. Whereas Stephen had published such fiction as Henry James's "Daisy Miller: A Study," Payn published more accessible works like the Sherlock Holmes stories (Annan, *Leslie Stephen*, 83).
8. LS/JPS, 16 April 1895. Other information in the last two paragraphs is from John Bicknell to me, 25 October 1994.
9. For a discussion of Virginia's inability to grieve, see Spilka, *Virginia Woolf's Quarrel with Grieving*.
10. In "The Symbol," probably Virginia's penultimate short fiction piece, a woman confesses that she had longed for her mother's death. Her mother had lingered before finally dying, as Julia did not, and Leslie Stephen did. Still, the tale encapsulates some of the guilt Virginia felt at each of her parents' deaths: The woman had "never told any one; for it seemed so heartless" (CSF, 289).
11. MB, 41. Virginia used the phrase "utmost intimacy" to refer to Stella's being alone in his study with Leslie and having to listen to his personal regrets and guilt, from which, of course, she could not absolve him (MB, 41). DeSalvo assumes that by "intimacy," Virginia (then twenty-five) meant that Stella was required to provide Leslie with sexual favors (*Virginia Woolf*, 67). DeSalvo never considers the passage (MB, 104) where Virginia speaks of her father's honor, his love for one woman, his chasteness.
12. MB, 40. For more details, see Hill, "Virginia Woolf and Leslie Stephen." In early drafts of "A Sketch of the Past" (MHP/A5c), Virginia was more explicit than in the final version, about her fury with her father and her inability to express it.
13. MHP/A5c. For a description of the physical symptoms of manic-depression, see Caramagno, *Flight of the Mind*, 11–18.
14. There are startling similarities between Styron's experiences and Woolf's. Regarding his mother's death when he was thirteen, Styron explains: "This disorder and early sorrow—the death or disappearance of a parent, especially a mother, before or during puberty—appears repeatedly in the literature on depression as a trauma sometimes likely to create nearly irreparable emotional havoc." Styron mentions Woolf as one whose symptoms replicate his own (*Darkness Visible*, 46–47, 79, 82). In "Manic-Depressive Psychosis," Caramagno objects to literary critics who either apply to Woolf procrustean Freudian models of neurosis or, like Trombley ("*All That Summer She Was Mad*") and Stemerick ("Virginia Woolf and Julia Stephen"), deny that she had any problem at all (*Flight of the Mind*, 1, 2, 36).
15. "At George" does not appear in the typescript of this passage, which led me to conclude erroneously that Stemerick added it. Stemerick's transcription is partially correct, for the manuscript version of this passage has a barely legible

scribbled phrase that does seem to be "atGeorge" (written as one word). Stemerick's transcription, however, is not entirely correct, for she does not preserve Virginia's blank space, ellipses, and question mark. See Stemerick, "Virginia Woolf and Julia Stephen," 61.

16. Bollas explains ego structure as "the trace of a relationship" (*Shadow of the Object*, 50).
17. All quotations in the preceding paragraph without parenthetical citations are from the unpublished version of "A Sketch of the Past" (MHP/A5c).
18. Physical irregularities (headaches, fevers, influenza) are associated with the timing of manic-depressive episodes, and depression can be manifested as a physical disorder (Caramagno, *Flight of the Mind*, 13).
19. "The role of creativity among [Holocaust] survivors may be compared with that of the bereaved in more usual circumstances, especially those who endure loss in childhood. . . . Through creativity, the artist may confront and attempt to master the trauma on his own terms and, in so doing, complete the work of mourning" (Aberbach, *Surviving Trauma*, 3). Kristeva also speaks of art as constituting both the art object and the artist (Meisel, "Interview with Julia Kristeva," 131–32).
20. PA, 16. The diaries and journals that Virginia kept between 1897 and 1909 have been edited by Leaska and published as *A Passionate Apprentice*. For detailed accounts of Virginia's stylistic development over these twelve years, see Leaska's excellent introductory and editorial comments.
21. PA, 10. In 1897 (aged fifteen) she read, among other books, *Essays in Ecclesiastical Biography*, Lockhart's ten-volume life of Scott, and Lowell's poems (all gifts from her father); Macaulay's five-volume *History*, Carlyle, Dickens, Eliot, Pepys, Trollope, Froude, Lowell, Trollope, James, Hawthorne, Thackeray, Locke, Mrs. Gaskell, Mary Mitford, Anthony Hope, Charlotte Brontë, a history of Rome, a life of Coleridge, Janet Ross's *Three Generations of English Women*, and Leslie Stephen's life of Fawcett.
22. Virginia's first record in her 1897 diary of accompanying Stella to the poorhouse was made on 25 February (PA, 43). DeSalvo discusses these diary entries but blurs the chronology in *Virginia Woolf*.
23. MB, 106–7. In the unpublished version of this analysis (MHP/A5c), Virginia was even more explicit about her father's disguising his own jealousy by accusing Jack of selfishness. She saw the problem as generic, faulting the Victorians in general for lacking a "truth telling machine."
24. Discounting the presence of Vanessa and (for two nights) Leslie Stephen, DeSalvo thinks that Virginia resisted going to Bognor because she realized that she alone "was being entrusted with seeing to it that Stella maintained her chastity until her wedding" (*Virginia Woolf*, 215). Quentin Bell insists "there is not the slightest reason to think that Virginia imagined any such thing" (interview, 1 April 1988).

25. VB/TS, 28 March 1897. The accident had occurred two days previously (VB/TS, 31 January 1897).
26. Quentin Bell blames the sisters' dislike of Brighton on the unbearable sweetness of their Fisher cousins (1:59). But I think there was a prior cause in the Stephen children's exiles there with their grandmother.
27. On the bipolarity of manic-depressives and the fact that their symptoms do not emerge as "full blown" until puberty, see Carmagno, *Flight of the Mind*, esp. 35–36.
28. Alan Bell and John Bicknell speculate that she might have died not from peritonitis but from a tubal pregnancy (Bicknell to me, 9 April 1994).
29. VD/VB, 16 June and 6 July 1942. Leaska writes that Stella's death "robbed" Virginia's vision of love of its "authenticity, making it like so much else during those early years, a chimera, something not to be trusted" (PA, xxx).
30. Her first ended in a miscarriage, and her second produced Laura (Annan, *Leslie Stephen*, 65, 71).
31. Stephen, *Mausoleum Book*, 96, 104. In 1900, he noted Virginia's bout with measles (107).
32. MB, 125; MHP/A5c. Virginia and Adrian later vowed to break that pattern (MB, 125). The Stephen sisters' force of character seems to have developed with their emerging outspokenness. Both sisters became openly and caustically critical of other people, in reaction to Victorian prohibitions. For a discussion of the denial mechanism, see Freud, *Ego and the Mechanisms of Defense*, esp. 87. On Virginia's private handwriting, see also MB, 122.
33. I examined the diaries before they were published and found this one nearly impossible to read. Leaska has deciphered it and describes it in PA, 160, note. He supposes that Virginia followed this odd procedure to "have a handsome leather binding" for her journal, but I think she mutilated the book as a desperate attempt to hide her writing from those who might be curious.
34. VWL, 1:28. DeSalvo maintains that the "peculiar words" were references to the pond's "duckweed," which was code for Gerald Duckworth and forced oral sex (*Virginia Woolf*, 257). According to DeSalvo's logic, slime cannot be actual slime but must be code for saliva. Such a "reading" deconstructs itself. DeSalvo's humorless allegorizing is totally at odds with "a person whose sense of humour is unimpaired by drowning or duckweed." For the longer version, see Woolf, "Terrible Tragedy in a Duckpond." Quotations are from 38, 41, 40.
35. Radin, *Virginia Woolf's "The Years,"* 95. The quotation comes from the holograph of *The Pargiters*, the first version of *The Years*.
36. For more details, see Willis, *Leonard and Virginia Woolf as Publishers*, 5–7.
37. For reproductions, see Gillespie, *Sisters' Arts*.
38. VB/Margery Snowden, 17 and Sunday, ? March 1903. There is some confusion about the year here. Virginia says that she was alone on a Sunday afternoon when "Nessa went to the Watts"—which sounds like an exhibition

(VWL, 1:71). But the letters quoted here are from Watts's home, Linnersease, and on black-bordered stationery. Possibly Vanessa and George visited Watts in 1904 during the period between Leslie Stephen's death and their subsequent visit to Manorbier and Italy (Chapter 3). Watts died on 1 July 1904.

39. VWL, 1:102. DeSalvo translates the metaphor as "Nessa was nothing but a cavity into which Jack could ejaculate his grief" (*Virginia Woolf*, 65).
40. MB, 154. In 1931, Virginia remembered being "dragged by my half-brothers against my will" to high-society parties (VWL, 4:297).
41. It is possible, however, to view Julia's indulgence of her Duckworth sons and her laissez-faire attitude toward boyish naughtiness as contributing factors in her sons' misbehavior. The *Hyde Park Gate News* offers several examples of Julia's indulgence. For example, she did not let George sit for exams lest the experience be too great a strain on his health (HPGN, 21 March 1992).
42. MHP/8, 13(a). About this time, Thoby quoted a few lines from *Lysistrata* to Clive Bell about a "certain youth—Melanion who fled from marriage" to a desert and mountain "so did he hate all the female tribe." Antifemale rhetoric enjoyed a certain caché at Cambridge. Thoby seems to have repeated misogynist rhetoric without a thought to his sisters or to the lessons of *Lysistrata*. He probably would not have understood why George's behavior was so distressing to Vanessa and Virginia.
43. MHP/A14. This version also ends with the phrase about George's being the sisters' "lover," making his unwanted attentions of a piece with his other exuberances.
44. Maynard Keynes responded to this memoir by suggesting that Virginia "continue to write about real people & make it all up" (D, 2:121).
45. FS/VB, 4 December 1918. As Olivier Bell pointed out (letter to me, 6 May 1991), incest was a crime and so was not something even members of the Memoir Club would joke about.
46. VB/VW, 8 September 1904. Leaska points out, "What Virginia wrote in 1897, however, does not convincingly support claims of sexual interference." In fact, the early diaries cast George (or "Georgie") in an "almost benign, light" (PA, xxxiv). DeSalvo, however, assumes that Virginia experienced a "pattern" of abuse lasting for sixteen years, perpetrated by various males in the family (*Virginia Woolf*, 101). But the fact that some Victorian men abused the image of female purity is hardly (despite DeSalvo's conviction) proof that the Stephens or even the Duckworths did.
47. MHP/8, 13(a). Fredegond Shove remembered her father and theirs "always talking about the making of the world—it was the only subject grand enough or great enough for two such wise *thin* men to be talking about at all" (FS/VB, 1925). Fredegond's mother was Florence Fisher (Virginia and Vanessa's first cousin). Her father was Frederic W. Maitland (Leslie Stephen's biographer and an eminent legal historian).

Chapter 3

1. Such behavior is typical of manic-depressives (Goodwin and Jamison, *Manic-Depressive Illness*, 302).
2. VWL, 1:90, 75. Bollas calls this pose "the mother's infant as sexual object" (*Shadow of the Object*, 82).
3. MB, 133; VB/MS, 18 April 1903. Leslie Stephen was actually seventy-one when he died on 22 February 1904, as he was born on 28 November 1832.
4. Some of these drawings are illustrations in *Selected Letters of Leslie Stephen*.
5. Virginia's handwriting, then and later, bore a striking resemblance to her father's.
6. TS/CB, ? April 1904. In his next letter, Thoby was "grieved that you discover so much ferocity in my letter." After parading his own opinions, Thoby concluded, "I am afraid that you have not studied" (TS/CB, ? April 1904).
7. Thoby had written to Clive asking that he find a cheap room for him alone in Paris (TS/CB, late April 1904).
8. Manic-depressive women are more likely than men to have hallucinations, which are "chiefly auditory and visual" and occur less frequently than do delusions (Goodwin and Jamison, *Manic-Depressive Illness*, 252, 263–64).
9. Bell, 1:89. On the absurdity of trying, as Panken (*Virginia Woolf and the "Lust of Creation"*) and others do, to demystify these hallucinations, see Caramagno, *Flight of the Mind*, 47. Unlike depressive episodes, manic episodes generally have no basis in "reality."
10. Thoby wrote to Clive about coming home to "fogs, rain & cold & misery in general—an exam next Monday" without mentioning Virginia's breakdown. Thoby's quarantine story must have sounded hollow to Clive, for Thoby also claimed that "not being infectious," he could break the quarantine by going to the New Gallery. This is the first mention in his letters to Clive (that is, the surviving ones) of either of his sisters (TS/CB, ? and ? May 1904; VB/Sir William Rothenstein, 6 July 1904, Harvard Theatre Collection.)
11. Regina Marler offers a comparable explanation: "The fact that VB was hated and feared by VW during this episode . . . is not insignificant. Observing VB's glee at Sir Leslie's death, and painfully aware of her own dependence on VB's care and love, it seems natural that in a state in which repressed emotions are made manifest, VW would strike out against her beloved sister" (letter to me, 7 January 1992).
12. Misreading a pronoun reference in a letter from Violet to Vanessa written some forty years later, King jumps to the totally implausible conclusion that Virginia's problem was that she wanted a baby "to nurse or mother the wounded child within herself" (*Virginia Woolf*, 92 and note).
13. Stemerick, however, explains Virginia's instability as "a legitimate reaction to

a social value system which established certain goals as desirable while it simultaneously deprived her of the means of achieving those goals, then her rage and anti-social acts make sense" ("Virginia Woolf and Julia Stephen," 54).

14. VBL, 29, note; VB/VW, 1 November 1904. Vanessa also decorated Virginia's room, choosing for her a couch with a chintz cover and having curtains made (VB/VW, 5 and 25 November 1904).
15. VWL, 1:151. Clearly, Jack's concern was preserving what Virginia called his "thickskulled proprieties!!!" by wanting her excerpts from her parents' letters to each other to be not "too intimate." Instead, DeSalvo asserts, without citing any evidence, that Stella had received "compromising letters from Leslie," which Jack was afraid Virginia might excerpt and publish (*Virginia Woolf*, 63). But there is nothing to suggest that there was any sexual relation between Leslie Stephen and Stella.
16. TS/LS, 16 November 1904; TS/CB, 18 November 1904. Thoby's next sentence shows as much concern for his pamphlet against compulsory chapel at Cambridge as for Leonard Woolf: "My poor pamphlet has had a sad history." To Lytton, he also wrote, "This unfortunate pamphlet is causing me the devil of a bother. They are making a prophet of me in my own despite" (TS/LS, 19 November 1904; LWL, 50, note).
17. Frederic Spotts remarks that Leonard's academic career reveals "as much about the times—the rapid changes in late Victorian society—as about Leonard. A generation earlier a preparatory school education for a boy who was not only middle class but also poor and Jewish would have been out of the question" (LWL, 5).
18. Perhaps Virginia's letter to Violet Dickinson (22 October 1904), describing "the Quaker trumpeting like an escaped elephant on the stairs" (VWL, 1:145), was further developed in a "Life" that made Violet "howl." Virginia also experimented with this new genre in a comic life of her aunt Mary Fisher (VWL, 1:163, note).
19. Janet Vaughan, in Stape, *Virginia Woolf*, 9. Virginia judged Will in much the same way that Leslie Stephen had judged Will's father for failing to appreciate his wife, Addy, and being so dull that his intellectual "abilities had to be taken on faith" (*Mausoleum Book*, 76).
20. Vanessa's especially apt allusion is to the nursery rhyme "They come as a boon and a blessing to men, / The Pickwick, the Owl, and the Waverly Pen" (VBL, 27 and note).
21. VB/VW, 5 December 1904; LS/JPS, 14 April 1881. Julia's letters do not survive. Bicknell speculates that they were among the many papers destroyed in the Woolf's Mecklenburgh Square House in the 1940 bombing (letter to me, 25 November 1994).

22. For an account of these publications and the many ways in which they anticipated Virginia Woolf's later writing, see Rosenbaum, "Virginia Woolf: Beginnings."
23. See Gillespie, *Sisters' Arts*, 21–62. Gillespie reproduces the drawings and purposively deemphasizes Fry's influence and psychological and biographical issues in order to focus on the sisters' artistic sharing.
24. Maitland, *Life and Letters of Leslie Stephen*, 477, 325, 438. Maitland also reclothed Virginia's memory of Leslie Stephen's rages regarding the account books: "Stephen rarely thought about money: so rarely that when he was called out of the eighteenth century [his writing] to face the domestic finance of the nineteenth or twentieth, there would be thunderings and lightnings and the gloomiest vaticinations. The doors of the workhouse would yawn before his eyes; and an hour after he would be making to a friend, who was out of health, an offer far too generous to be accepted" (477–78). Virginia focused on similar contradictions in Jonathan Swift.
25. King erroneously asserts that Leonard Woolf, too, was present (*Virginia Woolf*, 106), but Woolf was in Ceylon from 1904 to 1911.
26. VWL, 1:201. This debate suggests how insulated the British were, despite visits to Paris, from the avant-garde painting that was already replacing Impressionism.
27. Virginia's "A Dialogue upon Mount Pentelicus" was thought to be an incomplete essay until Rosenbaum reordered the typescript, found that it formed a narrative, and published it in the *Times Literary Supplement*, 11–17 September 1987, 979, and in the *Charleston Newsletter* 19 (September 1987): 23–32. Susan Dick republished the story with notes in CSF, 63–68.
28. Spalding writes that Vanessa's London doctor later confirmed this episode as a breakdown (*Vanessa Bell*, 59). Spalding attributes it to years of anxiety over Virginia's mental health, compounded by the conflicts created by Clive Bell's proposal. Something of what was going on in the household after their return is indicated in a letter from Vanessa Bell to Margery Snowden. She wrote that the new nurse did not know anything about "the Goat or the workings of this *very odd family." She must have added a note with an asterisk at the bottom of the letter clarifying what she meant by "very odd," but that part of the letter has been torn off (VB/MS, 16 November 1906).
29. Saxon Sydney-Turner/LW, 30 November 1906, LWP. Trekkie Ritchie Parsons says that the loss of his father, with whom he identified and felt "intellectually *d'accord*," had been devastating to Leonard (interview, 25 July 1993).
30. VB/CB, 29 January 1905. Virginia's own comment, "The Watts show is atrocious: my last illusion is gone," suggests that she was not so loyal as Vanessa implied (VWL, 1:174).
31. V. Woolf, "Friendships Gallery," 273, 279, 276, 272. The "Life" was written

early in 1907 when Vanessa married. Manic-depressives tend to exaggerate the intensity of their interactions with others (Goodwin and Jamison, *Manic-Depressive Illness*, 302).

32. Fogel begins his treatment of Woolf's debt to James with her visit to Rye and her decision to begin her novel. James was keenly aware of Virginia as an emerging writer, and she of him as a fading figure of greatness (*Covert Relations*, chaps. 3, 4, 5). Dates for the composition of *Melymbrosia* (later *The Voyage Out*) are taken from Heine, "Virginia Woolf's Revisions."
33. CB/VW, 11 August 1907, and 25 January 1908. Vanessa read *Les Liaisons* in April 1909 (VBL, 83). Years later, Clive also undertook the "sexual education" of Vanessa's daughter Angelica by giving the pubescent girl the "rather curious choice of *Les Liaisons dangereuses*" (Garnett, *Deceived with Kindness*, 139).
34. Vanessa's judgment was mimetic: "Of course I can only talk as do laymen about a portrait & tell you whether I think it like or not" (VB/VW, 20 April 1908).
35. CB/VW, 13 and 18 April 1908. Clive also wrote to Lytton that Vanessa and Mrs. Raven-Hill "talk bawdy when they must and seem to enjoy themselves" (CB/LS, 23 April 1908).
36. Clive jocularly talked about his affair in a 1921 Memoir Club paper. Virginia found his revelations "a surprise to me. She [the mistress] coincided with his attachment to me then." She endorsed the bifurcated nature of Clive's attentions with the phrase "But she was a voluptuary" (D, 2:89). For more about Clive's affair, see Dunn, *Very Close Conspiracy*, 111.
37. Bollas explains that when a child is "left to work on a life problem that is beyond his capability, he often assumes the problem to be unresolvable and it therefore becomes an inevitable part of his sense of identity. . . . In moods[,] the individual will remain in contact with the child self who endured and stored the unrepresentable aspects of life experience" (*Shadow of the Object*, 115, 111–12). Marler argues that Virginia, Vanessa, and Clive were playing roles in a revenge tragedy (VBL, 49).
38. CB/VW, 7 May 1908. However hypocritical Clive was about his emotions, he was not so about Virginia's talent: He wrote to Lytton that "I go to Kew with Virginia and talk about her genius in which somehow I believe rather strongly" (4 June 1908).
39. CB/VW, 12 July 1908. He also wrote to Lytton, in ways calculated to suggest intimacy with Virginia, that Young's overtures "provoked strange virile jealousies and infinite bitterness" in himself, "not but what they met with the coldest of responses" (CB/LS, 9 August 1908).
40. VB/VW, 31 July 1908. Clive was simultaneously writing to Lytton, hoping that Virginia would "be allowed" to see Lytton's poems, which were a "revelation" and not only for lustfulness (CB/LS, 30 July 1908). In 1907, the classi-

cist Walter Headlam had proposed, but Virginia refused. (Sixteen years her senior, Headlam died in 1908.)

41. VWL, 1:346; CB/VW, 30 July 1908. Virginia was conscious of her audience. Later in the year, she spoke of Julian's charms to the maternal Madge Vaughan (VWL, 1:372, 395).
42. The "chaste & maidenly" quotation is from the same letter (25 August 1908) but not printed in the published version.
43. On the centrality of Moore's book to Bloomsbury thought, see Rosenbaum, *Victorian Bloomsbury*.
44. This heading appears in the original diary along with the notes (PA 383) on the versos of 135–40 (BL).
45. Leaska says that she wrote the sketch "apparently at the beginning of the journey," which would have been in Siena (PA, 382). Much later, Clive Bell wrote, "And I have by me a picture post-card from Siena, written many years later. [It carries] an inky cross against a spot on the fortezza and beneath this legend—'Here Clive quarrelled with his sister-in-law'" (*Old Friends*, 93, reprinted in Stape, *Virginia Woolf*, 93–112).
46. The date was 15 January 1909. Clive's notes on the society are at King's College, Cambridge.
47. LS/LW, 19 February 1909, LWP. The published version of this letter (LWL, 147) omits the comments about virginity.
48. VWE, 1:268. At 2 percent interest, this would earn Virginia an extra £50 a year. Caroline Emelia left Vanessa and Adrian each £100 (VWE, 1:269, note).
49. Spalding, *Vanessa Bell*, 82. This painting was praised by Walter Sickert, probably the foremost English painter at the time (VBL, xxxi).
50. Nine years later, this courtship still held enough interest for Fredegond Shove to tell Vanessa, "I want a complete account (sometime) of Virginia's relations with Hilton Young—she was too modest to give a proper one herself" (FS/VB, ? December 1918).
51. VB/VW, 11 May 1909. In SS-T/LW, 17 August 1909, Saxon named all the operas they had seen and would see.
52. Bell, *Ad Familiares*, 14–15. In this privately printed collection, given to friends as Christmas presents, Clive gathered various poems that he had written over the years.
53. Bell, *Old Friends*, 100; also see VWL, 1:361. Olivier Bell believes that Virginia did not buy the desk but had it made. It was later passed on to Quentin. After her marriage to Quentin, Olivier Bell writes, "As it was too high to sit at, I cut about six inches off the legs. We sold it at the Sotheby Sale in aid of Charleston (21 July 1980, cat. no 327, illustrated). We rather think VW first had it at Fitzroy Square, and Q doesn't think in fact she used it much" (letter to me, 6 May 1991). Quentin Bell writes, "I have seen VW at work and she was quite comfortably seated" (letter to me, 15 March 1995).

Chapter 4

1. MRF. In an article in *Vogue*, Fry termed Durbins a model of "genuine domestic architecture" in which people "let their houses be the direct outcome of their actual needs" (VD, 190–94).
2. Even though his brother, Fitzjames, was an Apostle, Leslie's "claims, alas! if they were considered, were not considered to be sufficient" (Stephen, *Some Early Impressions*, 36–37). The odds were against Leslie's being elected, since Apostles were mostly from King's or Trinity College and Leslie was a member of Trinity Hall College. Lowes Dickinson (Goldie) was an Apostle with Fry. Later, E. M. Forster, G. E. Moore, Lytton Strachey, Bertrand Russell, Desmond MacCarthy, Saxon Sydney-Turner, Maynard Keynes, and Leonard Woolf were Apostles. As Woolf explained, "Of the ten men of Old Bloomsbury only Clive, Adrian and Duncan were not Apostles" (LWA, 2:11). Thoby had not been an Apostle either.
3. RF/Robert Trevelyan, 29 July 1897. They were staying at the Hôtel de la Place, Veules, Seine Inférieure.
4. Julian Fry remembered of his mother, "I was fond of her but I never felt safe with her" (Rosenbaum, "Conversation with Julian Fry," 130).
5. After that dinner, Vanessa also visited Roger and Helen Fry one summer; aside from Helen's illness, theirs seemed a "strange alive household" (MRF).
6. See Leaska, Afterword to *Pointz Hall* and "Biographer's Dilemma."
7. Virginia's first published article was in the 14 December 1904 issue of the *Guardian*, which also called attention to "Mr Roger Fry's charming drawing of St John's College, Oxford" in a New English Art Club exhibition (VWE, 1:xii). Fry's "Essay in Aesthetics" was first published in Desmond MacCarthy's *New Quarterly* 2 (April 1909), where Virginia certainly might have read it. If she did so, it made no impact at the time. But she probably heard Fry lecture later in 1909. At any rate, Virginia knew of Fry but seems not to have known him or read his works.
8. MRF. For a detailed description of this visit, see Spalding, *Roger Fry*, 126–27.
9. The conventionality of Julia's sister Mary and her husband, Herbert Fisher, had long been a joke with Leslie and his children. During one of Julia's extended stays at Brighton, Leslie wrote to her that "after lunch I walked in the gardens with Ginia & Adrian" where they "met Herbert Fisher. The little ones carefully avoided their [Fisher] cousins" (LS/JPS, 27 October 1891).
10. RF/Joseph Hodges Choate, 17 July 1909. Anticipating legal troubles, Fry kept a copy of his letter.
11. RF/Lionel Cust, 17 February 1910. Fry also kept a copy of this letter. Cust, coeditor with Fry of the *Burlington Magazine*, was also Keeper of the King's Pictures.
12. A letter from Vanessa to Clive on a Saturday when he was away suggests,

however, that the sisters went their own independent ways. Vanessa wrote, "I have hardly seen her this morning as I went off to paint—but she said she felt better" (VB/CB, spring 1910).

13. In a portion of a letter omitted from RFL, 1:321, Roger explained that Helen's delusions made life with her at home impossible for the children (RF/John G. Johnson, 6 May 1909, John G. Johnson Collection, Philadelphia Museum of Art).
14. For details, see Bell, 1:35–36. J. K. Stephen's father, Sir James Fitzjames Stephen, remained true to the family's resolution and refused to acknowledge his son's madness.
15. Helen's symptoms were often like Virginia's: sleeplessness, obsessions, wild excitement, depression, hallucinations, resentment of Roger and her nurses. She became finally incoherent and violent and remained in The Retreat, York, until her death in 1937. Now we know that she suffered from chronic paranoid schizophrenia (D, 5:270, note). (Laura Stephen was incarcerated in another Yorkshire institution until her death in 1945.)
16. VBL, 90. Dr. Silas Weir Mitchell (1829–1914) was an American neuropsychologist who introduced "rest cures" for nervous diseases and also wrote *Diseases of the Nervous System, Especially of Women* (1881). For a description of Mitchell's debilitating effect on another woman writer, Charlotte Perkins Gilman, see Lane, *To "Herland" and Beyond.*
17. VB/CB, 25?, 26?, and 24 June 1910; CB/VW, 27 June 1910. Virginia entered the nursing home on 30 June (Bell, 1:199).
18. Dunn sees this pattern as more mutually supportive and does not consider Vanessa's fondness for "picking apart" her sister (*Very Close Conspiracy*).
19. See Quentin Bell's "Letter" on his various names and on James King's confusion about whether Bell is Quentin or Christopher.
20. VB/CB, 9 September 1910. The Tate Gallery has dated this letter 1911, but the references to Jean and Cornwall clearly set it in 1910.
21. RF/G. L. Dickinson, 24 September 1910. Roger said he had to go to Hampton Court on Wednesday, 28 September 1910.
22. Morrell, *Ottoline*, 204. Lady Ottoline completed her memoirs up to 1918 using as her source her daily journal of those years. Thus her recollection of Virginia's opinion is contemporary and probably authentic. In the published memoirs, the remark appears on the same page as a reference to Roger Fry's preparation for the first Postimpressionist exhibition.
23. The letter from 10 Downing Street, dated 26 January 1911, says that Mr. Asquith, the prime minister, would be "glad" to appoint Fry to the keepership of the Tate Gallery. The salary would be £350 a year, increasing by £50 increments to £500 (FP).
24. Heine names the following versions: (1) The earliest version, 1907–1908; (2) a now lost "Valentine" version, finished early in 1910; (3) a 1910–1912 ver-

sion (*Melymbrosia*), of which Virginia finished eight chapters before going to Turkey in April 1911; and (4) a final, 1912–1913, complete rewriting. Virginia also made changes in the 1920 American edition that "reflected not what she had accomplished in publishing the novel in 1915, but the rejection of everyday detail and the development of new forms with which she was experimenting five years later" ("Virginia Woolf's Revisions," 400). The Uniform Hogarth Edition of 1929 returned to the 1915 version, but we do not know whether this was for reasons of convenience or copyright. In the Hogarth Definitive Collected Edition, Heine preserves, for bibliographical accuracy, the 1929 choice of the 1915 edition (401). According to Leonard Woolf, Virginia burned five or six versions of *The Voyage Out* (Stape, *Virginia Woolf*, 149–50).

25. Roger's letters to Clive about these arrangements are dated ? January and 11 February 1911. The first is published in RFL, 1:339–40. The date in VWL, 2:78, note, seems to be incorrect.
26. He also offered an introduction to meet a man that Roger described to Clive as "a great Bergsonite so you must either avoid that subject or be prepared to listen" (RF/CB, 26 January 1911). McLaurin sees Fry as "close to Bergson's ideas" in distinguishing between the practical and speculative ways of seeing (*Virginia Woolf*, 113). The letter to Clive, however, suggests that Fry himself had reservations about Bergson's antirationality.
27. CB/VW, 17 April 1911. "In your bad books" probably meant "in your debt."
28. VWL, 1:465. That Virginia was exposed to the physical circumstances of miscarriage is indicated by a letter that she wrote years later to Ethel Smyth, in which she described what happened when she neglected her work on *The Waves*: "I break the membrane, and the fluid escapes—a disgusting image, drawn I think from the memory of Vanessa's miscarriage" (VWL, 4:185).
29. Unpublished letters suggest that Roger and Vanessa may have become lovers before the trip. Earlier in 1911, Vanessa called Roger "my dear" and told him that she "wanted" him and that they must take a holiday together (VB/RF, 7 and 8 February 1911). Although Clive seems to have meant Roger when he wrote about Vanessa's "paramour," Virginia seems not to have believed him.
30. Given Virginia's habit throughout her biography of Fry of deleting herself as a presence, I think it safe to assume that the "someone reading a book" to whom she referred was herself (Woolf, *Roger Fry*).
31. Fry's terminology may have been misleading, as he never devalued literature itself. Instead, he condemned "literary" approaches to painting.
32. For more evidence of Fry's "impact," see *Melymbrosia*, 71, and Heine, "Virginia Woolf's Revisions," 420–32. Heine finds evidence in *Melymbrosia* that Virginia may have dropped the "art/nature conflict" as too "old-fashioned, after Post-Impressionism" (letter to me, 30 May 1991). Before going to

Turkey, Virginia had finished revising eight chapters of the version of *The Voyage Out* published as *Melymbrosia* (VWL, 1:461 and note). Because the letters were edited before Heine had sorted out the chronology, that note does not distinguish between Virginia's finishing eleven chapters of the 1909 version and eight chapters of the 1911 version.

33. Because she was trying to provide a stable home for Julian and Pamela Fry, Joan had little use for Vanessa. Years later, Julian remembered that Bloomsbury gave his aunt Joan a hard time: "I don't think any of the Bloomsbury people had any idea that religious inspiration could be genuine." Asked who was most intolerant, Julian replied, "Well, Clive, Vanessa, and Virginia to some extent. Leonard, no. He was much too nice a person" (Rosenbaum, "Conversation with Julian Fry").
34. In fact, Virginia looks so old in Figure 17 that the two-volume English Triad Paladin Edition of Quentin Bell's biography of her uses this 1911 portrait for the cover of the second volume, which spans 1912 to 1941; Vanessa Bell's 1912 portrait of her is on the cover of the first volume.
35. VWL, 1:466. Vanessa's reply to Virginia's depressed letter—if there was one—no longer exists.
36. For a description, see Pamela Fry Diamand, "Durbins," in Lee, *Cézanne in the Hedge*, 51–62. Photographs are in RFL, 1.
37. Quentin Bell explains that Virginia's flirtation with Clive made her "aware of her own normal proclivities, in making her feel the need, which she had not hitherto felt, for a man" (1:135). Those critics who see only Virginia's lesbianism may object to this hypothesis, but her letters from 1908 until her own marriage in 1912 exhibit a fascination with heterosexual relationships that is most obviously interpreted as a "need" for a man. One could read this "fascination" as merely an attempt to respond to Vanessa's prodding, but it seems to me to be a genuine interest.
38. These remarkable paintings are now in storage at the Tate Gallery.
39. For a reproduction, see Spalding, *Roger Fry*, 148.
40. Andrew McNeillie says the review, in the *Nation*, 14 October 1911, is "almost certainly" by Virginia Stephen [Woolf] (VWE, 1:xviii).
41. He confessed to Virginia that he was "selfish, jealous, cruel, lustful, a liar & probably worse still." For her part, she had confessed the day before to being "vain and egoist untruthful" (LWL, 169).
42. CB/MM, 14 March 1912. This is the only reference I know of to Virginia's actually seeing a psychologist.
43. For the dating of this and other canceled passages in *Melymbrosia*, see Heine, "Virginia Woolf's Revisions," 430.
44. Ka Cox/James Strachey, 17 and 22 April 1912. Leonard Woolf does not mention these events in his diary.

45. Trekkie Ritchie Parsons told me that Leonard was so passionately in love with Virginia that he simply adopted her value system, including its anti-Semitism, without realizing how deeply he was offending his family (interview, 25 July 1993).
46. Ka Cox/James Strachey, Wednesday, ? ? 1912, possibly from an envelope postmarked "Ju 16 12."
47. This is the version now published as *Melymbrosia.* As Heine writes, Virginia's talk of finishing "unknowingly misled her biographers" ("Virginia Woolf's Revisions," 425).
48. Bell, 2:40. Six years later, Virginia remembered the original rather than the revised date (D, 1:181). VB/VW, 14 August 1912; Bell, 2:40; Dunn, *Very Close Conspiracy*, 184; VB/MS, 20 August 1912.
49. Gordon suggests that because this "twinge of anger" follows an announcement that Leonard is "writing the first chapter of his new great work, which is about the suburbs," Woolf's anger was over the novel (*Virginia Woolf*, 150–55). But Virginia does not seem to have suspected then how much Leonard's novel would expose the Bloomsbury world. The context of the sentence suggests instead that the anger she did not want to visit on Leonard was anger at a sexual dimension of life to which he had suddenly exposed her.
50. For a discussion of the germination of *Night and Day*, see Heine, "Virginia Woolf's Revisions," 435.
51. VB/VW, 14 September 1912. Roger's undated letter to Duncan Grant about the poster is printed in RFL, 1:360; the unpublished letter with Vanessa's addition is in the Charleston Papers. Since Vanessa wrote to Roger on 27 August (FP) about Virginia's refusal to eat and to Leonard on 29 August (LWP) that Roger "looks on the matter as settled if you will really do it," it seems probable that Roger wired Leonard in response to Vanessa's report of Virginia's troubles.
52. In the momentous spring of 1912, however, Leonard had taken a life class in painting, perhaps because his brother Philip was a painter (LWL, 171).

Chapter 5

1. Desmond MacCarthy, "Kant and Post Impressionism," *Eye-Witness*, 10 October 1912, 533–34. Rosenbaum discovered this essay.
2. *Nation*, 9 November 1912, 250. Goldsworthy Lowes Dickinson worried, as he wrote to Fry on 26 November 1912, that the Postimpressionist aesthetic, as Fry explained it in the *Nation*, "points to the complete elimination of everything Actual."

3. These attacks on the exhibition are from, respectively, *Daily Mail*, 5 October 1912; *Yorkshire Post*, 4 October 1912; and *Court Journal*,14 October 1912 (FP).
4. Furbank, *Growth of the Novelist*, 206. Fry seems to have eventually convinced Forster. An unpublished fragment (perhaps written retrospectively) in the Forster papers at King's College, Cambridge, shows Forster defining, as "essential" Bloomsbury, Fry's insistence that treatment was more important than subject matter.
5. Later, Virginia wrote that "it was due to Mr. Fry" that her eyes "began painfully to concentrate about the autumn of 1912, upon the canvases of Cézanne" (Review of *Vision and Design*, by Roger Fry, 28).
6. Fry financed the Omega Workshop through his inheritance from Joseph Storrs Fry, so for the most part "the Omega Workshops came out of the Fry chocolate factory" (Rosenbaum, "Coversation with Julian Fry," 132). Shareholders also made contributions; for example, Vanessa wrote to Fry that Clive "really wants to give [£100]—& also to make a slight grievance of it!" (VB/RF, 24 January 1913).
7. For some of this correspondence, see Hall, *Hidden Anxieties*.
8. VB/CB, 27 December 1912; VB/RF, Sunday, 29 December 1912. Clearly, the letter to Clive should have been dated 28 December (to support "last night"), or the letter to Roger should have been dated Saturday (CB/MM, 31 December 1912). The Woolfs' conversation with Vanessa raises the probability, I think, of one or both of them having read Marie Stopes.
9. VB/MS, 29 December 1912; VB/CB, 8 January 1913; LWD. The second Postimpressionist exhibition was to close at the end of 1912, but it continued until the end of January 1913 (VB/VW, 26 January 1913), with Sidney Waterlow replacing Leonard as secretary in the new year.
10. VB/LW, 20 January 1913; VB/VW, 2 February 1913. Vanessa may have been thinking of Helen Fry, whose symptoms in so many ways matched Virginia's. Helen's decline into apparently hopeless madness intensified after childbirth.
11. VB/LW, 20 January 1913; VB/VW, 6 February 1913. Leonard kept Vanessa's letters (LWP), but he does not seem to have saved his article. His diary shows him writing "art" on 4, 17, and 18 January. Clive's article, "Post-Impressionism and Aesthetics," articulated his position, so opposed to Leonard's.
12. Fry's obituary notice for the fashionable painter Sir Lawrence Alma-Tadema sparked this debate (*Nation*, 18 January 1913, 666–67). Sir Philip Burne-Jones and Sir William Richmond defended Alma-Tadema and attacked "Mr. Roger Fry's Criticism." Bernard Shaw replied with two letters, Clive Bell one, and Roger Fry two in his own defense. Leonard mentioned Alma-Tadema on 3 February 1913 (LWD). The citation for the quotation from Fry is *Nation*, 22 February 1913, 851; for that from Shaw, *Nation*, 1 March 1913, 889. Since Vanessa told Leonard that she could not "sign" his article, it may have been intended as another reply to "Mr. Roger Fry's Criticism." After Vanessa's dis-

missal of his ideas, Leonard seems to have destroyed this effort in aesthetics. My thanks to Leila Luedeking, curator of Modern Literary Collections at Washington State University, for this suggestion.

13. VB/RF, 6 February 1913; VB/VW, 6 February 1913. The battle between the arts was public as well as private. Roger Fry wrote to Lytton Strachey that Bruce Richmond, editor of the *Times Literary Supplement*, "publicly suggested the flogging of art critics meaning me" (RF/LS, 11 February 1913).
14. Because most of these elaborations and additions are in the first part of the novel, Virginia probably began the visualization process in 1911 after her talks with Roger in Turkey. Then, after her honeymoon, she applied the process to the whole novel.
15. In reference to the subtitle of Bollas's *The Shadow of the Object: Psychoanalysis of the Unthought Known*.
16. Fear of ridicule is especially acute with manic-depressives, who fear that criticism may expose real fissures in their personality (Caramagno, *Flight of the Mind*, 44–48).
17. VB/VW, 17 August 1913; VB/RF, ? ? 1913; VB/CB, 12 August (VBL, 140) and 19 August 1913. Regina Marler suspects that "VB and LW were in collusion regarding the transference of care for VW from VB to Leonard" (letter to me, 6 January 1992).
18. VB/CB, 23 August 1913. Vanessa also told Clive that she "couldn't have managed" without Roger, but whether she was talking about managing to get the Woolfs on the train or herself and her children on another train or both is not clear. The glimpse that Vanessa offers into Virginia's psyche is itself a refutation of Alexander's and King's simplistic theory that her only problem was not having a baby (Alexander, *Leonard and Virginia Woolf*; King, *Virginia Woolf*).
19. VB/LW, 25 August 1913; VB/VW, 27 August 1913. Vanessa may have remembered Roger's stories of Helen Fry. As he wrote to another friend, Helen's "absolute intransigence" involved insisting that there was nothing wrong with her and refusing to go to an institution (RF/Robert Trevelyan, 16 September 1909).
20. The dates of these entries are 24, 26, 27, 28, and 31 August 1913. Caramagno writes that such delusions "give form to the inexplicable feelings of emptiness and evil that manic-depressives have when biochemistry fails" ("Neuroscience and Psychoanalysis," 211). Virginia worried that her breakdowns would be interpreted as signs of a moral flaw, that she would be blamed for not being in control of herself, and that her jealousy or resentment would precipitate disasters for others. Again and again over the years, irrational guilt battled with her self-confidence (LWA, 2:53–54, 116–17).
21. Caramagno explains: "Manic-depression fabricates evidence justifying itself and fulfills its own prophecy" ("Neuroscience and Psychoanalysis," 211).

22. Veronal was a depressant used as an inducement to sleep. Barbital is the modern equivalent (succeeded by phenobarbital). Virginia took twenty-one times the normal dose, potentially lethal. My thanks to Bill Ross, associate director of the Louisiana Drug and Poison Control Center at Northeast Louisiana University, for providing this information.
23. Various literary critics have recently refused to believe that Virginia was so ill; some have even implied that Quentin Bell's biography misunderstood or patronized her and hence exaggerated her "madness." Other critics have acknowledged that Virginia was ill but have construed her illness as simply a logical reaction to the excesses of a patriarchal system (represented not only by the Duckworth brothers but by Leslie and Leonard as well). Given the distance that such theories have taken us from the facts that are available, it seems to me appropriate to cite, whenever possible, heretofore unpublished accounts by the people who were there.
24. Although Quentin Bell gives the date as 20 September, the diary has it as 17 September.
25. VB/LW, 27 September 1913. Vanessa was correct about the side effects of Virginia's overdose. According to Bill Ross, an overdose will produce withdrawal symptoms afterward, including delirium, hallucination, and various other psychotic responses.
26. Their visit created another difficult situation for Leonard to deal with: In a letter to Leonard on 25 October 1913, Dr. Savage claimed that he was not "professionally jealous," but he nonetheless seems to have been so.
27. VB/RF, Friday, 3 October 1913. This letter is dated by internal references to the Ideal Home Exhibition, before the "rumpus" errupted on 5 October (RFL, 2:371–72). Vanessa also refers to Roger's plans for a painting excursion with Henri Doucet in France.
28. RFL, 371–74; VBL, 143–49. For the "rumpus" that Lewis caused, see Rosenbaum, "Bloomsbury Criticisms and Controversies," in *Bloomsbury Group*, 329–61.
29. VB/LW, 21 November 1913; Ka Cox/James Strachey, 22 November 1913; VB/LW, 24 November 1913. The University of Sussex tentatively dated this last letter as 1912, but it seems to me to date from 1913 when Virginia was doing poorly on 22 November 1913. Vanessa may have been asked to come to see her then; hence she would have needed to explain on 24 November 1913 why she did not come. The editors of VWL write that "Vanessa, Janet Case and Ka Cox took it in turns to watch over Virginia during [Leonard's] absences" in 1914. I see no evidence either in Leonard's diary or in any of Vanessa's letters that Vanessa stayed alone with Virginia when Leonard was away. She visited in January 1914, but Leonard was there, as his diary entry shows. Vanessa visited on 18 March 1914 (VBL, 159), but Ka Cox also was staying with Virginia then (LWD). With few exceptions, Vanessa's letters to

Clive and Roger about Virginia's health cite either Leonard or Ka as her source.

30. Perhaps in this period, Vanessa advised Madge Vaughan that "it was better not" to write to Virginia (Madge Vaughan/LW, n.d.).
31. The letters from Marie Woolf are dated 11 and 16 December 1913. The last quotation is from Bella's unpublished letter of 12 December 1913 (LWP).
32. LWA, 2:54; Caramagno, *Flight of the Mind*, 63. Caramagno further points out that to explain their depression, patients often focus obsessively on a past event such as the loss of a loved one or some "unforgivable" sin. In the absence of such an incident, depressives often invent one ("Neuroscience and Psychoanalysis," 219).
33. Spotts quotes from a letter from Gerald Brenan written in 1967 in which he says that on their honeymoon, Leonard gave up all idea of "sexual satisfaction" (LWL, 162–63). It is hard to believe that Brenan could remember such a conversation after more than forty years and even harder to believe that Leonard had confided in Brenan. Virginia's references to the "proper business of bed," to the "probable fruits" of her "lost virginity," and to the servants' not minding "the sight of us naked" every morning also may make Brenan's story suspect (VWL, 2:5, 9, 11). But Virginia may have simply been striking a pose.

Chapter 6

1. VB/RF, ? ? 1914; VB/VW, 2 February 1913. Even though Virginia had written a comic biography of her aunt Caroline Emelia Stephen, she was never so dismissive as Vanessa.
2. VB/LW, 14 and 23 January 1914. In the second letter, Vanessa refers to charges from Dr. Belfrage (who, she thought, "does V no good") of more than £88. Medical expenses were eating into the capital that Virginia had inherited and Leonard had saved.
3. VB/CB, 22 January 1914; VB/LW, 26 January 1914. Virginia's letter of 11? February 1914 does say that Vanessa visited "the other day." Regina Marler suggested to me that there might have been another visit after the one on 29 and 30 January (letter, 7 January 1992). Leonard Woolf records only the one visit. If Vanessa made only one trip to Paris in January 1914, it is odd that she did not write to Duncan Grant about it until (probably) 25 March 1914 (VBL, 160–61). At any rate, Vanessa's children did not keep her from Paris.
4. Whatever she typed for Violet, it must have been an odd contrast to typing Lytton's *Ermyntrude and Esmeralda*, which Willis describes as "the erotic,

comic epistolary exchanges between two Victorian girls about bowwows and pouting pussies" (*Leonard and Virginia Woolf as Publishers*, 4).

5. Leonard quotes correctly but omits this summary of Goldwin Smith's opinion of the Stephen brothers: "They were both *hard* men, Leslie less hard, more genial than Fitzjames. They were both critics" (Haultain, *Goldwin Smith*, 39).
6. Leonard's review of Freud's *Psychopathology of Everyday Life* appeared in June. He remembered that "before writing it I read *The Interpretation of Dreams*, which had been published the previous year" (LWA, 2:120). His diary says he read it on 12 May 1914, but he could have read or skimmed it beforehand. Certainly he had heard of it.
7. Fry, "New Theory of Art," 938; Spalding, *Roger Fry*, 164–65. Fry also maintained that it was "good for the public to have the same thing said twice over differently" (RF/CB, n.d.).
8. Ka's letter to Leonard is dated 18 March; probably the "sad evening" was the night of 17 March. Vanessa was at Asheham with Virginia and Ka on 18 March and left on 19 March (LWD; and see VBL). Whether Vanessa arrived on 18 March is unclear. Apparently Ka, too, left on 19 March after Leonard returned.
9. In the last years of her life, Virginia read Vanessa's letters to Roger, with, I believe, devastating effects.
10. VB/CB, 20 August 1914. In an undated letter to Roger, Vanessa began their breakup with her determination to "look after myself more independently. I did manage to refuse to do things & see things when you suggested them sometimes which I hadn't been able to do."
11. LWA, 2:119. Spotts explains that Woolf's "research into women's wages in industry well illustrates Leonard's feminist sympathies, his nascent concern for social justice and his work for better minimum wages for women" (LWL, xi).
12. According to Bill Ross of the Louisiana Drug and Poison Control Center, this was more than five times the normal therapeutic dose and markedly toxic, possibly lethal, depending on body weight. Given her familiarity with regular doses, Virginia must have known that she was taking an extraordinary number of pills; thus this event hardly seems in any simple way an "accident."
13. VB/RF, ? August and Thursday, 17 September 1914. These last comments may refer to Virginia's postsuicidal state in 1913, but the letter can be dated from 1914, when 17 September was a Thursday.
14. VB/CB, 1 October 1914; SW/LW, 2 January 1915. An undated letter from Vanessa to Roger in the fall of 1914 also describes her anger at the Woolfs for wanting her to take on their part of the lease of Asheham. She did not feel she should fully reimburse them because the decorations and most of the furniture were hers. The Woolfs then rented to the Waterlows and, for a brief period in the spring, to Maynard Keynes (MK/LW, 24 March 1914).

15. In several of his essays on damaged villages and Quaker peace efforts, Fry expressed what we now identify as a quintessential World War I distrust of abstract ideals ("Friends' Work").
16. Grieving for Rupert, Ka Cox joined Roger's sister Margery in the Quaker relief work in Corsica. Vanessa did not see the reason for seeing his death as a tragedy. She wrote to Clive of her irritation with people who "are driven to talking about the waste & meaningless [*sic*] of life by Rupert's death. After all it's not the first time a young person has died" (VB/CB, ? May 1915).
17. VBL, 172–73, 174. Vanessa's letter of 9 April 1915 refers to a now lost letter from Roger whose contents can be inferred from Vanessa's answers.
18. She had begun the year asserting that "if the British spoke openly about W.C.'s, & copulation, then they might be stirred by universal emotions" (D, 1:5). "Rapid and pressured speech" is a feature characteristic of mania, occurring in 98 percent of patients (Goodwin and Jamison, *Manic-Depressive Illness*, 270).
19. Bollas explains that "when a person goes 'into' a mood, he becomes that child self who was refused expression in relation to his parents for one reason or another" (*Shadow of the Object*, 115).
20. In 1914, the Woolfs had lived on £443 for the whole year, but in 1915 their medical bills alone came to £500. For details of the Woolfs' finances, see Spater and Parsons, *Marriage of True Minds*, 75–90.
21. Leonard recorded the doses he administered and their effect; for example, "no sleep w avalin" (LWD, 6 April 1915). His diary is the source for the other information in this and the preceding paragraph.
22. For details of this ménage, see Spalding, *Vanessa Bell*, chap. 7.
23. LWD; VB/RF, 24 May 1915. For more details, see Spalding, *Roger Fry*, 197–98.
24. VB/RF, 9 and 14 June 1915. Vanessa's illness seems oddly like Leslie Stephen's; that is, he was competing with Maria Jackson's claims, whereas Vanessa was competing with Virginia's.
25. LWL, 213. Leonard wrote this letter on 12 July, saying that Virginia had turned against him "about 4 weeks ago" and that "although it is a little better lately, she is still very opposed."
26. LWD. Adrian had married Karin Costelloe in October 1914 (VB/FR, 22 October 1914).
27. Perhaps Virginia was jealous of Leonard's work with Margaret or viewed her political activism as a reproach to her.
28. M. L. Davies/LW, no date on any of these letters.
29. Goodwin and Jamison explain that the low self-esteem of depression and the distorted profusions of mania are "encapsulated," or sealed off, so that normalcy can return (*Manic-Depressive Illness*, 292).

30. Virginia mentions a letter from the beginning of the year that she apparently did not save (D, 1:7).
31. Later, Leonard wrote that "when I was away Vanessa or Ka came to Asham and stayed with Virginia" (LWA, 2:118); his diaries at the time, however, record only the visits I have mentioned. Dunn acknowledges frictions but still sees the Stephen sisters in essentially a mutually supportive relationship with each other (*Very Close Conspiracy*). In her edition of Vanessa's letters, Marler necessarily can only sample letters, but that sampling tends to obscure much of the side of this relationship that my citations uncover (Reid, Review of *Selected Letters of Vanessa Bell*).
32. Ian Parsons, Introduction to *The Wise Virgins*, xii. Parsons also notes that Leonard omitted all reference to *The Wise Virgins* in his index of the portion of his autobiography entitled "Beginning Again." Leonard later attributed the book's failure to the conjunction of its publication with the outbreak of war (LWA, 63).

Chapter 7

1. Goodwin and Jamison explain that the spouses of manic-depressives and close family members bear the "interpersonal consequences" of their loved ones' manic episodes (*Manic-Depressive Illness*, 308, 372).
2. She intuitively sensed what feminist psychotherapists theorize: that the source of creativity lies in the desire to restore oneness with the mother (Segal, "Psycho-Analytic Approach to Aesthetics," 197–99; Klein, *Love, Guilt, and Reparation*).
3. Spater and Parsons offer a table of the names of Bloomsbury people that the Woolfs saw, listed in Leonard Woolf's diary in the sample years of 1913, 1919, 1923, 1928, 1933, and 1938 (*Marriage of True Minds*, 127). Vanessa Bell was mentioned forty-three times in 1913, nine in 1919, and thirteen in 1923, after which she appears almost as often as in 1913. In 1914, 1915, and early 1916, Vanessa's name (abbreviated as "Van") appears even less frequently than in 1919.
4. See, for example, Woolf's review of a book by Arthur Symons in VWE, 2:67–70, which originally appeared in the *Times Literary Supplement* on 21 December 1916. Also see Fry's critique of art historians who try to judge art without "some understanding of the problems of the creator" ("Baroque," 146).
5. In August 1913, Vanessa wrote to Fry: "How awful to think of Hampton Court." He and Paul Nash worked on the canvas in 1914 until Nash became

an official war artist. Further information about the Mantegna work will appear in Reid, "My Mantegna: The Triumphs of Caesar and the Hubris of Roger Fry."

6. Dowling argues that Virginia Woolf was not then thinking that literature could "follow the Post-Impressionists to create an analogous form" (*Bloomsbury Aesthetics*, 98). I take Virginia's triumphal claim of a "rout" of Postimpressionism to mean that she was distancing herself from the rival terms of the *paragone* and preparing to appropriate what she could from the visual arts.
7. For the saga of the "servant question" over the years, see Bell, *Virginia Woolf*; Spater and Parsons, *Marriage of True Minds*; and, most recently, King, *Virginia Woolf*.
8. In the late revisions, after her conversations with Fry, Virginia added innumerable references to patterns in the novel (Heine, "Virginia Woolf's Revisions," 433; and, for example, VO, 374).
9. VWL, 2:87. Roger's letters of 21 June 1916 and 15 February 1917 are particularly insistent that Vanessa write to him.
10. VWL, 2:88. Roger thought that Ruskin was "too lazy to think and too credulous to doubt the value of his mental overflow—rather like the Freudian children who preserve their excretion" (RF/Lytton Strachey, 2 April 1927). In an unpublished letter to Charles Mauron (17 December 1933), Roger explained in detail his objection to Turner, who, he thought, lacked a sense of form. In 1918, Virginia mentioned to Vanessa a Turner sketch that she found "very beautiful" (VWL, 2:260).
11. Goodwin and Jamison cite studies stating that manic-depressives recovering from bouts with the illness may be more hypersensitive and dependent than before but that they also score "higher on measures of emotional strength, resiliency, and extraversion" (*Manic-Depressive Illness*, 289).
12. According to Bloom and Lazerson, sleeplessness can trigger the onset of manic-depressive episodes (*Brain, Mind, and Behavior*, 337).
13. Virginia considered Saxon an unlikely husband for young Barbara Hiles because of his "lack of virility" (D, 1:69), and she wrote to Leonard as her "passionate and ferocious and entirely adorable" Mongoose (VWL, 2:191). These do not sound like the words of someone who had no interest in sex.
14. VWL, 2:95. Segal describes the necessity of such a "reparative capacity" ("Psycho-Analytic Approach to Aesthetics," 201).
15. Lord Curzon replied that "each must take his chances in the trenches" (13 May 1916, FP; RF/VB, 9 May 1916).
16. VB/RF, 26 July 1916. Vanessa seems to have misdated her letter, as the Woolfs actually were with her from 21 to 23 July (VWL, 2:106–9; RF/VB, 29 and 3 July 1916).
17. Goodwin and Jamison explain that manic-depressives' worst fear is the recurrence of their irrational illness (*Manic-Depressive Illness*, 731).

18. VWL, 2:179 and note. Katherine's letters to Virginia are flattering, but her remarks about her to others are usually derogatory. Each accused the other of smelling bad (D, 1:58; Mansfield, *Collected Letters*, 2:77).
19. VWL, 1:174. For copies and discussion of these bookplates, see Gillespie, *Sisters' Arts*, 28–31.
20. VWL, 2:151. For more details, see Willis, *Leonard and Virginia Woolf as Publishers*, 14.
21. For a valuable analysis of modern poetry's debt to the hand-press tradition, see McGann, *Black Riders*.
22. Much of Katherine Mansfield's writing can be described as Impressionist. See Weisstein's analysis of "Her First Ball" as an importation into language of the methods and assumptions of Impressionism ("Literature and the Visual Arts," 263).
23. My thanks to Janice Stein for this concept.
24. First published on 8 December 1910, Bennett's article stated that it was "London and not the exhibition which is making itself ridiculous." Bennett ended his tribute to Fry's catalyzing influence by saying that the painters' example suggested that writers of his generation might have achieved only "infantile realisms" (*Books and Persons*, 280–85). Arnold Bennett visited the Omega with W. B. Yeats on 1 March 1917, and Roger listed Bennett as a person of "tastes and interests" similar to his own in May 1917 (RFL, 2: 404, 411). In 1918, Fry designed the sets for Bennett's play *Too Much Money* (RFL, 1:99).
25. Dora Carrington had fallen in love—hopelessly, of course—with Lytton Strachey. She had become a combination of companion, housekeeper, pupil, and (rarely) lover. Her deference to the senior members of Bloomsbury can be seen in a 1917 letter to Virginia: Carrington was delighted at her "humble woodcuts embelishing your Literary masterpiece. . . . But I feel they are poor feathers to adorn your hat."
26. JMK/VB, 28 July 1917; FMS/VB, 19 July 1917. In 1915, Vanessa had described Fredegond Shove as a "remarkable young woman, oddly like Virginia, but I hope saner" (VB/RF, 25 June 1915).
27. For an association between the divisions of the right and left brain and East and West, see di Virgilio, "Chirally Yours," esp. 224–25.
28. Katherine wrote for Virginia a description of Lady Ottoline's garden (VWL, 2:174). Apparently it was rather like the one she sent to Ottoline, in which she thought of different "*pairs* of people" pacing in the garden and wondered "*who* is going to write about that flower garden" (Mansfield, *Collected Letters*, 1:325; CSF, 297–98).
29. *Nation*, 22 February 1913, 851. Years later, Roger's cousin Robert Bridges, the poet laureate, heard one of Roger's lectures on the radio and wrote to ask whether Roger had read *Portrait of the Artist as a Young Man* (RB/RF, 26 September 1929). Perhaps Joyce had read Fry, although neither Aubert (*Aesthet-*

ics of James Joyce) nor Harkness (*Aesthetics of Dedalus and Bloom*) mentions Fry. Nor does Harper (*Aristocracy of Art in Joyce and Woolf*) mention Fry in her study of the aesthetics of both Joyce and Woolf.

30. This opposition also replicates the distinction Kristeva makes between the semiotic and symbolic (*Revolution in Poetic Language*, esp. 86–89).
31. Actually, Roger was invited to lecture to the Fabians on "Literature in Relation to Life," but he altered the emphasis to "Art and Life" (Edward R. Pease/RF, 18 June 1917). Since he talked about aesthetics whenever he was with Virginia, she probably had heard what Roger was going to say about the dissocation of the aesthetic sensibility before he gave his lecture in August 1917. He did not publish "Art and Life" as an essay until 1920 (VD, 1–11, 213).
32. Virginia's attitude perhaps explains why, although she was involved with the "grass roots" of the Labour Party and the Cooperative movement, Leonard regarded her as "the least political animal that has lived since Aristotle invented the definition" (LWA, 2:204–5). In another comparison of artists and "social reformers & philanthropists," Virginia concluded that "the only honest people are the artists" (D, 1:293).
33. RF/VB, 25 April 1917. Fry's immediate reference is to his review of "The Hugh Lane Bequest," with its detailed treatment of Renoir's *Les Parapluies* (*Nation*, 5 April 1917, 13–14). Overcommitted and in need of money, however, Roger usually wrote too rapidly and repetitively to follow Virginia's influence.
34. Duncan Grant, in Noble, *Recollections of Virginia Woolf*, 29–30. Half a year later, Virginia showed genuine psychological insight into similar behaviorial patterns manifested by the painter Mark Gertler: "I felt about him, as about some women, that unnatural repressions have forced him into unnatural assertions" (D, 1:158–59).
35. For more details, see Barbara Bagenal, in Noble, *Recollections of Virginia Woolf*, 179–81.
36. Pamela Fry Diamand once told me that Bloomsbury was anti-Semitic. When I reminded her that Leonard was Jewish and also part of Bloomsbury, she answered that the other members of the group talked about him behind his back as if he were not one of them. (Permission to quote her was granted by the late Mrs. Diamand.) A letter from Virginia to Jacques Raverat in 1924 hints at the problem: "What is my husband like? A Jew: very long nosed and thin; immensely energetic; But why I don't talk about him is that really you are Anti-Semitic, or used to be, when I was in the sensitive stage of engagement; so that it was then impressed upon me not to mention him" (VWL, 3:130). On Leonard Woolf and Bloomsbury and Virginia Woolf and anti-Semitism, see also J. Wilson, *Leonard Woolf*.
37. A few months later, she was reading Ezra Pound's memoir of Gaudier-Brzeska, who had joined the French army and was killed in June 1915 (D, 1:90).

38. This had been Leonard's word for Virginia's delusions during her 1915 breakdown. Now they both used it apparently to mean the positive illusions necessary to believe in themselves.
39. D, 1:79. Aldous Huxley in *Crome Yellow* and D. H. Lawrence in *Women in Love* (both 1921 novels) did more than gossip. Huxley satirized and Lawrence caricatured Lady Ottoline.
40. D, 1:116. See also Barbara Bagenal's recollections of an air raid that she experienced with the Woolfs, in Noble, *Recollections of Virginia Woolf*, 181–82.
41. D, 1:140. It is tempting to associate Woolf's "A Society," perhaps begun in 1920 and published in 1921, with *Lysistrata*. In Woolf's story, a society of women agrees not to bring another child into the world until they can determine how well a man's world has succeeded in producing good people and good books.
42. In "Cézanne in the Hedge," Bell tells of this painting's acquisition and arrival at Charleston. Richardson describes the purchase of this Cézanne and three other paintings in Paris as the beginning of Keynes's career as a serious collector (*Bloomsbury Iconography*, 308).
43. John Graham recalls asking several of "the principals," including Leonard Woolf, whom he had come to know rather well, how these aesthetic issues "filtered from Fry through VW's consciousness. No one could tell me anything about that: I drew a blank from Leonard, who said that VW did not discuss with him the lengthy conversations she had with Fry" (letter to me, 21 June 1995).
44. VB/VW, 3 and 9 July 1918; VWL, 2:258. Vanessa's *A Conversation* is one of the color illustrations in Spalding, *Vanessa Bell*.
45. Dora Carrington/VW, 21 August 1918. Virginia's letter to Duncan (VWL, 2:270–71) must have followed this one and so is more likely dated 22 August than 21 August, as in VWL, 2.
46. For a full discussion of this issue, see Zwerdling, *Virginia Woolf and the Real World*, 87–119.
47. VWL, 2:416. This comment was especially prescient, for *Night and Day*, with its sense of progression, order, and logic, does seem controlled more by the left brain. See Appendix B.
48. For a detailed discussion of the aesthetic and emotional significance of Vanessa's large oil painting (now in the Tate Gallery) of the same woman-and-tub motif, see Spalding, *Vanessa Bell*, 170–71. The titles used here are those on the backs of the prints in my collection.
49. In VB/RF, 4 January 1919, Vanessa quotes his Wuthering Heights allusion; VB/RF, 6 February 1919; RF/VW, 11 January 1919; VB/VW, 11 January 1919.
50. All quotations from Virginia Woolf in this paragraph are from "Modern Novels," first printed in *Times Literary Supplement*, 10 April 1919, 189.
51. Gillespie makes this connection: "The circles so common in Vanessa Bell's

decorative work are similar embodiments of wholeness" (*Sisters' Arts*, 78). Spalding discusses the importance of the circle in Bell's art in *Vanessa Bell*, 171.

52. For more details, see Spalding, *Vanessa Bell*, 169–74. Duncan's supposed adolescence (he was six years younger than Vanessa, a month short of thirty-four when Angelica was born) was taken by Vanessa as an excuse for his behavior (Garnett, *Deceived with Kindness* 38).
53. Virginia later wrote that the "best critics, Dryden, Lamb, Hazlitt, were acutely aware of the mixture of elements, and wrote of literature with music and painting in their minds" (CE, 2:242).
54. V. Woolf, "The Royal Academy," *Athenaeum*, 22 August 1919, 774–76; VWE, 3:92. Later, Fry made the same point in BBC broadcasts entitled "The Meaning of Pictures," published in *The Listener*, 2, 9, 16, 23, and 30 October and 6 November 1929. In the last of these lectures, Fry mentioned a painting called *The Doctor* that invited viewers to identify with the doctor and "feel we're doing good for the poor patients." He deplored such an "indulgence in this fictitious sense of one's moral worth."
55. Although the methods of neither "The Haunted House" nor the story "Sympathy" are precisely Postimpressionist, their concern with the power of perception and imagination is related to Virginia's other experiments from this period. Probably they were among the manuscript stories Virginia referred to in an October 1919 letter to Vanessa (VWL, 2:393).
56. Fry, "Modern French Art at the Mansard Gallery," 723–24. In his overview of the relationship between Woolf and Fry, Quick convincingly argues that Fry's "translation" of Survage into words is directly inspired by "Kew Gardens" ("Virginia Woolf, Roger Fry and Post-Impressionism," 556–57).
57. Fry was pleased by her praise and in 1920 sent her several prose poems in what he called "the Virginian manner," which the Woolfs did not publish (FP).
58. VWL, 2:385–86. One of these primitive paintings, which hung in the dining room at Monk's House, is in Alexander, *Leonard and Virginia Woolf*, plate 19.
59. D, 1:314. Rather than seem similarly self-interested, Virginia refused to review other Dorothy Richardson novels, for she was afraid that if Richardson was "good then I'm not" (D, 1:315).

Chapter 8

1. For information about how right-brain visual apprehensions balance left-brain verbal apprehensions, see Hoppe, "Split Brains and Psychoanalysis," 229, and Goodwin and Jamison, *Manic-Depressive Illness*, 79.

2. D, 2:10–11. Clive, writing to Vanessa, felt no qualms about appropriating Roger's ideas: "As for Roger & Virginia, God knows I am not their keeper" (CB/VB, 15 April 1920).
3. In her collection of Woolf's shorter fiction (CSF), Dick places "Blue & Green" at the close of the 1917–1921 stories, but there is no external or internal evidence for dating "Blue & Green" more precisely.
4. HGW/LW, 19 March 20; VW/Molly MacCarthy, ? August 1920, MFS, 180. One of these letters may have been misdated.
5. Fry, "Negro Sculpture at the Chelsea Book Club," 516; VD, 71, 70. Virginia made a similar attack on the English: "One cannot help exclaiming that English society is making it impossible to produce English literature" (*Times Literary Supplement*, 5 February 1920, 83; VWE, 3:177).
6. There are no letters from Vanessa to Virginia in the Berg Collection for 1920, and only a few for the next several years.
7. In a letter of 10 October 1920, Roger warned Vanessa that Berenson "makes the appreciation of art a matter of social distinction wh[ich] is intolerable." Six letters from Vanessa to Roger during this trip have survived (although two might date from a similar trip in 1921). There are none from Vanessa to Virginia in the Berg Collection; Virginia mentions postcards from Duncan (VWL, 2:432).
8. VWL, 2:438. This letter is tentatively dated "[1 August 1920]" in VWL. Since Roger's letter of 27 July 1920 seems to be a reply, Virginia's letter should be dated earlier.
9. VB/RF, 15 August 1920; VWL, 2:472, note. Usually the Bloomsbury Group thought rather poorly of Clutton-Brock. For example, Maynard Keynes called a notice in the *Times* of an Omega show "typical Clutton rubbish" (JMK/VB, 3 November 1918).
10. D, 2:47. For an examination of this longing, see Zwerdling, *Virginia Woolf and the Real World*, 87–119.
11. Virginia's letters and Desmond's replies were printed in the 9 and 16 October 1920 issues of the *New Statesman*.
12. The Hogarth Press's Definitive Edition of 1990 dramatically preserves the text's blank spaces.
13. As in Fry's "Spanish Pictures," 136–37.
14. Both appeared in the *Athenaeum*, Fry's essay on 11 July 1919, and Virginia Woolf's story on 22 October 1920. She did not write again for the *Athenaeum* until it merged with the *Nation*.
15. Bell's *Still Life on the Corner of a Mantelpiece* and Grant's *The Mantelpiece*, both now in the Tate Gallery, were painted in 1914. Vanessa's *Still Life* is reproduced in Gillespie, *Sisters' Arts*. Virginia planned the beginning of "Solid Objects" in 1918 (VWL, 2:299). Then Fry's essay "The Artist's Vision," published on 11 July 1919, seems to have been a catalyst for her return to the story.

16. Spalding, *Vanessa Bell*, 166. Maynard Keynes, for example, wrote to Vanessa regarding other Hogarth publications: "As usual the printer's ink has got smudged about on the pages" (JMK/VB, 30 November 1921).
17. RFL, 2:454, 453. The letter was printed in the *Burlington Magazine*, August 1919, 85.
18. D, 2:123. Both the Woolfs were learning enough Russian to help their friend Samuel Koteliansky ("Kot") with translations of classic Russian authors for the press. For details, see Spater and Parsons, *Marriage of True Minds*, 105.
19. VWL, 2:472. Arthur Clutton-Brock's slim volume *Simpson's Choice* had been Roger's first publication at the Omega Workshop. See also Willis, *Leonard and Virginia Woolf as Publishers*.
20. A letter from Lawrence Binyon to Roger (16 November ?, FP) argues with him on this point.
21. Virginia's entire earnings from her books in 1921 were scarcely above £10 (LWA, 2:231).
22. VB/RF, 21 August 1921; VB/VW, ? August 1921: This letter is the only one from Vanessa to Virginia in 1920 and 1921 that has been preserved. Virginia alludes to at least one more letter in VWL, 2:491, but clearly Vanessa had all but ceased writing to her sister.
23. VWL, 2:495. For information about this printing task, see Willis, *Leonard and Virginia Woolf as Publishers*, 32. Since its inception, Virginia had courted Roger for the Hogarth Press. The considerable success of *Twelve Original Woodcuts* (1921) cemented this relation (D, 2:144; VWL, 2:497). He also published with them *Duncan Grant* (1923), *A Sampler of Castille* (1923), *The Artist and Psycho-Analysis* (1924), *Art and Commerce* (1926), and the preface to and translation of Charles Mauron's *The Nature of Beauty in Art and Literature* (1927). Hogarth also published posthumously Fry's translations of Mallarmé. These books amplify and illustrate his dialogue with Virginia about the nature of the arts.
24. VBL, 264. Regarding Virginia's habit of inventing a character for Vanessa, her daughter writes: "Virginia's egotism and Vanessa's passivity contributed towards a situation that was, like some illnesses, chronic" (Garnett, *Deceived with Kindness*, 23).
25. VWL, 2:525. Roger did not want to write himself and so asked André Gide to contribute (RFL, 2:530). He may also have suggested Virginia Woolf to Scott Moncrieff, who lists Virginia as among those who maintained that they did not know Proust well enough to write a tribute to him (Moncrieff, *Marcel Proust*, 3).
26. This unpublished lecture, as well as the Oxford lecture mentioned later, can be found in FP.
27. V. Woolf, in *Times Literary Supplement*, 20 July 1922, 465–66; the heavily revised version, first published in *The Moment*, is in VWE, 3:338–41. Fry's argu-

ment about how modern French artists created form, first printed in the *Catalogue of the Second Post-Impressionist Exhibit* (1912), was reprinted in 1920; see VD, 167. Gloversmith argues that Fry's statements about the autonomy of art were "as comprehensive as those of Ortega, and far more assured in formulation" ("Autonomy Theory," 158). Gloversmith also makes the point that theories about the autonomy of art are "almost self-deconstructing" (149).

28. RF/VW, 26 October 1922, unprinted portion of letter in FP. For the printed portion, see RFL, 2:529–30.
29. DC/VW, ? October 1922. Violet Dickinson wrote about how pleased Thoby would have been with Virginia's rendition of their experiences in Greece (VD/VW, March 1922).
30. Rose, *Woman of Letters*, 100, 102. This feminine form was no doubt what provoked the Irish novelist George Moore to say to Jacques-Emile Blanche, "That woman does not understand narrative" (Blanche, *More Portraits of a Lifetime*, 50).
31. V. Woolf, Introduction to *Mrs. Dalloway*.
32. The Woolfs' capital had been shrinking ever since Virginia's 1915 illness. Now Leonard's £500 salary ended their financial woes (Spater and Parsons, *Marriage of True Minds*, 84, 113–14).
33. D, 2:237. Virginia wrote to Raymond Mortimer, "I was miserable not to look in last night, but being sleepless the night before, had had a dose of chloral, which knocks me into complete stupidity next day—not a fit state for your party" (VW/RM, 1 March 1923?, LWP).
34. Brenan, *South from Granada*, 140–42; Stape, *Virginia Woolf*, 87. Brenan also gives an account of the Woolfs' leave-taking in *Personal Record*, 58.
35. RFL, 2:534. *Stimmung* means literally "tuning" or "pitch," as with a musical instrument; metaphorically, it means mood, temper, disposition, or, more abstractly, impression or atmosphere.
36. For the full run of photographs, see Heilbrun, *Lady Ottoline's Album*. For various testimonies to the liveliness of Virginia's conversation, see Noble, *Recollections of Virginia Woolf*, and Stape, *Virginia Woolf*.
37. When she investigated what turned out to be the sounds of Clive and Mary making love, for example, Virginia "thrust in my upper teeth" before going out into the hall (D, 2:225).
38. V. Woolf, "Jane Austen at Sixty," *Nation and Athenaeum*, 15 December 1923; revised and reprinted as "Jane Austen," CE, 1:148.
39. Roger inscribed a copy of the revised second edition of *Vision and Design*: "Virginia with love/ from Roger—Oct. 1923—." Virginia marked his analysis of Marchand and representation. This presentation copy of *Vision and Design* is in the Harry Ransom Humanities Center, University of Texas, Austin. Virginia's markings (it seems to me safe to assume they are hers) can be found on p. 281.

40. For details, see Spalding, *Vanessa Bell*, 195–97.
41. VB/RF, 25 June 1915; FC/VW, 12 October 1923. The letter from Virginia appears to have been lost.
42. The term was one that Roger Fry and Charles Mauron worked out together to integrate plastic and psychological values in art (Mauron, *Nature of Beauty*).

Chapter 9

1. See Figure 62 and, for a description, Spender, *World Within World*, 137–38.
2. Actually, the marriage of unlike elements was becoming a favorite metaphor for Virginia. Pamela Fry Diamand remembered Virginia's description of marbleized paper in terms of the courtship and marriage of different substances and colors (Broughton, "Pamela Fry Diamand"). Similarly, the manuscript of *To the Lighthouse* includes a sexual metaphor, the gratification of "bodily human love," for artistic creation (V. Woolf, *Original Holograph Draft*, 280). Also, Virginia's unidentified foreword to an exhibition of Vanessa's paintings (Berg, 73B6139) uses the following metaphor: "Greens, blues, reds and purples are here seen making love and war and joining in unexpected combinations of exquisite married bliss."
3. These comments can be found on the verso of p. 6 of the second notebook of "The Hours ? [Mrs. Dalloway]" (BL).
4. VWL, 3:135–36. Oppositions between masculine and feminine, left brain and right brain, are analogous to Bakhtin's (*Dialogic Imagination*) distinction between the monologic and dialogic imagination and Kristeva's (*Revolution in Poetic Language*) distinction between symbolic and semiotic figuration, phallocentric and gynocentric writing.
5. RF/VB, 18 November 1924; RF/LW, 13 September 1924. Bell excused his theft with a feeble acknowledgment: "I once heard Mr. Roger Fry trying to explain this to a roomful of psycho-analysts; and, following in his footsteps, I have attempted the same task myself" ("Dr. Freud on Art").
6. In 1924, Virginia was reviewing former essays for *The Common Reader* and apparently then, since she said she was "writing about Percy Lubbock's book," she made extensive revisions in this essay. See the revised version in VWE, 3:338–41.
7. From "Crepitation," one of Fry's replies to the controversy stirred up by his assessment of the work of the "Late Sir Lawrence Alma Tadema, O.M."
8. For a detailed treatment of the ways in which Virginia Woolf's narrative method subverts monologic, patriarchical narration, see Pearce, *Politics of Narration*, chaps. 8 and 9.

9. For a discussion of how the medical profession's reaction to postwar "shell shock" may have inspired the characters of Septimus and his doctors, see Thomas, "Virginia Woolf's Septimus Smith."
10. Gordon elaborates on this point in *Virginia Woolf*, 193.
11. Later, for example, Virginia told Vita that Madge Symonds Vaughan "is Sally in Mrs Dalloway" (Sackville-West, in Stape, *Virginia Woolf*, 36). For a map, see Beja, "London of *Mrs. Dalloway*."
12. Vanessa was Roger's confidante in the early stages of his relationship. She told him she was "entirely" sympathetic to his choice of Helen and not jealous (VB/RF, 20 May 1925), but her behavior sometimes suggested otherwise, as did Virginia's.
13. For this shift and its relation to Klein, see Abel, *Virginia Woolf*, 13–14, 68. Abel describes Karin Stephen as a "popularizer" of Klein (20).
14. For a discussion of this unpublished lecture, see Gillespie, *Sisters' Arts*, 46.
15. Virginia rejected Stein's *The Making of Americans* (Sprigge, *Gertrude Stein*, 143), but Stein's *Composition as Explanation* was published in November 1926 as one of the Hogarth Essays (D, 3:89, note).
16. Their conventionality was not passed on to their daughter, Dame Janet Vaughan, however. In 1967, she wrote to the editors of John Addington Symonds's letters about how Edmund Gosse had destroyed Symonds's papers "to preserve the good name of my grandfather J. A. S." A physician and educator, Dame Janet regretted the destruction of "all the case histories and basic studies of sexual inversion that J. A. S. is known to have made." She found "Gosse's smug gloating delight . . . nauseating" (Symonds, *Letters*, 2:382, note). My thanks to Elizabeth Richardson for pointing this out to me (letter, 20 August 1994). Janet Vaughan came to know the Woolfs better during the general strike of 1926 (Stape, *Virginia Woolf*, 9–12).
17. For a detailed treatment of the relation of their work and friendship, see Raitt, *Vita and Virginia*; also Reid, Review of *Vita and Virginia*.
18. Spotts argues that "even so, given his deep attachment, he must have been hurt by Virginia's sharing her affections with another person." He concludes that Leonard behaved "with the forbearance and grace of a saint" (LWL, 164). As Leonard Woolf's companion after Virginia's death, Trekkie Parsons gathered that Leonard was "rather impervious" to the affair and that Virginia simply liked to flirt, to be pursued, and to be courted. Also, Parsons thought that for Virginia, having an aristocratic lover was largely the fulfillment of a fantasy (interview, 25 July 1993).
19. Late in his life, Leonard told Trekkie Parsons that he had never had an affair with another woman because his infidelity might have upset Virginia's stability (Spotts, LWL, 163).
20. Bell, 2:120; interviews with Nigel Nicolson, 23 July 1993, and Quentin Bell, 26 July 1993. At our interview, Bell recalled that the argument lasted

throughout the afternoon, not the evening, as his biography implied. He remembered Julian as a member of the Labour Party, although because he was only seventeen, Julian may only have had Labour leanings then. Although Nicolson thought that his mother did not return to Charleston at all, Leonard's diary records meeting Vita for lunch there on 29 July 1928, after which she spent the night with the Woolfs at Monk's House.

21. See Bakhtin, *Dialogic Imagination*, esp. 3–40. As a "hole" in the narrative, "Time Passes" exemplifies what Pearce calls "Woolf's struggle with authority" (*Politics of Narration*, 129–43).
22. Keynes reported to Vanessa that Duncan sold ten paintings; Roger, seven; Vanessa, six; and Porter, five. "One of the papers has been saying that more pictures have been sold at this show than at the Royal Academy, which is not quite true but nearly" (JMK/VB, 3 June 1926).
23. Virginia erroneously wrote "son" instead of son-in-law and said that her great aunt was fifty when she took up photography. The latter error was her mother's (JDS, 214). My thanks to Elizabeth Richardson for correcting this point and identifying the "Miss Stephen" listed in Ritchie, *Letters*, 126.
24. Because it bears the roman numeral "II," it could not have been Roger's first Memoir Club paper given in 1920. I date it between 1925, when Virginia began to buy rather elegant clothes, and 1927, when Clive and Mary's affair ended. Virginia had her hair cut in February 1927; in this sketch, she may have short hair or long hair pulled back.
25. RF/VB, 18 May 1926. Rather than being published by Leonard, Roger's "Seurat" appeared in the *Dial.*
26. Poole writes that most of the men in Virginia's fiction "never look at the beauty of the created world of shapes and colour, they look through it with the hard stare of abstraction" (*Unknown Virginia Woolf*, 9). However derived from Leslie Stephen many of these characters seem, Leslie did have considerable skill in and appreciation of drawing.
27. Gerald Brenan remembered that Virginia was generally thought to be "wayward and capricious in her judgements—a being who lived in a fantasy world of her own and was incapable of real contact with other people." Brenan was surprised that her diary showed her "deeply devoted to her friends in spite of the malicious remarks she sometimes made about them" (*Personal Record*, 252–53.
28. I am indebted to my former student Deborah Wilson for her investigations of "The Fin and the Fish."
29. VB/VW, 31 December 1926. Roger was to arrive the next Monday.
30. Vanessa wrote to Virginia on 2 March 1927 that she had heard that they had decided to go to Greece instead of America.
31. RFL, 2:598. Haule includes both the typescript of this early version of "Time Passes" and Charles Mauron's "Le Temps passe." Haule feels that Woolf defied

Fry in her final revisions by diminishing the humanity of what Fry termed her "superb" characterization of Mrs. McNab (Introduction, 273). But Haule also says that Woolf's "reliance on the abstract representation of human disintegration and despair and her pointed explanations of the significance of insignificant labor [in the early version] are reduced in favor of a more controlled, more imagistic approach. This is, perhaps, the greatest testament to the impact of Roger Fry" (274). Woolf's evolving minimalist treatment of Mrs. McNab, then, is less a defiance of Fry's approval than a compliment to his aesthetic.

32. Julia's words, as quoted by Leslie Stephen in LS/JPS, 18 June 1881.
33. In 1917, the painting was the property of George Granville, fifth duke of Sutherland. Now the property of the current duke of Sutherland, it is currently on loan to the National Gallery of Scotland. My thanks to Aidan Weston-Lewis, assistant keeper, Italian and Spanish Art, for providing this information.
34. V. Woolf, "Notes for Writing. Holograph Notebook unsigned, dated March 1922–25," 11, Berg. For the diagram, see Haule, Introduction, 276, and for commentary, see Zwerdling, *Virginia Woolf and the Real World*, 200–201.
35. TL, 171. This phrase echoes Virginia's sense of Proust's achievement ("tough as catgut & as evanescent as a butterfly's bloom" [D, 3:7]), but she had admired the combination of fragility and hardness since her early twenties. In her 1904 record of Christmas in the New Forest, she wrote: "The stone window frames at Kings [House] are iridescent like a butterfly's wing when the sun shines through the stained glass" (PA, 215).
36. This pattern may also be described as "chiasmus/sequence opposition" (McCluskey, *Reverberations*). Of course, no artistic apprehension is absolutely static; our eyes move over a painting, seeing it over time as well as in space. Abel argues that Lily "seeks a mode of representation outside the father's symbolic universe" which is associated with the "prelinguistic experience of the mother" (*Virginia Woolf*, 47). For more on this novel's aesthetic achievement, see Hussey, *Singing of the Real World.*
37. VWL, 3:395, note. Meeting the Woolfs privately, Bennett found them "gloomy," but he said he "liked both of them in spite of their naughty treatment of me in the press" (Stape, *Virginia Woolf*, 29). Apparently he felt himself to be the aggrieved party.
38. RF/VW, 17 May 1927. For example, T. Semon Hsu wrote to Roger in 1927 from China, thanking Roger and Helen for entertaining him and introducing him to *To the Lighthouse*. He said he would like to "place observances at the shrine of this beautiful & sensitive writer" (FP).
39. The mention of Clive and the threat to broadcast news of Vita's sleeping with Mary Campbell are the passages omitted from VWL, 3:395 and 397 (V. Woolf, *Virginia Woolf Manuscripts*).

40. VBL, 322. They bought the car on 15 July 1927 (LWD). Vanessa bought hers around 1 July 1927 (VBL, 319).
41. LWL, 230, note, clarified by Quentin Bell to me (15 March 1995).
42. First printed as "Poetry, Fiction and the Future," *New York Herald Tribune*, 14 August 1927, and retitled "The Narrow Bridge of Art" in CE, 2:218–29.
43. For a detailed treatment of Bell's illustrations and the interrelation of text and image, see Gillespie, *Sisters' Arts*, 123–137.
44. D, 3:156–57. She, like Fry, left Leonard Woolf out of the portrait. Roger did so, one assumes, because Leonard's activist commitment to the world of politics alienated him from the core of the Bloomsbury circle. Virginia did so, presumably, because of her promise not to characterize him flippantly and to be more sympathetic to his feelings. Also, Leonard simply was an unlikely character in Orlando's milieu.
45. VWL, 3:457. Roger had complained about Virginia's "power of insisting on scenes & spreading black gloom" (RF/VB, 4 August 1927), perhaps between himself and Helen. Both Vanessa and Virginia complained about Helen's intention to buy a house near them.
46. VB/VW, 24 February 1928. In this period, Vanessa wrote several letters to Leonard asking for financial advice, and she always asked him to tell Virginia to write to her (LWP).
47. Interview with Quentin Bell, 26 July 1993. Virginia's response to Julian's poems is in VWL, 3:431–32 (dated 16 October 1927). In VB/JB, 25 October 1927, Vanessa told Julian that she had asked Virginia for her opinion of the poems.

Chapter 10

1. Later, Annan remarked that the 1920s, too, experienced a binary way of thinking "through the eyes of either Beatrice Webb or Virginia Woolf" (quoted in Michael Holroyd, Forward to Lee, *Cézanne in the Hedge*, 7). If so, Woolf represented the artistic and detached, a representation with which she was becoming irritated.
2. Sackville-West, *Letters*, 282, 283. Unless otherwise indicated, the details in the rest of this and the next two paragraphs are taken from "Vita Sackville-West's diary of her 1928 Journey with Virginia Woolf," in the Berg Collection of the New York Public Library, reprinted in Stape, *Virginia Woolf*, 34–36.
3. For a detailed treatment of the relationship between Vita and Virginia (including this trip) and *Orlando*, see Knopp, "'If I Saw You Would You Kiss

Me?'" Knopp does not, however, deal with the question of how Virginia and Vita "did not find each other out."

4. Vita gave Virginia permission to write, or take "vengeance," as she pleased (Sackville-West, *Letters*, 238).
5. Recent critics have been kinder to *Orlando* than Woolf herself was. For example, Edel finds it a better introduction to the art of biography than *Roger Fry* (*Literary Biography*, 98). Knopp implies that it outclasses Shakespeare ("'If I Saw You Would You Kiss Me?'"). And Sally Potter's movie *Orlando* has made it a success with much of the general public.
6. For a detailed discussion of the "third voice" or "Woolf's Other," see McGee, *Telling the Other*, esp. 105–9.
7. RF/VW, 3 November 1928. This portion of the letter is omitted from the published version.
8. For details, see Spater and Parsons, *Marriage of True Minds*, 115.
9. For a series of photographs that illustrate, among other things, Virginia's changing styles of dress, see Heilbrun, *Lady Ottoline's Album*.
10. Kennedy, *Boy at the Hogarth Press*, 34. Kennedy also says that Clive Bell was the "opposite" of Leonard Woolf. He preferred Leonard (39).
11. Brenan says that this topic, like "the Older Generation v. the Younger" or "even Men v. Women," was a standard subject of Bloomsbury conversation which, even though shopworn, stimulated "the most brilliant and fantastic conversation that one can hear anywhere in England" (*Personal Record*, 155–56).
12. D, 3:227, 228. The country seat of the Sidney family in Kent and home of the Renaissance poet Sir Philip Sidney, this grand house was celebrated in Ben Jonson's *To Penshurst* in 1616.
13. VB/CB, ? January 1930; VB/CB, 28 January 1930. By the end of 1930, the *Nation and Athenaeum* and the *New Statesman* merged, becoming the *New Statesman and Nation*.
14. Vanessa was worried that Virginia would discover Duncan and his lover (VBL, 348–53). By September, Clive did know about the affair, as his gossipy letter to Lytton of 6 September 1930 makes clear. For an insightful treatment of Vanessa's self-controlled and possibly self-deceptive reaction to Duncan's affair, see Spalding, *Vanessa Bell*, 235–39.
15. Also, Henry Duckworth's smile does resemble Virginia's, perhaps validating Lydia Keynes's theory that one of George's sons bore "a look of Virginia" (John Maynard Keynes/VB, 3 June 1926).
16. Interview with Quentin Bell, 26 July 1993. Whether or not they thought of such comments as part of an inoffensive game, the offhanded slurs did naturalize prejudices. For example, T. S. Eliot read Julian's poems and congratulated Clive on Julian's escaping "the Chicago Semite taint" (TSE/CB, 25 July

1930). Clive's letter to Lytton of 6 September 1930 called Duncan's lover "an east-side Jew boy from New York."

17. D, 4:6. In a marginal note added later, Virginia wrote that this new book was to be "Here and Now," an early title for *The Years* (Leaska, Introduction to *The Pargiters*, xv).
18. VB/CB, 11 February 1931, and 18 January 1931. Clive maintained that Virginia's complaints about young men arose from "jealousy about buggery" (CB/VB, 14 February 1931).
19. From a map in McQueeney, *Virginia Woolf's Rodmell*, drawn by the gardener's son. See also Spater and Parsons, *Marriage of True Minds*, 174, and Stape, *Virginia Woolf*, 19.
20. For a discussion of Woolf's voicing the "other" beyond historically inscribed sexual divisions, see McGee, *Telling the Other*, 110–15. The form of *The Waves* also fits the chiasmus pattern that McCluskey describes in *Reverberations*. In fact, its triangular pattern, peaking in chapter 5 with the farewell dinner, could be described as an iconic representation of the image of the fin.
21. Interview with Igor Anrep, 4 June 1994. Anrep was twenty-one when *The Waves* was published.
22. Notes by L. Patrick Wilkinson (FP). The Rembrandt remark is not in the written text of Fry's lecture. Although he did not write a book on Rembrandt as a dramatist, Fry did give a late lecture on that topic that he might have developed into a book had he lived. Two versions of this lecture can be found in the Fry papers. The more complete one, with interpolations from the other, was published by Pamela Fry Diamand as "Rembrandt: An Interpretation," *Apollo*, March 1962, 42–55.
23. The young man seems to have been George Bergen, who had almost taken Duncan away from Vanessa two years earlier (see note 14).
24. Woolf's essay "Poetry, Fiction and the Future" (*New York Herald Tribune*, 14 August 1927) contained the phrase "the narrow bridge of art."
25. D, 4:121. Leonard Woolf's description, now in LWP, of this attack makes "fainted" seem like an inappropriately mild verb. Virginia was suffering from menstrual pains, intestinal distress, heat, and a weak heart.
26. Sometime in this period, Virginia went through menopause "gently and imperceptibly" (VWL, 6:60).
27. See the entry in the *Oxford English Dictionary*.
28. Translated and abridged by Pamela Fry Diamand, this essay was published as "The Double Nature of Painting." The unpublished lecture is in the Fry papers at King's College, Cambridge.
29. This letter could date from 1933, although in *The Virginia Woolf Manuscripts*, it is placed with the 1929 letters.
30. WS/VW, n.d. Virginia's letter to Sickert apparently has not survived.

31. For a discussion of Fry's painting that particularly emphasizes his late work, see Morphet, "Roger Fry."
32. This foreword (Berg 73B6139) expands a notebook entry (in Berg 66B0233) in which Virginia follows comments on Sickert with comments on "Vanessa Bell's Exhibition." In these two contiguous sets of comments, she offers precisely opposite interpretations of the two artists. Of course, they were very different artists, but still Virginia's praise of Vanessa Bell's art for its wordlessness suggests her consciousness of the heresy of praising Sickert for his literary qualities.
33. According to the writer Rose Macaulay, one of Virginia's favorite topics of conversation was "Can there be Grand Old Women of literature, or only Grand Old Men? I think I shall prepare to be the Grand Old Woman of English letters. Or would you like to be?" (Noble, *Recollections of Virginia Woolf*, 200).

Chapter 11

1. For a discussion of the connections between Roger Fry and Leslie Stephen, see Leaska, Afterword to *Pointz Hall*, 454–58.
2. Barron, *Duncan Grant*, also makes this point.
3. CE, 2:240, 238, 243. This is the sort of perversity that Albright has in mind: "In the company of formalists [Woolf] will claim her art has direct human reference, while in the company of those who want simple representation of emotion she will defend formality" (*Personality and Impersonality*, 150). Gillespie sees *Walter Sickert* as "Woolf's culminating piece of formal art criticism" (*Sisters' Arts*, 8).
4. Quentin Bell is the source of information about Sickert's flirtatious relationship with Vanessa (interview, 26 July 1993). For a comparison of Vanessa Bell and Sickert, see Reed, "Apples," 20–24. Reed finds Bell's "own ideals of domesticity and female creativity" compatible with Fry's modernism and antithetical to Sickert's notion of the artist as a male voyeur. One wonders whether Virginia knew that Sickert was responsible for excluding women from the Camden Town Group of artists and whether Vanessa regretted encouraging Virginia to have a tête-à-tête with him.
5. Interviews with Igor Anrep, 16 July 1986 and 4 June 1994. Anrep also said, "There was an inquiry into the behavior of the sister [head nurse] of the ward, who tried to prevent Helen seeing Fry on moral grounds, as they were not married. But a nicer nurse called Helen from the waiting room to come while

they were trying to get the plaster [cast] off Roger" (permission to quote granted by Dr. Anrep).

6. D, 4:258. Although Virginia suspected that Margery's proprietary interests would extend that far, I have found no evidence that Margery Fry wanted to write a biography of her brother.
7. For the corrected dating of these photos (ca. 1934), see Richardson, *Bloomsbury Iconography*, 220. Both the *Bookman* cover and Figure 56 were taken by the some photographer, and in both Vita wears the same hat.
8. The label *absurdist* seems appropriate, given that the absurdist playwright Eugene Ionesco, along with the experimental writers Alain Robbe-Grillet and Nathalie Sarraute, later staged a famous reenactment in Paris and then in New York. Watts had intended to paint an allegorical rendering of the "Utmost for the Highest" before his death in 1904.
9. On 2 February 1933 (Chapter 10). See also Leaska, Introduction to *The Pargiters*, xvii.
10. VWL, 5:379, 380. For more on Virginia's double dealing, see King, *Virginia Woolf*, 522–23.
11. In 1936, when D. H. Lawrence's posthumous papers were published, another such baiting became public. This was Lawrence's satire of Roger Fry and Bloomsbury aesthetics: "They had renounced the mammon of 'subject' in pictures, they went whoring no more after the Babylon of painted 'interest,' nor did they hanker after the flesh-pots of artistic 'representation.' Oh purify yourselves, ye who would know the aesthetic ecstacy, and be lifted up to the 'white peaks of artistic inspiration'" (*Phoenix*, 565).
12. Various letters in his papers at the University of Sussex dealing with the disposition of Roger's funds establish that Leonard finished this duty by 25 March 1935.
13. Carolyn Heilbrun elaborates on these points in "Virginia Woolf in Her Fifties," in Marcus, *Virginia Woolf*, 236–53.
14. In June 1935, Virginia began collecting and ordering the Roger/Vanessa correspondence (VWL, 4:401, 406). In August, she collected more papers (D, 4:336). In September, she planned to "drop a few facts into my brain," began reading Roger's "schoolboy letters," bought a case to hold her loose pages of notes, and decided to "write in loose leaf books" (D, 4:341, 345, 344). In early October, she complained that "it will take me six months, only to read the letters at this rate" (VWL, 5:429). Eleven days later, she established a routine of reading "Roger between tea & dinner" and had "read all early R. letters [and] noted them" (D, 4:346).
15. Margery put Kenneth Clark in a similar position, for he wrote a short foreword with an honest assessment of Roger's painting that, according to Vanessa, angered Margery (VBL, 396).
16. VWL, 5:447. The editors of Virginia's letters confuse Julian's letter and mem-

oir. The letter arrived before 1 December 1935, but the more formal, lengthy memoir did not arrive until May 1936, when Virginia had it professionally typed (VWL, 6:33). Virginia copied and used the letter, but she seems to have ignored the forty-seven-page memoir in her biography. The Woolfs refused to print the memoir in their Hogarth Letters series.

17. D, 4:360. For example, although she had both Roger's handwritten and typed versions of the memoir about his father's becoming a Lord Justice, she retyped about eight pages of the handwritten one on blue chain-line paper that matches what she used especially in this period to take notes for Roger Fry (FP).
18. Radin, *Virginia Woolf's "The Years,"* xxii. Radin offers an excellent bibliographical and psychological treatment of Virginia's revisions in various versions of *The Years*.
19. Documentary evidence indicates that she had started but not finished extracting Roger's early letters to his parents and later letters to his daughter Pamela, to Basil Williams, to Goldsworthy Lowes Dickinson, to Helen Anrep, and to Vanessa Bell. My guess is that the "three stout volumes" were three of the following: MHP/B17a, MHP/17c, MHP/B17d, and MHP/B17e. Because Virginia shifted pages around in these volumes, it is impossible to be certain. For example, six pages of the transcripts of letters to Helen Anrep dating from this period were moved from MHP/B17e to MHP/B17f.
20. Quentin Bell narrates this incident and discusses his mother's character in his introduction to VBL.
21. Stansky and Abrahams, *Journey to the Frontier*, 278; also quoted in MFS, 196, note. The letter is dated 20 September 1936.
22. MHP/B17a, 17, dated October 1936. However "free" with the sequence of facts Virginia intended to be, though, a series of cards (impossible to date), now in the Fry Papers, do show her organizing material by some dates.
23. Y, 434. Radin prints the "two enormous chunks" that Virginia deleted. Leaska's edition of *The Pargiters* prints the essays that she had removed from *The Years*.
24. MB, 210. For example, in his next-to-last letter to her, Roger identified Edith Wharton as an "American snobbess" (RFL, 2:692). Roger had dined with Edith Wharton when he was the European curator for the Metropolitan Museum in New York (pocket diaries in FP).
25. D, 5:57 and note, 58. The joint project did not materialize. For more details, see Gillespie, *Sisters' Arts*, 172.
26. D, 5:91. Neither I nor any other Faulknerian I know has been able to find a record of his praise. Perhaps it was forwarded by word of mouth. There is, however, a record of Faulkner's awareness of Woolf's *Orlando* in Carvel Collins's notes from interviews with Faulkner's great-aunt Alabama McLean (Harry Ransom Humanities Research Library, University of Texas).

27. For a refutation of those critics who insist that Woolf was a consistent polemicist and radical feminist all her life, see Zwerdling, *Virginia Woolf and the Real World*, esp. 32–37. Such approaches, which tend to read "the whole career through the lenses of *Three Guineas*," oversimplify Woolf.
28. Julian died on 18 July 1937. Vanessa and the Woolfs heard the news on 20 July (D, 5:104). VWL 6:128, note, mistakenly gives the date as 21 July 1937.
29. Knowing of Julian Bell's death, one may remember Vanessa's comment (VB/CB, ? May 1915), about Rupert Brooke's death: "After all it's not the first time a young person has died." Vanessa's denial of feelings is compellingly treated by Dunn (*Very Close Conspiracy*), who also presents a moving picture of Virginia's selfless efforts to console Vanessa after Julian's death.
30. Vanessa made manuscript and typescript copies of the poems and letters in French from Roger to Josette Coatmellec, Roger's lover who killed herself in 1924. Vanessa also typed and translated Roger's sixty-three-page "L'Histoire de Josette," and Virginia took notes from that copy. These materials are in the Charleston Papers.
31. TG, 58–59. John Bicknell writes: "In 1936, when [he and Evangeline] got married 59 years ago today, and went to Ceylon to visit my parents for our honeymoon, we got about $5.00 to a pound; since a guinea was 21 shillings, I would guess that three guineas might have been worth somewhere between $16 [and] $18" (to me, 16 June 1995). Elizabeth Inglis sent me a page from the china and glass section of an Army and Navy Stores catalog from 1939. Then a set of a dozen after-dinner coffee cups, "richly gilt," or a set consisting of a china coffeepot with creamer and sugar bowl and six cups on a wooden tray was each £3.3, or 63 shillings, or 3 guineas.
32. D, 5:128, note. For corrections of Lehmann's version of his wrangles with Leonard, see LWL, 273.
33. VB/VW, 4 February 1938. This letter shows a very shaky hand, and it has no heading.
34. AS/VB, 28 July 1938, Berg. Igor Anrep remembers once asking Adrian why Virginia Woolf had not been analyzed. Adrian replied, "God help the analyst who undertook it" (interview, 4 June 1994).
35. These retypings are in MHP.
36. Leonard wrote that "Quentin volunteered before the war began—indeed last year—for anti-aircraft work, but was rejected owing to the fact that he had had trouble with his lungs. There was therefore no question, I think, of his being called up" (LWL, 245).
37. Vanessa's letters to Clive are more snide than those to Roger; nevertheless, there was enough material in the latter group to hurt Virginia. Several letters and diary entries establish that she had both sides of the correspondence, for example, VWL, 6:285, and D, 5:204.
38. Here Virginia echoes Roger's words from his introduction to Cameron, *Victo-*

rian Photographs. This exchange, however, probably affected Virginia's notes for an article on biography in which she wondered, "What sort of truth is possible?" and concluded, "Nothing is more dangerous than a gift for writing" ([Contemporary Criticism], Berg 66B0248).

Chapter 12

1. MB, 64. For similar comparisons, see Peter Jacobs, "'The Second Violin Tuning in the Ante-room': Virginia Woolf and Music," in Gillespie, *Multiple Muses*, 227–60.
2. VWL, 3:429; D, 4:245; D, 4:348; D, 5:91; D, 5:160. This pattern dates back to the composition of *The Voyage Out*.
3. Adjacent to Tavistock Square, Southampton Row becomes Woburn Place, but Virginia still called it Southampton Row.
4. Garnett, *Deceived with Kindness*, 144, 148; for more details, see chap. 11, "Bunny's Victory." Garnett feels that she was not well served by Vanessa's indulgence, as the title of her book makes clear. Vanessa's behavior toward her children, but especially toward Angelica, echoes (and magnifies) Julia's indulgence of her sons.
5. MFS, 201–2. Although Virginia did not allow him to help in her biography, Ben Nicolson did make use of his readings ("Post-Impressionism and Roger Fry").
6. For her version of the sitting, see Freund, *Gisèle Freund Photographer*, 96; and for the ordering of the pictures and further details, see Richardson, *Bloomsbury Iconography*, 301–2.
7. Her eightieth birthday party was on 29 October 1930, according to LWD. Leonard underestimated her age by a year or two in LWA, 2:375.
8. D, 5:229. Madge Vaughan was Symonds's older daughter.
9. RF/GLD, 28 May 1991, FP. In his collection of Fry letters, Denys Sutton omits the reference to "paederastia" and misattributes the recipient as Basil Williams (RFL, 1:147). Michael Halls, formerly the modern archivist at King's College, Cambridge, is responsible for the correct attribution and date.
10. Moving at this time was the only poor real-estate choice the Woolfs ever made, for they ended up paying the lease on both houses.
11. This is Berg 66B0204. One page is dated "Aug 23." Both Silver (*Virginia Woolf's Reading Notebooks*, 315) and Leaska (*Pointz Hall)*, date the entire book from 1939. The *Pointz Hall* and the *Roger Fry* fragments are very rough (and barely legible) outlines.
12. This is Berg 66B0214. The Fry notes are on 2–19 and 27–34. Embedded in

these Fry notes on 20–26 are the notes for Virginia's review, entitled "Gas at Abbotsford," of an edition of Scott's *Journal*. She had begun working on the review in September 1939 (VWL, 6:356; D, 5:237). The essay was published on 27 November 1939. The interjected Scott notes strongly suggest September 1939 as the date of composition for this rudimentary treatment of Fry's post–World War I life.

13. D, 5:236, 237. In 1937 her earnings were £2,466; in 1938, £2,972; and in 1939, £891 (LWA, 2:291).
14. For the quoted passages (in which I have not reproduced the textual variants), see V. Woolf, *Roger Fry*, 7–8, 11. This text revises and expands a handwritten treatment of Roger's sitting "for his portrait to a novelist" (Berg, 66B0214). That version can be dated from its inclusion with the materials for a review that Virginia was working on in September 1939 (see note 12). Therefore, I have dated this typed revision, entitled "Roger Fry; a series of impressions" (Berg 66B0227), from after September 1939. I plan to treat the bibliographical evidence concerning the composition of *Roger Fry* in a separate essay.
15. Nearly fifty quotations from the Fry/Bell correspondence, mostly from 1919, with a few excerpts from 1921, are in V. Woolf, *Roger Fry*, 218–26.
16. This letter was not "extracted" in the various notes that Virginia had taken over the years; instead, it seems that she discovered it only in 1939. In "'Virginia Is Anal'" (listed under Broughton), I discuss Virginia's reaction to reading (as I believe she did for the first time in December 1939) that rather startling description of herself.
17. The complete typescript is Berg, 66B0216–24.
18. For more of Vanessa's reaction, see VBL, 469.
19. He was the older brother of William Fisher, who had been flag commander of the *Dreadnought* back in 1911.
20. CB/VW, ? ? 1940. The situation, of course, was Roger's affair with Vanessa.
21. Olivier Bell states that "Virginia had noted a tiny, terrifying, and incomprehensible fragment of one of the world's decisive battles" ("Footnote," 21).
22. VWL, 6:426. Leaska speculates that Virginia's difficulties with this biography resulted from an internal emotional conflict between loyalty to her actual father and to Roger Fry as a "surrogate-father" ("Biographer's Dilemma," 30).
23. Like most letters to Virginia, these are in MHP.
24. D, 5:328. Igor Anrep remembers that Virginia could be "funny, gossipy" but also "bitchy" to his mother. She would end an amiable evening by saying something nasty and unprovoked to Helen that implied, "You silly woman, how could we listen to you?" (interview, 4 June 1994).
25. KF/VW, 26 October 1939, and 27 August 1940, MHP. DeSalvo argues that this correspondence shows Dame Furse as a pioneer in insisting on "lifting the curtain of secrecy surrounding sexuality and the treatment of children"

(Afterword). DeSalvo, apparently, did not read the letter in which Dame Katherine "took exception."

26. For an insightful study, based on documentary evidence, of Virginia's last writings, see Dick, "'The Writing "I" Has Vanished.'"
27. MB, 142, 143. An earlier version of this idea was "My relation with V. not scenic—too much a part of the structure" (MHP). This sentence was added after she completed the last section of her memoir (MB, 143–59). Since she began this section on 15 November 1940, the sentence about Vanessa was probably added in December 1940 or January 1941.
28. MB, 158. For the documentary argument that the memoir did end with this passage, see Broughton, "Virginia Woolf's Meta-Memoir."
29. Wilberforce, *Autobiography*, 168. This conversation seems to have taken place around New Year's Day, 1941.
30. After the death of her great-grandmother, Virginia's great-grandfather James Stephen had married a Wilberforce.
31. OW/LW, 29 March 1941, LWP. Octavia's recollection was that this conversation took place six to eight weeks before Virginia's death, when Octavia brought cream to the Woolfs. For mention of both the cream and Lord David's Cecil's failure to respond to Virginia's retort, see D, 5:354.
32. For evidence of the elaborate patterns of allusions in this novel, see Leaska, Introduction, Notes and References, and Afterword to *Pointz Hall.*
33. Adrian wrote to Vanessa on 19 February 1941 that he had been put in a "very swank" convalescent home in Ayrshire to get over his delirium (Berg).
34. Having established that Virginia described Leslie Stephen as both Godlike and childlike, DeSalvo argues that such a profile "is precisely the portrait of fathers within incestuous households," as if all persons who behave at those extremes are guilty of incest (*Virginia Woolf*, 125). DeSalvo's source is Moore, *Short Season Between Two Silences*, 16. Neither Moore nor DeSalvo quotes Wilberforce's record (*Autobiography*, 178) of what Virginia herself said she meant by not enjoying her body. Of course, Virginia may have been reluctant to explain what she actually meant, but the implications are certainly more complicated than DeSalvo's interpretation.
35. LWA, 2:433. Possibly, as Appendix A to the last volume of her letters argues, Virginia may have attempted suicide on Tuesday, 18 March. But given the concatenation of factors intensifying after that date, I think it is much more probable that this attempt took place on Tuesday, 25 March. See Appendix D.
36. Wilberforce quotes from her letter of 22 March in her *Autobiography*, 179–80. Unless otherwise noted, that letter is the source of information in the preceding three paragraphs. Nigel Nicolson also sees the sentence "I've lost the art" as the key to Woolf's suicide (VWL, 6:xvi).
37. Vanessa had written on 20 March after her visit with Virginia. Even in the

country, even in war, mail was promptly picked up and delivered several times a day.

38. The transcription in VBL of Vanessa's saying "believe in us" rather than "believe us" (as in VWL, 6:485, note) is correct. I also think that Vanessa wrote "Yours VB" rather than "Your," as in VBL, but the distinction is a minor one.
39. Regina Marler, however, finds the letter "more blunt than brutal" (VBL, 473, note).
40. Garrett, Introduction to *The Voyage Out*, vii.
41. VWL, 6:487. I see no reason to think that this assurance was anything but heartfelt. Kenney, however, writes, "'Methinks the lady doth protest too much'" ("Two Endings," 266). See also Lehmann, *Thrown to the Woolfs*, 101.
42. According to Jill Allbrooke and Christopher Fletcher of the British Library, there is no way to be certain when this issue, dated 29 March, actually arrived at the library, at newsstands, and in subscribers' mailboxes. However, because the paper is a "Weekend Review" and the "Coming Week" column begins with a lecture scheduled for Friday, 28 March, this issue would have been distributed between 25 and 27 March, with Thursday, 27 March, as the most likely day (Allbrooke to me, 13 May 1996).
43. Garnett, in Stape, *Virginia Woolf*, 171; Garnett, *Deceived with Kindness*, 151. In Stape's *Virginia Woolf*, Garnett's reference to "a few days" before Virginia's death suggests 24 or 25 March for this visit. But in Garnett's *Deceived with Kindness*, she specifically states that Virginia died three days after this visit (151). If that dating is correct, the Woolfs did visit on the late afternoon of the day when, by my reckoning, Virginia may have made an abortive suicide attempt. David Garnett dates the visit as "two days before" Virginia's suicide, or 26 March (Stape, *Virginia Woolf*, 159). Leonard's diary only records having tea with Angelica and Bunny on 11 March, but as he was under considerable pressure himself on 25 and 26 March, he may simply have omitted this detail.
44. Leila Luedeking agrees with me that these paragraphs on the Nazi destruction of Bloomsbury were written by Leonard. She suggests that they were probably written at the time of the bombings the previous fall and might even have been a collaborative effort by both Woolfs (letter to me, 16 May 1996).
45. Given my redating, this letter to Lehmann—and not the undated suicide letter to Leonard—was probably Virginia's last writing.
46. Information about Virginia's consultation with Octavia Wilberforce in the following paragraphs is from the last chapter of Wilberforce's *Autobiography*, entitled "Virginia Woolf's Last Year, 1940–41," 180–87; from Octavia's letter to Leonard of 29 March 1941 (LWL, 251–52); and from the notes she included in that letter (LWP).

47. My thanks to Ann Wilson and Angus Ross for helping me identify this tree.
48. When Vita visited Leonard about ten days after Virginia's death, Virginia's needlework and thimble were out in the sitting room as she had left them (Sackville-West, in Stape, *Virginia Woolf*, 81). I am assuming that she worked on the embroidery that night after promising Octavia that she would.
49. William Robson [coeditor with Leonard at the *Political Quarterly*] to Trekkie Parsons, 28 December 1975, recollecting a 1941 conversation with Leonard (LWP). In April 1941, Vanessa wrote to Vita that Leonard's "old friend Robson" had visited the weekend before Virginia's body was cremated, which seems to date this conversation as taking place on 19 or 20 April 1941 (collection and permission of Nigel Nicolson).
50. D, 3:235. The last time she visited Beatrice Webb, Virginia told the old woman with regret, "I have no living philosophy" (Webb, in Stape, *Virginia Woolf*, 33). On Virginia Woolf's beliefs, see Hussey, *Singing of the Real World*, 130–55. Leonard maintained a "profound disbelief in religion and its consolation," and after Virginia's death, he could conclude only that "she is dead and utterly destroyed" (William Robson to Trekkie Parsons, 28 December 1975, LWP).
51. Goodwin and Jamison insist that the "mortality rate for untreated manic-depressive patients is higher than it is for most types of heart disease and many types of cancer," and they protest "the erroneous but widespread belief that suicide is volitional" (*Manic-Depressive Illness*, 227). For firsthand testimony of how a manic-depressive can, at the same time, make seemingly volitional plans for suicide and be an involuntary victim of "nightmarish terrors," see Kauth, *Season in Hell*. When he decided to kill himself, Knauth recalled, "inside my head the clamorous voices were stilled at last" (3, 53).
52. LWA, 2:435; Louie Mayer, in Noble, *Recollections of Virginia Woolf*, 195; Sackville-West, in Stape, *Virginia Woolf*, 81. Louie's account varies slightly from Leonard's. He does not mention her or the duster. She says the letters in envelopes were on a "little coffee table." On the verso of the letter dated Tuesday, Leonard noted: "This is the letter left for me on the table in the sitting room which I found at 1 on March 28." His later recollection of the letters on the mantelpiece, therefore, seems to be a mis-remembering. My reconstruction is based on Leonard's recollections, Louie's account, and the evidence of folding and smudging on the suicide letters. It is impossible, however, to be certain about the precise sequence of these events.
53. Collection and permission of Nigel Nicholson.
54. Julian Bell [son of Quentin and Anne Olivier Bell and nephew of the first Julian Bell], "Monk's House and the Woolfs," 26–27. Bell also notes that Virginia was wearing a fur coat.
55. LWL, 258 and note, 259; also VBL, 479 and note.

56. For more details of Vanessa's last years, see Spalding, *Vanessa Bell*, 297–363.
57. From a letter to Vita, 14 June 1941 (collection and permission of Nigel Nicholson).
58. LWL, 472. For Spotts's treatment of Leonard's "Later Years," see LWL, 463–72.

Appendix A

1. Carder describes the "Powis Villas" as built in the 1850s (*Encyclopedia of Brighton*). One of these is plain but has "a good Doric porch," and four are "semi-detached villas with ironwork balconies and verandas." All have a good view over the town to the sea.
2. MJ/JPS, 11 February [1889?]. According to Stanley Weintraub, the Crown Prince was Queen Victoria's son-in-law, husband of Victoria, Princess Royal: "Frederick III of Germany was Crown Prince for nearly 30 years and Emperor for only 99 days, dying on 15 June 1888. . . . The 'organ' would have been the larynx, by the way: he refused surgery to remove it—which might have been fatal. . . . He came to London for the 1887 Jubilee already speechless" (letter to me, 4 March 1996).
3. When Julia and Leslie were in Switzerland, Maria wrote, "I only wish I could give you news of the dear children but I know Stella does that in detail they all seem so well" (MJ/JPS, 30 January 1989). Unfortunately, although Julia kept all her mother's letters, she seems to have discarded Stella's.

Appendix B

1. This gap between stimulus and response is caused by chemistry rather than neurotic conflict (Caramagno, *Flight of the Mind*, 72, 89; Broughton, Review of *The Flight of the Mind*).
2. From a biological perspective, "the patient has episodes of being inappropriately elated, unconcerned about important problems, overconfident, and hyperactive in both motor and speech patterns, with thoughts jumping quickly from one subject to another. Bursting with apparently endless energy, the manic patient has little need for sleep" (Bloom and Lazerson, *Brain, Mind, and Behavior*, 330).
3. For a study of the workings of the brain's hemispheres, see Springer and Deutsch, *Left Brain, Right Brain*, and Bloom and Lazerson, *Brain, Mind, and Behavior*, chap. 1.

Appendix C

1. The page numbers in this appendix refer to Leonardo da Vinci's "Paragone."
2. Mitchell suggests that the debate is ongoing; in fact, the "dialectic of word and image" may be a constant "in the fabric of signs that a culture weaves around itself." Mitchell argues that the word/image *paragone* has never been "just a contest between two kinds of signs, but a struggle between body and soul, world and mind, nature and culture" (*Iconology*, 43, 49).
3. Other Renaissance treatises served as resources: Cennino Cennini's (ca. 1400), Leone Battista Alberti's (1435), Averlino Filarete's (ca. 1460), and Piero della Francesca's (ca. 1485). Leonardo's work most resembled Alberti's, since both insisted on the scientific aspect of painting.
4. In *Laocoon*, Lessing acknowledged that painting uses natural objects and poetry uses artificial symbols and that painting is spatial and literature is temporal. But unlike Leonardo, Lessing saw literature's artificiality as proof of its superiority to painting. Mitchell (*Iconology*) sees Lessing's and Burke's distrust of the visual as an iconophobic reaction of the ruling classes to the prospects of mob rule. This association of the visual with the body and the "lesser" faculties was a bias that Virginia Woolf carried from her youth and modified only reluctantly.
5. For more on the modern arts trying to appropriate one another's qualities, see Steiner, *Colors of Rhetoric*.
6. cummings's own copy (now in the Harry Ransom Humanities Research Center at the University of Texas) of Fry's *Cézanne: A Study of His Development* is heavily underlined throughout, with marginal checks of especially important matters. In black ink and pencil, with some penciled lines inked over, cummings outlined the entire book on the front flyleaf.

Appendix D

1. For their discussion, see VWL, 6:489–91.
2. For a photograph of the undated letter, see VWL, 6:488.
3. Sally Brown, curator of Modern Manuscripts at the British Library, confirmed my belief that these two letters were indeed folded together. That folding would have been pointless, almost impossible, if the undated letter to Leonard was, as the editors of Virginia's letters believe, written on 28 March, the day of Virginia's suicide.

Bibliography

Abel, Elizabeth. *Virginia Woolf and the Fictions of Psychoanalysis*. Chicago: University of Chicago Press, 1989.

Aberbach, David. *Surviving Trauma: Loss, Literature and Psychoanalysis*. New Haven, Conn.: Yale University Press, 1989.

Albright, Daniel. *Personality and Impersonality: Lawrence, Woolf, Mann*. Chicago: University of Chicago Press, 1978.

Alexander, Peter F. *Leonard and Virginia Woolf: A Literary Partnership*. New York: St. Martin's Press, 1992.

Annan, Noel. *Leslie Stephen: The Godless Victorian*. Chicago: University of Chicago Press, 1984.

Aubert, Jacques. *The Aesthetics of James Joyce*. Baltimore: Johns Hopkins University Press, 1992.

Bakhtin, M. M. *The Dialogic Imagination*. Ed. Michael Holquist. Austin: University of Texas Press, 1981.

Barron, Wendy. *Duncan Grant and Bloomsbury*. London: Fine Art Society, 1975.

Battiscomb, Georgina. *Charlotte Mary Yonge: The Story of an Uneventful Life*. London: Constable, 1943.

Beja, Morris. "The London of *Mrs. Dalloway*." *Virginia Woolf Miscellany* 7 (Spring 1977): 4.

Bell, Anne Olivier. "A Footnote to Virginia Woolf's Diary." *Charleston Newsletter* 14 (March 1986): 20–21.

———. Letter. *Virginia Woolf Miscellany* 38 (Spring 1992): 2.

Bell, Clive. *Ad Familiares*. London: Pelican Press, 1917.

———. *Art*. London, 1914. Reprint. New York: Capricorn Books, 1958.

———. "Dr. Freud on Art." *Nation and Athenaeum*, 6 September 1924, 690–91.

———. *Old Friends: Personal Recollections*. London: Chatto & Windus, 1956.

———. "Post-Impressionism and Aesthetics." *Burlington Magazine* 22 (January 1913): 226–30.

———. *Potboilers*. London: Chatto & Windus, 1918.

Bell, Julian. "Monk's House and the Woolfs." In *Virginia Woolf's Rodmell: An Illustrated Guide to Sussex Village*, ed. Maire McQueeney, 20–27. Rodmell: Rodmell Village Press, 1991.

Bell, Quentin. "A Cézanne in the Hedge." In *A Cézanne in the Hedge and Other Memories of Charleston and Bloomsbury*, ed. Hugh Lee, 136–39. London: Collins and Brown, 1992.

———. *Elders and Betters*. London: Murray, 1995.

———. Introduction to *Jacob's Room*. Definitive Collected Edition. London: Hogarth Press, 1990.

———. Letter. *Charleston Magazine* 11 (Spring–Summer 1995): 50.

———. Review of *Virginia Woolf: The Impact of Sexual Abuse on Her Life and Art*, by Louise A. DeSalvo. *New York Review of Books*, 15 March 1990, 4.

———. *Virginia Woolf: A Biography*. New York: Harcourt Brace Jovanovich, 1972.

Bell, Quentin, and Angelica Garnett, eds. *Vanessa Bell's Family Album*. London: Jill Norman & Hobhouse, 1981.

Bell, Vanessa. *Selected Letters of Vanessa Bell*. Ed. Regina Marler. New York: Pantheon Books, 1993.

Bennett, Arnold. *Books and Persons: Being Comments on a Past Epoch, 1908–11*. 1917. Reprint. New York: Greenwood Press, 1968.

———. "Is the Novel Decaying?" *Cassell's Weekly*, 28 March 1923, 47.

———. *Our Women*. London: Cassell, 1920.

Bicknell, John W. "The Ramsays in Love." *Charleston Magazine* 9 (Spring–Summer 1994): 4–9.

Blanche, Jacques-Emile. *More Portraits of a Lifetime, 1918–1938*. Trans. and ed. Walter Clement. London: Dent, 1939.

Bloom, Floyd E., and Arlyne Lazerson. *Brain, Mind, and Behavior*. New York: Freeman, 1988.

Bollas, Christopher. *The Shadow of the Object: Psychoanalysis of the Unthought Known*. New York: Columbia University Press, 1987.

Brenan, Gerald. *Personal Record: 1920–1972*. New York: Knopf, 1975.

———. *South from Granada*. London: Folio Society, 1988.

Broughton, Panthea Reid. [See also Reid, Panthea.] "The Blasphemy of Art: Fry's Aesthetics and Woolf's Non-'Literary' Stories." In *The Multiple Muses of Virginia Woolf*, ed. Diane F. Gillespie, 36–57. Columbia: University of Missouri Press, 1993.

———. "Pamela Fry Diamand, 1902–1985: Recollections." *Charleston Newsletter* 19 (September 1987): 12–16.

———. Review of *The Flight of the Mind: Virginia Woolf's Art and Manic-Depressive Illness*, by Thomas C. Caramagno. *English Literature in Transition* 36 (Fall 1993): 498–501.

———. Review of *Julia Duckworth Stephen: Stories for Children, Essays for Adults*, ed. Diane F. Gillespie and Elizabeth Steele. *English Literature in Transition* 32 (1989): 125–28.

———. Review of *Selected Letters of Vanessa Bell*, ed. Regina Marler. *English Literature in Transition* 38 (Spring 1995): 208–11.

———. "'Virginia Is Anal': Speculations on Virginia Woolf's Writing Roger Fry and Reading Sigmund Freud." *Journal of Modern Literature*, Summer 1987, 151–57.

———. "Virginia Woolf's Meta-Memoir." Review of *Moments of Being*, by Virginia Woolf. 2nd ed. *Review* 10 (1988): 125–36.

Cameron, Julia Margaret. *Victorian Photographs of Famous Men and Fair Women by Julia Margaret Cameron: With Introductions by Virginia Woolf and Roger Fry*. London: Hogarth Press, 1926.

Caramagno, Thomas C. *The Flight of the Mind: Virginia Woolf's Art and Manic-Depressive Illness*. Berkeley: University of California Press, 1992.

———. "Manic-Depressive Psychosis and Critical Approaches to Virginia Woolf's Life and Work." *PMLA* 103 (January 1988): 10–23.

———. "Neuroscience and Psychoanalysis: The Mind/Brain Connection in Biographical Studies of Woolf, Dostoyevsky and Mishima." In *Biography East and West: Selected Conference Papers*, ed. Carol Ramelb, 206–14. Honolulu: University of Hawaii Press, 1989.

Carder, Timothy. *Encyclopedia of Brighton*. Lewes: East Sussex County Librarians, 1990.

Chodorow, Nancy. *The Reproduction of Mothering: Psychoanalysis and the Sociology of Gender*. Berkeley: University of California Press, 1978.

DeSalvo, Louise A. Afterword. *Tulsa Studies in Women's Literature* 9 (Fall 1990): 229–30.

———. *Virginia Woolf: The Impact of Childhood Sexual Abuse on Her Life and Work*. Boston: Beacon Press, 1989.

Dick, Susan. Introduction and Notes to *The Complete Shorter Fiction of Virginia Woolf*. London: Hogarth Press, 1989.

———. "'The Writing "I" Has Vanished': Virginia Woolf's Last Short Fictions." In

Essays for Richard Ellmann: Omnium Gatherum, ed. Susan Dick, Declan Kiberd, Dougald McMillan, and Joseph Ronsley, 134–45. Kingston, Ontario: McGill and Queen's University Press, 1989.

Dictionary of National Biography. London: Smith, Elder, 1886.

di Virgilio, P. S. "Chirally Yours: The Role of Visual Perception in Defining Right and Left Linear Grammars—Left and Right Branching as Signs of Cultural Identity." In *Biography East and West: Selected Conference Papers*, ed. Carol Ramelb, 215–30. Honolulu: University of Hawaii Press, 1989.

Dowling, David. *Bloomsbury Aesthetics and the Novels of Forster and Woolf*. London: Macmillan, 1985.

Dunn, Jane. *A Very Close Conspiracy: Vanessa Bell and Virginia Woolf*. London: Cape, 1990.

Edel, Leon. *Literary Biography*. London: Rupert Hart-Davis, 1957.

Fogel, Daniel Mark. *Covert Relations: James Joyce, Virginia Woolf, and Henry James*. Charlottesville: University Press of Virginia, 1990.

Forster, E. M. "Roger Fry: An Obituary Note." In *Abinger Harvest*, by E. M. Forster. New York: Harcourt, Brace, 1936.

Freud, Anna. *The Ego and the Mechanisms of Defense*. Trans. Cecil Baines. New York: International Universities Press, 1946.

Freud, Sigmund. *Civilization and Its Discontents*. Trans. James Strachey. New York: Norton, 1961.

———. *The Problem of Anxiety*. Trans. Henry Alden Bunker. New York: Norton, 1936.

Freund, Gisèle. *Gisèle Freund Photographer*. Trans. John Shepley. New York: Abrams, 1985.

Fry, Roger. "The Allied Artists at the Grafton Gallery." *Athenaeum*, 18 July 1919, 626–27.

———. "Art and Science." *Athenaeum*, 6 June 1919, 434–35.

———. *The Artist and Psycho-Analysis*. London: Hogarth Press, 1924.

———. "The Baroque." *Burlington Magazine* 39 (September 1921): 145–48.

———. *Cézanne: A Study of His Development*. New York: Macmillan, 1927.

———. *Characteristics of French Art*. London: Chatto & Windus, 1932.

———. "Crepitation." *Nation*, 22 February 1913, 851–52.

———. "The Double Nature of Painting." Trans. Pamela Fry Diamand. *Apollo* 89 (May 1969): 362–71.

———. "Explorations at Trafalgar Square." *Athenaeum*, 18 April 1919, 211.

———. "The Friends' Work for War Victims in France." *Charleston Magazine* 12 (Autumn–Winter 1995): 22–24.

———. "The Grafton Gallery—I." *Nation*, 19 November 1910, 331–32.

———. "The Grafton Gallery: An Apologia." *Nation*, 9 November 1912, 249–51.

———. "Jean Marchand." *Athenaeum*, 11 April 1919, 178–79.

———. *Last Lectures.* With an introduction by Kenneth Clark. Cambridge: Cambridge University Press, 1939.

———. *The Letters of Roger Fry.* Ed. Denys Sutton. London: Chatto & Windus, 1972.

———. "London Statues." *Nation and Athenaeum,* 6 June 1925, 293–95.

———. "London Statues." *Nation and Athenaeum,* 19 September 1925, 730–31.

———. "The Meaning of Pictures: I—Telling a Story." *Listener,* 2 October 1929, 429–32.

———. "Modern French Art at the Mansard Gallery." *Athenaeum,* 8 August 1919, 723–34.

———. "Negro Sculpture at the Chelsea Book Club." *Athenaeum,* 16 April 1920, 516.

———. "A New Theory of Art." *Nation,* 7 March 1914, 937–39.

———. "Quelques peintres français modernes." *Le Français: Journal de la Société des professeurs de français en Angleterre* 217 (October 1918), 218 (January 1919), and 219 (April 1919).

———. *A Sampler of Castille.* Richmond: Hogarth Press, 1923.

———. "The Scenery of 'La Boutique fantastique.'" *Athenaeum,* 13 June 1919, 466.

———. "Seurat." *Dial* 81 (September 1926): 224–32.

———. "Spanish Pictures." *New Statesman,* 6 November 1920, 136–37.

———. "The Toilet." *Listener,* 19 September 1934, 466–68.

———. *Transformations.* London: Chatto & Windus, 1926.

———. *Vision and Design.* 1920. Reprint. Oxford: Oxford University Press, 1981.

Furbank, P. N. *E. M. Forster: A Life.* Vol. 1, *The Growth of the Novelist (1879–1914).* New York: Harcourt Brace Jovanovich, 1977.

Garnett, Angelica. *Deceived with Kindness: A Bloomsbury Childhood.* London: Hogarth Press, 1984.

———. Introduction to *The Voyage Out.* Definitive Collected Edition. London: Hogarth Press, 1990.

Gillespie, Diane Filby. *The Sisters' Arts: The Writing and Painting of Virginia Woolf and Vanessa Bell.* Syracuse, N.Y.: Syracuse University Press, 1988.

———, ed. *The Multiple Muses of Virginia Woolf.* Columbia: University of Missouri Press, 1993.

Ginsberg, Elaine K., and Laura Moss Gottlieb, eds. *Virginia Woolf: Centennial Essays.* Troy, N.Y.: Whitson, 1983.

Glendinning, Victoria. *Vita: The Life of V. Sackville-West.* London: Weidenfeld and Nicolson, 1983.

Gloversmith, Frank. "Autonomy Theory: Ortega, Roger Fry, Virginia Woolf." In *The Theory of Reading,* ed. Frank Gloversmith, 147–97. Totowa, N.J.: Barnes and Noble, 1984.

Goodwin, Frederick K., and Kay Redfield Jamison. *Manic-Depressive Illness*. New York: Oxford University Press, 1990.

Gordon, Lyndall. *Virginia Woolf: A Writer's Life*. New York: Norton, 1984.

Graham, J. W., ed. *Virginia Woolf's "The Waves": The Two Holograph Drafts*. Toronto: University of Toronto Press, 1976.

Hall, Lesley. *Hidden Anxieties: Male Sexuality: 1900–1958*. London: Polity Press, 1991.

Harkness, Marguerite. *The Aesthetics of Dedalus and Bloom*. Lewisburg, Pa.: Bucknell University Press, 1984.

Harper, Margaret Mills. *The Aristocracy of Art in Joyce and Woolf*. Baton Rouge: Louisiana State University Press, 1990.

Haule, James M. Introduction to "'Le Temps passé' and the Original Typescript: An Early Version of the 'Time Passes' Section of *To the Lighthouse*." *Twentieth Century Literature* 29 (Fall 1983): 267–77.

Haultain, Arnold. *Goldwin Smith: His Life and Opinions*. London: T. Werner Laurie, 1913.

Heilbrun, Carolyn G., ed. *Lady Ottoline's Album: Snapshots and Portraits of Her Famous COntemporaries and of Herself Photographed for the Most Part by Lady Ottoline Marrell from the Collection of Her Daughter Julian Vinogradoff*. New York: Knopf, 1976.

Heine, Elizabeth. "Virginia Woolf's Revisions of *The Voyage Out*." In *The Voyage Out*, by Virginia Woolf, 399–463. Definitive Collected Edition. London: Hogarth Press, 1990.

Hill, Brian. *Julia Margaret Cameron: A Victorian Family Portrait*. New York: St. Martin's Press, 1973.

Hill, Katherine C. "Virginia Woolf and Leslie Stephen: History and Literary Revolution." *PMLA* 96 (May 1981): 351–62.

Holroyd, Michael. *Lytton Strachey: A Critical Biography*. Vol. 2, *The Years of Achievement (1910–32)*. London: Heinemann, 1968.

Holtby, Winifred. *Virginia Woolf: A Critical Memoir*. 1932. Reprint. Chicago: Cassandra Editions, 1978.

Hoppe, Klaus D. "Split Brains and Psychoanalysis." *Psychoanalytic Quarterly* 46 (1977): 220–44.

Hussey, Mark. *The Singing of the Real World: The Philosophy of Virginia Woolf's Fiction*. Columbus: Ohio State University Press, 1986.

Hyman, Virginia R. "Reflections in the Looking-Glass: Leslie Stephen and Virginia Woolf." *Journal of Modern Literature* 10 (June 1983): 197–216.

Kennedy, Richard. *A Boy at the Hogarth Press*. London: Heinemann, 1972.

Kenney, Susan M. "Two Endings: Virginia Woolf's Suicide and *Between the Acts*." *University of Toronto Quarterly* 44 (Summer 1975): 265–89.

Keynes, Geoffrey. *The Gates of Memory*. Oxford: Oxford University Press, 1981.

King, James. *Virginia Woolf*. London: Hamish Hamilton, 1994.

Klein, Melanie. *Love, Guilt, and Reparation and Other Works*. New York: Dell, 1975.

———. *The Selected Melanie Klein*. Ed. Juliet Mitchell. Harmondsworth: Penguin Books, 1986.

Knauth, Percy. *A Season in Hell*. New York: Harper & Row, 1975.

Knopp, Sherron E. "'If I Saw You Would You Kiss Me?': Sapphism and the Subversiveness of Virginia Woolf's *Orlando*." *PMLA* 103 (January 1988): 24–34.

Kristeva, Julia. *Language: The Unknown*. Trans. Anne M. Menke. New York: Columbia University Press, 1989.

———. *Revolution in Poetic Language*. Trans. Margaret Waller. New York: Columbia University Press, 1984.

Lane, Ann J. *To "Herland" and Beyond: The Life and Work of Charlotte Perkins Gilman*. New York: Pantheon Books, 1990.

Lawrence, David Herbert. *Phoenix: The Posthumous Papers of D. H. Lawrence*. Ed. Edward D. McDonald. 1936. Reprint. Harmondsworth: Penguin Books, 1978.

Leaska, Mitchell. "A Biographer's Dilemma: Virginia Woolf and Roger Fry." In *A Cézanne in the Hedge and Other Memories of Charleston and Bloomsbury*, ed. Hugh Lee, 82–85. London: Collins and Brown, 1992.

———. Introduction, Notes and References, and Afterword to *Pointz Hall*. New York: University Publications, 1983.

Lee, Hugh, ed. *A Cézanne in the Hedge and Other Memories of Charleston and Bloomsbury*. London: Collins and Brown, 1992.

Lehmann, John. *Thrown to the Woolfs*. New York: Holt, Rinehart and Winston, 1978.

Leonardo da Vinci. "Paragone: Of Poetry and Painting." In *Treatise on Painting*, trans. and annotated A. Philip McMahon, 3–44. Princeton, N.J.: Princeton University Press, 1956.

Lessing, Gotthold Ephraim. *Laocoon: An Essay on the Limits of Painting and Poetry*. New York: Noonday Press, 1957.

Lowell, James Russell. *Letters of James Russell Lowell*. Ed. Charles Eliot Norton. New York: Harper, 1893.

MacCarthy, Desmond. "Kant and Post Impressionism." *New Witness*, 10 October 1912, 533–34.

———. "Readers' Reports: *Another Country*, by H. du Coudray." *Life and Letters*, August 1928, 221–22.

Maitland, Frederic William. *The Life and Letters of Leslie Stephen*. London: Duckworth, 1907.

Malcolm, Janet. "A House of One's Own." *New Yorker*, 5 June 1995, 58–79.

Mansfield, Katherine. *The Collected Letters of Katherine Mansfield*. 3 vols. Ed. Vincent O'Sullivan, with Margaret Scott. Oxford: Clarendon Press, 1984, 1987, and 1993.

Marcus, Jane. "Virginia Woolf and Her Violin: Mothering, Madness and Music." In *Virginia Woolf: Centennial Essays*, ed. Elaine K. Ginsberg and Laura Moss Gottlieb, 27–49. Troy, N.Y.: Whitson, 1983.

———, ed. *Virginia Woolf: A Feminist Slant*. Lincoln: University of Nebraska Press, 1983.

Marsh, Jan. *Pre-Raphaelite Women: Images of Femininity in Pre-Raphaelite Art*. London: Weidenfeld and Nicolson, 1987.

Mauron, Charles. *The Nature of Beauty in Art and Literature*. Trans. and preface by Roger Fry. London: Hogarth Press, 1927.

McCluskey, Kathleen. *Reverberations: Sound and Structure in the Novels of Virginia Woolf*. Ann Arbor, Mich.: UMI Research Press, 1986.

McGann, Jerome. *Black Riders: The Visible Language of Modernism*. Princeton, N.J.: Princeton University Press, 1993.

McGee, Patrick. *Telling the Other: The Question of Value in Modern and Postcolonial Writing*. Ithaca, N.Y.: Cornell University Press, 1992.

McLaurin, Allen. *Virginia Woolf: The Echoes Enslaved*. Cambridge: Cambridge University Press, 1973.

McQueeney, Maire, ed. *Virginia Woolf's Rodmell: An Illustrated Guide to a Sussex Village*. Rodmell: Rodmell Village Press, 1991.

Meisel, Perry. "Interview with Julia Kristeva." Trans. Margaret Waller. *Partisan Review* 51 (Winter 1984): 128–32.

Meredith, George. *The Letters of George Meredith: Collected and Edited by His Son*. New York: Scribner, 1912.

Mitchell, W. J. T. *Iconology: Image, Text, Ideology*. Chicago: University of Chicago Press, 1986.

Moncrieff, C. K. Scott, ed. *Marcel Proust: An English Tribute*. London: Chatto & Windus, 1923.

Moore, Madeline. *The Short Season Between Two Silences: The Mystical and the Political in the Novels of Virginia Woolf*. Boston: Allen & Unwin, 1984.

Morphet, Richard. "Roger Fry: The Nature of His Painting." *Burlington Magazine* 122 (July 1980): 478–86.

Morrell, Ottoline. *Ottoline: The Early Memoirs of Lady Ottoline Morrell*. Ed. Robert Gathorne-Hardy. London: Faber & Faber, 1963.

Nicolson, Benedict. "Post-Impressionism and Roger Fry." *Burlington Magazine*, January 1951, 11–15.

Nicolson, Nigel. *Portrait of a Marriage*. New York: Atheneum, 1974.

Noble, Joan Russell, ed. *Recollections of Virginia Woolf by Her Contemporaries*. London: Sphere Books, 1989.

Oakley, David A., and Lesley C. Eames. "The Plurality of Consciousness." In *Brain and Mind*, ed. David A. Oakley, 217–51. London: Methuen, 1985.

Panken, Shirley. *Virginia Woolf and the "Lust of Creation": A Psychoanalytic Exploration*. Albany: State University of New York Press, 1987.

Parsons, Ian. Introduction to *The Wise Virgins*, by Leonard Woolf. New York: Harcourt Brace Jovanovich, 1979.

Pearce, Richard. *The Politics of Narration: James Joyce, William Faulkner, and Virginia Woolf*. New Brunswick, N.J.: Rutgers University Press, 1991.

Poole, Roger. *The Unknown Virginia Woolf*. Cambridge: Cambridge University Press, 1978.

Quick, Jonathan R. "Virginia Woolf, Roger Fry and Post-Impressionism." *Massachusetts Review*, Winter 1985, 547–57.

Radin, Grace. *Virginia Woolf's "The Years": The Evolution of a Novel*. Knoxville: University of Tennessee Press, 1981.

Raitt, Suzanne. *Vita and Virginia: The Work and Friendship of V. Sackville-West and Virginia Woolf*. Oxford: Clarendon Press, 1993.

Ramelb, Carol, ed. *Biography East and West: Selected Conference Papers*. Honolulu: University of Hawaii Press, 1989.

Ray, Man. *Self-Portrait*. Boston: Little, Brown, 1963.

Reed, Christopher. "Apples: 46 Gordon Square." *Charleston Newsletter* 23 (June 1989): 20–24.

Reid, Panthea. [See also Broughton, Panthea Reid.] Review of *Selected Letters of Vanessa Bell*, ed. Regina Marler. *English Literature in Transition* 38 (Spring 1995): 208–11.

———. Review of *Vita and Virginia*, by Suzanne Raitt. *English Literature in Transition* 38 (Winter 1995): 91–94.

Richardson, Elizabeth P. *A Bloomsbury Iconography*. Winchester: St. Paul's Bibliographies, 1989.

Ritchie, Annie Thackerray. *Letters of Annie Thackeray Ritchie*. Ed. Hester Ritchie. London: Murray, 1924.

Rose, Phyllis. *Woman of Letters: A Life of Virginia Woolf*. New York: Oxford University Press, 1978.

Rosenbaum, S. P. "Conversation with Julian Fry." *Modernist Studies* 3 (1979): 127–35.

———. *Victorian Bloomsbury: The Early Literary History of the Bloomsbury Group*. New York: St. Martin's Press, 1987.

———. "Virginia Woolf: Beginnings." In *Leon Edel and Literary Art*, ed. Lyall H. Powers, 43–53. Ann Arbor, Mich.: UMI Research Press, 1988.

———, ed. *The Bloomsbury Group: A Collection of Memoirs, Commentary and Criticism*. Toronto: University of Toronto Press, 1975.

Sackville-West, Vita. *The Letters of Vita Sackville-West to Virginia Woolf*. Ed. Louise DeSalvo and Mitchell A. Leaska. New York: Morrow, 1985.

Segal, Hanna. "A Psycho-Analytic Approach to Aesthetics." *International Journal of Psycho-Analysis* 33 (1952): 196–207.

Silver, Brenda. *Virginia Woolf's Reading Notebooks*. Princeton, N.J.: Princeton University Press, 1983.

Sitwell, Osbert. *Laughter in the Next Room*. Boston: Little, Brown, 1948.

Spalding, Frances. *Roger Fry: Art and Life*. London: Granada, 1980.

———. *Vanessa Bell*. London: Weidenfeld and Nicolson, 1983.

Spater, George, and Ian Parsons. *A Marriage of True Minds: An Intimate Portrait of Leonard and Virginia Woolf*. New York: Harcourt Brace Jovanovich, 1977.

Spender, Stephen. *World Within World*. New York: Harcourt, Brace, 1951.

Spilka, Mark. *Virginia Woolf's Quarrel with Grieving*. Lincoln: University of Nebraska Press, 1980.

Spotts, Frederic. Preface and Chapter Introductions to *The Letters of Leonard Woolf*. Ed. Frederic Spotts. San Diego, Calif.: Harcourt Brace Jovanovich, 1989.

Sprigge, Elizabeth. *Gertrude Stein: Her Life and Work*. New York: Harper, 1957.

Springer, Sally P., and Georg Deutsch. *Left Brain, Right Brain*. San Francisco: Freeman, 1981.

Stansky, Peter, and William Abrahams. *Journey to the Frontier: Two Roads to the Spanish Civil War*. Chicago: University of Chicago Press, 1966.

Stape, J. H. *Virginia Woolf: Interviews and Recollections*. Iowa City: University of Iowa Press, 1995.

Stein, Gertrude. *The Autobiography of Alice B. Toklas*. 1933. Reprint. New York: Vintage Books, 1960.

Stein, Janice. "A Thousand Appliances: Virginia Woolf and the Tools of Visual Literacy." Ph.D. diss., Louisiana State University, 1994.

Steiner, Wendy. *The Colors of Rhetoric: Problems in the Relation Between Modern Literature and Painting*. Chicago: University of Chicago Press, 1982.

Stemerick, Martine. "Virginia Woolf and Julia Stephen: The Distaff Side of History." In *Virginia Woolf: Centennial Essays*, ed. Elaine K. Ginsberg and Laura Moss Gottlieb, 51–80. Troy, N.Y.: Whitson, 1983.

Stephen, Julia Duckworth. *Julia Duckworth Stephen: Stories for Children, Essays for Adults*. Ed. Diana F. Gillespie and Elizabeth Steele. Syracuse, N.Y.: Syracuse University Press, 1987.

Stephen, Leslie. "Byron." In *Dictionary of National Biography*. London: Smith, Elder, 1886.

———. *Sir Leslie Stephen's Mausoleum Book*. Introduction by Alan Bell. Oxford: Clarendon Press, 1977.

———. *Selected Letters of Leslie Stephen*. 2 vols. Ed. John W. Bicknell. Columbus: Ohio State University Press, 1996.

———. *Some Early Impressions*. 1903. Reprint. London: Hogarth Press, 1924.

Styron, William. *Darkness Visible*. New York: Random House, 1990.

Suttie, Ian D. *The Origins of Love and Hate*. London: Kegan Paul, 1935.

Swanwick, Helena Maria. *I Have Been Young*. London: Gollancz, 1935.

Symonds, John Addington. *The Letters of John Addington Symonds*. Ed. Herbert M. Schueller and Robert L. Peters. Detroit: Wayne State University Press, 1968.

Thomas, Sue. "Virginia Woolf's Septimus Smith and Contemporary Perceptions of Shell Shock." *English Language Notes* 25 (1987): 49–57.

Trombley, Stephen. *"All That Summer She Was Mad": Virginia Woolf and Her Doctors.* London: Junction Books, 1981.

Watts, M. S. *George Frederic Watts.* Vol. 2, *The Annals of an Artist's Life.* London: Macmillan, 1912.

Weaver, Mike. *Whisper of the Muse: The Overstone Album & Other Photographs by Julia Margaret Cameron.* Malibu, Calif.: J. Paul Getty Museum, 1986.

Weisstein, Ulrich. "Literature and the Visual Arts." In *Interrelations of Literature,* ed. Jean-Pierre Barricelli and Joseph Gibaldi, 251–77. New York: Modern Language Association, 1982.

Wilberforce, Octavia. *The Autobiography of a Pioneer Woman Doctor.* Ed. Pat Jalland. London: Cassell, 1989.

Willis, J. H. *Leonard and Virginia Woolf as Publishers: The Hogarth Press, 1917–41.* Charlottesville: University Press of Virginia, 1992.

Wilson, Duncan. *Leonard Woolf: A Political Biography.* New York: St. Martin's Press, 1978.

Wilson, Jean Moorcroft. *Leonard Woolf: Pivot or Outsider of Bloomsbury.* London: Cecil Woolf, 1994.

———. *Virginia Woolf: Life and London, a Biography of Place.* London: Cecil Woolf, 1987.

———. *Virginia Woolf and Anti-Semitism.* London: Cecil Woolf, 1995.

Woolf, Leonard. *An Autobiography.* 1960, 1961, 1964, 1967, 1969. Reprint (5 vols. in 2). Oxford: Oxford University Press, 1980.

———. *Fear and Politics: A Debate at the Zoo.* London: Hogarth Press, 1925.

———. "The Future of British Broadcasting." *Political Quarterly* 2 (April–June 1931): 172–85.

———. *The Letters of Leonard Woolf.* Ed. Frederic Spotts. San Diego, Calif.: Harcourt Brace Jovanovich, 1989.

———. "Looking Back." Interview with Malcolm Muggeridge. London: BBC Television, 1967.

———. *Quack, Quack!* London: Hogarth Press, 1935.

———. *The Wise Virgins: A Story of Words, Opinions and a Few Emotions.* 1914. Reprint. New York: Harcourt Brace Jovanovich, 1979.

Woolf, Virginia. [Works not mentioned in Abbreviations of Published Sources]

———. "'Anon' and 'The Reader.'" Ed. Brenda R. Silver. *Twentieth Century Literature,* Fall–Winter 1979, 356–441.

———. "Asheham Diary." Ed. and with an introduction by Anne Olivier Bell. *Charleston Magazine* 9 (Spring–Summer 1994): 27–35.

———. "Character in Fiction." *Criterion,* July 1924, 409–30.

———. *A Cockney's Farming Experiences and the Experiences of a Paterfamillas.* Ed. Suzanne Henig. San Diego: San Diego State University Press, 1972.

———. *Contemporary Writers*. London: Hogarth Press, 1965.

———. "Four Hidden Letters." Ed. Joanne Trautmann Banks. *Charleston Magazine* 19 (Autumn–Winter 1994): 25–31.

———. "Friendships Gallery." Ed. Ellen Hawkes. *Twentieth Century Literature*, Fall–Winter 1979, 270–302.

———. Introduction to *Mrs. Dalloway*. New York: Modern Library, 1928.

———. *Melymbrosia: An Early Version of "The Voyage Out."* Ed. Louise A. DeSalvo. New York: New York Public Library, 1982.

———. *The Pargiters: The Novel-Essay Portion of "The Years."* Ed. Mitchell A. Leaska. London: Hogarth Press, 1978.

———. *Recent Painting by Vanessa Bell, with a Foreword by Virginia Woolf*. London: London Artists' Association, 1930.

———. Review of *Vision and Design*, by Roger Fry. *Woman's Leader*, 18 February 1921, 43. Reprint. "Two Contributions by Virginia Woolf to the Suffragette Press." Ed. B. J. Kirkpatrick. *Charleston Magazine* 12 (Autumn–Winter 1995): 25–29.

———. *Reviewing, with a Note by Leonard Woolf*. London: Hogarth Press, 1939.

———. *Roger Fry: A Series of Impressions*. Ed. Diane F. Gillespie. London: Cecil Woolf, 1994.

———. "Terrible Tragedy in a Duckpond." *Charleston Magazine* 1 (Spring–Summer) 1990): 37–42.

———. *Virginia Woolf: To the Lighthouse, The Original Holograph Draft*. Transcribed and ed. Susan Dick. Toronto: University of Toronto Press, 1982.

———. *Virginia Woolf & Lytton Strachey: Letters*. Ed. Leonard Woolf and James Strachey. London: Hogarth Press, 1956.

———. *The Virginia Woolf Manuscripts*. Woodbridge, Conn.: Research Publications International, 1993.

———. *The Virginia Woolf Manuscripts from the Monk's House Papers at the University of Sussex and Additional Manuscripts at the British Library, London Microform*. Brighton: Harvester Press Microform, 1985.

———. *Virginia Woolf's Pointz Hall: The Earlier and Later Typescripts of "Between the Acts."* Ed. Mitchell A. Leaska. New York: University Publications, 1983.

———. *Virginia Woolf's "The Waves": The Two Holograph Drafts Transcribed and Edited by J. W. Graham*. Toronto: University of Toronto Press, 1976.

Zwerdling, Alex. *Virginia Woolf and the Real World*. Berkeley: University of California Press, 1986.

Index

NOTE: In this index, VW refers to Virginia Woolf, JPS refers to Julia Prinsep Stephen (VW's mother), LS refers to Leslie Stephen (VW's father), LW refers to Leonard Woolf (VW's husband), RF refers to Roger Fry, and VB refers to Vanessa Bell (VW's sister).